GRAVITATIONAL WAVES

Related Titles from the AIP Conference Proceedings Subseries on Astronomy and Astrophysics

522 Cosmic Explosions: Tenth Astrophysics Conference
Edited by Stephen S. Holt and William W. Zhang, June 2000, 1-56396-943-2

516 26th International Cosmic Ray Conference: ICRC XXVI, Invited, Rapporteur, and Highlight Papers
Edited by Brenda L. Dingus, David B. Kieda, and Michael H. Salamon
May 2000, 1-56396-939-4

515 GeV-TeV Gamma Ray Astrophysics Workshop: Towards a Major Atmospheric Cherenkov Detector VI
Edited by Brenda L. Dingus, Michael H. Salamon, and David B. Kieda, May 2000, 1-56396-938-6

510 The Fifth Compton Symposium
Edited by Mark L. McConnell and James M. Ryan, March 2000, 1-56396-932-7

499 Small Missions for Energetic Astrophysics: Ultraviolet to Gamma-Ray
Edited by Steven P. Brumby, December 1999, 1-56396-912-2

493 General Relativity and Relativistic Astrophysics: Eighth Canadian Conference
Edited by C. P. Burgess and R. C. Myers, November 1999, 1-56396-905-X

478 COSMO-98: Second International Workshop on Particle Physics and the Early Universe
Edited by David O. Caldwell, May 1999, 1-56396-853-3

476 3 K Cosmology: EC-TMR Conference
Edited by Luciano Maiani, Francesco Melchiorri, and Nicola Vittorio, May 1999, 1-56396-847-9

456 Laser Interferometer Space Antenna: Second International LISA Symposium on the Detection and Observation of Gravitational Waves in Space
Edited by William M. Folkner, December 1998, 1-56396-848-7

433 Workshop on Observing Giant Cosmic Ray Air Showers from $>10^{20}$ eV Particles from Space
Edited by John F. Krizmanic, Jonathan F. Ormes, and Robert E. Streitmatter, June 1998, 1-56396-788-X

To learn more about these titles, or the AIP Conference Proceedings Series, please visit the webpage **http://www.aip.org/catalog/aboutconf.html**

GRAVITATIONAL WAVES

Third Edoardo Amaldi Conference

Pasadena, California   12–16 July 1999

EDITOR
Sydney Meshkov
California Institute of Technology, Pasadena

Melville, New York
AIP CONFERENCE PROCEEDINGS ■ 523

Editor:

Sydney Meshkov
California Institute of Technology
LIGO 18-34
1201 E. California Blvd.
Pasadena, CA 91125
USA

E-mail: syd@ligo.caltech.edu

The article on pp. 401–402 was authored by a U. S. Government employee and is not covered by the below mentioned copyright.

Authorization to photocopy items for internal or personal use, beyond the free copying permitted under the 1978 U.S. Copyright Law (see statement below), is granted by the American Institute of Physics for users registered with the Copyright Clearance Center (CCC) Transactional Reporting Service, provided that the base fee of $17.00 per copy is paid directly to CCC, 222 Rosewood Drive, Danvers, MA 01923. For those organizations that have been granted a photocopy license by CCC, a separate system of payment has been arranged. The fee code for users of the Transactional Reporting Service is: 1-56396-944-0/00/$17.00.

© 2000 American Institute of Physics

Individual readers of this volume and nonprofit libraries, acting for them, are permitted to make fair use of the material in it, such as copying an article for use in teaching or research. Permission is granted to quote from this volume in scientific work with the customary acknowledgment of the source. To reprint a figure, table, or other excerpt requires the consent of one of the original authors and notification to AIP. Republication or systematic or multiple reproduction of any material in this volume is permitted only under license from AIP. Address inquiries to Office of Rights and Permissions, Suite 1NO1, 2 Huntington Quadrangle, Melville, N.Y. 11747-4502; phone: 516-576-2268; fax: 516-576-2450; e-mail: rights@aip.org.

L.C. Catalog Card No. 00-103506
ISBN 1-56396-944-0
ISSN 0094-243X
Printed in the United States of America

CONTENTS

Preface xiii
Sponsors xv
Committees xvii

PART I. INVITED LECTURES

Amaldi Meeting Introduction 3
    B. C. Barish
Edoardo Amaldi, Scientist 5
    U. Amaldi
Bars in Action 32
    E. Coccia

Astrophysical Sources

Compact Binary Mergers and Accretion-Induced Collapse: Event Rates 41
    V. Kalogera
Instabilities in Stiff Stellar Cores: The Gravitational Radiation Reaction 51
    J. L. Houser
Gravitational Waves from the r-Modes of Rapidly Rotating Neutron Stars 55
    B. J. Owen
Gravitational Waves from Low-Mass X-Ray Binaries: A Status Report 65
    G. Ushomirsky, L. Bildsten, and C. Cutler
Gravitational Waves from Inspiral into Massive Black Holes 76
    S. A. Hughes
Radiation Reaction Force on a Compact Body Spiraling
into a Supermassive Black Hole 82
    Y. Mino
Self-Force Approach for Radiation Reaction 86
    L. M. Burko
Are Pre-Big-Bang Models Falsifiable by Gravitational Wave Experiments? ... 90
    C. Ungarelli and A. Vecchio
Is There a Signature in Gravitational Waves from Structure
Formation of the Universe? 94
    O. D. Miranda, J. C. N. de Araújo, and O. D. Aguiar

Status of Interferometers

The Status of LIGO 101
    M. W. Coles

*Italicized name indicates author(s) who presented the paper.

Status of the VIRGO Experiment 110
    F. Marion for the VIRGO Collaboration
The Status of GEO600 119
    H. Lück, P. Aufmuth, O. S. Brozek, K. Danzmann, A. Freise,
    S. Goßler, A. Grado, H. Grote, K. Mossavi, V. Quetschke,
    B. Willke, K. Kawabe, A. Rüdiger, R. Schilling, W. Winkler,
    C. Zhao, K. A. Strain, G. Cagnoli, M. Casey, J. Hough, M. Husman,
    P. McNamara, G. P. Newton, M. V. Plissi, N. A. Robertson, S. Rowan,
    D. I. Robertson, K. D. Skeldon, C. I. Torrie, H. Ward, B. F. Schutz,
    I. Taylor, and B. S. Sathyaprakash
TAMA Project: Design and Current Status 128
    M. Ando, K. Tsuboni, and the TAMA Collaboration
Status of the Australian Consortium for Interferometric Gravitational Astronomy 140
    D. McClelland, M. B. Gray, D. A. Shaddock, B. J. Slagmolen,
    S. M. Scott, P. Charlton, B. J. Whiting, R. J. Sandeman, D. G. Blair,
    L. Ju, J. Winterflood, D. Greenwood, F. Benabid, M. Baker,
    Z. Zhou, D. Mudge, D. Ottaway, M. Ostermeyer, P. J. Veitch,
    J. Munch, M. W. Hamilton, and C. Hollitt

Overviews

Suspensions 153
    S. Braccini
VIRGO Suspension R&D: Fused Silica and Creep 162
    L. Gammaitoni, *J. Kovalik*, F. Marchesoni, *M. Punturo*,
    and G. Cagnoli
Interferometer Configurations—An Overview 173
    K. A. Strain
Energetic Quantum Limit in Large-Scale Interferometers 180
    V. B. Braginsky, M. L. Gorodetsky, *F. Y. Khalili*, and K. S. Thorne

Advanced Configurations

A Power-Recycled Michelson Interferometer with Resonant Sideband Extraction 193
    M. B. Gray, D. A. Shaddock, and D. E. McClelland
The Polarization Sagnac Interferometer as a Candidate Configuration for an Advanced Detector 200
    P. T. Beyersdorf, R. L. Byer, and M. M. Fejer
Transfer Functions for Fields in a 3-Mirror Nested Cavity 204
    M. Rakhmanov
Signal Extraction and Length Sensing for LIGO II RSE 208
    J. Mason and P. Willems

*Italicized name indicates author(s) who presented the paper.

Lasers and Optics

The GEO 600 Stabilized Laser System and the Current-Lock Technique 215
    B. *Willke*, O. S. Brozek, K. Danzmann, C. Fallnich, S. Goßler,
    H. Lück, K. Mossavi, V. Quetschke, H. Welling, and I. Zawischa

The Influence of X-Ray Damage on High Purity Sapphire Optical Absorption and Investigation on the Origin of the Residual Absorption @1064 nm 222
    F. Benabid, M. Notcutt, *L. Ju*, and D. G. Blair

LISA

Technology of Free Fall for LISA 231
    S. Vitale and R. Dolesi

Deep Surveys of Massive Black Holes with LISA 238
    A. Vecchio

Supermassive Black Holes as Gravitational Wave Sources for LISA 248
    O. Blaes

The Angular Resolution of Space-Based Gravitational Wave Detectors 255
    T. A. Moore and R. W. Hellings

Bar Antennae

An Optical Transduction Chain for the AURIGA Detector 261
    L. Conti, F. Marin, M. De Rosa, G. A. Prodi, L. Taffarello,
    J. P. Zendri, M. Cerdonio, and S. Vitale

MiniGRAIL, A 65-cm Spherical Antenna 268
    A. de Waard and G. Frossati

Detection of Cosmic Rays by NAUTILUS 275
    P. Astone, M. Bassan, P. Bonifazi, P. Carelli, E. Coccia, V. Fafone,
    S. D'Antonio, S. Frasca, A. Marini, E. Mauceli, G. Mazzitelli,
    Y. Minenkov, I. Modena, *G. Modestino*, A. Moleti,
    G. V. Pallottino, V. Pampaloni, M. A. Papa, G. Pizzella,
    F. Ronga, R. Terenzi, M. Visco, and L. Votano

Niobe: Improved Noise Temperature and Background Noise Suppression 283
    M. E. Tobar, C. R. Locke, I. S. Heng, E. N. Ivanov, and *D. Blair*

Suspensions and Thermal Noise

Mechanical Loss Factors of Materials and Suspension Systems for Advanced Gravitational Wave Detectors 293
    S. Rowan, A. Alexandrovski, G. Cagnoli, M. M. Fejer, E. K. Gustafson,
    J. Hough, S. McIntosh, P. Sneddon, and R. Route

*Italicized name indicates author(s) who presented the paper.

Active Seismic Isolation for Enhanced LIGO Detectors 300
    J. *Giaime*, B. Lantz, D. DeBra, J. How, C. Hardham,
    S. Richman, and R. Stebbins

Effect of Optical Coating and Surface Treatments on Mechanical Loss in Fused Silica 306
    A. M. Gretarsson, *G. M. Harry*, S. D Penn, P. R. Saulson,
    J. J. Schiller, and W. J. Startin

Suspension Design for GEO 600—An Update 313
    N. A. Robertson, G. Cagnoli, J. Hough, M. E. Husman,
    S. McIntosh, D. Palmer, M. V. Plissi, D. I. Robertson,
    S. Rowan, P. Sneddon, K. A. Strain, C. I. Torrie,
    and H. Ward

New Seismic Attenuation System (SAS) for the Advanced LIGO Configurations (LIGO2) 320
    A. Bertolini, G. Cella, E. D'Ambrosio, *R. DeSalvo*, V. Sannibale,
    A. Takamori, and H. Yamamoto

Reducing Low-Frequency Residual Motion in Vibration Isolation to the Nanometre Level 325
    J. Winterflood, Z. B. Zhou, and *D. G. Blair*

Inertial Control of the VIRGO Superattenuator 332
    G. *Losurdo* for the Pisa and Florence VIRGO Groups

Measurements of Mechanical Q in Levitated Paramagnetic Crystals 338
    S. J. *Augst* and R. W. P. Drever

Signal Processing and Data Analysis

Validation of Data in Operating Resonant Detectors 345
    G. A. Prodi, L. Baggio, M. Cerdonio, V. Crivelli Visconti,
    V. Martinucci, A. Ortolan, L. Taffarello, G. Vedovato,
    S. Vitale, and J. P. Zendri

SNEWS: The SuperNova Early Warning System 355
    K. Scholberg

Removing Instrumental Artifacts: Suspension Violin Modes 362
    S. *Mukherjee* and L. S. Finn

Search for Gravitational Radiation with the Allegro and Explorer Detectors 369
    P. Astone, M. Bassan, P. Bonifazi, P. Carelli, E. Coccia,
    C. Cosmelli, V. Fafone, S. Frasca, K. Geng, W. O. Hamilton,
    W. W. Johnson, E. Mauceli, M. P. McHugh, S. Merkowitz,
    Y. Minenkov, I. Modena, G. Modestino, A. Moleti, A. Morse,
    G. V. Pallottino, M. A. Papa, *G. Pizzella*, N. Solomonson,
    R. Terenzi, M. Visco, and N. Zhu

Gravitational Wave Detection—The Way Forward 376
    J. Hough

---

*Italicized name indicates author(s) who presented the paper.

PART II. CONTRIBUTED PAPERS

Advanced Configurations

All-Reflective Interferometry for Gravitational-Wave Detection 385
    S. Traeger, P. Beyersdorf, E. Gustafson, R. Beausoleil,
    R. K. Route, R. L. Byer, and M. M. Fejer

Lasers and Optics

GEO 600 Slave Laser Prototype II 389
    I. Zawischa, O. S. Brozek, V. Quetschke, C. Fallnich,
    B. Willke, K. Danzmann, and H. Welling
An Injection Locked Nd:YAG Laser for the Glasgow
10m Interferometric Gravitational Wave Detector 391
    D. A. Clubley, K. D. Skeldon, and G. P. Newton
The Laser System for the LISA Technology Demonstrator 393
    O. S. Brozek, M. Peterseim, K. Danzmann, I. Freitag,
    C. Fallnich, and H. Welling
Effect of Annealing on the Light Absorption in Sapphire 395
    A. L. Alexandrovski, M. M. Fejer, and R. K. Route

LISA

Double Test-Mass Control for LISA 399
    P. C. E. Roberts
Computing LISA Far-Field Phase Patterns 401
    E. Waluschka
LISA Observations of Massive Black Hole Binaries Using
Post-Newtonian Waveforms 403
    A. M. Sintes and A. Vecchio
Globular Cluster Distance Measurements with LISA 405
    M. Benacquista

Bar Antennae

High-Sensitivity Accelerometers for High-Performance
Seismic Attenuators 409
    A. Bertolini, R. DeSalvo, F. Fidecaro, M. Francesconi,
    V. Sannibale, and A. Takamori
Errors on the Inverse Problem Solution in a Noisy Spherical
GW Antenna 411
    J. A. Lobo and S. M. Merkowitz

The First Phase of the Brazilian Graviton Project:
The Mário Schenberg Detector 413
    O. D. Aguiar, N. S. Magalhães, J. C. N. de Araújo, O. D. Miranda,
    J. L. Melo, K. L. Ribeiro, L. A. de Andrade, K. B. M. Salles,
    S. R. Furtado, N. F. Oliveira Jr., W. F. Velloso Jr., C. Frajuca,
    R. M. Marinho Jr., and G. Frossati

An Advanced Inductive Transducer for Resonant Mass
Gravitational Wave Detectors 415
    G. M. Harry, I. Jin, H. J. Paik, T. R. Stevenson,
    and F. C. Wellstood

Perspectives on Transducers for Spherical Gravitational
Wave Detectors 417
    C. Frajuca, O. D. Aguiar, N. S. Magalhães, K. L. Ribeiro,
    and L. A. Andrade

The Auriga Ultracryogenic Test Facility:
A New Capacitive Resonant Transducer 419
    V. Crivelli Visconti, J. P. Zendri, L. Taffarello, G. A. Prodi,
    S. Vitale, M. Cerdonio, M. Bonaldi, P. Falferi, R. Mezzena,
    and A. Mattioli

Status Report of the Gravitational Wave Detector AURIGA 421
    J. P. Zendri, L. Baggio, M. Bonaldi, M. Cerdonio, L. Conti,
    V. Crivelli Visconti, P. Falferi, P. L. Fortini, V. Martinucci,
    R. Mezzena, A. Ortolan, G. A. Prodi, G. Soranzo, L. Taffarello,
    G. Vedovato, A. Vinante, and S. Vitale

Noise and Signal Reduction and Characterization
in the AURIGA Detector 423
    L. Baggio, M. Cerdonio, V. Martinucci, A. Ortolan, G. A. Prodi,
    L. Taffarello, G. Vedovato, S. Vitale, and J. P. Zendri

Recent Results of NAUTILUS 425
    P. Astone, M. Bassan, P. Bonifazi, P. Carelli, E. Coccia,
    S. D'Antonio, V. Fafone, A. Marini, E. Mauceli, Y. Minenkov,
    I. Modena, G. Modestino, A. Moleti, G. V. Pallottino, M. A. Papa,
    G. Pizzella, F. Ronga, R. Terenzi, M. Visco, and L. Votano

Improving the Sensitivity of Resonant Detectors with Advanced
Linear Capacitive Transducers 427
    M. Bassan and Y. F. Minenkov for the ROG Collaboration

Tests on a Prototype Spherical Detector 430
    P. Astone, M. Bassan, P. Bonifazi, P. Carelli, E. Coccia,
    S. D'Antonio, V. Fafone, G. Frossati, A. Marini, E. Mauceli,
    S. Merkowitz, Y. Minenkov, I. Modena, A. Moleti, G. Modestino,
    G. V. Pallottino, M. Papa, G. Pizzella, G. Raffone, F. Ronga,
    M. Schipilliti, R. Terenzi, M. Visco, and L. Votano

Suspensions and Thermal Noise

High-Sensitivity Measurement and Control of Thermal Noise in a Cavity 435
    A. Heidmann, P. F. Cohadon, Y. Hadjar, and M. Pinard

The Maraging Steel Blades of the Virgo Superattenuator 437
    S. Braccini, C. Casciano, F. Cordero, F. Corvace,
    M. De Sanctis, R. Franco, F. Frasconi, E. Majorana,
    R. Passaquieti, G. Paparo, P. Rapagnani, F. Ricci,
    A. Solina, and R. Valentini

The ANU Thermal Noise Experiment 439
    M. B. Gray, B. J. J. Slagmolen, K. G. Baigent,
    and D. E. McClelland

Design of an Isolation System to a Medium-Size Spherical
Resonant Gravitational Wave Detector 441
    W. F. Velloso Jr., J. L. Melo, and O. D. Aguiar

Development of a Double Pendulum for Gravitational Wave Detectors 443
    M. Beilby, G. Gonzalez, M. Duffy, A. Stuver, and J. Poker

Bi-Filar Pendulum Mode Q Factor for Silicate-Bonded Pendulum 445
    K. Tokmakov, V. Mitrofanov, V. Braginsky, S. Rowan, and J. Hough

Prototype of the Suspension Last Stages for the Mirrors
of the Virgo Interferometric Gravitational Wave Antenna 447
    A. Bernardini, L. Brocco, E. Majorana, P. Puppo, P. Rapagnani,
    F. Ricci, and G. Testi

Signal Processing and Data Analysis

Toward Gravitational Wave Detection 451
    L. S. Finn, G. Gonzalez, J. Hough, M. F. Huq, S. Mohanty,
    J. Romano, S. Rowan, P. R. Saulson, and K. A. Strain

Separation of the Gravitational-Wave Signals and the Solar
Oscillation Signals 459
    X.-T. Ni and X. Xu

Data Archiving and Distribution of the VIRGO Antenna
for Gravitational Wave Detection 461
    F. Barone, A. Eleuteri, F. Garufi, and L. Milano

The Environment Monitoring of the VIRGO Antenna
for Gravitational Wave Detection 463
    A. Anastasio, F. Barone, A. Eleuteri, F. Garufi, and L. Milano

A Neural Network Approach for the Noise Identification
and Data Quality of the VIRGO Antenna 465
    F. Barone, A. Ciaramella, A. Eleuteri, F. Garufi, L. Milano,
    and R. Tagliaferri

Gravitational Wave Signal Detection with Neural Networks
for the VIRGO Antenna 467
    F. Barone, A. Eleuteri, F. Garufi, L. Milano, and R. Tagliaferri

LIGO End-to-End Simulation Program 469
    B. Bhawal, G. Cella, M. Evans, S. Klimenko, E. Maros,
    S. D. Mohanty, M. Rakhmanov, R. L. Savage Jr.,
    and H. Yamamoto

The Logging and Data Retrieve System for the GW Detector AURIGA....... 471
    A. Ortolan, L. Baggio, M. Cerdonio, V. Martinucci,
    G. A. Prodi, L. Taffarello, S. Vitale, G. Vedovato,
    and J. P. Zendri

An Efficient Matched Filtering Algorithm for the Detection of Continuous Gravitational Wave Signals............................. 473
    P. R. Williams and B. F. Schutz

A Likelihood-Based Scheme for Coincidence Analysis..................... 477
    S. Mukherjee and S. D. Mohanty

A Robust Test for Detecting Non-Stationarity........................... 479
    S. D. Mohanty

Program... 481

Author Index... 489

PREFACE

This volume contains the Proceedings of the Third Edoardo Amaldi Conference on Gravitational Waves, which was held at the California Institute of Technology, Pasadena, July 12 through 16, 1999.

The Amaldi Conference had been held twice before, in Frascati, Italy (1994) and at CERN, Geneva, Switzerland (1997), but took on a new significance after it was designated as the cornerstone meeting for the recently formed Gravitational Wave International Committee (GWIC). GWIC includes representatives from the five major interferometer projects – ACIGA, GEO, LIGO, TAMA, and VIRGO; the five major acoustic detector projects – ALLEGRO, AURIGA, EXPLORER, NAUTILUS, and NIOBE; and the space-based interferometer project LISA.

Gravitational wave scientists, two hundred twenty-six of them, gathered at Caltech to attend the Third Edoardo Amaldi Conference. Almost half of these scientists were from outside the United States, with the largest number, forty, coming from Italy. They were presented with a variety of stimuli, ranging from visual to gustatory to scientific, and in all cases, their response was enthusiastic.

The conference events began on Sunday evening, July 11, with a welcoming reception and registration at the Athenaeum Rathskeller. To start the Conference on Monday morning, introductions and welcomes were offered by LIGO Director Barry Barish and Caltech President David Baltimore. Ugo Amaldi, Edoardo Amaldi's son and himself a noted high-energy experimentalist, presented an insightful and loving portrait of his father.

The Conference was organized so that there were three purely plenary sessions. The other sessions were mixtures of plenary talks chosen by the session convenors, followed by a number of talks chosen, again, by the convenors, from abstracts submitted to the Conferences. Altogether there were sixty-two talks. A total of ninety-two abstracts were submitted to the Conference. Many of these were presented at a very popular Poster Session, on Tuesday evening, July 13.

As may be gleaned from a perusal of the program and of the papers presented in these Proceedings, every aspect of gravitational wave detection research, both theoretical and experimental, was touched on during the Conference. This was made possible by the hard work of the session convenors. Many thanks are due to them and to the members of the International Advisory Committee and the Program/Organizing Committee for their expert advice.

A featured social activity was a trip to the new Getty Museum on Wednesday afternoon. Later that evening, Kip Thorne gave a beautiful public lecture at the Beckman Auditorium entitled, "Probing Black Holes and the Dark Side of the Universe with Black Holes." Nine hundred forty-three fortunate listeners attended and kept Kip answering questions far into the night.

The success of the Amaldi Conference was crucially dependent on the financial support provided by the Caltech Department of Physics, Mathematics, and Astronomy, the US National Science Foundation, the Italian National Institute for Research in Nuclear and Subnuclear Physics (INFN), Lightwave Electronics Corporation, and the Parsons Corporation. We are grateful for their help.

Warm appreciation is due to Jake Robinson, who constructed the Amaldi Conference web page, and to Barbara Kratochwill of Caltech for her service in operating and maintaining the Amaldi Conference web page. Jeanne Rostant, of CERN, and Eugenio Coccia provided useful advice based on their work on earlier Amaldi conferences. Susan Davis, of Caltech, offered analogous useful advice based on her experience in organizing an Econometrics conference at Caltech.

We are pleased to thank Valeda Mercier and Veronica Kondrashov, who were crucial in helping to organize and coordinate all phases of the Conference, from arranging the trip to the Getty Museum, to ordering Conference bags and folders, and to choosing menus for the Cocktail Party, Poster Session and Conference Banquet.

Special thanks are due Veronica Kondrashov for her skill and patience in producing these Proceedings.

The impact of the Third Edoardo Amaldi Conference won't be known for some time. However, whenever you put two hundred twenty-six eager, creative, interested scientists together, interacting at breakfast, lunch, dinner and social occasions, good things are sure to follow.

For the Organizing Committee,
Sydney Meshkov,
Chair

SPONSORED BY

Department of Physics, Mathematics, and Astronomy - California Institute of Technology
NSF – US National Science Foundation
INFN - Istituto Nazionale di Fisica Nucleare
Parsons Corporation
Lightwave Electronics

THE GRAVITATIONAL WAVE INTERNATIONAL COMMITTEE (GWIC)

INTERNATIONAL ADVISORY COMMITTEE

P. Astone	(Roma)
P. Bender	(JILA)
B. Bertotti	(Pavia)
D. Blair	(UWA)
V. Braginsky	(Moscow)
A. Brillet	(CNRS)
R. Byer	(Stanford)
N. Cabibbo	(Roma)
J. Centrella	(Drexel)
M. Cerdonio	(Padova)
T. Damour	(Paris)
R. Drever	(Caltech)
J. Faller	(JILA)
S. Finn	(Penn Sate)
G. Frossati	(Leiden)
M. Fujimoto	(NAO)
A. Giazotto	(INFN Pisa)
W. Hamilton	(LSU)
J. Hough	(Glasgow)
Y. Kozai	(NAO)
K. Kuroda	(ICRR)
J. Pacheco	(Nice)
G. Pizzella	(Roma)
N. Robertson	(Glasgow)
A. Rüdiger	(MPI Garching)
J. Sandeman	(ANU)
B. Schutz	(Potsdam)
K. Tsubono	(Tokyo)
P. Van Amersfoort	(NIKHEF)
S. Vitale	(Trento)
H. Ward	(Glasgow)

PROGRAM/ORGANIZING COMMITTEE

S. Meshkov, Chair	(Caltech)
B. Allen	(UWM)
B. Barish	(Caltech)
J. Camp	(Caltech)
E. Coccia	(Roma)
M. Coles	(Livingston)
G. Gonzalez	(Penn State)
N. Mavalvala	(Caltech)
F. Raab	(Hanford)
G. Sanders	(Caltech)
P. Saulson	(Syracuse)
D. Shoemaker	(MIT)
R. Stebbins	(JILA)
K. Thorne	(Caltech)
G. Veneziano	(CERN)
R. Weiss	(MIT)

CONFERENCE SECRETARIAT

V. Kondrashov	(Caltech)
V. Mercier	(Caltech)

PART I
INVITED LECTURES

Amaldi Meeting Introduction

Barry C. Barish

California Institute of Technology
Pasadena, CA 91125

Welcome to Caltech and the 3rd Edoardo Amaldi Conference on Gravitational Waves. Obviously, something must be very interesting to bring more than 250 scientists from around the world to Pasadena in July for this particular meeting. In fact, July in Southern California does have many attractions, in addition to the good weather and cool nights. For this conference, we have arranged a visit to the new Getty Museum on our excursion day. This is meant to make your stay more pleasant, but is not the real reason we have gathered here. This meeting addresses the detection of gravitational waves, a much-anticipated event.

Einstein predicted Gravitational Waves in 1916, as a result of his theory of gravity. This new description, involving curved space-time, solved the fundamental problem in Newtonian gravity of instantaneous action at a distance. Other effects of general relativity have been observed, including the recent advances in gravitational lensing that are becoming so important in astrophysics. Gravitational waves, themselves, have been indirectly observed in the beautiful experiment of Hulse and Taylor, by measuring the gradual speed up of the period of the binary neutron star system PS 1913+16, and showing that it precisely agrees with the effect predicted from the radiation of gravitational waves.

The experimental challenge of detecting gravitational waves has been with us since Einstein introduced the concept. Unfortunately, the effects due to gravitational waves are so small that a self-contained laboratory experiment appears unfeasible. Instead, the present attempts are to detect gravitational waves resulting from some astrophysical or cosmological effect. Possible sources include supernovae, binary inspiral of compact objects like neutron stars or black holes, spinning neutron stars, or even gravitational waves that were emitted in the first instants of the early Universe.

Sensitive techniques have been developed over the past three decades, first using resonant bars and more recently, suspended mass interferometers. Vigorous research has been done to develop these techniques and large ambitious experiments are now being implemented. With the world-wide effort and the very large interferometers being developed the community has grown and there is a growing need for this emerging field to organize itself for collaboration, exchange of information and to represent the field.

For these reasons, the gravitational wave community organized itself a couple years ago forming the Gravitational Wave International Committee (GWIC, for short) in order to help foster the international aspects of this emerging field. One of the first tasks was to create a conference specifically for this field. Rather than create a new conference, GWIC decided to 'evolve' a successful existing topical workshop, the Amaldi meeting that had been held twice, once at Frascati, Italy and then at CERN in Switzerland.

GWIC has become an important committee for the field of gravitational waves. It has representatives from all major projects in the world - resonant bars, interferometers and also the space based initiative. The committee is developing a formal tie with IUPAP (the International Union of Physics and Applied Physics), which is the highest body of scientists representing international community. The relationship is through an Associated Commission - AC2 that covers general relativity; as well as being linked to PaNAGIC, a newly formed committee in IUPAP to represent particle astrophysics and related fields. We expect to seek IUPAP sponsorship of the next Amaldi meeting, which will be in Perth, Australia in 2001.

This Amaldi meeting at Caltech represents the first in the series under the auspices of GWIC and has been organized as a true international conference. To accomplish that, we have followed IUPAP guidelines for international conferences in terms of distribution of delegates, speakers, etc.

The program of the meeting covers astrophysics sources, resonant bar detectors (coincidences), suspended mass interferometers (new generation becoming operational over the next few years), and space based experiments (LISA) (which hopefully will become a reality over the coming decade). It covers technical details, as well as data analysis approaches and techniques. Syd Meshkov has put the conference together very capably with advice from an international organizing committee.

Lastly, I would like to comment that GWIC has kept the title Eduardo Amaldi Conference in tribute to the fact that Eduardo Amaldi was such an important figure in 20th century experimental physics, and that he spent the later part of his career pioneering this new field and developing the techniques to search for gravitational waves. We are fortunate today to have his son Ugo Amaldi, an old friend, here to talk and give us a tribute to his father.

Hopefully, this will be the first in a long series of successful Amaldi meetings sponsored by GWIC that will trace the progress in this exciting field as gravitational waves are directly detected and become a new tool for studying fundamental questions in physics and astronomy.

Edoardo Amaldi, Scientist

Ugo Amaldi

CERN, 1211 Geneva 23, Switzerland and
TERA Foundation, Via Puccini, 11, 28100 Novara, Italy

INTRODUCTION

Edoardo Amaldi was born in Piacenza (Italy) in 1908 and died in Rome on December 5th, 1989. His scientific activity, spanning sixty years, has covered many fields of physics. Atomic physics (for about five years, starting in 1929), nuclear physics (for ten years, from 1934 to 1944), cosmic rays (fifteen years, from 1945 to 1961), particle physics around particle accelerators (ten years) and the search for gravitational waves. This was the longest lasting (twenty years) of his many scientific commitments since he launched the first project in 1970, and when he died of an heart attack at the seat of the *Accademia dei Lincei*, of which he was the President, gravitational waves were still his only scientific interest.

In parallel with these scientific activities, Amaldi devoted much of his time to the management of science and to the definition of postwar European science policy, to the education of university and high school students, to the defense of human rights and to disarmament and arms control. Along the years he also wrote on the history of this century physics and on some of the physicists he knew, so that a volume of more than six hundred pages has recently been published (1). All these items will be touched upon in the last four Sections, while the rest of this paper focuses on his scientific activities.

Many more details on all the subjects can be found in Ref. 1 and in the biography written by Carlo Rubbia for the Royal Society (2). Comments on Edoardo Amaldi's life and activities appear in the volumes printed for his 60th (3) and 70th birthday (4) and in the proceedings of the First (5) and the Second (6) *Edoardo Amaldi Conference on Gravitational Waves*, published by World Scientific as Vols. 1 and 4 of the *Edoardo Amaldi Foundation Series*.

In 1979 Piero Angela, the best known Italian science journalist, recorded an interview of more than twelve hours during which Amaldi told many anecdotes and expressed his opinion, on the development of physics research, University teaching, the spin-offs of science to society, the nuclear bomb and the worldwide effort on disarmament. The book *"Interview on Matter: from Nuclei to Galaxies"* is a lively and balanced representation of his attitudes and ideas (7). After his death, this interview and the Italian translation of Rubbia's biography were published under the title *"Edoardo Amaldi, Scientist and European Citizen"* (8).

THE YOUTH YEARS

Edoardo Amaldi was born in a small town, seventy kilometers south of Milan, where his parents had a very simple country house to which he would come back every year towards the end of August, even for a very short stay. His father Ugo was a well-known mathematician, over the years professor at the Universities of Cagliari, Modena, Padua and Rome. He was an expert in 'infinite' Lie groups, i.e groups having an infinite number of generators. He was the first to classify them, as Elie Cartan had done with the 'finite' Lie groups. Ugo was a teacher of exceptional clarity, as testified also by Emilio Segrè, who had him as a professor in Rome around 1926. Much later Segrè wrote, addressing Edoardo (9): "In those halcyon years in Via Panisperna ... once in a while Majorana stunned us with his groups. We turned to your father for help, because my experience as a student had convinced me that if there were any hope of mastering group theory, it was through his teaching. Contrary to our rather utilitarian approach, typical of young experimental physicists, he told us that we should really study also Lie groups. He was a prophet, and would rejoice seeing physicists using SU(3) and beginning to appreciate what was one of his chief interests."

FIGURE 1. Edoardo Amaldi climbing at the age of 17.

Ugo Amaldi devoted much time, during almost fifty years, to the writing of high school textbooks on geometry and algebra in collaboration with one of the most influential Italian mathematician of this century, Federigo Enriques. Between 1905 and 1960 most Italian high school girls and boys, myself included, studied using their books.

Edoardo attended high school in Padua and entered University in Rome, where his father was appointed mathematics professor in 1925. He was fond of sports during all his life, in particular tennis, skiing and climbing (Fig. 1). He continued to climb well into his sixties and at that time he also used to run regularly. To indicate one of the many sides of his very precise character, it is worthwhile mentioning that, when he was around seventy, he would run five kilometres almost every morning and he was satisfied when the time he had taken was no more than twice the world record.

In Rome, Edoardo enrolled as a student in engineering, but two years later he changed to physics studies after an inspired address made by Senator Orso Mario Corbino, Director of the Physics Institute. Corbino told the engineering students that Enrico Fermi, a young brilliant scientist, had been appointed professor of theoretical physics, that Franco Rasetti, an excellent experimentalist, had been chosen as Fermi's assistant and that physics would soon see a new renaissance. By deciding for this change of faculty Edoardo followed Emilio Segrè, who had made the same move a couple of years before.

FIGURE 2. From left to right: Edoardo Amaldi, Franco Rasetti and Emilio Segrè, during a Sunday walk at the beginning of the thirties.

Thus the 'Rome group' was formed. Close to the group, but not really belonging to it, was Ettore Majorana: in 1929 Edoardo took his *Laurea* on the same day as Ettore, of whom he was the closest friend. (Much later he wrote a scientific biography of Majorana (10), who disappeared in 1938 – most probably by committing suicide).

The Rome group, as discussed by Gerald Holton (11), was the first group of physicist to work as such, dividing the responsibilities, sharing the work and enjoying vacations and sports together (Fig. 2). Since then, the group is known in Italy as 'i ragazzi di Via Panisperna' (the boys of Via Panisperna), because the Physics Institute where Fermi and collaborators worked, from 1926 to 1936, was located in Via Panisperna, in the centre of Rome.

It is worth mentioning that the building of Via Panisperna still exists (Fig. 3) and, after having being part for many decades of the Italian Ministry of the Interior, it will be soon restored and transformed into a museum.

FIGURE 3. The building of the Physics Institute in Via Panisperna.

Initially, the Rome group was specialised in atomic physics and in 1931 Amaldi spent a year in Lipsia working under Peter Debye on diffraction of X rays by liquids (12). Back in Rome, he discovered with Segrè the 'swollen' atoms (13), now known as 'Rydberg atoms', i.e. gas atoms whose external electrons have such a large principal quantum number that their orbit includes other atoms of the gas. Fermi soon published the theory of the phenomenon (14) and this would have probably been the subject of many other papers, if the discovery of the radioactivity induced by neutrons, at the beginning of 1934, had not induced a complete change in the direction of the research done in Rome.

THE NEUTRON WORK

The stream of discoveries initiated by the first Fermi paper on neutron radioactivity is described by Amaldi in a long review paper published in 1984 in *Physics Reports* under the title '*From the discovery of the neutron to the discovery of nuclear fission*' (15). Typical of his extremely careful way of documenting every detail, it contains 924 references and notes. In the report, the discussion on the radioactivity induced by neutrons is preceded by three chapters: *'The discovery of the neutron opens the great season of nuclear physics'*, *'Other highlights on the same period of nuclear physics'* and *'Beta decay discloses the existence of a new particle and a new interaction'*.

In the last one, Amaldi describes the proposal of Pauli of a 'neutron' emitted by the radioactive nucleus together with an electron and adds a very interesting note (number 277 of Ref.15). "The name *neutrino* (a funny and grammatically incorrect contraction of 'little neutron' in Italian: neutronino) entered the international terminology through Fermi, who started to use it sometime between the conference in Paris in July 1932 and the Solvay Conference in October 1933, where Pauli used it. The word came out in a humorous conversation at the Istituto di Via Panisperna. Fermi, Amaldi and a few others were present and Fermi was explaining Pauli's hypothesis about his 'light neutron'. For distinguishing this particle from the Chadwick neutron Amaldi jokingly used this funny name, - says Occhialini, who recalls of having shortly later told around this little story in Cambridge". I have known this for a long time and I have even asked Beppo Occhialini about it, getting a sharp answer as if everybody should know how the name neutrino came about.

The theory of beta decay was published by Fermi in *La Ricerca Scientifica* in December 1933 and explained to his collaborators at the beginning of 1934. Fifty years later, in the Physics Report, Amaldi writes: "The first time I heard about this paper - that much later became known as the first step in the theory of weak interactions - was from Enrico Fermi himself, one evening between Christmas 1933 and the beginning of 1934, at the Hotel Oswald in Selva, Val Gardena. A few physicists of the University of Rome were spending their Christmas vacation in this beautiful village in the Dolomites, and one evening, after a full day of skiing, Fermi invited us to his room for explaining the essence of the paper he had sent for publication some time before. Since in the room there was at most one chair, Fermi sat with croutched legs in the middle of his bed, while Segrè, Rasetti and I sat around him, on the edge of the bed, with our necks twisted trying to see what he was writing on a piece of paper leaning on his knees" (16). A detail is obviously not reported here, which he told me a couple of times: this vacation was memorable for him because not only was he taught how nuclei beta decay, but he also learned from Ginestra that she was expecting their first baby.

The neutron work was performed in Rome between 1934 and 1936, when the group of Via Panisperna dispersed. The details can be found in the already quoted *Physics Report*, which deals with the many experiments leading to four major discoveries:

(i) the radioactivity induced by neutrons,
(ii) the radioactivity induced by neutrons slowed down by collisions with light nuclei, in particular hydrogen,
(iii) the $1/v$ law, with which slow neutrons are absorbed in nuclei with the emission of gamma rays,
(iv) the large absorption cross-section of cadmium for slow neutrons, the existence of strong selective absorption bands and the effect of chemical bonds on the phenomenon.

The two first papers on the discovery of the radioactivity induced by neutrons are signed by Fermi (17). Already in the second one he acknowledges the contribution of

Amaldi and Segrè in carrying out the experiment. Since then, the Rome group worked actively together and by summer 1934 fifty new radioactive isotopes had been found and three articles had been published in *La Ricerca Scientifica* (18), where at the time Ginestra – who was an astronomer before becoming in 1933 Edoardo's wife - worked as deputy editor. Through her, the reprints were made available within days, so that they could be sent by mail to a selected list of prominent physicists: the Rome group was the first to use 'preprints' to spread its results rapidly.

FIGURE 4. Edoardo Amaldi in his youth.

The group work "was organised in a very efficient way. Fermi, helped by Rasetti, did a good part of the measurement and calculations, Segrè secured the substances to be irradiated and the necessary equipment and later became involved in most of the chemical work. I took care of the construction of the Geiger Müller counters and of what we now call electronics. The division of the activities, however, was not rigid at all and each of us participated in all phases of the work. We immediately realised that we needed the help of a professional chemist. Fortunately we succeeded almost immediately in convincing Oscar D'Agostino, (who) had held a fellowship in Paris in the laboratory of Madame Curie." (19)

In summer 1934, a manuscript summarising the work done in Rome was brought by Amaldi (Fig. 4) and Segrè to Lord Rutherford in Cambridge (20). At the first encounter, Segrè asked whether it would be possible to obtain speedy publication in the Proceedings of the Royal Society. Many years later he wrote: "I imprudently recommended prompt publication, whereupon he answered, whether in jest or annoyance I could not tell 'What do you think I am the President of the Royal Society for?'" (21). Edoardo adds: "Unfortunately our understanding of Rutherford English at the time was imperfect and we could not follow most of his remarks, many of which must have been humorous because he laughed from time to time and only then took the pipe out of his mouth." (22)

The way to the second discovery was opened when, on October 18th, 1934, Edoardo Amaldi started a systematic investigation to clarify the miraculous properties of some wooden tables, on which the induced radioactivity was larger than the one measured with the same apparatus mounted on a marble table. On October 22nd, Fermi, being alone, instead of using a lead absorber decided, without any conscious argument in mind, to use a paraffin wedge. The letter written the same evening in Amaldi's apartment was sent the day after to *La Ricerca Scientifica,* but I do not have any direct recollection of the facts, since I was at the time two months old. Emilio Segrè described the scene as follows: "Fermi dictated while I wrote. He stood by me; Rasetti, Amaldi and Pontecorvo paced the room excitedly, all making comments at the same time. The din was such that when we left, Amaldi's maid discreetly asked whether the evening guests were tipsy. Ginestra Amaldi handed the paper to her boss at La Ricerca Scientifica the following morning" (21).

The $1/v$ law was suspected since the beginning, but the proof came from experiments done in Rome and elsewhere. The Rome group was the one which performed the first successful mechanical experiment in which the moderator was moving with respect to the source, so as to determine the relevant neutron velocities (21). The fourth discovery, related to the existence of strong absorption lines, started with the observation of the anomalous absorption of cadmium by the Rome group and continued with the observations made by Bjerge and Westcott (24) and Moon and Tillman (23), who noticed that the absorption coefficient for slow neutrons of certain elements would differ according to the element used as detector. In 1936 Amaldi and Fermi performed the systematic investigation of 11 elements in all possible combinations with 7 detectors, whose induced activity was quantitatively measured (26).

The framework of this work was described by Edoardo in introducing the relevant papers in the *Enrico Fermi Collected Papers* (27), where one can also find the English translation of all the articles published in Italian by the Rome group. Amaldi wrote: "After the summer vacation of 1935, Fermi and I found ourselves alone in Rome. Most of the group had dispersed by now. The general atmosphere in Italy was chiefly to blame for this as the Country prepared for the Ethiopian war. Rasetti had gone to the United States … . Segrè too… . D'Agostino no longer worked with us. Pontecorvo … for a few months worked with Wick. Upon resuming Fermi and I … went to work with even greater energy than in the past as if by our own more intensive efforts we wanted to compensate for the loss of manpower in our group. We had prepared a systematic plan of attack which we jokingly summarised by saying that we would measure the absorption coefficient of all 92 elements combined in all possible ways with the 92 elements used as detectors."

As a result of this systematic search, Amaldi and Fermi also discovered the effect of the chemical binding forces. The corresponding papers (28) conclude the intellectual adventure of the boys of Via Panisperna.

FIGURE 5. Photo taken at the Copenhagen Conference of June 1936. On the first row from the left: W. Pauli, P. Jordan, W. Heisenberg, M. Born. L. Meitner, O. Stern, J. Franck. Standing from the left: N. Bohr, L. Rosenfeld, E. Amaldi, G. Wick. Amaldi was 28.

All the work done in Rome was summarised by Edoardo at the International conference on "*Probleme der Atomkernphysik*", held in Copenhagen from 14 to 20 June, 1936 (Fig. 5). At the conference Edoardo Amaldi realised that, in order to be competitive, more intense neutron sources were mandatory and, in particular, that the best experiments would be done around cyclotrons and other accelerators already existing in a few laboratories. Thus, the group decided that an accelerator was needed in Rome and that direct information should be collected on the matter.

In summer 1936 Amaldi left for the United States, sharing his time between Columbia University and the Carnegie Institution of Washington, where Merle Tuve (a school-mate of Ernest Lawrence) and collaborators had constructed a 'sectionalised accelerating tube' powered by a Van der Graf electrostatic accelerator. He learned about accelerators and also had the time to perform some experiments. As a result, a paper was published on the neutron yields from artificial sources (29).

PHYSICS IN THE WAR YEARS

In November 1936, to get the funds needed for the accelerator, Fermi wrote a proposal to the Italian Ministry of Interior together with Domenico Marotta, at that time Director of *Istituto Superiore di Sanità* (ISS). (The Italian *Higher Health Institute* is roughly equivalent to the American *National Institutes of Health* while its Physics Laboratory, where the accelerator had to be installed, can be compared with the *National Bureau of Standards*, though on a much smaller scale). Waiting for the funds, a prototype 200 KeV Crocroft-Walton was constructed; it accelerated particles in June 1937 (30). Shortly afterwards Edoardo was appointed to the chair of Corbino, who had prematurely died. The funds for the accelerator were available and the construction had just started when, at the end of 1938, Fermi left for the States directly after receiving in Stockholm the Nobel prize for the discovery of the radioactivity induced by neutrons. With this momentous event, the Rome group ceased to exist. Still, there was time for Amaldi and Rasetti to discover the now much used property of gadolinium of emitting electrons when capturing neutrons (31) and to complete, with the physicists of the Physics Laboratory of ISS, the 1.1 MeV Crockroft-Walton (32).

Fermi's departure changed the life of Edoardo Amaldi. He began to realise that, being the only member of the group in Rome, he had to modify his plans and devote much of his time to organise the research of other people. Thus, he proceeded to direct the experimental work around the Crockroft-Walton, which was installed in the building of the ISS in Rome (financed by the Rockfeller Foundation), not far from new Physics Institute of the University, where the Fermi group had moved from Via Panisperna in 1936. Then, in 1939, he made a long trip in the States, which he recounted in a moving unfinished Italian manuscript. This has been published and commented by Gianni Battimelli and Michelangelo De Maria (33) who, together with Gianni Paoloni and Lucia Orlando, have set up and take care of the *Amaldi Archive* at the Department of Physics of the University *La Sapienza* in Rome (34).

This trip had the purpose of collecting direct information in Berkeley, Columbia, Ann Arbour and elsewhere, so as to be in a position to construct in Italy an even more powerful machine, a cyclotron. This accelerator was never built, as the many others that Edoardo Amaldi proposed and staunchly defended from 1939 to 1950. All the work and ingenuity gone in these failed attempts have been described by Battimelli and Gambaro in a contribution having the self-explanatory title *'From Via Panisperna to Frascati: the accelerators which where never built'* (35). These failures never discouraged Amaldi. On the contrary, very characteristically they pushed him to pursue on a different and much wider basis the same purpose, so that in the fifties he became one of the founding fathers of CERN and of the Frascati Laboratories.

During the long tour in the States, Amaldi quietly enquired whether it was possible for him to find a position and emigrate. Indeed, before leaving, he had requested a passport for Ginestra - who was now expecting my brother Francesco - myself and my sister Paola. The passport was refused. On September 1st, 1939, two days before Edoardo arrived in Berkeley, Germany invaded Poland. In the following days the war broke out and he decided to leave with the next ship for Europe. Edoardo did not want to be separated from his family and this was his main motivation. But in taking the ship he also had his mind on two conversations he had had with Werner Heisenberg and Felix Bloch, who had both told him on two different occasions of the responsibilities he would have to accept, if he was going back to Rome, to save, at least in part, Italian physics.

He wrote: "The day after (the conversation with Bloch) on October 4, 1939, I boarded the ship 'Vulcania' which sailed to Naples, where I arrived on October 14. In very few periods of my life, may be in none, I felt so anxious as during these ten days of navigation. I was coming back knowing that our group was definitely destroyed, and I had no hope of leaving fascist Italy for a few years, in an Europe raged by war, in which within few months even our country will have been thrown and, even worst, thrown on the wrong side. The considerations by Felix Bloch were also coming back to my mind as a way to give a meaning to my life, but the way was different from all those I had always imagined and far from my aspirations and capacities" (33).

Back in Rome, he concentrated on the use of the Crockroft-Walton in collaboration with physicists from Istituto di Sanità, studying the fission processes induced by high-energy neutrons. This scientific activity was interrupted for about six months when, in May 1940, he was mobilised in the Italian Army and sent as an officer to North Africa. About these experiments he wrote (36): "We found an interesting increase in the cross-section and together with Gian Carlo Wick – our theoretician – I corresponded with Niels Bohr, who suggested a transparent interpretation of our results. When we understood that the United States were close to entering the war, we summarised our results in a paper written in English and sent it to Physical Review (37), where it appeared after the Japanese attack on Pearl Harbour. At that time we held a small meeting (G. Bernardini, B.N. Cacciapuoti, B. Ferretti, G.C. Wick, E. Amaldi and a few others), in which we decided to stop any work on fission and to employ our limited manpower on completely different problems. The reason for such a decision was that we had arrived at the conviction that almost any problem related to the investigation of fission could become of interest for the construction of weapons and we did not want to become involved in this kind of work."

Thus, the Crockroft-Walton was used to study neutron-proton and neutron-deuteron collisions. Despite the war and the occupation of Rome, experiments were also performed on the elastic scattering of fast neutrons by medium and heavy nuclei (38, 39). The most interesting result concerns the first experimental test of the 'optical theorem', a direct consequence of the wave nature of the neutrons absorbed and scattered by nuclei.

While experiments were proceeding at the *Istituto Superiore di Sanità*, at the Physics Institute of the University cosmic rays were studied under the guidance of Gilberto Bernardini, so that in Rome there were two main research programmes: nuclear physics under Amaldi and cosmic rays under Bernardini. Giuseppe Cocconi would often visit and Marcello Conversi, Ettore Pancini and Oreste Piccioni were junior assistants. Their famous experiment was completed in 1946, one year after the Allies entered Rome, but the apparatus had been kept running with the contribution of everybody even during the times of the nazi occupation. Amaldi wrote: "Immediately after the arrival of the Allies we brought back to the Institute from the Liceo Virgilio (which was close to the Vatican and thus protected from air-raids) the equipment that Conversi and Piccioni had built and used for measuring the mean life of mesons. The measurement had been carried almost without interruption also during the months of occupation. To keep this experiment in operation, at any cost, had become, for all of us, a kind of symbol of our will of cultural and scientific continuity" (40).

COSMIC RAYS

In fall 1946, Ginestra and Edoardo Amaldi spent almost three months in the States visiting the most important laboratories and universities. Edoardo could thus describe in many seminars and private conversations the research carried out in Italy during the war and learn the fantastic developments of American nuclear physics. During his stay, he also had the satisfaction to receive, from Rome, news of the unexpected results of the Conversi-Pancini-Piccioni experiment, which demonstrated that mesotrons were not strongly interacting, and to present them on various occasions. Fermi offered him a chair in Chicago, but Amaldi decided to come back to Italy and to devote himself to what he later called 'the years of reconstruction' (40). Such a well planned programme had in fact already started one year before, with the creation of the '*Centro di Studio della Fisica Nucleare e delle Particelle Elementari*', of which he was director and Bernardini deputy director, and he did not want to leave this endeavour unfinished.

Back in Rome, while organising the work of the Centre and of its numerous new collaborators, he decided to abandon neutron physics and to devote himself to the study of cosmic rays, for which no accelerator was needed. At the same time, Bernardini conceived and directed the construction of the Testa Grigia Laboratory in the Alps at an altitude of 3500 m (Fig. 6). Many Italian physicists have worked there for almost twenty years.

FIGURE 6. Edoardo Amaldi, Gilberto Bernardini and Ettore Pancini in front of the cosmic ray laboratory 'Testa Grigia'.

As their first cosmic ray experiment, Amaldi and Giuseppe Fidecaro constructed an hodoscope of more than hundred long Geiger counters and, running it in Rome, proved that a few GeV cosmic muons only have elastic electromagnetic interactions with nuclei (41). An underground experiment performed in a cavern constructed for a hydro-electric power plant under the Gran Sasso showed the same to be true in the case of inelastic collisions of 10 GeV muons (42).

As a consequence of this work, the idea came to use muons for studying the structure of nuclei, and the formula for Coulomb elastic and inelastic scattering of muons were computed using an independent particle model for the target nucleus (43). As recalled by Fidecaro (44), even a proton form factor was introduced without naming it; this work was recognised by Robert Hofstadter when receiving the Nobel prize for his electron scattering experiments on nuclei and nucleons. (45).

In 1950 Amaldi moved to the technique of nuclear emulsions, brought to the Rome Institute a few years before by Bernardini, who introduced G. Cortini and A. Manfredini to it. The newly formed group took part in the expeditions organised on the initiative of C.F. Powell in 1952 and 1953 for the prolonged exposure of 'stripped' emulsions using high altitude balloons. A section of the report *'Personal Notes on Neutron Work in Rome in the 30s and Post-war European collaboration in High-Energy Physics'* is devoted to this period (46). Here Amaldi describes the first international expedition of 1952 – which had its bases in Naples and Cagliari and was already under the sponsorship of CERN - and the second one from Sardinia in 1953. He writes: "Eighteen laboratories from European countries in addition to one from Australia took part in it. 25 balloons were launched and over 1000 emulsions were exposed. ... They were successfully recovered at sea on account of the employment of a seaplane and a corvette (Pomona) generously placed at the disposal of the expedition by the Italian Airforce and the Italian Navy. ...The development required special care and, in particular, the construction of special developing systems at the University of Bristol, Padua and Rome, where the emulsions of the whole expedition were processed" (47). The 'generous' offer was of course the result of the influence that he had already acquired in Italy, of many discussions with the military authorities and some lobbying.

The first results were presented in July 1953 at the conference of Bagnères de Bigorre in the French Pyrenées Alps, where the Rome group introduced the logarithm of the tangent of the production angle (later called 'rapidity') as an appropriate random variable to describe high-energy events (48). The paper was presented by the young Moreno and in the discussion Amaldi replied to Bhabha, Kaplon and Peters. In spite of the efforts of Bruno Rossi, who gave the concluding speech, the situation of the V-particles observed among cosmic ray events was very confusing. To indicate this state of affairs, on the first page of the conference proceedings the editors wrote: "The particles described in this conference are not entirely fictitious and every analogy with particles really existing in nature is not purely coincidental".

The final results of the emulsion research were published in the following years in many papers signed with G. Baroni, C. Castagnoli, G. Cortini, C. Franzinetti and A. Manfredini. Their main results concerned the decay properties of what are now known as K-mesons (49). Edoardo Amaldi became at the time one of the world experts in the field of properties of the K-mesons and of the so-called *tau-theta puzzle*. Later on, he wrote a much quoted review article (50). In 1955 the group published an *"Unusual event produced by cosmic rays"*, which describes an event consisting of two interconnected stars, which are very probably due to the annihilation of a cosmic antiproton, but was very cautiously interpreted by the authors (51).

PHYSICS WITH PARTICLE ACCELERATORS

This paper marks another change in the scientific interests of Edoardo Amaldi: from cosmic rays to experiments on fundamental particles produced by accelerators. Indeed, its publication gave him the opportunity, during a long stay in Berkeley, to initiate a collaboration with the group of E. Segrè and O. Chamberlain. Together with C. Wiegand and T. Ypsilantis, and in competition with the group led by Edward Lofgren, Segrè and Chamberlain were preparing an experiment to observe the creation of antiprotons with the recently completed Bevatron. The story of this competition has been told by Heilbron (52), who writes: "In September (1955 Lofgren's) group was to begin a four-day hunt for the antiproton. Instead it ceded its time to Segrè's group, which immediately got its first antiproton counts. For the next month the entire research effort at the Bevatron went to confirming and extending the counts. ... Segrè's group fell back on a technique they had tried unsuccessfully during the summer: exposure of a stack of emulsion to the momentum-selected beam. ... By November 14, 1955, enough emulsion had been searched to have yielded a half-dozen annihilations events (but) the film showed nothing. ... The Berkeley physicists grew worried. The situation was classified as 'fluid and somewhat disturbing'. Amaldi's experienced scanner then went to work on emulsions exposed at the Bevatron. Their first find, 'Faustina', ... resembled in several ways the odd star that Amaldi's group had found in the film of the second Sardinian expedition, which strengthened their conviction that both events recorded the end of an antiproton".

Figure 7. Emilio Segrè, Edoardo Amaldi and Bruno Rossi discussing physics.

The first antiproton star observed was published in 1956 (53). About these events Emilio Segrè wrote in his autobiography: "At the time of the antiproton experiment, Amaldi and his wife Ginestra were at our home in Lafayette as our guests. He and I established a collaboration for the study of photographic emulsions exposed at Berkeley, taking advantage of the numerous well-trained scanners available in Rome.

When Amaldi returned to Italy, some Italian newspapers wrote inappropriate comments and tried to ascribe to him a part he had not played. This misreporting could have had unpleasant consequences, but Amaldi set things straight and we kept calm" (54). After this first exposure, a collaboration was organised with the Berkeley emulsion group headed by W.H. Barkas, which led to various common papers. Independently, the Rome group studied the annihilation in flight of 700 MeV/c antiproton collecting a statistics of 650 events (55) and observing a first event in which an anti-sigma was produced (56).

Quite surprisingly, in the years from 1956 to 1958 Edoardo Amaldi found the time to write for the *Handbuch der Physik* a very thick treatise on "*The Production and Slowing down of Neutrons*" (57). Having in mind the research and science policy activities he had in parallel, this is a really impressive piece of work. Down to minor details it covers - with more than one thousand references, 250 figures and 115 tables – neutron research from the first experiments to the last results published in 1958. Later Ugo Fano who had written with two collaborators for the same *Handbuch* a much thinner book on the interactions of gamma rays with matter – told me that he could not understand how Edoardo had managed alone such an enterprise. I could only answer that I had heard from Edoardo himself that with this book he wanted to complete and conclude the work initiated 35 years before.

More or less at the same time, he was fascinated by the Dirac hypothesis on the existence of magnetic monopoles and started a series of experiment at the CERN protonsynchrotron with emulsions (58) and later at the newly built CERN *Intersecting Storage Rings* (ISR). As founder, with the CERN Director General Viki Weisskopf, of the *European Committee for Future Accelerators* (ECFA) and its first Chairman from 1963 to 1969, he had been one of the main promoters of the parallel construction by CERN of the Superprotonsynchrotron (SPS) and of the ISR, a completely novel proton-proton collider. The monopole experiment - in collaboration with the group of Luke Yuan of Brookhaven - made use of the high energy electron and gamma detector built by Pierre Darriulat, Luigi Di Lella, Leon Ledermann and collaborators for other purposes. The aim was to search for events in which virtual monopole-antimonopole pairs would annihilate producing, according to a suggestion by Ruderman and Zwanziger, about 137 gamma rays (59). The interest in Dirac hypothesis also brought him to write a review paper with Nicola Cabibbo (60).

Another series of experiments was performed in the sixties with the external electron beam of the Frascati 1 GeV electronsynchrotron in collaboration with Gherardo Stoppini and junior physicists from the Rome Institute (61). The idea was to study pion electroproduction close to threshold, a difficult experiment - due to the low counting rate - from which, with a relatively simple theoretical interpretation, one could deduce the axial form factor by applying the partial conservation of the axial current. Adopting the same approach used for the Dirac monopole, he eventually wrote on the subject a long report with Fubini and Furlan, who had given original theoretical contributions to the interpretation of the data (62).

THE SEARCH FOR GRAVITATIONAL WAVES

The interest of Edoardo Amaldi for the search of gravitational waves arose in 1961, while following a course given by Joseph Weber at the Varenna International School of the Italian Physical Society. He took the occasion of the next trip to the States to visit Weber's laboratory. I still remember his enthusiasm about what he saw, immediately followed by suggestions on how one could improve the techniques used by Weber, whom he continued to appreciate along the years for his pioneering effort, even if he did not always agree with some of his later claims.

Back in Rome, he talked to many junior colleagues, trying to convince them to start a new experimental activity. Among the others, he discussed the subject with Giorgio Ghigo and Guido Pizzella, who was his assistant. In 1968 William Fairbank spent a sabbatical year in Rome, guest of Giorgio Careri. Thus Amaldi learned of Fairbank's intention to build a resonance antenna running at low temperature. These conversations convinced him even more of the need to start such an activity in the Rome Institute, but no collaborator materialised.

Pizzella writes that in the middle of 1969, due to some misunderstandings in connection with the course he was in charge of as assistant, he went to see Amaldi and communicated him that he would soon leave for the States. "I still remember his expression while I was explaining my decisions. I saw in his eyes perplexity but also understanding and, I think, sorrow. He said nothing. I left for one year and came back as though nothing had happened. ... In the following twenty years we never came back on this episode. ... In the period spent in the States I had thought about the future and concluded that the most beautiful thing I could do was to engage myself in an important experiment of fundamental physics. Thus, on September 3, 1970, the day after my arrival, I went to his office, located on the corner of the second floor and I told him (I was still his assistant): *'Professore*, I would like to make an experiment searching for gravitational waves'. Immediately his eyes gleamed as flares fixing me for a long time. Never I have seen so much enthusiasm and awareness concentrated in a single look. In that moment nothing was more important than the decision which was about to be taken. When we shall observe gravitational waves that look will still be there to shine light " (63).

The initial members of the group were, together with Guido Pizzella, Massimo Cerdonio and Ivo Modena – who were already well recognised experts in SQIDs (Cerdonio) and cryogenics (Modena). Soon they were joined by Gian Vittorio Pallottino, in charge of all the low noise electronics. In 1971 during a visit of Cerdonio, Marconero, Modena and Pizzella in the United States, a collaboration was established with the groups of Stanford (W. M. Fairbank) and of Louisiana (W.O. Hamilton), with the aim of constructing three 5 tons gravitational antennas cooled to 20 mK. The first experimental results of a triple coincidence experiment were published almost twenty years later. The paper is signed by thirty authors (64).

In 1974 the Italian programme was well advanced, when the already built cryostat collapsed, due to a technical fault. Then the decision was taken to move at slower pace and a 24 kg antenna was built and tested at 1.1K (65). It was found that the Q-value of an aluminium cylinder increases by orders of magnitude when cooled. In 1978 the 400 kilograms antenna Altair – mounted in Frascati at the *Istituto di Fisica dello Spazio*

Interplanetario of the Italian Research Council CNR - was cooled to 4.2K (66). It run intermittently for two years with a cooled FET amplifier having a sensitivity larger than the one of the more massive large room temperature antennas *a la* Weber. The longest run lasted more than one month.

FIGURE 8. The Rome group in 1973 in front of the first large cryostat.

Since in Frascati it was difficult to guarantee the needed liquid helium supply for a larger antenna, in 1978 Pizzella suggested to move the major centre of activity to CERN, where the facilities would be more than adequate. Initially Amaldi was cautious, probably because such an activity was outside the scope of the laboratory he knew so well and he did not want to use his influence to receive from CERN a special treatment. Eventually, he spoke to the Research Director General Leon Van Hove, who enthusiastically accepted the idea.

In 1979 in a meeting where Amaldi, Van Hove, Modena, Pallottino and Pizzella were present, an agreement was signed between CERN and the University *La Sapienza*. The 2300 kg antenna *Explorer* had its first run in 1986 at 4K and later reached 2K (67). Explorer was followed by *Nautilus*, initially mounted at CERN and cooled to the record temperature (for such a large mass) of about 100 mK, and later transported to Frascati.

In parallel Massimo Cerdonio, who held a Physics chair at the University of Padua, initiated a new group and constructed *Auriga*, which is now running at the INFN National Laboratory of Legnaro (Padua).

Since 1980, INFN has financed the *La Sapienza* activities and those other Italian groups – of the Universities of Tor Vergata, Trento and L'Aquila – which later joined the first one and formed together the *Centro Interuniversitario Ricerche sulla Gravitazione (CIRG) Edoardo Amaldi*. In 1990 Nicola Cabibbo, at that time President of INFN, wrote the Introduction to the volume describing the activities of CIRG (68). One reads: "In Italy Edoardo Amaldi was the first to understand the relevance of these studies, to promote them, to follow constantly their developments and, among other things, to argue with prophetic intuition (gauge theories were not yet accepted) that these studies have to be connected with those related to particles and fundamental interactions. Thus Amaldi managed to convince both INFN to take a responsibility in this search and CERN to host it".

I have personally followed quite closely the CERN part of this activity since in 1973 I moved from *Istituto Superiore di Sanità* to CERN. Either directly or by phone - during the long conversations we had between Rome and Geneva every Sunday night - I heard Edoardo speaking enthusiastically of any little progress and describing, without complaining, the next recovery steps planned after each one of the many setbacks. He was very conscious of all the difficulties they still had to face and told me more than once: "I'll never see 'them' and I am not sure that Guido will see them".

To conclude this short description of an activity well known to the attendees of this *Third Edoardo Amaldi Conference on Gravitational Waves*, I reproduce few lines written by Guido Pizzella few months after his death. "The contribution of Edoardo Amaldi to the Rome gravitational wave experiment had many facets. He brought the weight of his experience and of his deep physical insight, in particular in the interpretation of the experimental data collected during the long way that is slowly taking to the final goal. He played a major role in the installation of the large cryogenic antennas at CERN. His contagious enthusiasm was essential for keeping the research group united for so many years. He taught to his co-workers how to tackle difficult problems, find out the most important point, pursue it with strong determination to the very end, in a simple and pragmatic way, and never be discouraged. Certainly, when gravitational waves will be discovered all of us will be in great debt with him".

SCIENCE POLICY

In describing Edoardo Amaldi's rich scientific activity I have briefly mentioned his crucial influence on the post-war development of physics in Italy and in Europe. While referring to Refs. 1-6 for the details, here I want to recall the main facts.

In 1946, with Gilberto Bernardini he founded in Rome the *Centro di Studio della Fisica Nucleare e delle Particelle Elementari*. In 1951 this Center joined with similar institutions of Milan, Padua and Turin, becoming the nucleus of the *Istituto Nazionale di Fisica Nucleare* (INFN). Bernardini – who was spending half of his time in Columbia and later in Urbana – was INFN first President and Amaldi its second one (1961-1965). The construction of the Frascati 1 GeV electronsynchrotron under Giorgio Salvini was decided and completed during Bernardini's chairmanship (69). The Ada and Adone programmes belong to the Amaldi presidency (70).

Interested in applications, and not only in fundamental research, since 1946 Amaldi has promoted the peaceful use of nuclear energy by contributing to the creation of CNEN, later called ENEA - the Italian Atomic Energy Commission - of which he was Vice-president from 1956 to 1960.

In the years between 1948 and 1950 the idea of an international laboratory devoted to accelerator physics was being discussed more or less independently in many European circles. Amaldi was particularly worried by the increasing gap between

American and European research potentialities and by the 'brain drain', which would make the reconstruction of European science even more difficult. In 1977 he wrote (71): "It is impossible to establish the beginning of the first stage (of the future CERN) since around 1948 the idea of starting an international collaboration among European countries in the field of nuclear physics and elementary particles was making its first appearance, in more or less nebulous form, in many places. ...I remember that in the years 1948-1950 the various aspects of the problem - including energy and cost of machines - were examined in Rome in frequent discussions between Ferretti and myself and in letters exchanged with Bernardini, who in those years was at Columbia University and thus had the opportunity of contributing to stimulate the interest of Rabi on the subject".

FIGURE 9. Pierre Auger and Edoardo Amaldi at one of the first CERN Councils.

The process was initiated in June 1950 by a proposal made by Isidor Rabi to the Florence General Assembly of UNESCO. "In the official statement approved unanimously by the General Assembly along the same lines, neither Europe nor high-energy physics were mentioned. But the specific case was clearly intended by many people taking part in the Assembly, in particular by Auger, who was Director of Natural Sciences of UNESCO, and by Rabi, with whom I had discussed the subject at length a few days before" (71). Pierre Auger moved very rapidly and already at the beginning of 1951 Italy, and then France and Belgium, put at his disposal funds to act. In February 1952 an Agreement was signed and Edoardo Amaldi was nominated Secretary General of the provisional organisation. In 1952 the newly proposed strong focusing principle was chosen for the future CERN protonsynchrotron (PS), allowing to increase its energy from the initially planned 10 GeV to more than 20 GeV. The

Geneva site was chosen in February 1953 and temporary offices were occupied in October 1953.

Amaldi spoke of his last months as Secretary General at a Symposium held at Fermilab in May 1985. "On 12th August 1954 France completed its ratification procedure of the CERN convention and the situation appeared sufficiently promising to dare starting, on 13 August 1954, major excavation work on the Meyrin site for the foundations of the permanent buildings, in particular the one to house the synchrocyclotron. I had taken this decision, with the agreement of the council, under the pressure of the Geneva weather, months before all ratification formalities were completed. (Sir Ben) Lockspeiser (President of the Council) was clearly pleased and commented 'Now we have another task – keeping Amaldi out of jail'. ... The provisional organisation came to an end while the permanent organisation was not yet in existence. All the assets of the provisional organisation were suddenly master-less. For eight days I had the honour, as secretary general, of having sole responsibility for all the properties and liabilities on behalf of a new born organisation. Then, at the first meeting of the permanent council, assembled in Geneva on 7 October 1954, the secretary general presented his final report. (Felix) Bloch was nominated Director General, and CERN entered its final permanent form" (72).

What he does not say is that he had been offered to be the first CERN director and that he had refused. In his introduction to the Second Amaldi Conference Llewellyn Smith mentions the reasons for such a decision: "…he wished to return to research and also to put a definite end to a suspicion that some of the fathers of CERN were motivated by personal ambitions…" (6).

He also played a crucial role in initiating the European collaboration in space research. In 1958, after the launch of the Soviet Sputnik of October 4th, 1957, the United States established NASA and it appeared immediately clear to him that Europe had to take part in the scientific competition. I vividly remember when, coming back from a train trip, at dinner he told the family "I have being thinking today and I am convinced that now, after CERN, we must create *Euromoon*". He began by discussing the matter with Luigi Crocco from Princeton University in July 1958. Then he contacted his old friend Pierre Auger and he talked about it with I. Rabi (in September), H.S.W. Massey (in November), F. Perrin (in November), Th. Von Karman, chairman of the NATO Advisory Group for Aeronautical Research and Development (in January 1959). Then, in May, he wrote to a group of prominent European scientists and science managers a letter, which was published in December under the title '*Let us create an European Organisation for Space Research*' (73).

Within one year a conference was held at CERN and in 1961 a commission began working on a European programme. Between 1960 and 1964 the European Space Research Organisation (ESRO) and the European Launcher Development Organisation (ELDO) were created. He always expressed very strongly his opposition to this split, and, above all, to the creation of ELDO, which he considered to be rather a gift to the militaries and space industries than a contribution to space research (74). Eventually, the need for a reorganisation was recognised by everybody and in 1971 the two organisations were replaced by the European Space Agency (ESA), where he was Chairman of the Scientific Programme Committee and of the Space Science Advisory Committee.

The success of CERN can be gauged by the fact that since 1988 the number of American experimental high-energy physicists working at CERN is larger than the number of European physicists working in USA (Fig. 10). One can say that - at least in this field - the 'brain drain' has been stopped.

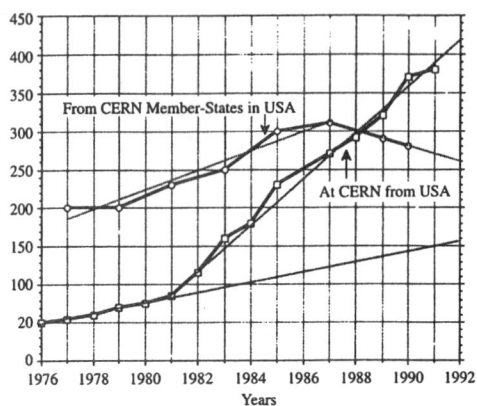

FIGURE 10. Since 1988 there have been more American experimenters working with CERN accelerators than European high-energy physicists working in the United States.

Carlo Rubbia wrote in his biography (75): "His commitment to CERN and ESO was inspired by two basic considerations, in addition to genuine scientific curiosity: the independence of science from military purposes and the danger to scientific liberty when military interests intervened; and the notion that ventures in fundamental science warranted the consistent collaboration of all European nations. Amaldi was a convinced European before many others. To compete with the United States and the Soviet Union, national differences had to be set aside and forces had to be combined. With the great success of CERN and of the *Ariane* programme and Europe's current highly competitive position in space technology today, the importance of Edoardo Amaldi's determined initiatives can be fully appreciated".

TEACHER AND AUTHOR OF PHYSICS BOOKS

Amaldi was a good speaker and an outstanding teacher. He was capable of catching and keeping the full attention of the hundreds of students who followed his course, usually held at eight o'clock in the morning. Every year many thousands Italian physics and engineering students, in *La Sapienza* and in other Universities, learned optics, electricity, classical electromagnetism and relativity from his treatise, that completed Gilberto Bernardini's one on mechanics and thermodynamics. The first version of his book was published in 1937, the last one fifty years later with Bizzarri and Pizzella (76).

Even more influential have been the high school books he wrote with Ginestra who, having graduated as astronomer, discovered her vocation when in 1937 she wrote with Laura Fermi 'Alchimia del tempo nostro' (*Alchemy in our times*) (77), a popular book describing atomic and nuclear physics to the layman. After the war, she became the most known Italian science writer and, later, the responsible and presenter of many radio and television science programmes. She had the gift of explaining in simple and appealing terms the most difficult subjects, not only in physics but also in biology and the other natural sciences. Her best known books (*Questo nostro mondo* and *Materia e antimateria*) were translated into English (78), French, Spanish and Russian.

FIGURE 11. Ginestra Giovene Amaldi and Edoardo Amaldi during the ceremony of an award.

In 1947 Ginestra and Edoardo published a volume of elementary physics for the 'commercial' high schools, followed in 1952 by three volumes for the 'scientific' high schools. The publisher was Zanichelli (Bologna), the same one which had been publishing, since the beginning of the century, the books of geometry and analysis by Ugo Amaldi and Federigo Enriques.

Many different editions of physics books for the many types of Italian high schools were published between 1947 and 1972 because the work was well planned: Ginestra would write by day what both of them had decided together the evening before. In the evening, the new text was carefully checked and this was often causing discussions: typically Edoardo would find the presentation too simplistic and thus not complete enough and Ginestra would argue that the book had to be written for everybody and not only for the best students. As a matter of fact, these were the only occasions in which I heard my parents involved in heated discussions.

In 1971, when she was 60, Ginestra had to undergo a brain surgery as a consequence of a sudden aneurysm. She recovered after a long coma and within about one year she was talking and speaking again, but a partial paralysis obliged her to use a wheel chair. What is worst, she had lost her wonderful taste for writing. Edoardo took care of her for almost twenty years with great affection, but many of his activities had to slow down – and even be interrupted for a long time – also because he decided not to travel any more.

In 1981 we started to rewrite together the three volumes course for the 'scientific' high schools, which was becoming obsolete. The new edition was published in 1984 and has been followed by many others, but since 1990 I am the only author. A few weeks before his death the publisher brought him a golden fountain pen. After reading aloud what was written on it (*un milione*) Amaldi said: "I do not think I'll see the second million (copies)".

CIVIL COMMITMENT

Edoardo Amaldi never took part in programmes aimed at the development of nuclear weapons, but he knew that, if in 1939 he had stayed in the United States, he would have most probably chosen to collaborate, because of the concern that a nuclear programme was underway in Germany. He thus decided already in 1946 to battle against all links between scientific research and military applications and to contribute as he could to the reduction of the number of nuclear weapons. He was already active in this direction when, after the publication in July 1955 of the 'Russell-Einstein Appeal', he received a letter from Russell, inviting him to take part in a meeting to be held in the Canadian township of Pughwash. He could not go because of other commitments, but he did attend the meeting of 1958 and was co-opted onto the Pughwash Movement's Continuing Committee in 1962, as soon as it was set up.

He was a member for ten years until 1972, when he had to reduce his travelling, and, in particular, he was very active in the organisation of the 1965 Pughwash Conference in Venice and then in drafting a strong statement in support of the non-proliferation treaty. In Italy he battled, with Francesco Calogero and others, to convince all the parties to vote in favour of the treaty. When Calogero, having replaced him in the Committee, was elected Pughwash Secretary General in 1989, Amaldi put some offices of the *Accademia dei Lincei*, of which he was President, at the disposal of the Movement.

This was a practical consequence of his deep conviction that Academies should engage in socially important issues. In 1976 he had tried to introduce in the Academy a School of disarmament, but his attempt failed since not everybody shared his views. But as Vice-president, and later President, he succeeded in creating both a group on disarmament and arms control and a committee for the defence of civil rights, with aims similar to those of the committees existing since long in the framework of the American National Academy of the Sciences. He was thus ready to take on himself to treat such themes at regular meetings of the world academies, thus realising the proposal brought to him in 1989 by Wolfang Panofsky for the American Academy, at which, as Panofsky says, 'he caught fire'. The first meeting had to be held in Rome and was almost organised when he died. But the ball was launched and, after that, many meetings have followed with the participation of more and more Academies. In 1999 the *12th Amaldi Conference of the Academies* on Disarmament and Arms Control took place in Mainz.

Finally, let me mention the *International School on Disarmament and Research Conflicts* (ISODARCO), that came to light during a conversation between Carlo Schaerf, who was at the time in Stanford, and Amaldi. They were in Varenna for the 1962 physics summer school and they thought that residential courses - as the one they were following - could offer to a wide number of scientists technical expertise related to the nuclear arms race and international security. Amaldi was initially director of ISODARCO, but a couple of years later he became the president. Schaerf has been and still is the main mover. The first summer course was held in June 1966 in Frascati and the first winter course took place in Trento in 1988. Thirty-two courses have been held till now and, since 1988, a course takes place in China every two years.

THE MANY FACETS

To conclude, I quote from the last sentences of the biography written by Carlo Rubbia (79).

FIGURE 12. Edoardo Amaldi in his home office at the age of 80. On the left one can see his preferred photo of Enrico Fermi and on the background his beloved mountains.

"Edoardo Amaldi was present and active in some of the great turning points of European science from 1930s onward. ...To stress the importance of (the period of

Via Panisperna) when Amaldi was still a young man does not of course belittle his later achievements in the study of cosmic rays, particle physics and gravitational waves."

"Although Edoardo Amaldi did not like to be congratulated by colleagues on his appointment to important functions, because he considered himself above all as a man of science, there can be no doubt that he was one of Europe's great post-war scientific statesmen. It is above all as a man of vision, as a man guided by powerful and timely ideas, as a man determined to build new scientific institutions, that Edoardo Amaldi has left an indelible trace on history."

"Amaldi was inspired by two clear principles. First he was convinced that science should not be pursued for military purpose. Amaldi spent much of his life ensuring basic research in his field received the funds and the facilities which it needed in order to prosper, yet he was emphatic that it should not become militarised. This does not mean that he thought that science had no social significance. On the contrary he was well aware that 'big science' in particular stimulates industry, and trained highly qualified specialists who could be useful in many sectors of the economy. Secondly, Edoardo Amaldi was a dedicated European."

"He realised very early that no single European nation could hope to hold its own scientifically and technologically. To compete they had to collaborate... (T)hese were not simply pious hopes or abstract ideals. Amaldi was determined to see them put into practice. This is clearly demonstrated by the key role he played in the launching of CERN and ESO."

"Historian, teacher, man of ideals, all these facets of Edoardo Amaldi's life have been referred to but it is also because of his human qualities that he will be remembered. His exemplary coolness of mind, his total sincerity, his respect for the deserving, his unwavering intellectual and moral honesty, and his constant affection to his family were the hallmarks of this remarkable man".

REFERENCES AND NOTES

1. Amaldi E., *Essays and Recollections on 20th Century Physics*, G. Battimelli and G. Paoloni eds., Singapore – New Jersey – London – Hong Kong: World Scientific, 1999.
2. Rubbia C., "Edoardo Amaldi", *Biogr. Mem. Fellows R. Soc.* **37,** 3-31 (1991).
3. *Evolution of Particle Physics*, M. Conversi ed., New York and London: Academic Press, 1970.
4. *Perspectives of Fundamental Physics*, C. Schaerf ed., Chur – London – New York: Harwood Academic Publishers, 1979.
5. Note by the Editors, "Edoardo Amaldi's Life and Activity", in *Gravitational Wave Experiments,* E. Coccia, G. Pizzella, and F. Ronga eds., Singapore – New Jersey – London – Hong Kong: World Scientific, xvii-xxi, 1995.
6. Llewellyn Smith, C., "An Introduction to Edoardo Amaldi and CERN", in *Gravitational Waves,* E. Coccia, G. Pizzella, and G. Veneziano eds., Singapore – New Jersey – London – Hong Kong: World Scientific, 1-23, (1998).
7. Angela, P., *Amaldi – Intervista sulla Materia dal Nucleo alla Galassia,* Roma-Bari: Laterza, 1980.
8. Rubbia, C., and Angela, P., *Edoardo Amaldi – Scienziato e Cittadino d'Europa,* Milano: Leonardo Periodici, 1992.
9. Segrè, E., in Ref. 3, p. XXII.

10. Amaldi, E., *La Vita e l'Opera di Ettore Majorana*, Roma: Acc. Naz. Lincei, 1966. For the English translation see Ref. 1 and E. Amaldi, "Ettore Majorana, Man and Scientist", in *Strong and Weak Interactions – Present Problems*, A. Zichichi ed., New York and London: Academic Press, (1966), p. 10.
11. Holton, G., "Fermi's Group and the Recapture of Italy's Place in Physics", in *The Scientific Imagination*, Case Studies, Cambridge: Cambridge University Press, Part 2, Section 5, 1978.
12. Amaldi, E., *Phys. Zeit.* **32,** 914 (1931).
13. Amaldi, E., and Segrè, E., *Nature* **133**, 141 (1934).
14. Fermi, E., *Nuovo Cimento* **11,** 191 (1934).
15. Amaldi, E., "From the Discovery of the Neutron to the Discovery of Nuclear Fission", *Phys. Reports 111(1-4)*, 1-334, (1984).
16. Ref. 13.
17. Fermi, E., *Ric. Scient.* **5(1),** 283 (1934). *Ibid.* **5(1),** 330 (1934).
18. Amaldi E., D'Agostino, O., Fermi, E., Rasetti, F., and E. Segrè, *Ric. Scient.* **5(1),** 452 (1934). *Ibid.* **5(1),** 652 (1934). *Ibid.* **5(2),** 21 (1934).
19. Ref. 13, p. 124.
20. Amaldi, E., D'Agostino, O., Fermi E.,. Rasetti, F., and Segrè, E., *Proc. Roy. Soc.* **A146,** 483 (1934).
21. Segrè E., *Enrico Fermi Physicist*, Chicago: The University of Chicago Press, pp. 93-95, 1970.
22. Ref. 13, p. 132.
23. Amaldi, E., D'Agostino O., Fermi E., Rasetti F. , Segrè E., *Ric. Scient.* **6(1),** 581 (1935).
24. Bjerge, T., and Westcott, D.P., *Proc. Roy. Soc. Lond.* **A150,** 709 (1935).
25. Moon, B.P., and Tillman, J.R., *Nature* **136,** 106 (1935).
26. Amaldi, E., and Fermi, E., *Ric. Scient.* **6(2),** 443 and 774 (1935). *Ibid.* **7(1),** 56, 223, 310, 393 and 454 (1936).
Amaldi, E., and Fermi, E., *Phys. Rev.* **50,** 899 (1936).
27. *Enrico Fermi Collected Papers*, Vol. 1: Italy 1921-1938, Vol. 2: USA 1939-1954, Rome and Chicago: Accademia dei Lincei and University Chicago Press, 1962 and 1965.
28. Amaldi, E., and Fermi, E., *Ric. Scient.* **7(1),** 223, 393, 454 (1936).
Amaldi, E., and Fermi, E., *Phys. Rev.* **50,** 899, (1936).
29. Amaldi, E., Hafstadt, L.R., and Tuve, M.A., *Phys. Rev.* **51,** 896 (1937).
30. Amaldi, E., Fermi, E., and Rasetti, F., *Ric. Scient.* **8(2),** 40 (1937).
31. Amaldi, E., and Rasetti, F., *Ric. Scient.* **10,** 623 (1939).
32. Ageno, M., Amaldi, E., Bocciarelli, D., and Trabacchi, G.C., *Rend. Ist. Sup. Sanità* 3, 201 (1940).
33. Amaldi, E., *Da Via Panisperna all'America*, G. Battimelli, and G. Paoloni eds., Roma: Editori Riuniti, 1997.
34. The Amaldi archive contains almost 30 000 letters, together with Amaldi's papers and documents.
35. Battimelli, G., and Gambaro, I., "Da Via Panisperna a Frascati, gli Acceleratori mai Realizzati", *Quaderno di Storia della Fisica*, **N. 1**, Bologna: Soc. It. Fisica, 1997.
36. Ref. 2, p.10.
37. Ageno, M., Amaldi, E., Bocciarelli, D., and Trabacchi, G.C., *Phys Rev.* **60,** 67 (1941).
38. Ageno, M., Amaldi, E., Bocciarelli, D., and Trabacchi, G.C., *Nuovo Cimento* **1**, 253 (1943). *Ibid.* **3**, 203 (1946).
Ageno, M., Amaldi, E., Bocciarelli, D., and Trabacchi , G.C., *Phys. Rev.* **71,** 30 (1947).
39. Amaldi, E., and Cacciapuoti, B.N., *Phys. Rev.* 739 (1947).
40. Amaldi, E., "The Years of Reconstrucion", in Ref. 4, p. 389.
41. Amaldi, E., and Fidecaro, G., *Phys. Rev.* **81,** 338 (1951).
42. Amaldi, E., Castagnoli, C., Gigli, A., Sciuti, S., *Nuovo Cimento* **9,** 453 and 969 (1952). *Proc. Phys. Soc.* **A65,** 556 (1952).
43. Amaldi, E., Fidecaro, G., and Mariani, F., *Nuovo Cimento* **7,** 553 (1950). *Ibid.* **7,** 757 (1950).
44. Fidecaro, G., "From Cosmic Rays to Physics with Accelerators", in *The Restructuring of Physical Sciences in Europe and the United States 1945-1960*, M. De Maria, M. Grilli and F. Sebastiani eds., Singapore – New Jersey – London – Hong Kong: World Scientific, p. 663, (1989).

45. Hofstadter, R., "The Electron-scattering Method and its Application to the Structure of Nuclei and Nucleons", in *Nobel Lectures - Physics 1942-1962*, Amsterdam: Elsevier, 560, (1964).
46. Amaldi, E., "Personal Notes on Neutron Work in Rome in the 30s and Post-war European collaboration in High-Energy Physics", in *History of Twentieth Century Physics*, LVII Course of the Enrico Fermi Int. School -Varenna, New York: Academic Press, 1977, pp. 294-351.
47. Ref. 44, p. 294.
48. *Determination of the Energy of the Primary of a Jet*, reported by D. Moreno, in the typed proceedings of *Congrès International sur le Rayonnement Cosmique*, Université de Toulose, 1953, p. 189.
49. Amaldi, E., Baroni, G., Cortini, G., and Franzinetti, C., *Suppl. Nuovo Cimento* **12**, 181 (1954). Amaldi, E., Cortini , G., and Manfredini, A., *Suppl. Nuovo Cimento* **12**, 205 (1954).
50. Amaldi. E., "Report on the tau-mesons", *Suppl. Nuovo Cimento* **4(2)**, 179 (1956).
51. Amaldi, E., Castagnoli, C., Cortini, G., Franzinetti, C., and Manfredini, A., *Nuovo Cimento* **1**, 492 (1955)
52. Heibron, J.L., "The Detection of the Antiproton", in *The Restructuring of Physical Sciences in Europe and the United States 1945-1960*, M. De Maria, M. Grilli and F. Sebastiani eds., Singapore – New Jersey –London – Hong Kong: World Scientific, pp. 161-217, 1989.
53. Chamberlain, O., Chupp, W., Goldhaber, G., Segrè, E., Wiegand, C., Amaldi, E., Baroni, G., Castagnoli, C., Franzinetti, C., and Manfredini, A., *Phys. Rev.* **101**, 909 (1956).
54. Segrè, E., *Enrico Fermi Physicist*, Chicago: The University of Chicago Press, 258, (1970).
55. Amaldi, E., Baroni, G., Bellettini, G., Castagnoli, C., Ferro-Luzzi, M. and Manfredini, A., *Nuovo Cimento* **5**, 977 (1957).
56. Amaldi, E., Barbaro-Galtieri, L., Baroni, G., Castagnoli, C., and Ferro-Luzzi, M., *Nuovo Cimento* **16**, 392 (1960).
57. Amaldi, E., "*The Production and Slowing down of Neutrons*", Vol. XXXVII/2 of *Handbuch der Physik*, S. Flügge ed., Berlin – Gottingen – Heidelberg: Springer Verlag, 1959, pp. 1-659.
58. Amaldi, E., Baroni, G., Manfredini, A., Bradner, P., Hoffmann, L., Vanderhaege, G., *Nuovo Cimento* **28**, 773 (1963).
59. Dell, G. F., Uto, H., Yuan, L.C.L., Amaldi, E., Beneventano, M., Borgia, B., Pistilli, P., and Sestili, I., *Lett. Nuovo Cimento* **15**, 269 (1976).
60. Amaldi, E., and Cabibbo, N., "On the Dirac Magnetic Monopoles", in *Aspects of Quantum Theory*, A. Salam and P. Wigner eds., Cambridge – New York – New Rochelle – Melbourne – Sidney: Cambridge University Press, 1972.
61. Amaldi, E., Balla, M., Borgia, B., Di Giorgio, G.V., Giazotto, A., Giorgi, M., Pistilli, P., Serbassi, S., and Stoppini, G., *Nuovo Cimento* **1**, 247 (1969).
62. Amaldi, E., Fubini, S., and Furlan, G., *Pion-Electroproduction*, Berlin – Heidelberg – New York: Springer-Verlag, Springer Tracts in Modern Physics, 1979, pp. 1-162.
63. Pizzella, G., "Quattro Episodi", in *Sapere*, Roma: Edizioni Dedalo, December 1990, p. 30. The Italian text reads as follows "*Di colpo i suoi occhi si accesero come fari e mi fissò a lungo. Mai ho visto tanto entusiasmo e consapevolezza concentrati in uno sguardo. In quel momento non c'era nessun'altra cosa che contasse di più di questa decisione che stava per essere presa. Quando scopriremo le onde gravitazionali quello sguardo starà ancora lí a far luce*"
64. Amaldi, E., Agular, O., Bassan, M., Bonifazi, P., Carelli, P., Castellano, M.G., Cavallari, G., Coccia, E., Cosmelli, C., Fairbank, W.M., Frasca, S., Foglietti, W., Habel, R., Hamilton, W.O., Henderson, J., Johnson, W., Mann, A.G, McAshan, M.S, Michelson, P.F., Modena, I., Moskowitz, B.E., Pallottino, G.V., Pizzella, G., Price, J.C., Rapagnani, R., Ricci, F., Somonson, N., Stevenson, T., Taber, R.C., Xu, B.X., *First Gravitational Wave Coincidence Experiment between Three Cryogenic Resonant-mass Detectors, Louisiana-Roma-Stanford,* Astronomy and Astrophysics **216**, 325 (1989).
65. Amaldi, E., Cosmelli, C., Bonifazi, P., Bordoni, F., Ferrari, V., Giovanardi, U., Vannaroni, G., Pallottino, G.V., Pizzella, G., and Modena, I., *Nuovo Cimento* **1C**, 341, (1978).
66. Amaldi, E., Cosmelli, C., Frasca, S., Modena, I., Pallottino, G.V., Pizzella, G., Ricci, F., Bonifazi, P., Bordoni, F., Ferrari, V., Giovanardi, U., Iafolla, V., Pavan, B., Ugazio S., and Vannaroni, G., *Nuovo Cimento* **1C**, 497 (1978).

Amaldi, E, Modena, I., Pallottino, G.V., Pizzella, G., Ricci, F., Bonifazi, P., Bordoni, F., Fuligni, F., Giovanardi, U., Iafolla, V., and Ugazio, S., *Lett. Nuovo Cimento*, **28**, 362 (1980).

67. Amaldi, E., Astone, P., Bassan, M., Bonifazi, P., Castellano, M.G., Cavallari, G., Coccia, E., Cosmelli, C., Frasca, S., Majorana, E., Modena, I., Pallottino, G.V., Pizzella, G., Rapagnani, P., Ricci, F., Visco M., and Zhu Ning, *Europhys. Lett.* **12**, 5 (1990).
68. Cabibbo N., "Prefazione", in *Il Centro Interuniversitario Ricerche sulla Gravità 'Edoardo Amaldi'*, Rome: Universitalia, 1991, p. 5.
69. Salvini, G., *The Electron Synchrotron and the Birth of the National Laboratories of Frascati*, in *The Restructuring of Physical Sciences in Europe and the United States 1945-1960*, M. De Maria, M. Grilli and F. Sebastiani eds., Singapore – New Jersey – London – Hong Kong: World Scientific, 1989, p. 663.
70. Amaldi, E., *The Bruno Touschek Legacy*, CERN Yellow Report 81-19, 1981. This report in reproduced in Ref.1.
Bernardini, C., "Ada: the Smallest Electron-positron Ring", in *The Restructuring of Physical Sciences in Europe and the United States 1945-1960*, M. De Maria, M. Grilli and F. Sebastiani eds., Singapore – New Jersey – London – Hong Kong: World Scientific, 1989, p. 444.
Amman, F., "The early Times of Electron Colliders", *Ibid.*, p. 449.
71. Ref. 44, p.336.
72. Amaldi, E., "The History of CERN during the Early 1950s", in *Pions to Quarks*, L.M. Brown, M. Dresden and L. Hoddeson eds., Cambridge – New York – New Rochelle – Melbourne – Sidney: Cambridge University Press, 1989, pp. 508-518.
73. Amaldi, E., *Créons une Organization Européenne pour la Recherche Spatiale*, L'Expansion de la Récherche Scientifique, N. 4, Décembre 1959, pp. 6-8.
74. Amaldi, E., "Science, Technology and the arms Race: European Perspectives", in *Proceedings of the Niels Bohr Symposium, 12-14 November 1985*, Cambridge: American Academy of Arts and Sciences, 1987.
75. Ref. 2, p. 20.
76. Amaldi, E., Bizzarri R., and Pizzella, G., *Fisica Generale: Elettromagnetismo, Relatività e Ottica*, Bologna: Zanichelli, 1986.
77. Amaldi G., and Fermi, L., *Alchimia del Tempo Nostro*, Milano: Ulrico Hoepli Editore, 1937.
78. Amaldi, G., *The Nature of Matter*, Chicago: University of Chicago Press, 1966.
79. Ref. 2, pp. 30-31.

Bars in action

Eugenio Coccia

Dipartimento di Fisica, Università di Roma "Tor Vergata"
and INFN, Sezione di Roma 2, Via Ricerca Scientifica 1, 00133 Rome, Italy
coccia@roma2.infn.it

Abstract.
We report on the status of the five resonant-mass detectors of gravitational waves operating today in Australia, Italy and USA. These bar detectors are in continuous observational mode with burst sensitivity h$\simeq 4 \times 10^{-19}$, or, in spectral units, 2×10^{-22} Hz$^{-1/2}$ over bandwidths of about 1 Hz, with a duty cycle mainly limited by cryogenic operations. The strongest potential sources of GW bursts in our Galaxy and in the local group are today monitored by such instruments.

With the formation of the IGEC (International Gravitational Event Collaboration), the activity on bar detectors passed a phase transition: from the occasional exchange of data between two groups to the systematic exchange of data between all the groups, following an agreed protocol.

In addition to the search for impulsive events, the data collected are being used to detect periodic waves over long time periods, to give new upper limits for the stochastic background of cosmological origin, and to study possible correlations with gamma ray bursts.

THE FIVE CRYOGENIC BAR DETECTORS

Cryogenic resonant-mass detectors were conceived in the '70s with the aim of improving the sensitivity of room temperature Weber detectors by many orders of magnitude, by reducing the temperature of the bar to or below helium temperature (4.2 K) and employing superconducting electronic devices in the readout system.

The first cryogenic detector was operated at the beginning of the '80 by the Fairbank group in Stanford, followed by the Rome group detector EXPLORER [1] located at CERN and by the LSU group detector ALLEGRO [2]. Only at the beginning of the '90s, however, the cryogenic detectors entered the continuous operational mode and hence the field of reliable instruments of physics. The Stanford detector, damaged by the 1989 earthquake, was shut down. Another detector, called NIOBE [3], started operating in Perth in 1993.

In the years 1982-1984 a feasibility study was conducted to establish the technical possibility of the cooling of a multiton Al 5056 bar to milliKelvin temperatures [4].

In 1995 and in 1997, respectively, the ultracryogenic detectors NAUTILUS [5], located at the INFN Frascati Laboratories, and AURIGA [6], located at the INFN Legnaro Laboratories, started taking data.

The principle of operation of resonant-mass detectors is based on the assumption that any vibrational mode of a resonant body that has a mass quadrupole moment, such as the fundamental longitudinal mode of a cylindrical antenna, can be excited by a GW with nonzero energy spectral density at the mode eigenfrequency.

The mechanical oscillation induced in the antenna by interaction with the GW, is transformed into an electrical signal by a motion or strain transducer and then amplified by an electrical amplifier. Unavoidably, Brownian motion noise associated with dissipation in the antenna and the transducer, and electronic noise from the amplifier, limit the sensitivity of the detector.

The sum at the output of the contributions due by the Brownian noise and by the electronic noise gives the total detector noise. This can be referred to the input of the detector (as if it were a GW spectral density) and is usually indicated as $S_h(f)$. This function has a resonant behaviour and can be characterized by its value at the detector resonance frequency f_0 and by its half height width. $S_h(f_0)$ can be written as:

$$S_h(f_0) = \frac{\pi}{2} \frac{kT}{M\nu_s^2 Q f_0} \qquad (1)$$

where T is the antenna temperature, M is the antenna mass, ν_s is the velocity of sound in the material used, Q is the quality factor of the mode.

The half height width of this function gives the bandwidth of a resonant detector:

$$\Delta f = \frac{f_0}{Q} \Gamma^{-1/2} \qquad (2)$$

We used the convenient dimensionless parameter Γ that gives the ratio of the wide band noise in the resonance bandwidth to the narrow band noise. Γ decreases as the noise temperature of the amplifier decreases and as the transducer efficiency increases.

These relations characterize completely the sensitivity of a resonant-mass detector. For instance, the minimum detectable (SNR=1) GW amplitude for a short burst signal lasting for a time τ_g can be written as [7]:

$$h_0^{min} = 2\tau_g^{-1}[\frac{S_h(f_0)}{2\pi \Delta f}]^{1/2} \qquad (3)$$

In table 1 we give a summary of the presently operating detectors. All the bars are equipped with a resonant transducer forming with the bar a system of two coupled oscillators. The frequencies of the two resulting normal modes are indicated with f_- and f_+. The vibration of the transducer modulates a dc electric field in the case of the capacitive transducers used by AURIGA, EXPLORER and NAUTILUS and a dc magnetic field in the case of the inductive transducer used by ALLEGRO. All these detectors use a dc SQUID amplifier. An active transducer using a microwave cavity is adopted by the Australian group for the NIOBE detector.

TABLE 1. Main features of the operating cryogenic detectors. T_{eff} expresses (in Kelvin) the minimum detectable energy innovation and determines the detector sensitivity to short (1 ms) bursts, indicated with h^{min}. We remark that $h = 6 \times 10^{-19}$ corresponds to the signal of a millisecond GW burst due to the total conversion of about $10^{-4} M_\odot$ in the Galactic Center.

	ALLEGRO	AURIGA	EXPLORER	NAUTILUS	NIOBE
material	Al5056	Al5056	Al5056	Al5056	Nb
lenght (m)	3.0	2.9	3.0	3.0	2.75
M (kg)	2296	2230	2270	2260	1500
f_- (Hz)	895	912	905	908	694
f_+ (Hz)	920	930	921	924	713
$Q \times 10^6$	2	3	1.5	0.5	20
T (K)	4.2	0.25	2.6	0.1	5
$S_h(f)^{1/2}(Hz^{-1/2})$	$1 \cdot 10^{-21}$	$2 \cdot 10^{-22}$	$6 \cdot 10^{-22}$	$2 \cdot 10^{-22}$	$8 \cdot 10^{-22}$
Δf (Hz)	$\simeq 1$	$\simeq 1$	$\simeq 1$	$\simeq 1$	$\simeq 1$
T_{eff} (mK)	10	2	10	2	3
h^{min}	$6 \cdot 10^{-19}$	$4 \cdot 10^{-19}$	$6 \cdot 10^{-19}$	$4 \cdot 10^{-19}$	$6 \cdot 10^{-19}$
duty cycle (%)	95	75	75	75	75
latitude	30°27'00"N	44°21'12"N	46°27'00"N	41°49'26"N	31°56'00"S
longitude	268°50'00"	11°56'54"	6°12'00"	12°40'21"	115°49'00"

RECENT RESULTS

All the resonant mass detector groups formed the International Gravitational Event Collaboration (IGEC) on July 4th 1997 at CERN, during the Second Amaldi Conference, and agreed on a data exchange protocol for searching short gw bursts. The main goal of IGEC [8] is to standardize and simplify the data exchange between the groups and to mantain a continuous discussion on the analysis procedures. Analysis of the 1997-1998 data of all the detectors, released under the IGEC protocol, is in progress. Each group provided a file containing a list of events above a threshold corresponding roughly to a Fourier component of the gw burst amplitude $H \simeq 10^{-22}$ Hz^{-1} or a strain amplitude for a conventional 1ms burst $h \simeq 10^{-18}$. This threhshold corresponds roughly to $\simeq 10^{-3}$ solar masses converted in a GW burst at the Galactic Center.

For each event the following information is provided:

- UTC event time (year, month, day, hour, minute, second);

- amplitude of GW burst candidate as Fourier trasform of strain amplitude of the wave;

- detector noise level (SNR in amplitude).

The results will be reported in the forthcoming GWDAW conference (Rome, December 1999).

Other recent analysis using the data of the bar detectors, just published or in progress, are the following:

- A search for coincidences among ALLEGRO and EXPLORER, using six month of data taken in 1991 [9], resulting in a new upper limit to GW bursts (see the contribution of G. Pizzella)

– A search for coincidences among EXPLORER , NIOBE and NAUTILUS [10]. The analysis is based on the comparison of candidate event lists recorded by the detectors in the period December 1994 through October 1996. Due to the different periods of data taking it was not possible to search for triple coincidences. The results have been: a weak coincidence excess with respect to the accidental ones between EXPLORER and NAUTILUS and no coincidence excess between EXPLORER and NIOBE.

– The first crosscorrelation analysis between two cryogenic detectors for the measurement of the GW stochastic background [11]. The results obtained with the data of EXPLORER and NAUTILUS, tuned in February 97 at 907 Hz gives a value of $\Omega(f) \leq 60$.

– Searches for monochromatic signals with the ALLEGRO data and with the data of EXPLORER are almost completed. The present sensitivity of bar detectors allows the detection at $SNR = 1$ of a continuous GW signal around 1 kHz of amplitude $h \simeq 5\,10^{-26}$ with an observation time of 100 days.

– Searches for coincidences among EXPLORER and the detectors of gamma-ray bursts BATSE and Beppo-SAX have been performed [12].

– The passage of cosmic rays (extensive air showers) has been observed to give delta-like excitations to the NAUTILUS resonant mode [13] (roughly one per week at the reported NAUTILUS sensitivity, vetoed by a system of cosmic ray detectors; see the contribution of G. Modestino). The correlation is very significant, and the result is in agreement with the thermoacoustic conversion model. It is remarkable that the detected energy changes in the vibrational status of NAUTILUS are of the order of 10 mK, or 10^{-6} eV.

– Last hour result. A record sensitivity is presented at this conference by the AURIGA group, the younger of the community: a detector noise temperature of 1 mK on long term periods (see the contribution of G. Prodi).

PRESENT DEVELOPMENTS

The crucial point in the present research and development work is to get a larger bandwidth. The use of a high-Q resonant-mass seems to imply that this type of detector has intrinsically a very narrow bandwidth. This is not the case. Both the strain signal and the thermal Brownian motion noise in the antenna exhibit the same resonant response near the mode eigenfrequency. Thus the signal-to-noise ratio is not bandwidth limited by the antenna thermal noise. The significant bandwidth limitation comes from the transducer and amplifier.

Near quantum-limited SQUID amplifiers and their coupling with high Q mechanical and electrical resonators are under study in the ALLEGRO [14], AURIGA [15], and ROG [16] groups. The work necessary to integrate such sensitive devices into the real detectors without degrading their performance is in progress: it appears possible to reach an effective bandwidth of the order of 50 Hz.

A different approach is the parametric transducer of the NIOBE group, based on the microwave technique, which shows a lot of room for improvement and can reach in principle a quantum limited sensitivity [17].

Also promising appears the new optical transducer developed by the AURIGA group [18].

For more information on the present readout activity, see the following contributions at this conference:

ALLEGRO and developments in the USA: Z. Allen (Multimode transducer using a double SQUID amplifier) and G. Harry (Advanced inductive transducer)

AURIGA: L. Conti (optical transduction chain).

EXPLORER and NAUTILUS: M. Bassan (Advanced transducer and SQUID)

NIOBE: D. Blair (Improved microwave transducer)

SPHERICAL DETECTOR

It is assumed now in our community that the next generation of resonant-mass detectors will be of spherical shape. A single sphere is capable of detecting gravitational waves from all directions and polarizations and is capable of determining the direction information and tensorial character of the incident wave.

A sphere will have a larger mass than the present bars (with the same resonant frequency), translating into an increased cross section and improved sensitivity. Omnidirectionality and source direction finding ability make a spherical detector an unique instrument for gravitational wave astronomy with respect to all present detectors. The measurement of the polarization states and the scalar-tensor discrimination open new possibilities in the study of gravitational physics. Finally, the different features and technology makes a spherical detector complementary to an interferometer. It emerges that an observatory composed of both a sphere and an interferometer will have unprecedented sensitivity and signal characterization capabilities. Studies and measurements essential to define a project of a large spherical detector, 40 to 100 tons of mass, cooled to 10 mK [19], have been made in USA, Italy, Netherland and Brasil (see [20] and references therein).

Burst sensitivity $h \simeq 10^{-22}$, or, in spectral units, 10^{-24} $Hz^{-1/2}$ over a bandwidth of about 50 Hz, can be reached [21].

In the last few years significant barriers towards the realization of a spherical antenna have already been overcome. Cooling large masses to ultra-low temperatures for long periods of time is possible and was demonstrated by the operation of the 2.3 ton AURIGA and NAUTILUS antennas at 100 mK. The 5 quadrupole modes of a real spherical mass are independent and have the required high mechanical Q at ultra-low temperatures [22]. Practicality of the truncated icosahedral symmetry for the positioning of the transducers was demonstrated and coupling of 6 resonant transducers with the 5 quadrupole modes of a spherical mass is understood [23]. The possibility of obtaining large pieces of material suitable for use as spherical antennas was investigated. Large pieces of Al5056 with high quality factors can be obtained by means of explosive bonding [22], and CuAl alloys, which can be cast in large pieces, were also found to have high quality factors [24].

Small (about 70 cm diameter) spherical detectors are in preparation at Leiden University, in San Paulo State, and Frascati INFN Labs (see the contributions of A. de Waard and M. Visco at this conference).

CONCLUSIONS

With the formation of the IGEC, the activity on bar detectors passed a phase transition: from the occasional exchange of data between two groups to the systematic exchange of data between all the groups, following an agreed protocol. We can confidently say that today the strongest sources in our Galaxy and in the Local Group will not pass unnoticed to at least two resonant-mass detectors. This fact is extremely important as the search for GW is based on the technique of coincidences among two or more detectors.

To fully exploit the potentialities of bar detectors, R&D programs devoted to the development of transducer-amplifier chains are in progress.

Bars and possible future spherical detectors can join with confidence the world wide gravitational wave observatory in formation at the beginning of the new millennium with the first generation large interferometric detectors.

REFERENCES

1. Astone P. et al. (ROG collaboration), *Phys. Rev D* **47** 2 (1993).
2. Mauceli E. et al.(ALLEGRO collaboration), *Phys. Rev D* **54** 1264 (1996).
3. Blair D.G. et al.(NIOBE collaboration), *Phys. Rev Lett.* **74** 1908 (1995).
4. Coccia E. and Niinikoski T.O., *Journal of Physics E* **16** 695 (1983).
5. Astone P. et al. (ROG collaboration), *Astroparticle Phys.* **7**, 231-243 (1997).
6. Prodi G. et al. (AURIGA collaboration), in *Gravitational Waves* Proc. of the 2nd Amaldi Conference, CERN (Geneva), 1997 E. Coccia, G. Pizzella and G. Veneziano (eds) (World Scientific, Singapore, 1999).
7. Pizzella G., *Class. Quantum Grav.* 1481-1485 (1997).
8. For more information on IGEC: http://igec.lnl.infn.it/igec.
9. Astone P. et al. (ALLEGRO and ROG collaborations), *Phys. Rev. D* **59**, 122001 (1999).
10. Astone P. et al. (NIOBE and ROG collaboration), *Astroparticle Phys.* **10**, 83-92 (1999).
11. Astone P. et al. (ROG collaboration), *Astron. Astrophys.* **351**, 811-814 (1999).
12. Astone P. et al. (ROG collaboration), *Astron. Astrophys. Suppl. Series* **138**, 603 and 605 (1999).
13. Astone P. et al. (ROG collaboration), *Phys. Rev. Lett.* 84, 14 (2000).
14. Jin I. et al. *IEEE Trans. Appl. Sup.* **7** 2742 (1997).
15. Falferi P. et al., *Appl. Phys. Lett.* **78** 3859 (1998).

16. Carelli P. et al., *Appl. Phys. Lett.* **72** 115 (1998).
17. Tobar M., Physica B Proceedings Supplement of LT22, Helsinki 1999, in press (1999).
18. Conti L. et al., Proc. 34th Rencontres de Moriond, Les Arcs 1999, in press.
19. Frossati G. and Coccia E., *Cryogenics (ICEC Suppl.)* **35** 9 (1994).
20. See for instance: Velloso, W.F., Aguiar, O.D., Magalhaes, N.S. Omnidirectional Gravitational Radiation Observatory, Sao Jose dos Campos, Brasil 1996, World Sci., Singapore, 1997.
21. Coccia E. et al. *Phys. Rev. D* **57**, 2051 (1998)
22. Coccia E. et al., *Phys. Lett.* A 219, 263 (1996).
23. Merkowitz S.M. and Johnson W.W., *Phys. Rev. D* **53** 5377 (1996).
24. Frossati G. et al. in *Proceedings of the International Conference on Gravitational Waves: Sources and Detectors*, edited by I. Ciufolini and F. Fidecaro (World Scientific, Singapore, 1997).

ASTROPHYSICAL SOURCES

Compact Binary Mergers and Accretion-Induced Collapse: Event Rates

Vassiliki Kalogera

Harvard-Smithsonian Center for Astrophysics
Cambridge, Massachusetts 02138

Abstract. This paper is a brief review of the topic of binary systems as sources of gravitational-wave emission for both LIGO and LISA. In particular I review the current estimates of the associated Galactic event rates and their implications for expected detection rates. I discuss the estimates for (i) the coalescence of close binaries containing neutron stars or black holes, (ii) white dwarfs going through accretion-induced collapse into neutron stars, and (iii) detached but close binaries containing two white dwarfs. The relevant uncertainties and robustness of the estimates are addressed along with ways of obtaining conservative upper limits.

INTRODUCTION

An important factor in the design and development of gravitational wave observatories is the prospect for detection of astrophysical systems known or expected to be sources of gravitational radiation. At an initial level, even qualitative knowledge of the source properties dictates the frequency range of operation and the desired sensitivity levels of the instruments. More detailed understanding of signal characteristics, such as waveforms and polarization, allows the development of optimized data analysis techniques, followed by early testing and calibration of the system based on model data. Hence, studies of several different astrophysical sources of gravitational radiation is an integral part of the collective effort for the *direct* detection of gravitational waves in the near future.

The inspiral of close binary compact objects, neutron stars (NS) or black holes (BH), driven by gravitational wave emission is considered one of the major sources for ground-based laser interferometers, such as LIGO, VIRGO, GEO600, and TAMA300. At present only systems containing two neutron stars (NS) have been detected, PSR B1913+16 being the prototypical NS–NS system [1]. This binary radio pulsar has provided striking empirical confirmation of general relativity with the measurement of orbital decay due to gravitational radiation [2]. As this decay proceeds, both the amplitude and characteristic frequency of the gravitational-wave signal increase. Although for PSR B1913+16 the frequency will not enter the LIGO window for another $\sim 3 \times 10^8$ yr, the expectation is that similar systems in other galaxies, well ahead in their inspiral phase, should be detectable now. In addition to NS–NS systems, and based on theories of binary evolution, BH–NS and BH–BH binaries are also expected to exist in galaxies, although they have not been discovered yet.

Another type of gravitational-wave source is provided by hot, young NS formed through the collapse of massive stars or the collapse of white dwarfs driven beyond the Chandrasekhar limit by accretion from a close binary companion (accretion-induced collapse). A variety of physical phenomena (e.g., rotational instabilities) can induce a quadrupole moment in proto-neutron stars and cause them to emit gravitational waves (see contributions in these proceedings by S. Hughes and J. Houser).

An assessment of detection prospects for the various sources requires estimates of (i) the signal strength, and hence the maximum distance out to which sources could be detected given the instrument sensitivity, and (ii) the source formation rate out to that maximum distance, extrapolated from Galactic formation rate estimates. In this paper I review current estimates of Galactic event rates for the coalescence of binary compact objects (NS–NS, BH–NS, BH–BH) and the formation of young NS in accretion-induced collapse of white dwarfs. A critical discussion of the various uncertain factors involved in these estimates is presented. At the end of the paper, I review briefly current predictions for close binaries as gravitational-wave sources expected to be detected by the future space laser interferometer LISA.

COALESCING BINARIES AND LIGO

Estimates of formation rates for *coalescing* binary compact objects (systems with tight enough orbits that will merge due to gravitational radiation within a Hubble time) can be predicted theoretically, based on our current understanding of binary evolution models. For NS–NS systems, we can also obtain empirical rate estimates based on the observed sample. In what follows I critically review the current coalescence-rate estimates, addressing the various uncertainties involved. A discussion of ways to obtain limits on the NS–NS coalescence rate is also included.

Given the expected strength of the gravitational-wave signal of double NS coalescence, the maximum distance out to which it could be detected by the LIGO-II interferometers has been estimated to be ~ 450 kpc [3]. A Galactic NS–NS coalescence rate of $\sim 10^{-6}$ yr^{-1} is then required for a detection rate of $2-3$ events per year. The corresponding estimates for the coalescence of two $10\,\mathrm{M}_\odot$ BH are ~ 2000 kpc and $\sim 10^{-8}$ yr^{-1} (these distance estimates take into account cosmological corrections for a flat universe and $H_0 = 65\,\mathrm{km\,s^{-1}\,Mpc^{-1}}$ [?]).

Theoretical Estimates

Theoretical calculations of the formation rate of coalescing binaries are possible, given a sequence of evolutionary stages followed by primordial binaries. Over the years a relatively standard picture has been formed based on the consideration of NS–NS binaries [5], although more recently variations of it have also been discussed [6]. In all versions of their formation path the main picture remains the same. The initial binary progenitor consists of two binary members massive enough to eventually collapse into NS or BH. Its evolution involves multiple phases of stable or unstable mass transfer, common-envelope evolution, and accretion onto neutron stars, as well as two supernova explosions.

Such theoretical modeling has been undertaken by various authors by means of population syntheses. This provides us with *ab initio* predictions of the coalescence rate. The evolution of an ensemble of primordial binaries with assumed initial properties is followed through specific evolutionary stages until a coalescing binary is formed. The changes in the properties of the binaries at the end of each stage are calculated based on our current understanding of the various processes involved: wind mass loss from massive hydrogen- and helium-rich stars, mass and angular-momentum losses during mass transfer phases, dynamically unstable mass transfer and common-envelope evolution, effects of highly super-Eddington accretion onto neutron stars, and supernova explosions with kicks imparted to newborn neutron stars. Given that several of these phases are not very well understood, the results of population synthesis are expected to depend on the assumptions made in the treatment of the various processes. Therefore, exhaustive parameter studies are required by the nature of the problem.

Recent studies of the formation of binary compact objects and calculations of coalescence rates (see [7], [8], [9], [10]) have explored the model parameter space and the robustness of the results at different levels of (in)completeness. Almost all have studied the sensitivity of the coalescence rate to the average magnitude of the kicks imparted to newborn neutron stars. The range of predicted Galactic NS–NS rates from *all* these studies obtained by varying the kick magnitude within reasonable ranges is $< 10^{-7} - 5 \times 10^{-4} \, \text{yr}^{-1}$. This large range indicates the importance of supernovae (two in this case) in the evolution of binaries. Variations in the assumed mass-ratio distribution for the primordial binaries can *further* change the predicted rate by about a factor of 10, while assumptions of the common-envelope phase add another factor of about $10 - 100$. Variation in other parameters typically affects the results by factors of two or less. Results for BH–NS and BH–BH binaries lie in the ranges $< 10^{-7} - 10^{-4} \, \text{yr}^{-1}$ and $< 10^{-7} - 10^{-5} \, \text{yr}^{-1}$, respectively when the kick magnitude to both NS and BH is varied. Other uncertain factors such as the critical progenitor mass for NS and BH formation lead to variations of the rates by factors of $10 - 50$.

It is evident that recent theoretical predictions for the coalescence rates cover a very wide range of values (typically 3-4 orders of magnitude). We note, however, that binary properties other than the coalescence rate, such as orbital sizes, eccentricities, center-of-mass velocities, are much less sensitive to the various input parameters and assumptions; the latter affect more severely the absolute normalization (birth rate) of the population. Given these results it seems fair to say that population synthesis calculations have quite limited predictive power and provide fairly loose constraints on coalescence rates.

Empirical Estimates

In the case of NS–NS binaries, there is another way to estimate their coalescence rate, using the properties of the observed coalescing NS–NS (two systems: PSR B1913+16 and PSR B1534+12) and models of selection effects in radio pulsar surveys. For each observed object, a scale factor is calculated based on the fraction of the Galactic volume within which pulsars with properties identical to those of the observed pulsar could be detected, in principle, by any of the radio pulsar surveys, given their detection thresholds. This scale factor is a measure of how many more pulsars like those detected in the coalescing NS–NS systems exist in our galaxy. The coalescence rate can then be calculated based on the scale factors and estimates of detection lifetimes summed up for the observed systems. This basic method was first used by Phinney [11] and Narayan et al. [12] who estimated the Galactic rate to be $\sim 10^{-6}\,\mathrm{yr}^{-1}$.

Since then, estimates of the coalescence rate have decreased significantly primarily because of (i) the increase of the Galactic volume covered by radio pulsar surveys with no additional coalescing NS–NS binaries discovered [13], (ii) the increase of the distance estimate for PSR B1534+12 based on measurements of post-Newtonian parameters [14] (iii) changes in the lifetime estimates for the observed systems [15], [16]. On the other hand, in these recent studies a upward correction has been added to account for the population of pulsars too faint to be detected by the surveys. The most recently published study [16] gives a lower limit of $2 \times 10^{-7}\,\mathrm{yr}^{-1}$ and a "best" estimate of $\sim 6 - 10 \times 10^{-7}\,\mathrm{yr}^{-1}$ which agrees with other recent estimates of $2 - 3 \times 10^{-6}\,\mathrm{yr}^{-1}$ [14], [17]. Additional uncertainties (typically by factors $\lesssim 10$) arise from the estimates of pulsar ages and distances, the pulsar beaming fraction, the spatial distribution of NS–NS binaries in the Galaxy, the form of the faint end of the luminosity function and the small number of objects in the observed sample.

Despite all these uncertainties the empirical estimates of the NS–NS coalescence rate appear to span a range smaller than two orders of magnitude, which is relatively narrow compared to the range covered by the theoretical estimates.

Small-Number Sample

One important limitation of empirical estimates of the coalescence rates is that they are derived based on *only two* observed NS–NS systems, under the assumption that the observed sample is representative of the true population, particularly in terms of their radio luminosity. Therefore, assessing the effect of small-number statistics on the results of the above studies is necessary. Assuming that NS–NS pulsars follow the radio luminosity function of young pulsars and that therefore their population is dominated in number by low-luminosity pulsars, it can be shown that the current empirical estimates most probably underestimate the true coalescence rate. If a small-number sample is drawn from a parent population dominated by low-luminosity (hence hard to detect) objects, it is statistically more probable that the sample will actually be dominated by objects from the high-luminosity end of the population. Consequently, the empirical estimates based on such a sample will tend to overestimate the detection volume for each observed system, and therefore underestimate the scale factors and the resulting coalescence rate.

This effect can be clearly demonstrated with a Monte Carlo experiment [18] using simple models for the pulsar luminosity function and the survey selection effects. As a first step, the average observed number of pulsars is calculated given a known "true" total number of pulsars in the Galaxy (thick-solid line in Figure 1). As a second step, a large number of sets consisting of "observed" (simulated) pulsars drawn from a Poisson distribution of a given mean number ($< N_{obs} >$) with assigned luminosities according to the assumed luminosity function are realized using Monte Carlo methods. Based on each of these sets, one can estimate the total number of pulsars in the Galaxy using empirical scale factors, as is done for the real observed sample. The many (simulated) 'observed' samples can then be used to obtain the distribution of the estimated total Galactic numbers (N_{est}) of pulsars. The median and 25% and 75% percentiles of this distribution are plotted as a function of the assumed number of systems in the (fake) 'observed' samples in Figure 1 (thin-solid and dashed lines, respectively).

It is evident from Figure 1 that, in the case of small-number observed samples (less than ~ 10 objects), it is highly probably that the estimated total number, and hence the estimated coalescence rate, is underestimated by a significant factor. For a two-object sample, for example, the true rate maybe higher by more than a factor of ten. This correction factor associated with the faint-end of the luminosity

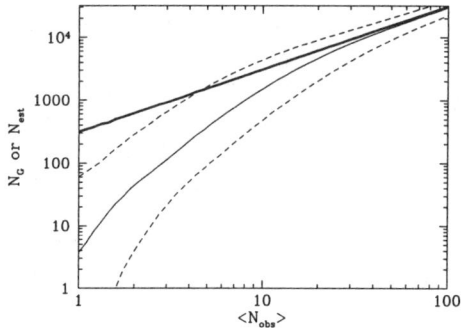

FIGURE 1. Bias of the empirical estimates of the NS–NS coalescence rate because of the small-number observed sample. See text for details.

function should be applied to the estimated NS–NS coalescence rate in place of the factor of ~ 10 used so far from a direct extrapolation of the luminosity function.

Limits on Coalescence Rates

One way to circumvent the uncertainties involved in the estimates of the NS–NS coalescence rate is to focus on obtaining upper or lower limits to this rate. Depending on how their value compares to the value of $\sim 10^{-6}\,\text{yr}^{-1}$ needed for a few LIGO-II events per year, such limits can provide us with valuable information about the prospects of gravitational-wave detection.

Bailes [19] used the absence of any young pulsars detected in NS–NS systems and obtained a rough upper limit to the rate of $\sim 10^{-5}\,\text{yr}^{-1}$, while recently Arzoumanian et al. [16] reexamined this in more detail and claimed a more robust upper limit of $\sim 10^{-4}\,\text{yr}^{-1}$.

An upper bound to the rate can also be obtained by combining our theoretical understanding of orbital dynamics (for supernovae with neutron-star kicks occurring in binaries) with empirical estimates of the birth rates of *other* types of pulsars related to NS–NS formation [20]. Binary progenitors of NS–NS systems experience two supernova explosions when the neutron stars are formed. The second supernova explosion (forming the neutron star that is *not* observed as a pulsar) provides a unique tool for the study of NS–NS formation, since the post-supernova evolution of the system is simple, driven only by gravitational radiation. There are three possible outcomes after the second supernova: (i) a coalescing binary is formed (CB), (ii) a wide binary (with a coalescence time longer than the Hubble time) is formed (WB), or (iii) the binary is disrupted (D) and a single pulsar similar to the ones seen in NS–NS systems is ejected. Based on supernova orbital dynamics we can calculate the probability branching ratios for these three outcomes, P_CB, P_WB, and P_D. For a given kick magnitude, we can calculate the maximum ratio $(P_\text{CB}/P_\text{D})^{\max}$ for the complete range of pre-supernova parameters defined by the necessary constraint $P_\text{CB} \neq 0$ (Figure 2). Given that the two types of systems have a common parent progenitor population, the ratio of probabilities is equal to the ratio of the birth rates (BR_CB/BR_D).

We can then use (i) the absolute maximum of the probability ratio ($\simeq 0.26$ from Figure 2) and (ii) an empirical estimate of the birth rate of single pulsars similar to those seen in NS–NS systems based on the current observed sample to obtain an upper limit to the NS–NS coalescence rate. The selection of this small-number sample involves some subtleties [20], and the analysis shows $BR_\text{CB} \lesssim 1.5 \times 10^{-5}\,\text{yr}^{-1}$ [20]. Note that this number could be increased because of the small-number sample and luminosity bias affecting this time the empirical estimate of BR_D by a factor of $2-6$.

This is an example of how we can use observed systems other than NS–NS to improve our understanding of their coalescence rate. A similar calculation can also be done using the wide NS–NS systems instead of the single pulsars [20].

Conclusions

A comparison of the various results on the NS–NS coalescence rate indicates that theoretical estimates based on modeling of NS–NS formation have a rather limited predictive power. The range of predicted rates exceeds 3 orders of magnitude and most importantly includes the "critical" value of $10^{-6}\,\mathrm{yr}^{-1}$ required for a LIGO-II detection rate of 2-3 events per year. This means that at the two edges of the range the conclusion swings from no detection to many per month. In other words no firm conclusions can be drawn from these estimates about the detection prospects of NS–NS coalescence. Empirical estimates, on the other hand, derived based on the observed sample appear to be more robust (estimates are all within a factor smaller than 100). Given those we would expect a LIGO-II detection rate of a few events per year up to even a few tens of events per year.

For coalescence rates of BH–NS and BH–BH systems we have to rely solely on our theoretical understanding of their formation. As in the case of double NS, the model uncertainties are significant and the ranges extend to more than 2 orders of magnitude. However, the requirement on the Galactic rate so that the detection rate is a few events $(2-3)$ per year is less stringent for the BH binaries, only $\sim 10^{-8}\,\mathrm{yr}^{-1}$. Therefore, even with the pessimistic estimates for BH–BH coalescence rates ($\lesssim 10^{-7}\,\mathrm{yr}^{-1}$), we would expect at least a few or even up to 10 detections per year, which is quite encouraging. We note that a very recent examination of dynamical BH–BH formation [21] in globular clusters leads to detection rates as high as a few per day.

Our expectations for the detectability of BH–NS coalescence could be improved significantly if we actually detect one or more such systems in the near future. Current pulsar surveys, such as the Multibeam Parkes Survey, are considerably more sensitive than previous searches to distant and faint pulsars in close binaries. This high sensitivity is the combined result of long integration times, rapid sampling, and the incorporation of acceleration searches in the data analysis techniques. A

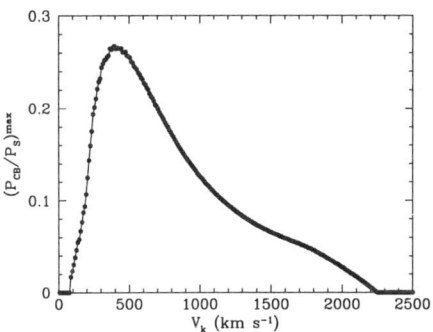

FIGURE 2. Maximum probability ratio for the formation of coalescing NS–NS binaries and the disruption of binaries as a function of the kick magnitude at the second supernova.

candidate NS–NS system (PSR J1811-1736) has been already discovered [22] and it would not be surprising if in the next few years close BH–NS systems are discovered.

YOUNG NS IN ACCRETION-INDUCED COLLAPSE

Hot, young NS formed in supernovae are susceptible to a range of instabilities each of which has a different physical origin: e.g., convection, bar-mode (dynamical or secular), r-mode instabilities. These instabilities can induce a quadrupole moment to the proto-NS and cause it to emit gravitational waves. The strength of the emission varies significantly with the physical mechanism and its calculation depends on model assumptions made (for a review see [23]). A mechanism which recently has attracted attention is related to the exponential growth of rotational mode (r-mode) instabilities in fast spinning ($\lesssim 10-15\,\text{ms}$), hot ($> 10^9\,\text{K}$) proto-NS (see contribution in these proceedings by B. Owen, [24], and [25]). Other mechanisms involve the centrifugal hang-up of a rapidly spinning collapsing core at early or late stages of the collapse [23]. Such rapidly rotating proto-NS can be formed either during the collapse of massive stars, provided that the stellar core is not well coupled to the envelope and is spinning fast just prior to collapse (see [26]), or perhaps in accretion-induced collapse (AIC) of fast spinning white dwarfs (WD) in close binaries that accrete at appropriate mass transfer rates.

Although the details of the the growth of the r-mode instability and the processes (hydrodynamic or gravitational radiation) for the removal of the excess angular momentum of rapidly rotating cores are not yet fully understood, it is interesting to consider these possibilities, primarily because the signal strength could be high enough for such sources to be detected by the LIGO-II interferometers out to $\sim 20\,\text{Mpc}$ for the r-mode instability [27] and late centrifugal hang-up at $\sim 20\,\text{km}$, and out to $\sim 100\,\text{Mpc}$ for early centrifugal hang-up at $\sim 100\,\text{km}$ [23]. Adopting the extragalactic extrapolation used by Phinney [11], we find that for a detection rate of a few events per year Galactic AIC rate of $\sim 10^{-4}\,\text{yr}^{-1}$ and $\sim 10^{-5}\,\text{yr}^{-1}$ are required, respectively.

As in the case of binary compact objects, formation of accreting WD that are expected to go through AIC can be studied via binary population synthesis techniques and theoretically predicted event rates can then be obtained. The accuracy and robustness of these results are actually significantly better than for binary NS and BH because they solely depend on our understanding of WD formation, which is considerably better than that of NS and BH formation with birth kicks. The most recent theoretical study of accreting WD formation [28] includes the most up-to-date picture for the conditions (WD mass and mass transfer rate) necessary for AIC to occur. The predicted AIC rates for a wide variety of models lie in the range $8 \times 10^{-7} - 8 \times 10^{-5}\,\text{yr}^{-1}$.

Recently an alternative way of addressing the question of AIC event rate has been explored by [29]. They use the measured abundances of neutron-rich nuclei (e.g., ^{62}Ni, ^{66}Zn, ^{87}Rb, ^{88}Sr) to set an *upper limit* on the AIC rate. This study includes a detailed investigation of the effect of a number of uncertain factors (equation of state, neutrino transport, etc.) and the conclusion is that the upper limit lies in the range $10^{-7} - 10^{-4}\,\mathrm{yr}^{-1}$.

Both the actual estimates and the empirical upper limits from the above studies appear to be in agreement and this gives us confidence that our understanding of the frequency of AIC events is relatively good. Comparison with the required Galactic rates for a few events detected per year indicates that it may be difficult to actually detect gravitational waves from rapidly rotating proto-NS. However, the upper end of our estimates is high enough to justify further consideration of AIC as candidate sources of gravitational radiation.

CLOSE BINARIES AND LISA

The construction of a space laser interferometer, LISA, is also being planned for the next one or two decades. The frequency range of its operation is expected to be $\sim 0.1 - 10^3\,\mathrm{mHz}$, clearly different than that of the ground-based interferometers that are limited by seismic noise at such low frequencies. As a consequence, LISA will be sensitive to very different types of gravitational-wave sources and allow us to explore different regimes of gravity and astrophysical processes.

Although the primary source-target for LISA are supermassive black holes in the centers of galaxies, LISA will also be sensitive to emission from a large population of relatively close binaries in our Galaxy [30]. For frequencies exceeding $\sim 0.1\,\mathrm{mHz}$, this emission is dominated by close binaries consisting of two white dwarfs. At frequencies up to $\sim 3-4\,\mathrm{mHz}$ the emission from a large collection of WD–WD binaries is blended and appears as a continuous background. For higher frequencies up to $\sim 30\,\mathrm{mHz}$, individual sources could be detected, and at even higher frequencies the extragalactic background becomes detectable but at a much lower power (a factor of ~ 10).

It is only very recently that WD–WD binaries have been discovered [31], despite the large expected number in the Galaxy. Their identification requires challenging optical observations that explain the small number of objects detected. With such a limited observed sample estimates of their true number in the Galaxy, and hence of the strength of the associated GW background, rely on purely theoretical calculations of their formation (population synthesis). As in the case of AIC models, the results of these studies are quite reliable since our understanding of the formation process for WD–WD binaries is better than that of NS and BH. This is clearly suggested by a comparison of the results from a number of different recent studies and for different model assumptions [32], [33]. The formation rate estimates span a relatively narrow range (factor of 2) from $5 \times 10^{-3}\,\mathrm{yr}^{-1}$ to $0.1\,\mathrm{yr}^{-1}$. The predicted GR background is largely insensitive to model variations with the most significant factor being that of mass-ratio distribution in the primordial binaries.

REFERENCES

1. Hulse, R.A., and Taylor, J.H., *ApJ* **195**, L51 (1975).
2. Taylor, J.H., and Weisberg, J.M., *ApJ* **253**, 908 (1982).
3. Gustafson, E., Shoemaker, D., Strain, K., and Weiss, R., *LSC White Paper on Detector Research and Development* (LIGO-Project document, September 11, 1999).
4. Finn, L.S., *private communication* (1999).
5. van den Heuvel, E.P.J., *Structure and Evolution of Close Binary Systems*, Dordrecht: Kluwer Academic Publishers, 1996, pp. 100-100.
6. Brown, G.E., *ApJ* **440**, 270 (1995).
7. Lipunov, V.M., et al., *MNRAS* **288**, 245 (1997).
8. Fryer, C.L., et al., *ApJ* **496**, 333 (1998).
9. Portegies-Zwart, S.Z., and Yungel'son, L.R., *A&A* **332**, 173 (1998).
10. Brown, G.E., and Bethe, H., *ApJ* **506**, 780 (1998).
11. Phinney, E.S., *ApJ* **380**, 17 (1991).
12. Narayan, R., et al., *ApJ* **379**, 17 (1991).
13. Curran, S.J., and Lorimer, D.R., *MNRAS* **276**, 347 (1995).
14. Stairs, I.H., et al., *ApJ* **505**, 352 (1998).
15. van den Heuvel, E.P.J., and Lorimer, D.R., *MNRAS* **283**, 37 (1996).
16. Arzoumanian, Z., et al., *ApJ*, **520**, 696 (1999).
17. Evans, T., et al., to appear in the proceedings of the XXXIVth Rencontres de Moriond on "Gravitational Waves and Experimental Gravity", Les Arcs, France, (1999).
18. Kalogera, V., et al., in preparation (1999).
19. Bailes M., *Compact Stars in Binaries*, Dordrecht: Kluwer Academic Publishers, 1996, pp. 213-223
20. Kalogera, V., and Lorimer, D.R., *ApJ* **530**, in press (2000) [astro-ph/9907426].
21. Portegies-Zwart, S.F., and McMillan, S.L.W., *ApJ Letters*, submitted [astro-ph/9910061].
22. Manchester, R.N., et al., to appear in *Pulsar Astronomy - 2000 and Beyond*, 1999.
23. Thorne, K.S., *Compact Stars in Binaries*, Dordrecht: Kluwer Academic Publishers, 1996, pp. 153-183.
24. Lindblom, L., et al., *PRL* **80**, 4843 (1998).
25. Andersson, N., et al., *ApJ* **510**, 846 (1999).
26. Spruit, H.C., and Phinney, E.S., *Nature* **393**, 139 (1998).
27. Owen, B., et al., *PRD* **58**, 084020(1998).
28. Yungel'son, L., and Livio, M., *ApJ* **497**, 168 (1998).
29. Fryer, C.L., et al., *ApJ* **516**, 892 (1999).
30. Hils, D., et al., *ApJ* **360**, 75 (1990).
31. Marsh, T.R., et al., *ApJ* **275**, 828 (1995).
32. Nelemans, G., et al., to appear in the proceedings of the XXXIVth Rencontres de Moriond on *Gravitational Waves and Experimental Gravity*, Les Arcs, France, (1999).
33. Webbink, R.F., and Han, Z., *Laser Interferometer Space Antenna*, New York: American Institute of Physics, 1999, pp.61-67.

Instabilities in Stiff Stellar Cores: The Gravitational Radiation Reaction

J.L. Houser[*][†]

[*]*Harvard-Smithsonian Center for Astrophysics, 60 Garden Street, Cambridge, MA, 02138*
[†]*LIGO Visiting Scientist, MIT, Cambridge, MA, 02139*

Abstract. Symmetry-breaking rotational instabilities may play a significant role in the dynamics of rapidly rotating compact objects. During the growth of such global non-axisymmetric instabilities, the external spacetime undergoes dynamical changes in response to the changing shape of the object and gravitational radiation is emitted. Using a three-dimensional Smoothed Particle Hydrodynamics code, the numerically generated gravitational wave signals of two pre-supernova stellar cores undergoing a dynamical instability will be compared. The first model assumes a purely Newtonian gravitational field, while the second model incorporates the General Relativistic "back reaction" into the hydrodynamical equations.

INTRODUCTION

An interesting possible source of gravitational radiation is the rotational instability that can occur during the gravitational collapse of a massive star's degenerate core, or during the accretion-induced collapse of a white dwarf. In these scenarios, the collapsing core and the accretion-formed neutron star may increase their rotation rates until sufficiently high to allow the development of a triaxial symmetry-breaking instability. Such global instabilities can develop via two different physical mechanisms and may potentially produce observable amounts of gravitational radiation [12,16]. The *secular* instability exists only in the presence of a dissipative mechanism such as viscosity or gravitational radiation, and develops on a timescale dependent on the triggering mechanism. The *dynamical* instability is driven by Newtonian hydrodynamical and gravitational forces and develops on a timescale of one rotation period [15].

The high rotational velocities and strong gravitational fields present in such compact objects demand the introduction of relativity into the time-dependent dynamics. General relativistic conservation laws guarantee that the source must lose angular momentum and energy consistent with the removal of these quantities due to the emission of gravitational radiation [10]. To study the effect of these losses on the system, the standard Newtonian potential, which conserves energy and angular momentum, is modified to include the relativistic back reaction potential [7].

METHOD

The Quadrupole Approximation was derived to compute gravitational wave generation in the limit of weak gravitational fields and nearly flat spacetime. In such a region of space, the metric can be written as $g_{\mu\nu} = \eta_{\mu\nu} + h_{\mu\nu}$ [10], which consists of a second rank tensor field $h_{\mu\nu}$ ($|h_{\mu\nu}| \ll 1$ and $\mu, \nu = t, x, y, z$) imposed on Minkowski, or flat ($\eta_{\mu\nu}$) spacetime. These restrictions enable a multipole expansion of the Newtonian potential $\Phi_{\text{Newtonian}}$ to be performed, yielding monopole, dipole, quadrupole, and octopole terms [10]. Conservation of mass and angular momentum requires there be no monopole or dipole radiation. Thus the first non-vanishing type of gravitational radiation is mass quadrupole. For such slow motion sources, the gravitational waveform profiles at a distance r from the source are written as

$$rh_{ij}^{\text{TT}} = 2\frac{G}{c^4}\ddot{I}_{ij}^{\text{TT}}, \qquad (1)$$

where TT stands for the transverse-traceless components of the metric perturbation h_{ij}, and I_{ij} is the source's trace-reduced quadrupole moment tensor [10].

For an observer located along the z axis ($\theta = 0$, $\phi = 0$) of a spherical coordinate system centered on the source's center of mass, the waveform amplitudes for the two polarization states take the simple form

$$h_{+,\text{axis}} = \frac{G}{rc^4}(\ddot{I}_{xx} - \ddot{I}_{yy}), \quad h_{\times,\text{axis}} = \frac{2G}{rc^4}\ddot{I}_{xy} \qquad (2)$$

where a dot indicates the time derivative d/dt. The total energy emitted by gravitational radiation is given by

$$\Delta E = \int \frac{G}{5c^5} \left\langle \dddot{I}_{ij}\dddot{I}_{ij} \right\rangle dt, \qquad (3)$$

where there is an implied sum on repeated indices, and the angle-brackets indicate an average over several characteristic wave periods.

This loss is incorporated into the quadrupole approximation by modifying the standard Newtonian potential to include a back reaction potential such that $\Phi = \Phi_{\text{Newtonian}} + \Phi^{\text{react}}$ [10]. The first term in the multipole expansion for the radiation reaction potential, Φ^{react}, is given by [6,7,10,12]

$$\Phi^{\text{react}} = \frac{1}{5}\frac{G}{c^5}\left\{x^2 I_{xx}^{(5)} + y^2 I_{yy}^{(5)} + z^2 I_{zz}^{(5)} + 2\left(xy I_{xy}^{(5)} + xz I_{xz}^{(5)} + yz I_{yz}^{(5)}\right)\right\}. \qquad (4)$$

Here, the superscript (5) indicates five time derivatives are to be taken.

The numerical method known as Smooth Particle Hydrodynamics (SPH) is used for the temporal evolution of an unstable core [3,9]. To incorporate the back reaction of the gravitational waves on the system, the SPH hydrodynamical equations are modified such that the net acceleration $\mathbf{a} = -\nabla\Phi = \mathbf{a}_0 + \mathbf{a}_{\text{gr}}$. Here, the acceleration $\mathbf{a}_0$ is calculated via the SPH representation of the fluid momentum conservation equation which uses only standard Newtonian forces [3]. The additive term $\mathbf{a}_{\text{gr}}$ is the back reaction acceleration and is calculated by taking the gradient of the Newtonian-type potential, Φ^{react} [7,17].

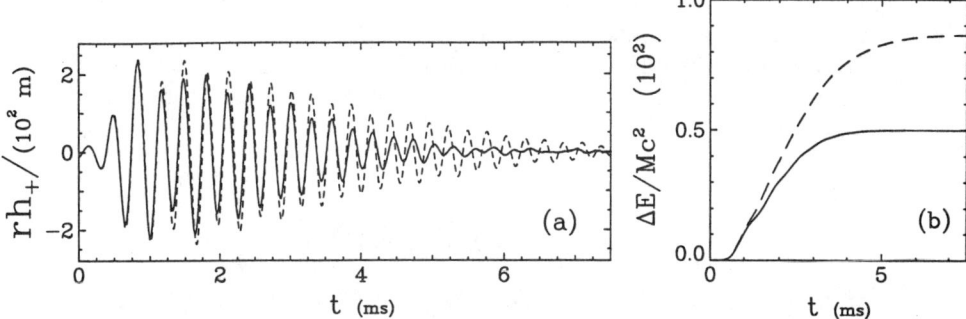

FIGURE 1. (a) rh_+ polarization state for an observer located on axis ($\theta = \phi = 0$), and (b) total rest mass energy radiated. The solid and dashed lines depict the time evolution for the relativistic and Newtonian models, respectively. Both models are assumed to have a total mass of $M = 1.4\,M_\odot$ and equatorial radius of $R_{eq} = 10$ km.

RESULTS

Two models of the dynamical instability are presented here: one which uses a purely Newtonian gravitational potential and one which includes the relativistic back reaction potential defined in Eq. (4). The initial model used in the simulations was a differentially rotating, axisymmetric equilibrium model which is dynamically unstable. This equilibrium model was produced using the self-consistent field method [11], and used an angular momentum distribution for a uniformly rotating, constant density, Maclaurin spheroid [4,5,13,14]. The resulting density and angular velocity profiles were converted into a particle model by using the "rejection method" for particle placement and rotational cooling to remove discretized density fluctuations [1,6].

When time evolved, the Newtonian and relativistic models produce identical gravitational waveforms until the growth of the primary instability, at ~ 0.44 ms (see Fig. 1a). After this time, the Newtonian simulation reaches a slightly higher peak wave amplitude and maintains an almost constant wave frequency throughout the evolution. In contrast, the relativistic model modulates between several frequency domains, eventually decreasing by ~ 700 Hz in approximately 4 ms [7,8]. The amplitude decrease in $rh_+(t)$ results from the inclusion of $\mathbf{a}_{gr}$, which behaves as a dissipative mechanism, removing mass and angular momentum from the system. Since this damping retards the growth of the instability, a more compact structure results. This in turn produces a weaker signal (see Fig. 1a), reducing the rest mass radiated in the form of gravitational waves to approximately 36% of that calculated via the Newtonian simulation (see Fig. 1b) [7].

This study yielded several findings of note. These are as follows: (1) The overall evolution of the relativistic unstable core remains structurally more compact throughout its temporal evolution. (2) The gravitational wave amplitude of the primary burst appears to remain unaffected by the inclusion of the general relativistic back reaction. Hence, $|h|_{max}$ remains unchanged from its Newtonian value [6,7]. (3) The frequency of the emitted waves generated by the relativistic model sweeps down from ~ 3600 Hz to ~ 3000 Hz during the evolution, thus bringing it closer to LIGO's sensitivity band $(10 - 1000$ Hz$)$. (4) The weaker signal produced by the relativistic model reduces the rest mass radiated in the form of gravitational waves.

ACKNOWLEDGEMENTS

This work was supported by NSF Cooperative Agreement No. PHY-9603177 in support of LIGO's Visitor's Program. Simulations were carried out at the Harvard-Smithsonian Center for Astrophysics (CFA) and at the Massachusettes Institute of Technology (MIT).

REFERENCES

1. Centrella, J.M. & McMillan, S.L.W., *Ap. J.* **416**, 719 (1993).
2. Durisen, R., Gingold, R., Tohline, J. & Boss, A., *Ap. J.* **305**, 281 (1986).
3. Hernquist, L. & Katz, N., *Ap. J. Suppl.* **70**, 419 (1989).
4. Houser, J.L., Centrella, J.M., & Smith, S.C., *Phys. Rev. Letters* **72**, 1314 (1994).
5. Houser, J.L. & Centrella, J.M., *Phys. Rev. D.* **54**, 7278 (1996).
6. Houser, J.L., *MNRAS* **299**, 1069 (1998).
7. Houser, J.L., in preparation (2000).
8. Lai, D. & Shapiro, S., *Ap. J.* **442**, 259 (1995).
9. Lucy, L., *A.J.* **82**, 1013 (1977).
10. Misner, C., Thorne, K. & Wheeler, J., *Gravitation*, Freeman: New York, 1973.
11. Ostriker, J. & Mark, J., it Ap. J. **151**, 1075 (1968).
12. Schutz, B., in *Gravitation in Astrophysics: Cargèse 1986*, edited by B. Carter & J.B. Hartle, Plenum Press: New York, 1986.
13. Smith, S.C & Centrella, J.M., in *Approaches to Numerical Relativity*, edited by R. d'Inverno, Cambridge University Press: New York, 1992.
14. Smith, S.C., Houser, J.L., & Centrella, J.M., *Ap. J.* **458**, 236 (1996).
15. Tassoul, J.,*Theory of Rotating Stars*, Princeton University Press: Princeton, 1978.
16. Thorne, K., in *Proceedings of the Snowmass 95 Summer Study on Particle and Nuclear Astrophysics and Cosmology*, ed. E. W. Kolb and R. Peccei, World Scientific: Singapore, 1996.
17. Thorne, K.S., *Rev. of Mod. Phys.* **52**, 285 (1980).
18. Tohline, J., Durisen, R. & McCollough, M., *Ap. J.* **298**, 234 (1985).

Address for correspondence: Janet L. Houser, Harvard-Smithsonian Center for Astrophysics, 60 Garden Street, Mail Stop 33, Cambridge, MA 02138, USA.

Email: jhouser@cfa.harvard.edu

Gravitational waves from the r-modes of rapidly rotating neutron stars

Benjamin J. Owen

Max Planck Institut für Gravitationsphysik (Albert Einstein Institut)
Am Mühlenberg 1, 14476 Golm bei Potsdam, Germany

Abstract. Since the last Amaldi meeting in 1997 we have learned that the r-modes of rapidly rotating neutron stars are unstable to gravitational radiation reaction in astrophysically realistic conditions. Newborn neutron stars rotating more rapidly than about 100 Hz may spin down to that frequency during up to one year after the supernova that gives them birth, emitting gravitational waves which might be detectable by the enhanced LIGO interferometers at a distance which includes several supernovae per year. A cosmological background of these events may be detectable by advanced LIGO. The spins (about 300 Hz) of neutron stars in low-mass x-ray binaries may also be due to the r-mode instability (under different conditions), and some of these systems in our galaxy may also produce detectable gravitational waves—see the review by G. Ushomirsky in this volume. Much work is in progress on developing our understanding of r-mode astrophysics to refine the early, optimistic estimates of the detectability of the gravitational waves.

I THE R-MODE INSTABILITY

The r-mode instability has been the subject of about thirty papers over the past two years. I will not be able to do them all justice here.[1] Instead I will summarize the most important (as I see them) results with a direct impact on gravitational-wave detection, beginning with the basic model worked out in 1998 and ending with the latest (end of 1999) developments in this rapidly changing field.

The reason for the excitement is a version of the CFS instability—named for Chandrasekhar, who discovered it in a special case [2], and for Friedman and Schutz, who investigated it in detail and found that it is generic to rotating perfect fluids [3]. The CFS instability allows some oscillation modes of a fluid body to be driven rather than damped by radiation reaction, essentially due to a disagreement between two frames of reference.

The mechanism can be explained heuristically as follows. In a non-rotating star, gravitational waves radiate positive angular momentum from a forward-moving

---

[1] For a recent review with more emphasis on completeness, see Friedman and Lockitch [1].

mode and negative angular momentum from a backward-moving mode, damping both as expected. However, when the star rotates the radiation still lives in a non-rotating frame. If a mode moves backward in the rotating frame but forward in the non-rotating frame, gravitational radiation still removes positive angular momentum—but since the fluid sees the mode as having negative angular momentum, radiation drives the mode rather than damps it. Another example of such an effect due to a disagreement between frames of reference is the well-known Kelvin-Helmholtz instability, which leads to rough airplane rides over the jet stream and pounding surf on the California coast.

Mathematically, the criterion for the CFS instability is

$$\omega(\omega + m\Omega) < 0, \tag{1}$$

with the mode angular frequency ω (in an inertial frame) in general a function of the azimuthal quantum number m and rotation angular frequency Ω. For any set of modes of a perfect fluid, there will be some modes unstable above some minimum m and Ω. However, fluid viscosity generally grows with m and there is a maximum value of Ω (known as the Kepler frequency Ω_K) above which a rotating star flies apart. Therefore the instability is only astrophysically relevant if there is some range of frequencies and temperatures (viscosity generally depends strongly on temperature) in which it survives.

The r-modes are a set of fluid oscillations with dynamics dominated by rotation. They are in some respects similar to the Rossby waves found in the Earth's oceans and have been studied by astrophysicists since the 1970s [4]. The restoring force is the Coriolis inertial "force" which is perpendicular to the velocity. As a consequence, the fluid motion resembles (oscillating) circulation patterns. The (Eulerian) velocity perturbation is

$$\delta\vec{v} = \alpha\Omega R(r/R)^m \vec{r} \times \vec{\nabla} Y_{mm}(\theta, \phi) + O(\Omega^3), \tag{2}$$

where α is a dimensionless amplitude (roughly $\delta v/v$) and R is the radius of the star. Since $\delta\vec{v}$ is an axial vector, mass-current perturbations are large compared to the density perturbations. The Coriolis restoring force guarantees that the r-mode frequencies are comparable to the rotation frequency,

$$\omega + m\Omega = \frac{2}{m+1}\Omega + O(\Omega^3). \tag{3}$$

It was not until the time of the last Amaldi Conference in mid-1997 that Andersson [5] noticed that the r-mode frequencies satisfy the mode instability criterion (1) for all m and Ω, and that Friedman and Morsink [6] showed the instability is not an artifact of the assumption of discrete modes but exists for generic initial data. In other words, *all* rotating perfect fluids are subject to the instability.

II DRIVING VS. DAMPING

The universe is inhabited not by balls of perfect fluid, but by stars subject to internal viscous processes which tend to damp out oscillation modes. To evaluate the stability of modes in realistic neutron stars, we must compare driving and damping timescales.

In the small-amplitude limit, a mode is a driven, damped harmonic oscillator with an exponential damping timescale [7]

$$\frac{1}{\tau} = -\frac{1}{2E}\frac{dE}{dt} = -\frac{1}{2E}\left[\left(\frac{dE}{dt}\right)_G + \sum_V \left(\frac{dE}{dt}\right)_V\right] = \frac{1}{\tau_G} + \sum_V \frac{1}{\tau_V}. \quad (4)$$

Here E is the energy of the mode in the rotating frame and dE/dt is the sum of contributions from gravitational radiation (subscript G) and all viscous processes (subscript V). The mode is stable if the damping timescale τ is positive, unstable if τ is negative. The gravitational radiation timescale τ_G depends on the rotation frequency Ω, and the viscous timescales generally depend also on the temperature T. Therefore we define a critical frequency Ω_c such that

$$\frac{1}{\tau(\Omega_c, T)} = 0 \quad (5)$$

and decide if a given mode is astrophysically interesting by examining the curve $\Omega_c(T)$.

Neutron stars are complicated objects, but a simple model suffices to estimate the most important driving and damping timescales in the very young ones. When hotter than 10^9K (younger than about a year), most of the star is a ball of ordinary, barotropic (equation of state independent of temperature) fluid. Given a putative equation of state, the gravitational radiation timescale τ_G can be calculated by standard multipole integrals [8], although the r-modes are nonstandard in that the leading-order (in Ω) contribution is not from the mass multipoles but from the mass-current multipoles [9]. Viscous damping is due both to shearing of the fluid and to compression and rarefaction of individual fluid elements (bulk viscosity). The shear viscosity is stronger (timescale is shorter) at lower temperatures (like everyday experience with motor oil) and can be calculated from neutron-neutron scattering cross-sections [10]. The bulk viscosity is a weak nuclear interaction effect and thus is much stronger at higher temperatures. Compression and rarefaction of the fluid by the mode disturbs the density-dependent equilibrium $p + e \leftrightarrow n$, generating neutrinos which efficiently carry energy away [11]. As the star cools the viscous mechanisms change (see Sec. V), but this model is good enough for a first look.

The net damping timescale of the most unstable ($m = 2$) r-mode can be written in terms of fiducial timescales (written with tildes)

$$\frac{1}{\tau(\Omega, T)} = \frac{1}{\tilde{\tau}_G}\frac{\Omega^6}{(\pi G \bar{\rho})^3} + \frac{1}{\tilde{\tau}_S}\left(\frac{10^9 \text{K}}{T}\right)^2 + \frac{1}{\tilde{\tau}_B}\left(\frac{T}{10^9 \text{K}}\right)^6 \frac{\Omega^2}{\pi G \bar{\rho}}, \quad (6)$$

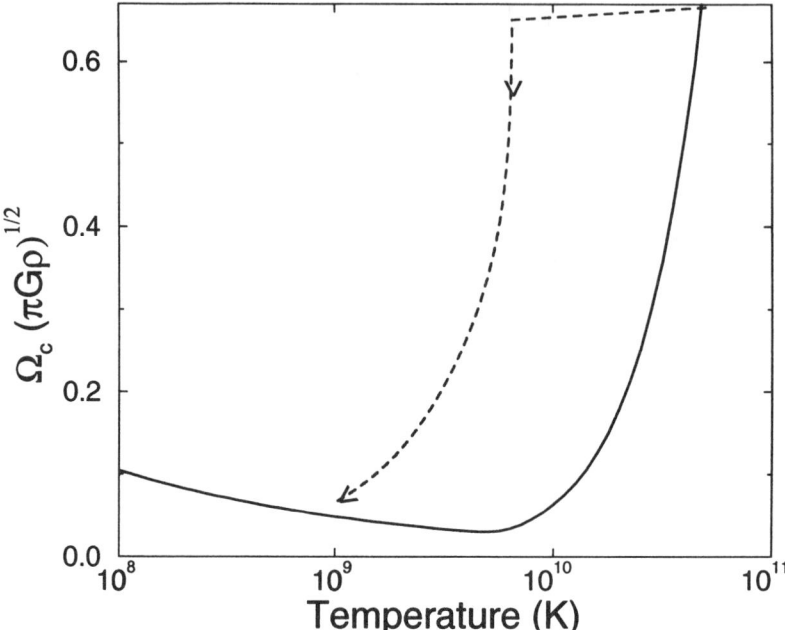

FIGURE 1. The solid curve is critical angular velocity Ω_c as a function of temperature T, assuming ordinary fluid viscosity given by Lindblom, Mendell, and Owen [13]. Above this curve, the fastest-growing ($m = 2$) r-mode is unstable. The dashed curve is the evolutionary track of a neutron star in the first year of its life if the r-modes begin at amplitude $\alpha = 10^{-5}$ and saturate at $\alpha = 1$, although it is not very sensitive to these values.

where $\bar{\rho}$ is the mean density of the equilibrium star. The numerical values of the fiducial timescales (for a simplistic equation of state) have been evaluated as [9,12,13]

$$\tilde{\tau}_G = -3.3 \text{ s}, \quad \tilde{\tau}_S = 2.5 \times 10^8 \text{s}, \quad \tilde{\tau}_B = 2.0 \times 10^{11} \text{s}. \tag{7}$$

The numbers change by factors of two or so for different neutron-star models, but the curve plotted in Fig. 1 and its conclusion are very robust. The r-modes are unstable in realistic neutron stars over an interesting range of frequencies and temperatures. Neutron stars born rotating at or near the Kepler frequency will spin down and emit gravitational radiation in the process; the question now is how much.

III SPINDOWN-COOLING MODEL

To detect gravitational waves from the r-modes, we need to know how the modes grow beyond the limits of perturbation theory and spin down the neutron star as it cools during the first year of its life. Even in the simple approximation of an ordinary fluid ball, this involves nonlinear hydrodynamics and radiation reaction, which are both tricky subjects and could take years to explore properly. In the meantime we make do with a simple model [15] developed to make the first, rough estimates of detectability.

In this model, we consider three coupled systems—a uniformly rotating fluid background (with angular velocity Ω), the most unstable r-mode (with dimensionless amplitude α), and the rest of the universe. The two systems in the star (mode and background) couple to each other by viscosity and nonlinear fluid effects. The mode couples to the universe by gravitational radiation; the background does not. The mode's energy evolves by gravitational radiation and viscous damping; the behavior of the background is determined by conservation of energy and angular momentum. Although some of the mode's energy goes into heating the star, the standard neutrino cooling law $(T/10^9 \text{K}) \sim (1 \text{ yr}/t)^{1/6}$ is all but unaffected.

If a star is born spinning at Ω_K at temperature 10^{11}K (and recent observations [14] suggest that some stars are), the evolution falls into three distinct phases. The *growth phase* begins when the r-modes go unstable of order one second after the supernova. During this phase a small initial perturbation α grows exponentially on a timescale of order one minute while Ω remains almost constant (the mode is too small to emit much angular momentum). In this regime linearized hydrodynamics (which is all we know at the moment) is a good approximation. Within at most a few minutes after the supernova, α becomes so large that nonlinear hydrodynamic effects can no longer be neglected. Previous studies of other modes [16] indicate that the main effect might be a saturation of the mode amplitude at some constant value, which we can treat as a phenomenological parameter. In this *saturation phase*, the star spins down very rapidly ($d\Omega/dt \sim \Omega^7$) and emits gravitational radiation of strain amplitude

$$h(t) = 4 \times 10^{-24} \left(\frac{\Omega}{\pi G \bar{\rho}}\right)^3 \left(\frac{20 \text{ Mpc}}{D}\right) \alpha_{\text{max}}, \tag{8}$$

at a detector at distance D (normalized here to the distance at which we expect several events per year). As the star spins down, the gravitational radiation gets much weaker (recall $1/\tau_G \sim \Omega^6$). Also, viscous damping becomes stronger, especially since other mechanisms come into play—for instance, when the neutrons become superfluid after cooling to about 10^9K. Thus, within a year the star has moved along a track such as that in Fig. 1 and entered the *decay phase*, where the r-modes are stabilized by viscosity and α slowly dies away without changing Ω much. The final spin frequency Ω_{end} is in practice another phenomenological parameter, since it depends on the more complicated viscous processes of cooler neutron stars as well as on α_{max}.

IV DETECTABILITY OF GRAVITATIONAL WAVES

Even at its strongest, an r-mode signal is below the strain noise of a gravitational-wave detector. But electromagnetic astronomers have been pulling faint pulsar signals out of noisy data for decades, and their data analysis techniques can be adapted for the r-modes.

Surprisingly, even the crude model of the source given in Sec. III is good enough to estimate the detectability of the gravitational waves. The quantity of interest is not the raw strain $h(t)$ but rather a *characteristic strain*

$$h_c(f) = h[t(f)]\sqrt{f^2/|df/dt|}, \tag{9}$$

where df/dt is the time derivative of the gravitational wave frequency. The optimal (filtered) signal-to-noise ratio is

$$(S/N)^2 = 2\int (d\ln f)(h_c/h_{\rm rms})^2, \tag{10}$$

where the rms strain noise is related to the detector's power spectral noise density by

$$h_{\rm rms} = \sqrt{f\, S_h(f)}. \tag{11}$$

Thus $(S/N)^2$ can be estimated by looking at a plot such as Fig. 2. For the r-modes, we find that [15]

$$h_c = 6 \times 10^{-22} \left(\frac{f}{1\text{ kHz}}\right)^{1/2} \left(\frac{20\text{ Mpc}}{D}\right) \tag{12}$$

with $(S/N) = 8$ for the projected LIGO-II (enhanced) noise curve as of 1998. This result is independent of much of the detailed physics of the source, including $\alpha_{\rm max}$.[2] However, it does depend on the detailed astrophysics through the final low-frequency cutoff, which does not change much even if the viscous damping changes by orders of magnitude.

The optimal signal-to-noise ratio is only an upper limit—it assumes matched filtering, which requires precise tracking of the signal phase. While our knowledge of astrophysics will never be good enough to track an r-mode signal to within one cycle out of 10^9, there are alternatives. The lower limit has been set by Brady and Creighton [17] using the simplest possible search algorithm, patterned on the techniques used to find pulsar signals in electromagnetic data. Assuming the supernova has been observed optically and a sky position is available, the Doppler shifts due to the Earth's motion can be removed to obtain a signal which is sinusoidal but for

---

[2] To my knowledge the argument was first made by R. D. Blandford in 1984 (but never published) that such a robust result holds for any system evolving mainly via gravitational radiation—like the r-modes in the saturation phase.

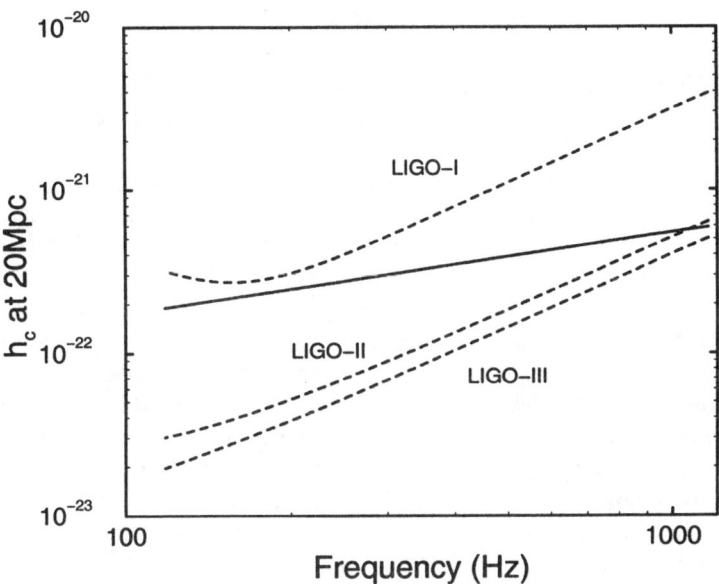

FIGURE 2. Characteristic signal amplitude h_c and noise $h_{\rm rms}$. The signal is from a neutron star with mass $1.4 M_\odot$ and radius 12.5 km at a distance of 20 Mpc (several events per year). The noise is for the three versions of LIGO used in Ref. [15]. Integrated (optimal) signal-to-noise ratio is 8; realistic signal-to-noise is somewhat less.

the (slow) intrinsic frequency evolution of the source. Even without any modeling of this evolution, i.e. by expanding

$$f(t) = f_0 \left(1 + \sum_k f_k t^k \right) \qquad (13)$$

for short Fourier transforms (integrating for year is computationally too expensive) and combining the transforms in some way for different trial values of the spindown parameters f_k, it is possible to obtain one fifth of the optimal signal-to-noise. This is in spite of the fact that the search is computationally limited by the requirement that data analysis keep pace with data acquisition and by the fact that the r-modes evolve so quickly that many terms f_k are needed. With constraints—even rough ones—from a physical model, the f_k are no longer all independent and the efficiency of data analysis could be increased. Since event rate goes roughly as $(S/N)^3$, it is important to beat the lower limit.

A stochastic background from the superposition of many faint r-mode signals out to cosmological distances will also exist. However, it is much fainter than a single signal and thus detectable only by (advanced) LIGO-III [15,18].

V OPEN QUESTIONS

We are now (at the end of 1999) in the midst of a renewed flurry of activity on r-mode astrophysics. Several effects neglected in the first simple scenario are being worked out. Some of them could damp the r-modes much more effectively than previously thought, pushing the detectability of the gravitational waves from LIGO-II to LIGO-III. However, this is far from certain and the astrophysicists are having an exciting time working it out. Here is a list of the effects that (I think) have the most direct impact on detection prospects.

Superfluid viscosity. One of the most eagerly awaited papers has been the calculation of the damping effects of "mutual friction", a process which paradoxically increases the viscous damping when a neutron star cools to a superfluid. At temperatures below about 10^9K this viscous mechanism was expected to dominate, and the big question was whether the damping was sufficient to stabilize the r-modes in stars older than about a year—especially the low-mass x-ray binaries (see the review by G. Ushomirsky in this volume). The answer [19] is a definite maybe. The damping timescale varies by several orders of magnitude, depending on a parameter of superfluid physics (the neutron-proton entrainment coefficient) which is yet poorly known. More work is in progress, but recently mutual friction has been upstaged by other issues.

Relativistic effects. Most r-mode calculations to date have assumed Newtonian gravity. Relativity was thought to simply multiply various numbers by redshift factors of order unity, but there are two important qualitative differences with Newtonian gravity. First there is the claim by Kojima [20] that the r-mode frequency becomes smeared over a finite bandwidth. This claim is contradicted, however, by Lockitch [21]. With Andersson and Friedman, he [22] finds that relativistic r-modes do however have an increased coupling to bulk viscosity similar to that of the "generalized r-modes" [23,24] of Newtonian stars, which are still unstable but less so.

Nonlinear fluid dynamics. At least two groups [25,26] are working on codes to numerically solve the fully nonlinear fluid equations for the r-modes and determine the saturation amplitude. However, the problem is complicated and the investment of coding and formalism is large, so expect results in a year or two at best. Order of magnitude arguments [27] have been made to claim that the r-modes saturate due to mode-mode coupling at a very small amplitude $\alpha \sim 10^{-5}$, which would render the signal undetectable. However, these arguments neglect the unique symmetries of the r-modes; and based on work on the g-modes of white dwarfs [28] it seems that the r-modes could indeed grow much larger. Semi-analytical analyses [29] of mode-mode coupling may give some indications about mode saturation while we wait for numerical results.

Magnetic fields. If the growth of the r-modes leads to substantial differential rotation, it could wind up magnetic field lines frozen into the fluid of a young neutron star, amplifying any seed field and saturating the r-modes at a small amplitude. Two recent papers [30,31] claim that the r-modes produce differential rotation, but this is a nonlinear effect which the authors have tried to treat with linear perturbation theory. Strictly speaking the gravitational radiation is also nonlinear (quadratic in α), but the canonical energy and angular momentum are global quantities whose perturbation can be derived self-consistently from a Lagrangian principle. It is not clear how to do this for local, dynamical quantities such as vorticity.

Crust formation. Perhaps the most important new result is that the formation of a solid crust (below about 10^{10}K) can act to strongly stabilize the mode. Bildsten and Ushomirsky [32] find that shear viscosity in the fluid boundary layer just below the crust decreases the damping timescale by 10^5–10^7. They conclude that the r-modes are completely suppressed in low-mass x-ray binaries and that the signal-to-noise ratio is reduced by three for newborn neutron stars. But it is not clear to me that this result is correct for newborn neutron stars. If an r-mode is already excited when the crust starts to form (of order a minute after the supernova), the intense and localized shear heating in the boundary layer can re-melt the crust if the pre-existing r-mode is strong enough. In this case, the outer layers stay in a self-regulating equilibrium at the melting temperature and the old model of the evolution is largely unaffected. I estimate that, in this case, "strong enough" means an r-mode amplitude of $\alpha = 10^{-3}$. This points out some interesting questions for future research: First, what is the initial value of α when the r-modes first go unstable? The first model [15] used gratuitously small values to make a point, but no one knows yet what are reasonable values. Also, what exactly is the melting temperature of a new crust? If it is 8×10^9K rather than 10^{10}K then the r-mode could have plenty of time to grow, and in astrophysics a 20% error is considered high precision.

Although I have skipped over many astrophysics issues, I realize even this short list may be bewildering to the experimenters and data analysts who are the main audience at the Amaldi Conference. If I had to distill my presentation into one sentence, I would say: Let the theorists argue for another two years; the r-modes are not as good a bet as binaries, but they may not be far behind.

VI ACKNOWLEDGMENTS

I am grateful to many colleagues for discussions of published and especially unpublished work which enabled me to give a good review: H. Asada, É. Flanagan, J.-A. Font, J. Friedman, J. Ipser, F. Lamb, Y. Levin, K. Lockitch, G. Mendell, S. Morsink, E. S. Phinney, L. Rezzolla, N. Stergioulas, K. Thorne, G. Ushomirsky, and especially L. Lindblom.

REFERENCES

1. Friedman, J. L., and Lockitch, K. H., gr-qc/9908083.
2. Chandrasekhar, S., *Phys. Rev. Lett.* **24**, 611–614 (1970).

3. Friedman, J. L., and Schutz, B. F., *Astrophys. J.* **222**, 281–296 (1978).
4. Papaloizou, J., and Pringle, J. E., *Mon. Not. Roy. Astron. Soc.* **182**, 423–442 (1978).
5. Andersson, N., *Astrophys. J.* **502**, 708–713 (1998).
6. Friedman, J. L., and Morsink, S. M., *Astrophys. J.* **502**, 714–720 (1998).
7. Ipser, J. R., and Lindblom, L., *Astrophys. J.* **373**, 213–221 (1991).
8. Thorne, K. S., Rev. Mod. Phys. **52**, 299-338 (1980).
9. Lindblom, L., Owen, B. J., and Morsink, S. M., *Phys. Rev. Lett.* **80**, 4843–4846 (1998).
10. Cutler, C., and Lindblom, L., *Astrophys. J.* **314**, 234–241 (1987).
11. Sawyer, R. F., *Phys. Rev. D* **39**, 3804–3806 (1989).
12. Andersson, N., Kokkotas, K. D., and Schutz, B. F., *Astrophys. J.* **510**, 846–853 (1999).
13. Lindblom, L., Mendell, G., and Owen, B. J., *Phys. Rev. D* **60**, 064006 (1999).
14. Marshall, F. E., Gotthelf, E. V., Zhang, W., Middleditch, J., and Wang, Q. D., *Astrophys. J.* **499**, L179–L182 (1998).
15. Owen, B. J., Lindblom, L., Cutler, C., Schutz, B. F., Vecchio, A., and Andersson, N., *Phys. Rev. D* **58**, 084020 (1998).
16. Detweiler, S. L., and Lindblom, L., *Astrophys. J.* **213**, 193–199 (1977).
17. Brady, P. R., and Creighton, T., to appear in *Phys. Rev. D* (gr-qc/9812014).
18. Ferrari, V., Matarrese, S., and Schneider, R., *Mon. Not. Roy. Astron. Soc.* **303**, 258–264 (1999).
19. Lindblom, L., and Mendell, G., submitted to *Phys. Rev. D* (gr-qc/9909084).
20. Kojima, Y., *Mon. Not. Roy. Astron. Soc.* **293**, 49–52 (1998).
21. Lockitch, K. H., Ph.D. Thesis, University of Wisconsin at Milwaukee (gr-qc/9909029).
22. Andersson, N., Lockitch, K. H., and Friedman, J. L., in preparation.
23. Lindblom, L., and Ipser, J. R., *Phys. Rev. D* **59**, 044009 (1999).
24. Lockitch, K. H., and Friedman, J. L., *Astrophys. J.* **521** 764–788 (1999).
25. Rezzolla, L., Shibata, M., Asada, H., Baumgarte, T. W., and Shapiro, S. L., to appear in *Astrophys. J.* (gr-qc/9905027).
26. Font, J. A., Stergioulas, N., and Kokkotas, K. D., submitted to *Mon. Not. Roy. Astron. Soc.* (gr-qc/9908010).
27. Phinney, E. S., at this conference; Lamb, F. K., Marković, D., Rezzolla, L., and Shapiro, S. L., in preparation.
28. Goldreich, P., and Wu, Y., *Astrophys. J.* **523**, 805–811 (1999).
29. Morsink, S. M., in preparation; Schenk, K., and Flanagan, É. É., in preparation.
30. Rezzolla, L., Lamb, F. K., and Shapiro, S. L., submitted to *Astrophys. J.* (astro-ph/9911188).
31. Levin, Y., and Ushomirsky, G., submitted to *Mon. Not. Roy. Astron. Soc.* (astro-ph/9911295).
32. Bildsten, L., and Ushomirsky, G., to appear in *Astrophys. J.* (astro-ph/9911155).

Gravitational Waves from Low-Mass X-ray Binaries: a Status Report[1]

Gregory Ushomirsky[†], Lars Bildsten[§], and Curt Cutler[‡]

[†]*Department of Physics and Department of Astronomy,
University of California, Berkeley, CA 94720*
[§]*Institute for Theoretical Physics and Department of Physics,
University of California, Santa Barbara, CA 93106*
[‡]*Max-Planck-Institut fuer Gravitationsphysik, Albert-Einstein-Institut,
Am Muehlenberg 1, D-14476 Golm bei Potsdam, Germany*

Abstract. We summarize the observations of the spin periods of rapidly accreting neutron stars. If gravitational radiation is responsible for balancing the accretion torque at the observed spin frequencies of ≈ 300 Hz, then the brightest of these systems make excellent gravitational wave sources for LIGO-II and beyond. We review the recent theoretical progress on two mechanisms for gravitational wave emission: mass quadrupole radiation from deformed neutron star crusts and current quadrupole radiation from r-mode pulsations in neutron star cores.

I SPINS OF ACCRETING NEUTRON STARS

Gravitational wave emission from rapidly rotating neutron stars (NS) has attracted considerable interest in the past several years. In addition to radiation from the spindown of newborn NSs (see the review by B. Owen in this volume), it has long been suspected [1,2] that rapidly accreting NSs, such as Sco X-1, may be a promising class of gravitational wave (GW) emitters. However, firm observational evidence of fast spins of these neutron stars had been missing until recently.

NSs in low-mass X-ray binaries (LMXBs) have long been thought to be the progenitors of millisecond pulsars [3]. However, directly measuring their periods has proved elusive, probably because of their rather low magnetic fields. With the launch of the *Rossi X-ray Timing Explorer*, precision timing of accreting NSs has opened new threads of inquiry into the behavior and lives of these objects. *RXTE* observations [4] have finally provided conclusive evidence of millisecond spin periods of NSs in about one-third of known Galactic LMXBs. These measurements

---

[1] Much theoretical progress has been made in understanding GW emission from LMXBs in the six months since the Amaldi conference in July 1999. Rather than just transcribe the talk given by one of us, we review the situation as of December 1999. Because of space limitations, this review is far from complete.

CP523, *Gravitational Waves: Third Edoardo Amaldi Conference*, edited by S. Meshkov
© 2000 American Institute of Physics 1-56396-944-0/00/$17.00

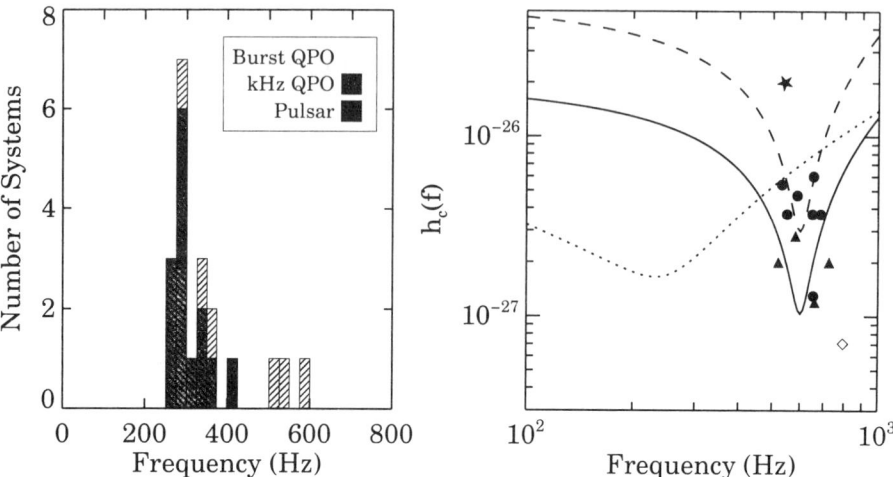

FIGURE 1. Left (a) The distribution of spins of NSs in LMXBs, summarizing Tables 1-4 of [4]. Right (b) Characteristic signal amplitude h_c for several known LMXBs (symbols, see text) compared with the sensitivity $h_{3/yr}$ of LIGO-II in broadband (the dotted line) and narrowband (the solid line) configuration (provided by K. A. Strain on behalf of the LIGO Scientific Community). The dashed line shows LIGO-II sensitivity for a two-week integration (see text).

are summarized in Fig. 1a. Altogether, there are seven such NSs with firmly established spin periods, by either pulsations in the persistent emission (discovered by Wijnands & van der Klis in the millisecond X-ray pulsar SAX J1808.4-3658; [5]) or oscillations during type I X-ray bursts (burst QPOs, first discovered in 4U 1728-34 by Strohmayer et al. [6]). There are an additional thirteen sources with twin kHz QPOs for which the spin may be approximately equal to the frequency difference [4]. A striking feature of all these neutron stars is that their spin frequencies lie within a narrow range, 260 Hz $< \nu_s <$ 589 Hz. The frequency range might be even narrower if the burst QPOs seen in KS 1731-260, MXB 1743-29, and Aql X-1 are at the first harmonic of the spin frequency, as is the case with the 581 Hz burst oscillations in 4U 1636-536 [7]. These NSs accrete at diverse rates, from $10^{-11} M_\odot$ yr^{-1} to the Eddington limit, $\dot{M}_{\rm Edd} = 2 \times 10^{-8} M_\odot$ yr^{-1}. Since disk accretion exerts a substantial torque on the NS and these systems are very old [8], it is remarkable that their spin frequencies are so similar, and that none of them are near the breakup frequency of ≈ 1.5 kHz.

One possible explanation, proposed by White & Zhang [9], is that these stars have reached the magnetic spin equilibrium (where the spin frequency matches the Keplerian frequency at the magnetosphere) at nearly identical frequencies. This requires that the NS dipolar B field correlate very well with $\dot{M}$ [9,10]. However, there are no direct B field measurements for LMXBs, and in the strongly magnetic binaries, where the B field *has* been measured directly, such a correlation is not observed. More importantly, for 19 out of 20 systems, there must be a way of hiding persistent pulses typically seen from magnetic accretors. These difficulties led Bildsten [11] to resurrect the conjecture originally due to Papaloizou & Pringle [12] and Wagoner [1] that gravitational radiation can balance the torque due to accretion. The detailed mechanisms will be discussed in the following sections.

Regardless of the detailed mechanism for GW emission, if gravitational radiation balances the accretion torque, then it is easy to estimate the GW strength. As noted by Wagoner [1], in equilibrium the luminosities in GWs and in X-rays are both proportional to the mass accretion rate $\dot{M}$, so the characteristic strain amplitude h_c depends on the X-ray flux F_x at Earth and the spin frequency

$$h_c = 4 \times 10^{-27} \frac{R_6^{3/4}}{M_{1.4}^{1/4}} \left(\frac{300 \text{ Hz}}{\nu_s}\right)^{1/2} \left(\frac{F_x}{10^{-8} \text{ erg cm}^{-2} \text{ s}^{-1}}\right)^{1/2}. \quad (1)$$

In Fig. 1b we show h_c for Sco X-1 (marked with a star), a few other bright LMXBs with the spin inferred from kHz QPO separation (thick dots) and burst QPO frequency (triangles), and the millisecond X-ray pulsar SAX J1808.4-3658 (open diamond). The dotted line shows LIGO-II sensitivity $h_{3/\text{yr}}$ (i.e., h_c detectable with 99% confidence in 10^7 s, provided the frequency and the phase of the signal are known in advance [13]) in the broadband configuration, while the solid line shows $h_{3/\text{yr}}$ for the narrowband configuration. However, the frequency and the phase are known precisely only for the SAX J1808.4-3658 millisecond X-ray pulsar [14]. For other sources, Brady & Creighton [13] showed that the number of trials needed to guess the poorly known orbital parameters or to account for the torque noise due to $\dot{M}$ variations lowers the effective sensitivity by roughly a factor of two.

While the average $\dot{M}$ certainly correlates with the X-ray brightness, current observations unfortunately do not let us robustly infer the *instantaneous* torque [4]. Even though $\dot{M}$ varies on a timescale of days, torque noise leads to frequency drift only on a timescale of weeks. The accretion torque is $N_a = \dot{M}(GMR)^{1/2}$, and the total time-averaged torque is zero due to equilibrium with GW emission. Assume that N_a flips sign randomly on a timescale $t_s \approx$ few days. The spin frequency Ω will experience a random walk with step size $\delta\Omega = (N_a/I)t_s$, where I is the NS moment of inertia. After an observation time t_{obs}, the drift is $\Delta\Omega = (t_{\text{obs}}/t_s)^{1/2}\delta\Omega$. This will exceed a Fourier frequency bin width, i.e., $\Delta\Omega \gtrsim 2\pi/t_{\text{obs}}$ only after

$$t_{\text{obs}} = \frac{21 \text{ days}}{M_{1.4}^{1/3} R_6^{1/3}} \left(\frac{1 \text{ day}}{t_s}\right)^{1/3} \left(\frac{10^{-8} M_\odot \text{ yr}^{-1}}{\dot{M}}\right)^{2/3} \quad (2)$$

Hence, on a timescale of tens of days, the intrinsic GW signal is coherent. The dashed line in Fig. 1b shows the LIGO-II sensitivity for a two-week integration in a narrowband configuration. This suggests that the way to detect GWs from LMXBs may be short integrations [13].

Currently, there are two classes of theories for GW emission from NSs in LMXBs. The presence of a large-scale temperature asymmetry in the deep crust will cause it to deform [11]. The resulting "mountains" will give the rotating star a time-dependent mass quadrupole moment. Alternatively, unstable r-mode pulsations (see a review by B. Owen in this volume) of a suitable amplitude in the NS liquid core can emit enough gravitational radiation to balance the accretion torque [11,15].

II DEFORMATIONS OF ACCRETING NS CRUSTS

The crust is a ≈ 1 km layer of crystalline "ordinary" (albeit neutron-rich) matter that overlies the liquid core composed of free neutrons, protons, and electrons. The crust's composition varies with depth in a rather abrupt manner. As an accreted nucleus gets buried under an increasingly thick layer of more recently accreted material, it undergoes a series of e^- captures, neutron emissions, and pycnonuclear reactions [16–18], resulting in layered composition. In Fig. 2a, we show schematically two such compositional layers (light and dark shading) sandwiched between the liquid core and the ocean. Since an appreciable fraction of the pressure is supplied by degenerate electrons, e^- captures induce abrupt density increases. In the outer crust, these density jumps are as large as $\approx 10\%$, while in the inner crust the density contrast is $\lesssim 1\%$. At $T = 0$, the e^- captures occur when the electron Fermi energy E_F is greater than the mass difference between the e^- capturer and the product of the reaction. In the absence of other effects, this depth is the same everywhere, and such an axisymmetric capture boundary (the dashed line in Fig. 2a) does not create a mass quadrupole moment.

However, in accreting NSs the crustal temperatures are high enough (in excess of 2×10^8 K) that e^- capture rates become temperature-sensitive [19]. Bildsten [11] pointed out that if there is a lateral temperature gradient in the crust (the arrow in Fig. 2a), then regions of the crust that are hotter undergo captures at a lower density than the colder regions. The capture boundary becomes "wavy" (the solid line in Fig. 2a), with captures proceeding a height Δz_d higher on the hot side of the star, and Δz_d lower on the cold side. Such a temperature gradient, if misaligned from the spin axis, will give rise to a nonaxisymmetric density variation and a nonzero quadrupole moment Q_{22} [11].

The required quadrupole moment Q_{eq} such that GW emission is in equilibrium with the accretion torque is

$$Q_{eq} = 3.5 \times 10^{37} \text{g cm}^2 M_{1.4}^{1/4} R_6^{1/4} \left(\frac{\dot{M}}{10^{-9} M_\odot \text{ yr}^{-1}} \right)^{1/2} \left(\frac{300 \text{ Hz}}{\nu_s} \right)^{5/2}, \qquad (3)$$

The range of $\dot{M}$'s in LMXBs is $\approx 10^{-11} - 2 \times 10^{-8}$ $M_\odot$ yr^{-1}, requiring $Q_{22} \approx 10^{37} - 10^{38}$ g cm^2 for $\nu_s = 300$ Hz [11]. Can temperature-sensitive e^- captures sustain a quadrupole moment this large? The quadrupole moment generated by a temperature-sensitive capture boundary is $Q_{22} \sim Q_{fid} \equiv \Delta\rho \Delta z_d R^4$, where $\Delta\rho$ is the density jump at the electron capture interface. Q_{fid} is the quadrupole moment that would result if the crust did not elastically adjust (or just moved horizontally)

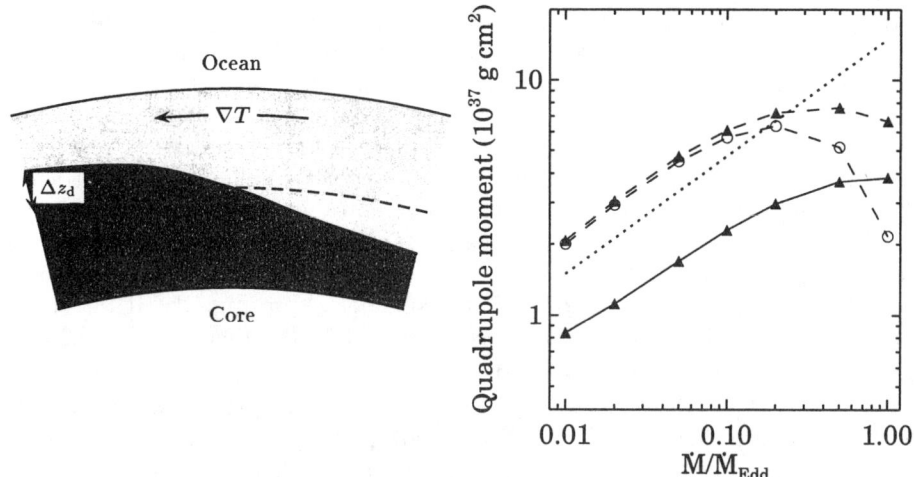

FIGURE 2. Left (a) A cartoon description of how a transverse temperature gradient leads to a varying altitude for the e⁻ captures. Right (b) The quadrupole Q_{22} due to a single capture layer in the inner crust as a function of $\dot{M}$. The solid and dashed lines denote the results for several NS models for a 10% composition asymmetry, while the dotted line is the relation given by Eq. (3), i.e., the quadrupole necessary for spin equilibrium at $\nu_s = 300$ Hz as a function of $\dot{M}$.

in response to the lateral pressure gradient due to wavy e⁻ captures. Using this estimate, Bildsten [11] argued that a single wavy capture boundary in the thin outer crust could generate Q_{22} sufficient to buffer the spinup due to accretion (Eq. [3]), provided that temperature variations of $\approx 20\%$ are present in the crust.

However an important piece of physics is missing from this estimate: the shear modulus μ. If $\mu = 0$, the crust becomes a liquid and cannot support a non-zero Q_{22}. Ushomirsky, Cutler, & Bildsten [20] recently calculated of the elastic response of the crust to the wavy e⁻ captures. They found that the predominant response of the crust to a lateral density perturbation is to sink, rather than move sideways. For this reason, Q_{22} generated in the *outer* crust [11] is much too small to buffer the accretion torque. However, a single e⁻ capture boundary in the deep *inner* crust can easily generate an adequate Q_{22}. Because of the much larger mass involved by captures in the inner crust, the temperature contrasts required are $\lesssim 5\%$, or only $\approx 10^6 - 10^7$ K, not $\approx 10^8$ K as originally postulated [11].

What causes the lateral temperature asymmetry, and can it persist despite the strong thermal contact with the almost perfectly conducting core? In LMXBs, the crusts are composed of the compressed products of nuclear burning of the accreted material. The exact composition depends on the local accretion rate, which could have a significant non-axisymmetric piece due to, e.g., the presence of a weak B field. Moreover, except in the highest accretion rate LMXBs, nearly all of the nuclear burning occurs in type I X-ray bursts. Burst QPOs (see Sec. I) provide conclusive evidence that bursts themselves are not axisymmetric. Until the origin of this symmetry breaking is clearly understood, it is plausible to postulate that these burst asymmetries get imprinted into the crustal composition.

Ushomirsky et al. [20] showed that such a non-uniform composition leads directly to lateral temperature variations δT. Horizontal variations in the charge-to-mass ratio Z^2/A (which determines the crustal conductivity) and/or nuclear energy release modulate the radial heat flux in the crust and set up a nonaxisymmetric δT. The δT's required to induce a $Q_{22} \approx Q_{\rm eq}$ can easily be maintained if there is a $\approx 10\%$ asymmetry in the nuclear heating or Z^2/A. So long as accretion continues, these δT's persist despite the strong thermal contact with the isothermal NS core.

The e^- capture Q_{22} calculations [20] are summarized in Fig. 2b. If the size of the heating or Z^2/A asymmetry is a constant fixed fraction, then for $\dot{M} \lesssim 0.5\dot{M}_{\rm Edd}$ the *scaling* of $Q_{22}(\dot{M})$ is just that needed for all of these NSs to have the same spin frequency (the normalization is proportional to the magnitude of the asymmetry, but the scaling is fixed by the microphysics). For $\dot{M} \gtrsim 0.5\dot{M}_{\rm Edd}$, in order to explain the spin clustering at *exactly* 300 Hz, this mechanism requires that the crustal asymmetry correlate with $\dot{M}$. Alternatively, if the asymmetry is the same as in the low $\dot{M}$ systems, then one would expect the bright LMXBs to have higher spins, a possibility that cannot be ruled out by current observations (Sec. I).

So long as crustal deformations are due to shear forces only, the crustal Q_{22} is limited by the yield strain $\bar{\sigma}_{\rm max}$ to be less than [20]

$$Q_{\rm max} \approx 10^{38} \text{ g cm}^2 \left(\frac{\bar{\sigma}_{\rm max}}{10^{-2}}\right) \frac{R_6^{6.26}}{M_{1.4}^{1.2}}. \quad (4)$$

Q_{22}'s needed to buffer the accretion torque require strains $\bar{\sigma} \approx 10^{-3} - 10^{-2}$ at ≈ 300 Hz, with $\bar{\sigma} \gtrsim 10^{-2}$ in near-Eddington accretors. Estimates for the yield strain of the neutron star crust range anywhere from 10^{-1} for perfect one-component crystals to 10^{-5}. Hence $\bar{\sigma} \gtrsim 10^{-2}$ is probably higher than yield strain, though this conclusion is based on extrapolating experimental results for terrestrial materials by > 10 orders of magnitude. Such high strains are perhaps the biggest problem with the crustal Q_{22} mechanism. At high pressures ($\gg$ shear modulus) terrestrial materials tend to deform plastically rather than crack, and so the crusts of accreting NSs may be in a state of continual plastic flow. If accretion continually drives the crust to $\bar{\sigma}_{\rm max}$, this leads to a natural explanation for spin similarities near $\dot{M}_{\rm Edd}$.

However, many fundamental issues remain unanswered. First, the calculation [20] is only good up to an overall prefactor set by the density of capture layers in the deep crust. We thus need an exploratory calculation of both the composition of the products of nuclear burning in the upper atmosphere over the entire range of $\dot{M}$ in LMXBs, and their detailed nuclear evolution under compression in the crust. Knowledge of the composition is also necessary for a robust calculation of the shear modulus, which is clearly the crucial number to know when computing the elastic response of the crust. Recent results [21,22] indicate that inner crusts of NSs are composed of highly nonspherical nuclei and may be more like liquid crystals (solids that provide no elastic restoring force for certain kinds of distortions) rather than simple Coulomb solids. Such improved calculations have implications far beyond the problem of the crustal quadrupole moment. The shear modulus of the crust affects the maximum elastic energy that can be stored in the crust, and hence the energetics of pulsar glitches and starquakes, as well as the models of magnetic field evolution that depend on crustal "plate tectonics" (see [23]). It even has bearing on the stability of r-modes in neutron stars (Sec. III). In addition, much work needs to be done on understanding what sets the shear strength $\bar{\sigma}_{\max}$ of multicomponent crystals, likely with defects and highly nonspherical nuclei, or what happens when $\bar{\sigma}_{\max}$ is exceeded and viscoelastic flow ensues.

III R-MODES IN ACCRETING NS CORES

Bildsten [11] and Andersson, Kokkotas, & Stergioulas [15] pointed out that the r-mode instability (see the review by B. Owen in this volume for the introduction and notation) may also explain the spins of NSs in LMXBs, and, if so, produce GW signal detectable by LIGO-II. An accreting NS is spun up (along a line in (ν_s, T) plane marked with an arrow in Fig. 3) until it reaches the r-mode instability line (the solid line in Fig. 3). At that point (marked by a thick dot in Fig. 3) the r-mode amplitude needed to balance the accretion torque is rather small. The NS can then hover at the instability line, with $1/\tau_G + 1/\tau_V = 0$, and the r-mode amplitude such that it balances the accretion torque. However, at $T = \text{few} \times 10^8$ K, the r-mode−accretion equilibrium spin frequency would be ≈ 150 Hz, rather than ≈ 300 Hz, resulting in an apparent disagreement with the observed spins of LMXBs. Bildsten [11] and Andersson et al. [15] speculated that including other sources of viscosity, e.g., superfluid mutual friction, is likely to raise the instability curve, resulting in equilibrium frequencies closer to the canonical 300 Hz. Finally, the narrow range of the observed spin frequencies would presumably arise because of the similar core temperatures of the accreting NSs (shown by the shaded box in Fig. 3).

Recent theoretical work brought up several challenges to this scenario. Levin [24] and Spruit [25] showed that steady-state equilibrium between accretion and r-modes is thermally unstable for normal fluid cores. In a normal fluid (i.e., not superfluid), the shear viscosity scales as T^{-2}, so the increase in the core temperature due to viscous heating decreases the shear viscosity. The smaller shear viscosity increases the growth rate of the r-mode, leading to an unstable runaway. Using a phenomenological model of nonlinear r-mode evolution [26], Levin [24] showed that in this case, instead of just hovering near the instability line, the r-mode grows rapidly until saturation, heats up the star, and spins it down and out of the instability region in less than 1 yr. Therefore, if NSs in LMXBs have normal fluid cores, we would not expect to see any of them with ≈ 300 Hz spins.

The unstable regime for r-modes in normal fluid NSs (above the solid line in

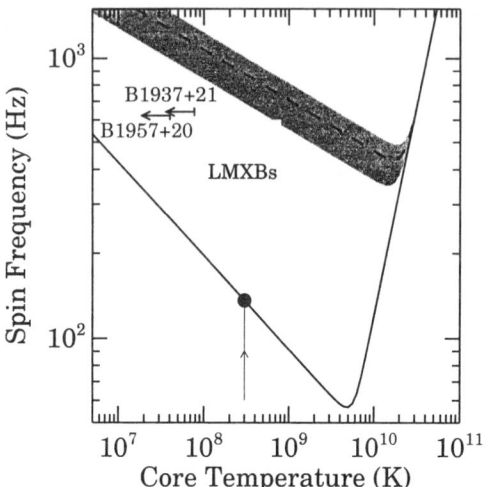

FIGURE 3. Critical spin frequencies of the r-mode instability. The solid line is the critical frequency set only by shear and bulk viscosities in the core (same as Fig. 1 of B. Owen's review in this volume). The dashed line also includes viscous boundary layer damping, while the shading around it displays the effect of core superfluidity.

Fig. 3) encompasses much of the parameter space occupied by NSs in LMXBs and newborn NSs. Because of the large torques exerted by the unstable r-modes, we would not expect to see any NSs in this region. In addition, the existence of two 1.6 ms radio pulsars (the spins and upper limits on core temperatures of which are shown by arrows in Fig. 3) means that rapidly rotating NSs are formed in spite of the r-mode instability [15]. While it is not clear whether their *current* core temperatures place these pulsars within the r-mode instability region, normal-fluid r-mode theory says that they were certainly unstable during spinup.

Superfluid r-mode calculations have been eagerly awaited, as they could resolve these conflicts. However, Lindblom & Mendell [27] showed that, for most values of the neutron-proton entrainment parameter, the superfluid dissipation is not competitive with gravitational radiation. Only over about 3% of the possible entrainment parameter values is mutual friction strong enough to compete with gravitational radiation. The r-mode instability line in this case is an approximately horizontal line (see Fig. 8 of [27]) separating the unstable spin frequencies ($\nu_s > \nu_{\rm crit}$) from the stable ones ($\nu_s < \nu_{\rm crit}$). If the superfluid entrainment parameter has a value such that $\nu_{\rm crit} \approx 300$ Hz, then the LMXB spin frequencies could still be understood in terms of the r-mode instability and the special nature of the NS superfluid.

Before learning about these results, Brown & Ushomirsky [28] ruled out such a simple superfluid equilibrium *observationally* for a subset of LMXBs. In steady state, the shear in the r-mode deposits ≈ 10 MeV of heat per accreted baryon into the NS core. When the core is superfluid, Urca neutrino emission from it is suppressed, and this heat must flow to the NS surface and be radiated thermally. In steadily accreting systems (such as Sco X-1) this thermal emission is dwarfed by the accretion luminosity of $GM\dot{M}/R \approx 200$ MeV per accreted baryon. However, in *transiently* accreting systems, such as Aql X-1, when accretion ceases, the r-mode heating should be directly detectable as enhanced X-ray luminosity from the NS surface. For Aql X-1 and other NS transients, Brown & Ushomirsky [28] showed that, if the superfluid r-mode equilibrium prevails, then the quiescent luminosity should be about 5−10 times greater than is actually observed.

A possible resolution of this conundrum has been recently proposed by Bildsten & Ushomirsky [29]. All but the hottest ($\gtrsim 10^{10}$ K) NSs have solid crusts. The r-mode motions are mostly transverse, and reach their maximum amplitude near the crust-core boundary. The fluid therefore rubs against the crust, which creates a thin (few cm) boundary layer. Because of the short length scale, the dissipation in this boundary layer is very large. The damping time due to rubbing is [29]

$$\tau_{\rm rub} \approx 100 \text{ s } T_8 \frac{M_{1.4}}{R_6^2} \left(\frac{1 \text{ kHz}}{\nu_s}\right)^{1/2}, \qquad (5)$$

substantially shorter than the viscous damping times due to the shear and bulk viscosities in the stellar interior, as well as the mutual friction damping time for most values of the superfluid entrainment parameter.

The critical frequency for the r-mode instability in NSs with crusts is shown by the dashed line in Fig. 3 for the case where all nucleons are normal, and the dark shading around it represents the range of frequencies when either neutrons or all nucleons are superfluid. The crust-core rubbing raises the minimum frequency for the r-mode instability in NSs with crusts to $\gtrsim 500$ Hz for $T \approx 10^{10}$ K, nearly a factor of five higher than previous estimates. This substantially reduces the parameter space for the instability to operate, especially for older, colder NSs, such as those accreting in binaries and millisecond pulsars. In particular, the smallest unstable frequency for the temperatures characteristic of LMXBs is $\gtrsim 700$ Hz, safely above all measured spin frequencies. This work resolves the discrepancy between the theoretical understanding of the r-mode instability and the observations of millisecond pulsars and LMXBs, and, along with observational inferences [28], likely rules out r-modes as the explanation for the clustering of spin frequencies of neutron stars in LMXBs around 300 Hz.

To summarize, a significant role of *steady-state* r-modes in LMXBs has probably been ruled out, both on theoretical grounds [29] (unless crust-core coupling is much stronger than was estimated), and observationally [28] (for Aql X-1 in particular). However, stochastically excited r-modes that decay rapidly may still play a significant role in accreting systems, as even a very small amplitude ($\alpha \lesssim 10^{-5}$) can balance the accretion torque at $\dot{M}_{\rm Edd}$. In addition, the issues of crustal shear modulus and the structure of the crust-core boundary, highlighted in Sec. II, are of paramount importance for r-modes as well. Crustal quadrupoles [11,20] can explain the spins of LMXBs and remain a viable source of continuous GWs, but the strains at $\dot{M}_{\text{Edd}}$ are rather high. The crustal breaking strain $\bar{\sigma}_{\text{max}}$ is not likely to be understood theoretically any time soon, and detection of GWs from LMXBs with LIGO-II type instruments will surely teach us many new things about NSs.

ACKNOWLEDGMENTS

GU acknowledges support from the Fannie and John Hertz Foundation.

REFERENCES

1. R. V. Wagoner, ApJ, **278**, 345 (1984).
2. K. S. Thorne, in *Three Hundred Years of Gravitation*, edited by S. W. Hawking and W. Israel (Cambridge University Press, Cambridge, 1987), Chap. 9, p. 330.
3. D. Bhattacharya, in *X-ray Binaries*, edited by W. H. G. Lewin, J. van Paradijs, and R. E. Taam (Cambridge University Press, 1995), p. 233.
4. M. van der Klis, in *Proceedings of the Third William Fairbank Meeting* (1999), astro-ph/9812395.
5. R. Wijnands and M. van der Klis, Nature, **394**, 344 (1998).
6. T. E. Strohmayer *et al.*, ApJ, **469**, L9 (1996).
7. M. C. Miller, ApJ, **515**, L77 (1999).
8. J. van Paradijs and N. White, ApJ, **447**, L33 (1995).
9. N. E. White and W. Zhang, ApJ, **490**, L87 (1997).
10. M. C. Miller, F. K. Lamb, and D. Psaltis, ApJ, **508**, 791 (1998).
11. L. Bildsten, ApJ, **501**, L89 (1998).
12. J. Papaloizou and J. E. Pringle, MNRAS, **184**, 501 (1978).
13. P. Brady and T. Creighton, Phys. Rev. D, (1999), in press. gr-qc/9812014.
14. D. Chakrabarty and E. H. Morgan, Nature, **394**, 346 (1998).
15. N. Andersson, K. D. Kokkotas, and N. Stergioulas, ApJ, **516**, 307 (1999).
16. K. Sato, Prog. Theor. Physics, **62**, 957 (1979).
17. P. Haensel and J. L. Zdunik, A&A, **227**, 431 (1990).
18. O. Blaes, R. Blandford, P. Madau, and S. Koonin, ApJ, **363**, 612 (1990).
19. L. Bildsten and A. Cumming, ApJ, **506**, 842 (1998).
20. G. Ushomirsky, C. Cutler, and L. Bildsten, MNRAS, submitted (2000).
21. C. J. Pethick and D. G. Ravenhall, Annu. Rev. Nucl. Part. Sci., **45**, 429 (1995).
22. C. J. Pethick and A. Y. Potekhin, Phys. Lett. B, **427**, 7 (1998).
23. M. Ruderman, ApJ, **382**, 587 (1991).
24. Y. Levin, ApJ, **517**, 328 (1999).
25. H. C. Spruit, A&A, **341**, L1 (1999).
26. B. J. Owen *et al.*, Phys.Rev.D, **58**, 084020 (1998).
27. L. Lindblom and G. Mendell, Phys. Rev. D, (1999), submitted. gr-qc/9909084.
28. E. F. Brown and G. Ushomirsky, ApJ, (2000), submitted. astro-ph/9912113.
29. L. Bildsten and G. Ushomirsky, ApJ, (2000), in press. astro-ph/9911155.

Gravitational waves from inspiral into massive black holes

Scott A. Hughes

Theoretical Astrophysics, California Institute of Technology, Pasadena, CA 91125

Abstract. Space-based gravitational-wave interferometers such as LISA will be sensitive to the inspiral of stellar mass compact objects into black holes with masses in the range of roughly 10^5 solar masses to (a few) 10^7 solar masses. During the last year of inspiral, the compact body spends several hundred thousand orbits spiraling from several Schwarzschild radii to the last stable orbit. The gravitational waves emitted from these orbits probe the strong-field region of the black hole spacetime and can make possible high precision tests and measurements of the black hole's properties. Measuring such waves will require a good theoretical understanding of the waves' properties, which in turn requires a good understanding of strong-field radiation reaction and of properties of the black hole's astrophysical environment which could complicate waveform generation. In these proceedings, I review estimates of the rate at which such inspirals occur in the universe, and discuss what is being done and what must be done further in order to calculate the inspiral waveform.

One of the most exciting sources that should be measured by the space-based gravitational-wave interferometer LISA (the Laser Interferometer Space Antenna [1]) is the inspiral of a "small" ($1-10\,M_\odot$) compact body into a massive ($10^{5-7}\,M_\odot$) black hole. Such massive black holes reside at the cores of galaxies; the smaller compact bodies will become bound to the hole and spiral into them after undergoing interactions with stars and other objects in the environment of the galactic center. The measurement of gravitational waves from such inspirals will make possible very high precision tests of general relativity, and probe the nature of the galactic core's environment.

To set the stage for understanding why these inspirals are such interesting objects, consider the following estimates: the orbital energy of a small body in an equatorial, prograde orbit of a Kerr black hole is

$$E^{\rm orb} = \mu \frac{1 - 2v^2 + qv^3}{\sqrt{1 - 3v^2 + 2qv^3}}, \qquad (1)$$

where $v \equiv \sqrt{M/r}$ and $q \equiv a/M$. (I use units in which $G = c = 1$ throughout.) The orbital frequency of the small body is

$$\Omega = \frac{M^{1/2}}{r^{3/2} + aM^{1/2}} . \qquad (2)$$

Radiation reaction carries orbital energy away from the system, causing the orbit to shrink. Eventually it shrinks enough that the body reaches the innermost stable circular orbit (ISCO). Orbits inside this radius are dynamically unstable; further radiative evolution tends to push the body into the black hole.

Post-Newtonian theory allows us estimate the rate at which the system loses energy as a power series in the quantity $u \equiv (M\Omega)^{1/3}$ (which is roughly the orbital speed) and the black hole's spin a. Reference [2] gives the energy loss in such a post-Newtonian expansion:

$$\frac{dE}{dt} = -\frac{32}{5}\left(\frac{\mu}{M}\right)^2 u^{10} \left[f_{\text{Schw.}}(u) + f_{\text{spin}}(a,u)\right] . \qquad (3)$$

The prefactor $-32/5(\mu/M)^2 u^{10}$ is the result one gets applying the quadrupole formula to a binary described with Newtonian gravity; the function $f_{\text{Schw.}}(u)$ is a (rather high-order) correction appropriate for zero-spin black holes, and $f_{\text{spin}}(a,u)$ is a correction incorporating information about the hole's spin. (Note that this formula is only appropriate for $\mu \ll M$: it does *not* incorporate any finite mass ratio corrections.)

Equations (1)–(3) can be used to estimate the time it takes for a small body to spiral from radius r_1 to r_2, and the number of gravitational-wave cycles it emits in that time:

$$T = \int_{t_1}^{t_2} dt = \int_{r_1}^{r_2} \frac{dt}{dr} dr = \int_{r_1}^{r_2} \frac{dE/dr}{dE/dt} dr , \qquad (4)$$

$$N_{\text{cyc}} = \int_{t_1}^{t_2} f_{\text{gw}} dt = \int_{r_1}^{r_2} \frac{\Omega}{\pi} \frac{dt}{dr} dr = \int_{r_1}^{r_2} \frac{\Omega}{\pi} \frac{dE/dr}{dE/dt} dr . \qquad (5)$$

(On the last line, I have assumed that the bulk of the radiation comes out in the quadrupole $m = 2$ mode, so that $f_{\text{gw}} = 2\Omega/2\pi$.) Consider now the inspiral of a $5\,M_\odot$ body into a rapidly spinning ($a \simeq M$) $10^6\,M_\odot$ black hole. The small body spirals from a radius of $8M$ (in Boyer-Lindquist coordinates) to the ISCO in one year, emitting around 5×10^5 gravitational-wave cycles as it does so. The gravitational-waves that it emits lie in the frequency band $3 \times 10^{-3}\,\text{Hz} \lesssim f \lesssim 3 \times 10^{-2}\,\text{Hz}$ — the band to which LISA is most sensitive. Indeed, careful analyses of the detectability of the signal by LISA [3] indicate that such an inspiral should be detected out to a distance of roughly 1 Gigaparsec with amplitude signal-to-noise ratio of around 10 to 100 (depending on factors such as the mass of the small body, the mass of the black hole, and the black hole's spin). The fact that such a large number of cycles are emitted indicates that details of the waveform can in principle be measured to very high precision. *Detection of extreme mass-ratio inspirals by LISA offers the possibility of very high precision measurements of the characteristics of extreme strong-field regions of spacetime.*

The high precision tests that LISA will be able to make should allow us to directly map the characteristics of the massive body's spacetime metric and confirm that they in fact exhibit the Kerr metric. Most likely, the way that this will be done will be to measure the multipole moments of the massive body. Fintan Ryan [4] has shown that the gravitational waves which are emitted as a small body spirals into a massive compact object contain a "map" of the massive body's spacetime. By measuring the gravitational waves and decoding the map, one learns the mass and current multipole moments which characterize the massive body. All multipole moments of Kerr black holes are parameterized by the holes' mass M and spin a:

$$M_l + iS_l = M(ia)^l. \tag{6}$$

For Kerr black holes, knowledge of the moments $M_0 \equiv M$ and $S_1 \equiv aM$ determines *all* higher moments. This is one way of stating the "no-hair" theorem: The macroscopic properties of a black hole are entirely determined by its mass and spin. (I neglect the astrophysically uninteresting possibility of charged holes.) *By measuring gravitational waves from extreme mass ratio inspiral and thereby mapping the massive body's spacetime, LISA will test the no-hair theorem for black holes, determining whether the massive body has multipole moments characteristic of the Kerr metric, or whether the body is something more exotic, such as a boson star.*

The waves emitted by extreme mass ratio inspiral are thus likely to be directly measureable by LISA, and are likely to be extremely interesting. One might next wonder whether they occur often enough to be interesting. This question has been examined in some detail by Martin Rees and Steinn Sigurdsson [5,6]. They consider the scatter of stellar mass black holes in the central density cusp of galaxies into tightly bound orbits of the galaxies' central black hole. Occasionally, such a scattering event will put the stellar mass hole into an orbit which is so tightly bound that its future dynamics are driven by gravitational wave emission, and it becomes an interesting source for LISA. They find that the rate of such events is likely to lie in the range

$$\frac{1 \text{ event}}{\text{year Gpc}^3} \lesssim \mathcal{R} \lesssim \frac{1 \text{ event}}{\text{month Gpc}^3}. \tag{7}$$

Obviously, there are large uncertainties in this calculation. The low mass end of the massive black hole population (which is most relevant to LISA observations) is not as well constrained as the population of very massive black holes ($M \gtrsim 10^8 \, M_\odot$), and there are uncertainties in the rate at which stellar mass black holes are "fed" into the central hole to produce extreme mass ratio binaries. However, the lower end of the rate estimate (7) is based on very conservative estimates. We may rather robustly estimate that the rate measured by LISA will be several events per year out to a Gigaparsec [7].

The waves that LISA will measure from these sources will come from orbits that are rather eccentric and inclined with respect to the black hole's equatorial plane [8]. To best interpret the measured waves (and, indeed, in order to improve the odds of seeing the waves at all in the detector's noisy data stream), it will be necessary to have some theoretical modeling of the orbit and the waves that it emits as gravitational radiation reduces the orbit's energy and angular momentum. We expect that the radius, inclination angle, and eccentricity of the orbit will change as radiative backreaction drives the system's evolution. Some means of understanding these changes, in detail, is needed in order to model the waves' evolution accurately.

Before discussing work in radiation reaction, it is necessary to take a sanity check. Relativity theorists often work in a very idealized universe: their extreme mass ratio binary is likely to consist of a big black hole, a small body, and gravitational waves. In the real astrophysical world, there will be complications to this pretty (but highly idealized) picture. One should worry whether the complications render the relativity theorist's modeling invalid.

Perhaps the most important such complication arises from interaction between the small inspiraling body and material accreting onto the massive black hole. Recently, Narayan has analyzed this interaction and concluded that, in almost all cases, accretion induced drag is unlikely to significantly influence extreme mass ratio inspiral [9]. This conclusion is based on the fact that in the majority of cases, the rate at which the central black hole accretes gas from its environment is rather low (several orders of magnitude less than the Eddington rate [10]). For these "normal" galaxies, much evidence [9,10] suggests that the gas accretes via an advection dominated accretion flow (ADAF). Narayan's calculation [9] shows that the timescale for ADAF drag to change the orbit's characteristics (*e.g.*, the orbital angular momentum) is many (9 to 16) orders of magnitude longer than the timescale for radiation reaction to change the orbit's characteristics. Thus, the relativity theorist's idealized view of an extreme mass ratio binary is probably quite accurate: radiation reaction is likely the most important factor driving the evolution of extreme mass ratio binaries.

When the mass ratio is extreme, one can analyze the spacetime of the binary using a perturbative expansion: the spacetime metric can be written as a "background" from the central object (which I will assume from now on is a Kerr black hole), plus a perturbation due to the inspiraling body:

$$g_{\alpha\beta} = g_{\alpha\beta}^{\text{Kerr}}(M, a) + h_{\alpha\beta}(\mu) \ . \tag{8}$$

The evolution of the perturbation $h_{\alpha\beta}(\mu)$ should then describe the dynamical evolution of the system. To linear order in the mass ratio μ/M (which should be adequate for extreme mass ratios), this evolution can be described using perturbation techniques, such as the Teukolsky equation[1] [11].

When the mass ratio is extreme, the effects of radiation reaction are gentle enough that the system's evolution is adiabatic: the radiation reaction timescale is much longer than the orbital timescale. At any given moment, the trajectory of the small body is very nearly a geodesic, parameterized by the three constants of Kerr

---

[1] The Teukolsky equation actually describes the evolution of a curvature quantity related to the perturbation.

orbital motion: the energy E, the (z-component of) angular momentum L_z, and the "Carter constant" Q. Gravitational-wave emission causes these three constants to change on the radiation reaction timescale. In this adiabatic limit, the evolution of the system can be understood in terms of the evolution of the quantities (E, L_z, Q).

It is well known that gravitational waves carry energy and angular momentum. One might think that they carry "Carter constant" as well, and that therefore one might be able to deduce the effects of radiation reaction by measuring the flux of radiation at infinity and down the event horizon. By measuring the amount of E, L_z, and Q carried in the flux one should be able to deduce how much E, L_z, and Q are lost from the orbit. This would then allow one to figure out orbits of Kerr black holes radiatively evolve.

This approach does not work. One can deduce the change in the orbit's E and L_z by examining radiation fluxes, but one cannot so deduce the change in Q:

$$\delta E_{\text{orbit}} = -\delta E_{\text{radiated}},$$
$$\delta L_{z,\text{orbit}} = -\delta L_{z,\text{radiated}},$$
$$\delta Q_{\text{orbit}} \neq -\delta Q_{\text{radiated}}. \tag{9}$$

The change δQ turns out to depend explicitly on the local radiation reaction force, $f^\mu = dp^\mu/d\tau$, which the small body experiences due to radiative backreaction (see [14]). The properties of this force (and programs to calculate it) are described elsewhere in this volume [12]. Here, it is sufficient to note that an understanding of the radiation reaction force for gravitational radiation reaction lies some time in the future, so that we cannot yet evolve generic Kerr black hole orbits.

There are special cases where evolution of the Carter constant is not such a nasty impediment. One case is the evolution of equatorial orbits. Equatorial orbits have $Q = 0$; and, one can easily show that an orbit which starts off equatorial remains equatorial. In this case, one need only evolve the energy and angular momentum. The local radiation reaction force is not needed in this case. Another case is the evolution of circular, non-equatorial orbits. (For non-zero spin, "circular" means "constant Boyer-Lindquist coordinate radius".) Such orbits have recently been shown to remain circular under adiabatic radiation reaction [13]. Thus, in an adiabatic evolution, a system which is initially circular and inclined will remain circular and inclined: the system evolves through a sequence of orbits changing only its radius and inclination angle. By imposing "circular goes to circular", one can write down a simple rule relating the change of the Carter constant to the flux of energy and angular momentum: $\dot{Q} = \dot{Q}(\dot{E}, \dot{L}_z)$.

Recently, I have examined the evolution of these circular orbits under adiabatic radiation reaction, using a flux-measuring formalism based on the Teukolsky equation. The formalism and results are presented at length in [14]. Some highlights of the results are presented in Figure 1.

Consider the left panel of Figure 1. The horizontal axis is orbital radius r; the vertical axis is inclination angle ι. This figure shows the direction, in (r, ι) phase space, in which radiation reaction tends to push the orbit. This particular analysis

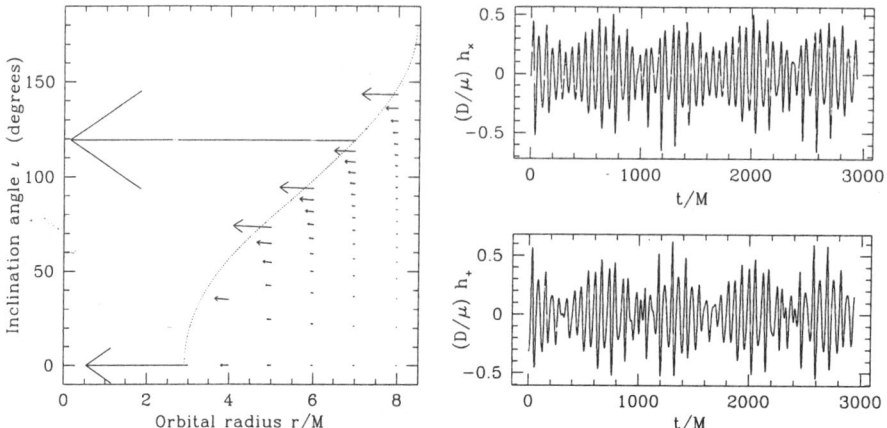

FIGURE 1. The left panel illustrates the effect of radiation reaction on circular, inclined orbits. It is for a black hole with $a = 0.8M$. The tail of each arrow represents a particular orbit with radius r and inclination angle ι. The arrow indicates the direction in which radiation reaction drives the orbit; the size of the arrow indicates how quickly it is so driven. The right panel is a snapshot of the gravitational waveform emitted during an inspiral into a hole with $a = 0.95M$. At this moment, the small body is spiraling through $r = 7M$, $\iota = 62.43°$.

is for a black hole with $a = 0.8M$. Notice that the orbits are in the strong-field of the hole; the dotted line indicates the maximum inclination angle which the orbit can have and remain stable. Orbits tilted beyond this line are dynamically unstable to small perturbations and plunge into the hole. The tail of each arrow represents a particular orbit. The direction of the arrow gives the direction in which that orbit tends to evolve due to gravitational-wave emission; the arrow's length indicates the relative rate of evolution. In all cases, the direction of the arrow is such that the inclination angle *increases*: radiation reaction tends to make tilted orbits more inclined. This is exactly what one would have predicted by extrapolating from post-Newtonian theory [15]. The rate at which this inclination angle increases is rather slow — the aspect of the arrows in Figure 1 is nearly flat. Indeed, the value of $d\iota/dt$ in this strong-field regime is roughly 3 times smaller than what post-Newtonian theory predicts[2]. An interesting feature of Figure 1 is the very long arrow at $r = 7M$, $\iota \simeq 120°$. This orbit lies extremely close to the marginally stable orbit: it is at $\iota = 119.194°$; the marginally stable orbit is at $\iota = 119.670°$. This orbit is barely dynamically stable, so a small push has drastic effects.

The right panel of Figure 1 shows a portion of the gravitational waveform emitted during inspiral. The central black hole in this case has $a = 0.95M$; it lies at

---

[2] Recent analyses are showing that in the extreme strong-field of rapidly rotating holes, the inclination angle changes rather more dramatically, and in the opposite direction: radiation reaction tends to *decrease* the inclination angle. This work is in progress [16].

luminosity distance D from the detector. The waveform here is shown as the small body passes through $r = 7M$, $\iota = 62.43°$. Note the low-frequency modulation of both polarizations. This is due to the frame dragging induced precession of the orbital plane — Lense-Thirring precession. Note also the many sharp, short-timescale features present in the two polarizations. When the spin is high, many harmonics of the small body's fundamental orbital frequencies contribute to the gravitational waveform. This leads to a rather complicated structure; the energy spectrum corresponding to these waveforms extends to rather high frequencies. Accurate measurement of such complicated waveforms will be quite a challenge. However, the payoff is likely to be immense.

I thank Kip Thorne and Steinn Sigurdsson for help and advice in writing this talk; I also thank Ramesh Narayan for allowing me to use a preliminary draft of Ref. [9]. I am indebted to many people who helped me construct my radiation reaction code, including (but not limited to) Sam Finn, Daniel Kennefick, Yuri Levin, Amos Ori, Sterl Phinney, and "The Capra Gang": Lior Burko, Patrick Brady, Éanna Flanagan, Eric Poisson, and Alan Wiseman. This research was support by NSF Grants AST-9731698 and AST-9618537, and NASA Grants NAG5-6840 and NAG5-7034.

REFERENCES

1. K. Danzmann et al., *LISA — Proposal for a Laser-Interferometric Gravitational Wave Detector in Space*, Max-Planck-Institut für Quantenoptik, Report MPQ 233 (1998).
2. Y. Mino, M. Sasaki, M. Shibata, H. Tagoshi, and T. Tanaka, Prog. Theor. Phys. Supplement No. 128 (1997).
3. L. S. Finn and K. S. Thorne, in preparation.
4. F. D. Ryan, Phys. Rev. D **56**, 1845 (1997).
5. S. Sigurdsson and M. J. Rees, Mon. Not. R. Astron. Soc. **284**, 318 (1997).
6. S. Sigurdsson, Class. Quantum Grav. **14**, 1425 (1997).
7. S. Sigurdsson, private communication.
8. D. Hils and P. L. Bender, Astrophys. J. **447**, L7 (1995).
9. R. Narayan, Astrophys. J. submitted; also astro-ph/9907328.
10. R. Narayan, R. Mahadevan, and E. Quataert, in *Theory of Black Hole Accretion Disks*, edited by M. A. Abramowicz, G. Bjornsson, and J. E. Pringle (Cambridge University Press, Cambridge, 1998).
11. S. A. Teukolsky, Astrophys. J. **185**, 635 (1973).
12. Y. Mino, these proceedings; L. Burko, these proceedings.
13. D. Kennefick and A. Ori, Phys. Rev. D **53**, 4319 (1996); F. D. Ryan, Phys. Rev. D **53**, 3064 (1996); Y. Mino, unpublished Ph.D. thesis, Kyoto University (1996).
14. S. A. Hughes, Phys. Rev. D, in press; also gr-qc/9910091.
15. F. D. Ryan, Phys. Rev. D **52**, R3159 (1995).
16. S. A. Hughes, in preparation.

Radiation Reaction Force on a compact body spiralling into a Supermassive Black Hole [1]

Yasushi Mino[2]

Department of Physics, Faculty of Science, University of Tokyo,
7-3-1 Hongo, Bunkyo-ku, Tokyo, 113-0033 JAPAN

I APPROACH

In this short talk, we briefly discuss the recent development in an analytic method to derive the gravitational evolution of a compact object rotating around a supermassive black hole.

The supermassive black hole is considered to make a dominant contribution to the gravitational potential, and we take the black hole spacetime as a background. We treat the compact object as a point source on the background black hole spacetime, and we consider to derive the reaction force acting on the particle with the well-known black hole perturbation technique.

$$g_{\mu\nu} = g_{\mu\nu}^{BH} + h_{\mu\nu} \tag{1}$$

$$T_{\mu\nu} = \hat{T}_{\mu\nu} \int d\tau \delta\left(x - z(\tau)\right) \tag{2}$$

By using the point source, the metric perturbation can be derived by the Green function method. However we have a very serious problem of divergence when we derive the equation of motion of the point particle, therefore, it is necessary to regularize the gravitational potential in order to derive the physically meaningful the gravitational radiation reaction force.

II EQUATION OF MOTION

We review the discussion in deriving the equation of motion made by us [2][3]. The discussion begins with a delicate question about the validity of the linear

---

[1] This work was supported in part by Monbusho Grant-in-Aid for Scientific Research No.5427.
[2] e-mail:mino@utap.phys.s.u-tokyo.ac.jp
[3] The other derivation in an axiomatic manner was found to obtain the same result [3].

perturbation as the metric perturbation around the point source becomes divergent. By using the matched asymptotic technique, we consider to avoid this problem.

For the simple discussion, we assume that the compact object around the supermassive black hole is a star or a black hole making a Schwarzschild black hole-like geometry around it, and the typical length scale of the gravitational potential made by the particle (say $G\mu$) is enough smaller than the curvature length of the background spacetime (say L). [4]

$$G\mu \ll L \quad (3)$$

The self-gravity of the particle dominates around the particle, but the increment of the gravitational field made by the particle's self-gravity becomes smaller and the full gravitational field gets dominated by the background geometry as we go away from the particle. With this physical intuition, we first prepare the two metrics. One is the internal metric valid only around the particle, which is a Schwarzschild black hole plus its perturbation.

$$h_{ab}^{internal} = h_{ab}^{BlackHole} + \frac{1}{L} h_{ab}^{(1)} + \frac{1}{L^2} h_{ab}^{(2)} + \cdots . \quad (4)$$

(a,b are the coordinate indices of the black hole.) With the centrifugal radius r of the black hole spacetime, the convergent radius is $r \ll L$. The other is the external metric valid away from the particle, which is the background metric plus its perturbation.

$$H_{AB}^{external} = H_{AB}^{Background} + G\mu H_{AB}^{(1)} + (G\mu)^2 H_{AB}^{(2)} + \cdots . \quad (5)$$

(A,B are the coordinate indices of the background metric.) With 'the centrifugal radius' R from the position of the particle on the background metric, the convergent radius is $G\mu \ll R$.

The radia r and R scale in the similar manner and there is a region where both expansion of the metrics converges, which we call the overlapping region defined as

$$G\mu \ll R \sim r \ll L . \quad (6)$$

Taking into account the difference of the coordinate systems and the gauge condition in (4) and (5), we explicitly implement matching of these metrics.

We find that the linear pertrubation of the metric away from the particle could be effectively generated by the point source,

$$T_{\mu\nu}(x) = G\mu v_\mu v_\nu \int d\tau \delta(x - z(\tau)) . \quad (7)$$

In a sense, the point source of the linearized Einstein equation in our perturbation study does not represent a point like matter distribution, but, effectively works as a boundary condition for the metric perturbation away from the particle.

---

[4] This assumption is self-consistent in the sense that the scale of the particle's self-gravity can be defined only when the particle is placed at an almost flat background.

In constructing the metric, we put a gauge condition so that the black hole has no dipole perturbation, which corresponds to fixing the position of the particle in the background metric. Then we obtain the singularity-free equation of motion of the particle.

The resulting equation is expressed with the linear perturbation of the metric generated by the point source, though it diverges at the particle as we stated. The trick is that we could divide the linear metric perturbation into two parts; the instantaneous part and the tail part.

$$h_{\mu\nu} = h_{\mu\nu}^{inst.} + h_{\mu\nu}^{tail} \tag{8}$$

The instantaneous part is the part which directly propagates on the future light cone of the particle without scattering of the background curvature and which is diverge at the particle. The tail part is the rest of the metric perturbation which is generated by the background curvature scattering, and is regular around the particle. We found that the equation of motion is made only by the tail part of the metric perturbation, thus, it is regular though it uses the metric perturbation at the particle.

III POWER REGULARIZATION

The argument made in Ref. [2] is just a local one, therefore, the tail part of the metric perturbation is still unknown. Here we review the idea to derive the tail part proposed in Ref. [4] and give some comment of the remaining problem of this approach.

The local analysis in Ref. [2] is appropriate to analyze the divergent behavior around the particle, but, it is not sufficient to get the tail part of the metric perturbation which depends on the global structure of the background spacetime. We, therefore, consider to combine the well-known method to derive the metric perturbation of the black hole spacetime, such as Zerilli-Regge-Wheeler formalism [5] and Teukolsky formalism [6] with Chrzanovski formalism [7]. These formalisms make use of the Fourier-Harmonic mode expansion, and the radial equations are recently found to be solved in a quite systematic manner [8].

Our prescription to derive the finite radiation reaction force is to extract the divergent instantaneous part from the whole perturbation. We discuss a component of the local gravitational force, $\mathcal{F} = \hat{F} \cdot h^{tail}(z)$. acting on the particle at z. We define $\mathcal{F}^{reg.}$, $\mathcal{F}^{inst.}$ and $\mathcal{F}^{bare}$, at a field point $\mathbf{x}$ as,

$$\mathcal{F}^{reg.}(\mathbf{x}) = \hat{F} \cdot h^{tail}(\mathbf{x}), \quad \mathcal{F}^{inst.}(\mathbf{x}) = \hat{F} \cdot h^{inst.}(\mathbf{x}), \quad \mathcal{F}^{bare}(\mathbf{x}) = \hat{F} \cdot h(\mathbf{x}), \tag{9}$$

which we call the 'regularized' quantity, the 'instantaneous' quantity and the the 'bare' quantity respectively. We then have

$$\mathcal{F}^{reg.}(\mathbf{x}) = \mathcal{F}^{bare}(\mathbf{x}) - \mathcal{F}^{inst.}(\mathbf{x}), \tag{10}$$

and the regularization prescription

$$\mathcal{F} = \lim_{\mathbf{x} \to \mathbf{z}} \mathcal{F}^{reg.}(\mathbf{x}). \quad (11)$$

The essential difficulty of this extraction is the fact that the divergence of the bare quantity is obscure. This is because the whole metric perturbation is derived as an infinite summation of the Fourier-Harmonic series and each radial mode function is finite in the present formalism. The summation usually requires very complicated sum formula of Harmonic functons.

We suggest that this complication can be avoid by following two tricks. The first idea is to calculate the quantity of a special location of $\mathbf{x}$. We find that the radial direction is easier for the latter calculation. For the particle at $\mathbf{z} = \{t_0, r_0, \theta_0, \phi_0\}$ in the Boyer-Lindquist coordinates, we calcluate the bare quantity at $\mathbf{x}(\eta) = \{t_0, r_0(1+\eta), \theta_0, \phi_0\}$, where η is a parameter for regularization. The second trick is to expand the bare and instantaneous quantities by the power of $r = r_0(1+\eta)$. With these tricks, we could find out the power expansion of the regularized quantity, $\mathcal{F}^{reg.}$, which we could take the coincidence limit, $\eta \to 0$.

IV GAUGE PROBLEM

Though the extraction is possible by the power regularization prescription, we found that there still remains a divergence coming from the gauge difference. The instantaneous quantity can only be derived in the harmonic gauge condition, however the bare quantity can be derived in the Regge-Wheeler gauge condition [5] or in the radiation gauge condition [7]. We found that these difference of the gauge condition makes a residual divergence in the regularized quantity, $\mathcal{F}^{reg.}$

However the prescription argued in the previous section is still promising and the analytic derivation of the gauge transformation is under development in order to extract the residual divergence [9].

REFERENCES

1. Mino, Y., Sasaki, M., Shibata, M., Tagoshi, H., and Tanaka, T., *Suppl. Prog. Theore. Phys.***128**,1(1997).
2. Mino, Y., Sasaki, M., and Tanaka, T., *Suppl. Prog. Theore. Phys.***128**,373(1997).
3. Quinn, T. C. and Wald, R. M., *Phys. Rev.***D56**,3381(1997).
4. Mino, Y., and Nakano, H., *Prog. Theore. Phys.***100**,507(1998).
5. Zerilli, F. J., *Phys. Rev.***D2**,2141(1970).
6. Teukolsky, S. A., *Astrophys. J.***185**,635(1973).
7. Chrzanowski, P. L., *Phys. Rev.***D11**,2042(1975).
8. Mano, S., Suzuki, H., and Takasugi, E., *Prog. Theore. Phys.***95**,1079(1996).
9. Mino, Y., and Nakano, H., in preparation.

Self-force approach for radiation reaction

Lior M. Burko

Theoretical Astrophysics, California Institute of Technology, Pasadena, California 91125

Abstract. We overview the recently proposed mode-sum regularization prescription (MSRP) for the calculation of the local radiation-reaction forces, which are crucial for the orbital evolution of binaries. We then describe some new results which were obtained using MSRP, and discuss their importance for gravitational-wave astronomy.

The problem of including the radiation-reaction (RR) forces in the orbital evolution of a binary is a long-standing open problem. This problem is as yet unresolved even in the extreme mass-ratio limit, with the particle orbiting a non-rotating black hole, although there has been a remarkable progress obtained from various directions [1]. The conventional approach is to consider the fields in the far zone, and then use a balance argument to relate the far-zone fields to the local properties of the particle. The generic failure of such approaches [2] prompted the idea to calculate the *local* forces acting on the particle, including the RR forces. In the following we discuss the RR forces acting on a scalar point-like charge, but for electric or gravitational charges the basic ideas are similar. The RR force $^{RR}F^\mu$ which acts on a point-like scalar charge q is given by [3]

$$^{RR}F^\mu(\tau) = q^2 \left[\frac{1}{3} \left(\ddot{u}^\mu - u^\mu \dot{u}_\nu \dot{u}^\nu \right) + \frac{1}{6} \mathcal{R}^\mu + \int_{-\infty}^{\tau} \nabla^\mu G_R \, d\tau' \right], \quad (1)$$

where $\mathcal{R}^\mu = R^\mu_{\ \nu} u^\nu + R_{\nu\sigma} u^\nu u^\sigma u^\mu - \frac{1}{2} R^\nu_{\ \nu} u^\mu$, $R_{\nu\sigma}$ is the Ricci tensor, u^μ is the charge's 4-velocity, a dot denotes (covariant) derivative with respect to proper time τ, ∇_μ denotes covariant differentiation, and G_R is the retarded Green's function. The first term is a local Abraham-Lorentz-Dirac type damping force, the second is a local force, which couples to Ricci-curvature and preserves conformal invariance, and the third is the so-called "tail" term, which arises from the failure of the Huygens principle in curved spacetime. The greatest problems in the calculation of the RR forces lurk in the tail term, because it requires the knowledge of G_R along the entire past world line of the charge. In addition, the self field of any particle diverges at the position of the particle, and the calculation of the RR forces will have to handle the infinities connected with the self field by providing a regularization prescription.

Recently, Ori proposed to approach the RR problem via mode decomposition [4]. Ori observed, that the individual Fourier-harmonic modes of the self field are bounded, also for a point-like particle, although the sum over all modes diverges. This observation is very useful, because the calculation of the individual modes is relatively easy. This still leaves the second, harder problem of having a regularization prescription to handle the mode sum. Very recently, Ori suggested MSRP [5], which is very successful for the few simple cases to which it has already been applied. In what follows, we overview MSRP very briefly, and describe some of the recent results which were obtained using it.

The tail part of the RR force can be decomposed into stationary Teukolsky modes, and then summed over the frequencies ω and the azymuthal numbers m. This force equals then the limit $\epsilon \to 0^-$ of the sum over all ℓ modes, of the difference between the force sourced by the entire world line (the bare force $^{\text{bare}}F_\mu^\ell$) and the force sourced by the half-infinite world line to the future of ϵ, where the particle has proper time $\tau = 0$, and ϵ is an event along the past ($\tau < 0$) world line. Next, we seek a function h_μ^ℓ which is independent of ϵ, such that the series $\sum_\ell (^{\text{bare}}F_\mu^\ell - h_\mu^\ell)$ converges. Once such a function is found, the regularized self force is then given by $^{\text{tail}}F_\mu = \sum_\ell (^{\text{bare}}F_\mu^\ell - h_\mu^\ell) - d_\mu$, where d_μ is a finite valued function. MSRP [5] then shows, from a local integration of G_R, that $h_\mu^\ell = a_\mu \ell + b_\mu + c_\mu \ell^{-1}$. MSRP also provides an algorithm for the calculation of the functions a_μ, b_μ, c_μ and d_μ analytically. It has been conjectured, that for all orbits $a_\mu = 0 = c_\mu$. (It was found to be true for all the special cases calculated so far.) If this is indeed the case, the tail force is given by

$$^{\text{tail}}F_\mu = \sum_\ell (^{\text{bare}}F_\mu^\ell - b_\mu) - d_\mu. \qquad (2)$$

Note that b_μ is just the limit $^{\text{bare}}F_\mu^{\ell \to \infty}$, and that $^{\text{bare}}F_\mu^\ell$ can be computed using the Teukolsky formalism. Alternatively, b_μ can also be calculated analytically using MSRP. The only remaining problem then, is to calculate d_μ. Even though there is an algorithmic way to calculate d_μ, this calculation is by no means easy. Although MSRP has been developed as yet only for very simplified cases, the approach is likely to be susceptible of generalization also for more realistic cases. If robust, MSRP can be of the greatest importance for the calculation of templates for gravitational-wave detection.

Table 1 displays the values of the MSRP parameters for the cases which have already been calculated. Note that the motion is not necessarily geodesic. The data in Table 1 may suggest the conjecture that d_μ exactly equals the sum of the two local terms of Eq. (1) (or its analog for other charge types). If this hypothesis is proved to hold in general, then a truely remarkable thing happens: the full RR force can be calculated directly from Eq. (2) when d_μ is ignored. It should be emphasized that presently the support for this far-reaching hypothesis is only the special cases listed in Table 1.

Next, we present some results which were obtained by application of this new approach. For the case of a static, minimally-coupled, massless scalar charge in Schwarzschild, the self force is known to equal zero [11]. For a static electric charge q in Schwarzschild the self force is known to be purely radial and to be given by

TABLE 1. Values of the MSRP parameters. All spacetimes are spherically symmetric. For all the cases $a_\mu = 0 = c_\mu$. The charge's spin is s. A $*$ denotes the cases for which b_μ and d_μ were inferred indirectly, and a † denotes cases where the values of b_μ were corroborated numerically. Here, $I_a = {}_2F_1(1/2, 1/2; 1; v^2)$ and $I_b = {}_2F_1(1/2, 3/2; 1; v^2)$, $v^2 = -(d\varphi/dt)^2 r^2/g_{tt}$, a^μ is the 4-acceleration, $a^2 = \dot{u}_\alpha \dot{u}^\alpha$, and $k_\mu = \frac{q^2}{3}(\ddot{u}_\mu - u_\mu a^2)$.

Type of orbit	Spacetime	s	μ	$b_\mu / [-q^2/(2r^2)]$	d_μ
Static [5,6]	Minkowski	0, 1	t	0	0
Static [5,6]	Minkowski	0, 1	r	1	0
Static† [5,7]	Schwarzschild	0	r	$(r-M)/(r-2M)$	0
Circular† [5,6]	Minkowski	0	r	$(2I_a - I_b)/u^t$	0
Circular*† [5,6]	Minkowski	0	t	0	k_t
Circular*† [8]	Minkowski	1	t	0	$2k_t$
Circular† [5,9]	Schwarzschild	0	r	$(2I_a - \frac{r-3M}{r-2M} I_b)/(u^t \sqrt{-g_{tt}})$	0
Radial [10]	Minkowski	0	r	$1 - \dot{r}^2 + r\ddot{r}$	k_r
Radial [10]	$g_{tt}(r) g_{rr}(r) = -1$	0	r	$1 - \dot{r} u_r + r a_r$	$k_r + q^2 \mathcal{R}_r/6$
Radial [10]	$g_{tt}(r) g_{rr}(r) = -1$	0	t	$-\dot{r} u_t + r a_t$	$k_t + q^2 \mathcal{R}_t/6$
Static* [6]	Reissner-Nordström	1	t	0	$q^2 \mathcal{R}_t/3$

$f_r = q^2 M r^{-3} (1 - 2M/r)^{-1/2}$ [12]. These results were recovered using MSRP in [7]. Note that for these two simple static cases the solution for the modes can be obtained analytically. In general, however, this is not expected to be possible, and the solution can be obtained only numerically.

The case of a scalar charge q in uniform circular orbit around a Schwarzschild black hole was recently considered in [9]. The RR force was calculated numerically without any simplifying assumptions, such as far field or slow motion, and the solution is fully relativistic. Both the temporal and the azimuthal, dissipative components and the radial, conservative component of the RR force were computed. Figure 1 displays the behavior of the radial component of the RR force for both geodesic and non-geodesic orbits. In the slow motion and far field limits the force is repulsive, and behaves like ${}^{RR}F_r \approx q^2 M^2 \Omega^2 / r^2$. However, in strong fields the force grows faster, and for fast motion is changed from repulsive to attractive. This expression for the radial, conservative RR force may be very important for the detection of gravitational waves, and also for gravitational-waves astronomy. The conservative radial force causes an additional precession of the periastron of the particle's orbit, and thus induces a change in the frequency and phase of the emitted radiation [13]. Although the radial self force has been obtained only for the simple case of a point-like scalar charge, the result indicates that one can expect a non-zero periastron precession also for a small mass. However, the magnitude of the effect will very reasonably depend on the type of the charge. A large-magnitude effect can cause the entire search algorithm to fail in the very detection of the signal (depending also on the size of the template library), and a small effect will introduce errors in the parameters of the observed binary, namely, the wave form would fit the template of a system with parameters different from the parameter of the actual binary. Note that the conservative force depends not only on the

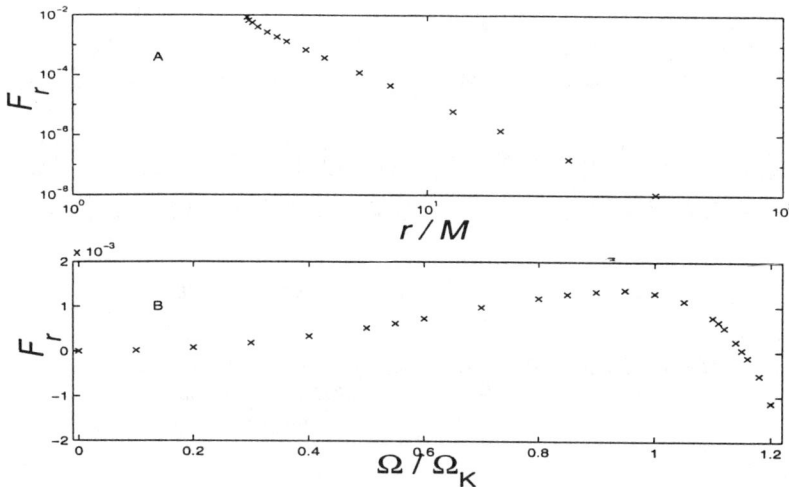

FIGURE 1. The radial RR force acting on a scalar charge in uniform circular orbit around a Schwarzschild black hole. Top panel (A): F_r as a function of r/M for geodesic orbits. Bottom panel (B): F_r as a function of the angular velocity Ω in units of the Keplerian angular velocity Ω_K, when the orbit is at $r = 4M$.

radiative modes of the field, but also on the non-radiative modes [13].

I thank Leor Barack and Amos Ori for discussions and for letting me use their results before their publication. This work was supported by NSF grants AST-9731698 and PHY-9900776 and by NASA grant NAG5-6840.

REFERENCES

1. See, e.g., B. F. Schutz, Class. Quantum Grav. **16**, A131 (1999).
2. S. A. Hughes, these proceedings; gr-qc/9910091.
3. T. C. Quinn and A. G. Wiseman, in preparation.
4. A. Ori, Phys. Lett. **A202**, 347 (1995); Phys. Rev. **D55**, 3444 (1997).
5. A. Ori, unpublished; For a brief outline of the proposed regularization method, see L. Barack and A. Ori, gr-qc/9911040.
6. L. M. Burko, unpublished.
7. L. M. Burko, Class. Quantum Grav. **17**, 227 (2000).
8. L. M. Burko, Am. J. Phys. (in press), also gr-qc/9902079.
9. L. M. Burko, in preparation.
10. L. Barack and A. Ori, in preparation.
11. A. I. Zel'nikov and V. P. Frolov, Sov. Phys. JETP **55**, 191 (1982); A. G. Wiseman, in preparation; A. E. Mayo, Phys. Rev. D **60**, 104044 (1999).
12. A. G. Smith and C. M. Will, Phys. Rev. D **22**, 1276 (1980).
13. L. M. Burko, in preparation.

Are pre-big-bang models falsifiable by gravitational wave experiments?

Carlo Ungarelli[1] and Alberto Vecchio

Max Planck Institut für Gravitationsphysik, Albert Einstein Institut
Am Mühlenberg 1, D-14476 Golm, Germany

Abstract. One of the most interesting predictions of string-inspired cosmological models is the presence of a stochastic background of relic gravitational waves in the frequency band accessible to Earth-based detectors. Here we consider a "minimal" class of string cosmology models and explore whether they are falsifiable by gravitational wave observations. In particular, we show that, the detectability of the signal depends crucially on the actual values of the model parameters. This feature will enable laser interferometers – starting from the second generation of detectors – to place stringent constraints on the theory for a fairly large range of the free parameters of the model.

MOTIVATIONS

One of the direct phenomenological predictions of inflationary cosmological models is the generation of a stochastic background of gravitational waves (GW's). For "slow-roll" inflationary models this prediction is hardly testable: since the spectrum is almost flat over the huge frequency band $\sim 10^{-16}\,\mathrm{Hz} - 1\,\mathrm{GHz}$, in the window $10\,\mathrm{Hz} - 1\,\mathrm{kHz}$, where ground-based GW experiments operate, the maximum value of the energy spectrum $\Omega_{\mathrm{gw}}(f) \equiv \rho_c^{-1} d\rho_{\mathrm{gw}}(f)/d\ln(f)$ compatible with the COBE measurements of the cosmic micro-wave background (CMB) temperature fluctuations at large scales – $h_0^2 \Omega_{\mathrm{gw}} \lesssim 10^{-14}$ – is well below the sensitivity expected for the third generation of detectors, $h_0^2 \Omega_{\mathrm{gw}} \gtrsim 10^{-11}$.

"Pre-big-bang" (PBB) models [1] represent an alternative to the standard "slow-roll" inflationary scenario. In a minimal PBB model, one assumes that the initial state is the perturbative vacuum of super-string theory, the flat 10-dimensions Minkowski space-time; the Universe goes first through an inflationary phase where the curvature and the string coupling increase, eventually reaches a "stringy epoch" where the curvature scale is of order one in units of the string length, and finally evolves into a typical decelerated radiation/matter dominated era. The inflationary PBB phase has a precise consequence on the structure of $\Omega_{\mathrm{gw}}(f)$, which affects the

---

[1] Present address: School of Computer Science and Mathematics, University of Portsmouth, Mercantile House, Hampshire Terrace, Portsmouth P01 2EG, UK

detectability of the stochastic background by laser interferometers, whose "science-runs" are expected to begin at the end of 2001: in the low frequency range, the spectrum is characterized by a steep power law, $\Omega_{\rm gw} \propto f^3$ [2]; indeed the COBE bound is easily evaded, and the spectrum can peak at frequencies of interest for GW experiments, while satisfying the existing experimental bounds [3,4].

For the rather general class of minimal PBB models, $\Omega_{\rm gw}(f)$ depends on two free parameters: (i) the red-shift $z_s \equiv f_1/f_s$ of the high-curvature phase, which fixes the "knee" frequency f_s between the low and high frequency regime ($f_1 \sim 10^{10}$ Hz is the cut-off frequency of the spectrum, whose exact value is irrelevant for the issues discussed here); (ii) the value g_s of the string coupling at the onset of the high-curvature phase, which determines the high-frequency slope of the spectrum. The maximum value of the spectrum compatible with the constraints due to pulsar timing data and to the abundance of light elements at the nucleosynthesis epoch is $h_0^2 \Omega_{\rm gw} \sim 10^{-7}$ [4]. $\Omega_{\rm gw}(f)$ strongly depends on z_s and g_s: they affect both the frequency behaviour and the peak value of the spectrum. Whether GW experiments will be able or not to detect a signal predicted by PBB models does depend on the actual value of the "true" parameters. Indeed, GW experiments represent one of the very few avenues where the cosmological models can be verified, and could open a new era for studies of the very-early Universe and the structure of fundamental fields at high energies. The emphasis of this contribution is not on finding the best theoretical scenario which would guarantee a detection. The goal of our analysis is to address to which extent GW observations can test PBB models, and more in detail to identify the the region of the free-parameter space that experiments can probe. A similar analysis, in the context of string cosmology models, was carried out in [5].

RESULTS

In order to address whether a background characterized by a given energy spectrum $\Omega_{\rm gw}(f; z_s, g_s)$ is indeed detectable or not, one needs to evaluate the signal-to-noise ratio (SNR) that can be obtained by cross-correlating the output of two detectors as a function of the model parameters, in our case z_s and g_s (see also [5]). The data analysis issues related to the detection of a stochastic background have been thoroughly discussed in [6], where the reader can find full details. Here we recall only the expression of the signal-to-noise ratio:

$$\text{SNR} \approx \frac{3H_0^2}{10\pi^2} T_{\rm obs}^{1/2} \left[\int_{-\infty}^{\infty} df \, \frac{\gamma^2(|f|)\Omega_{\rm gw}^2(|f|; z_s, g_s)}{f^6 S_1(|f|) S_2(|f|)} \right]^{1/2} \tag{1}$$

In Eq. (1) $T_{\rm obs}$ is the observation time – which is usually assumed to be a few months long – $H_0 = 100 \, h_0$ km sec^{-1} Mpc^{-1} is the Hubble constant, $S_{i=1,2}(f)$ is the noise spectral density of the i-th detector, and $\gamma(f)$ is the overlap reduction function. The spectrum $\Omega_{\rm gw}(f; g_s, f_s)$ for minimal PBB models is given in [4].

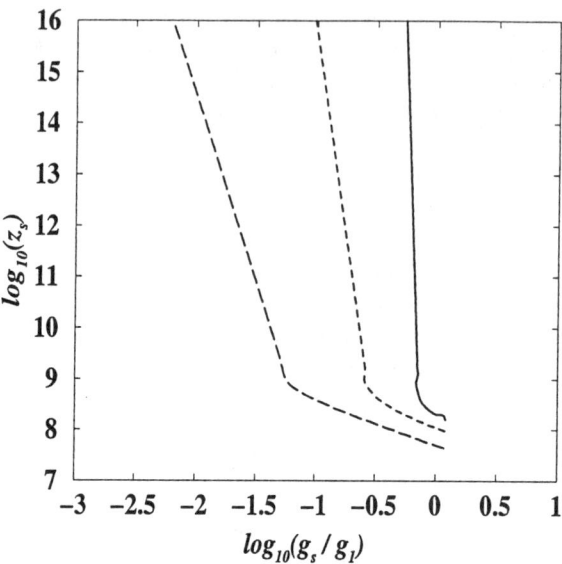

FIGURE 1. The signal-to-noise ratio (SNR) produced by the GW stochastic background predicted by minimal pre-big-bang models. The figure shows the contour plots of the SNR in the plane of the free-parameters of the model, f_s and g_s, obtained by cross-correlating the two LIGO interferometers. The curves corresponds to: SNR = 0.1 for the initial LIGO configuration (solid line); SNR = 5 for the enhanced LIGO configuration (short-dashed line); SNR = 10 for the advanced LIGO configuration (long-dashed line). We remind that the physical range over which the parameters are defined is: $0 < g_s/g_1 < 1$ and $1 < z_s < 10^{16}$.

Fig. 1 summarizes the results: it shows the contour plots of SNR for $T_{\rm obs} = 10^7$ sec which can be achieved by cross-correlating the two 4-km LIGO interferometers located in Hanford and Livingston. The contours are drawn in the relevant 2-dimensional parameter space: in order to highlight the physical content, our parameter choice corresponds to $\log_{10}(z_s)$ and $\log_{10}(g_s/g_1)$; here $g_1^2/4\pi = \alpha_{GUT}$, where α_{GUT} is the gauge coupling at the unification scale ($\alpha_{GUT} \sim 1/20$). The two free-parameters of the models are defined over the following range: $0 < g_s/g_1 < 1$ and $1 < z_s < 10^{16}$. In our analysis we have considered the estimated noise spectral density for the initial, enhanced and advanced LIGO configuration (the so-called LIGO I, II, and III, respectively). For initial LIGO, the minimum value of a detectable stochastic background is $h_0^2 \Omega_{\rm gw}^{\min} \sim 5 \times 10^{-6}$, cfr. also [6], and therefore the experiments can not provide any direct hint about PBB models: the maximum value of SNR, for fined-tuned parameters of the model, is just below one. However, the enhanced LIGO configuration will allow us to explore a relatively large parameter region: for SNR > 5, one can probe models where g_s differs by one order of magnitude with respect to its present value g_1 and about eight orders of magnitude in f_s; the SNR reaches a maximum ~ 100 when $g_s \simeq g_1$. The advanced LIGO configuration – which however requires major technological developments – will enlarge the g_s-range that one can test by more than one order of magnitude; for selected parameter values the experiment could reach SNR ≈ 1000. Notice that the experiments are very sensitive to g_s, and achieve the maximum SNR for $g_s \sim g_1$.

One might wonder whether other interferometers and/or resonant detectors could provide relevant information for the problem at hand. Unfortunately the location, orientation and sensitivity of VIRGO, GEO600 and TAMA are such that any search involving one of these three instruments will not allow to detect a stochastic background of PBB gravitons; just as reference, a VIRGO-GEO600 cross-correlation experiment will reach $h_0^2 \Omega_{\text{gw}}^{\min} \sim 8 \times 10^{-6}$ for an integration time of 4-months. Neither resonant antenna are suitable: two "bars" operating at the quantum limit with exactly the same resonance frequency would reach a maximum sensitivity $h_0^2 \Omega_{\text{gw}}^{\min} \sim 5 \times 10^{-6}$ (see also [7]), which is a factor ~ 10 worse than the one required for the most optimistic theoretical prediction; experiments involving hollow spheres [8], which are currently under study, could reach $h_0^2 \Omega_{\text{gw}}^{\min} \sim 10^{-7}$ for detectors operating at the quantum limit and co-located; this would produce a SNR ~ 1 for a PBB model whose parameters are such that the peak of $\Omega_{\text{gw}}(f)$ is right in the middle of the sphere frequency band.

CONCLUSIONS

Minimal PBB models predict a GW stochastic background which could be detectable by cross-correlating the two LIGO instruments – or, more in general, two interferometers at a distance < 3000 km and quasi-optimally oriented – with a sensitivity which is intermediate between the first and second stage. In the time frame 2004-2005, when the detectors are expected to operate in the so-called enhanced configuration, we will be able to place experimental constraints on PBB models for a fairly large portion of the free-parameter space. A more detailed analysis regarding the issues discussed here is currently in preparation.

REFERENCES

1. M. Gasperini and G Veneziano, *Astropart. Phys.* **1**, 317 (1993),
2. R. Brustein et al., *Phys. Lett.* B **361**, 45 (1995).
3. R. Brustein, M. Gasperini and G. Veneziano, *Phys. Rev.* D **55**, 3882 (1997).
4. A. Buonanno, M. Maggiore, C. Ungarelli, *Phys. Rev.* D **55**, 3330 (1997).
5. B. Allen and R. Brustein, *Phys. Rev.* D **55**, 3260 (1997).
6. B. Allen and J.D. Romano, *Phys. Rev.* D **59**, 102001 (1999).
7. S. Vitale, M. Cerdonio, E. Coccia and A. Ortolan *Phys. Rev.* D **55**, 1741 (1997).
8. E. Coccia, V. Fafone, G. Frossati, J. A. Lobo and J. A. Ortega *Phys.Rev.* D **57**, 2051 (1998).

Is there a signature in gravitational waves from structure formation of the Universe ?

Oswaldo D. Miranda, José C.N. de Araujo and Odylio D. Aguiar

Divisão de Astrofísica - Instituto Nacional de Pesquisas Espaciais [1]
Av. dos Astronautas 1758, São José dos Campos, SP, 12227-010, Brazil.

Abstract. We study the generation of a background of gravitational waves (GWs) produced from a cosmological Population of core-collapse supernovae in an explosive scenario of structure formation of the Universe. We show that these pre-galactic supernovae can produce a stochastic background of gravitational waves with maximum amplitude as high as $h_{BG} \sim 10^{-22}$ at frequency of ~ 50Hz. This background of GWs will probably be detected by VIRGO and LIGO.

INTRODUCTION

In the multicycle explosive scenario for the formation of the large-scale structures of the Universe (see [1,2]), the evolution of positive density perturbations, which collapsed after the recombination era, resulted in supernova explosions (of Population III objects) at high redshifts.

The shock wave generated by the explosion of these seed objects compressed the matter in front of it, resulting in a dense shell that eventually becomes gravitationally unstable, fragmenting into objects (or globules) which collapse, explode and form, in this way, a new generation of supernovae repeating the process in a cyclical form.

When the globules are formed, they are converted into stars which explode as supernovae at the final of their lives (we assume that these stars follow a like Salpeter initial mass function - IMF) . After the supernova event a neutron star or black hole can be produced as remnant. Considering that the remnants of the evolution of these pre-galactic objects are black holes (BHs), we obtain that both frequency and signal amplitude are dependent on the redshift when the explosive cycles occurred. We show that this scenario produces GWs that could be detected by both laser interferometer and resonant mass detectors.

---
[1] This work was supported by the Brazilian agency FAPESP (ODM thanks the grant 98/13735-7; JCNA thanks the grants 97/06024-4 and 97/13720-7).

It is worth mentioning that quite recently, in different scenarios, [3] and [4] study the cosmological background of GWs produced by supernova explosions in galaxies.

THE GRAVITATIONAL WAVES PRODUCTION

An ensemble of BH produces a background of GWs that can be calculated as follows

$$h_{BG}^2 = \frac{1}{\nu_{obs}} \int h_{BH}^2 dR_{BH} \tag{1}$$

(see, [5]), where dR_{BH} is the differential BH formation rate and h_{BH} is the dimensionless amplitude to the formation of each BH.

The collapse to black hole of a star produces a signal with frequency (see, e.g., [6])

$$\nu_{obs} = \frac{1}{5\pi M_r} \frac{c^3}{G}(1+z)^{-1} \simeq 1.3 \times 10^4 \text{Hz}\left(\frac{M_\odot}{M_r}\right)(1+z)^{-1}, \tag{2}$$

where G is the gravitational constant, c is the light velocity and the factor $(1+z)^{-1}$ takes into account the redshift effect on the emission frequency, that is, a signal emitted at frequency ν_e at redshift z is observed at frequency $\nu_{obs} = \nu_e(1+z)^{-1}$.

The collapse of a star, or star cluster, to form a black hole produces a dimensionless amplitude

$$h_{BH} = \left(\frac{15}{2\pi}\varepsilon\right)^{1/2} \frac{G}{c^2} \frac{M_r}{r_0} \simeq 7.4 \times 10^{-20} \varepsilon^{1/2} \left(\frac{M_r}{M_\odot}\right)\left(\frac{r_0}{1\text{Mpc}}\right)^{-1}, \tag{3}$$

where ε is the efficiency of generation of GWs, M_r is the mass of the remnant BH and r_0 is the distance to the source.

For the differential rate of black hole formation we have

$$dR_{BH} = \dot{\rho}_*(z) \frac{dV}{dz} \phi(m) dm dz, \tag{4}$$

where $\dot{\rho}_*(z)$ is the star formation rate (SFR) density (in $M_\odot \, \text{yr}^{-1} \, \text{Mpc}^{-3}$), $\phi(m)$ is the Salpeter IMF (that is, $\phi(m) \propto m^{-2.35}$) and dV is the comoving volume element.

Then, the dimensionless amplitude reads

$$h_{BG}^2 = \frac{(7 \times 10^{-21})^2 \varepsilon}{\nu_{obs}} \int \left(\frac{m}{M_\odot}\right)^2 \left(\frac{d_L}{1\text{Mpc}}\right)^{-2} \times \dot{\rho}_*(z) \frac{dV}{dz} \phi(m) dm dz, \tag{5}$$

where $M_r = 0.1\, m$ (see [4]) and $d_L = r_0(1+z)$ is the luminosity distance.

Another quantity of interest is the duty cycle that indicates whether the collective effect of the bursts of GWs, emitted in the collapse of a progenitor star with mass m, generate a continuous background. The duty cycle is defined as follows

$$D(z) = \int_{z_{c_f}}^{z_{c_i}} dR_{BH} \Delta \bar{\tau}_{GW}(1+z), \tag{6}$$

where $\Delta \bar{\tau}_{GW}$ is the average time duration of single bursts at the emission, which we assume to be $\sim 1\,\text{ms}$ (see, e.g., [4]).

TABLE 1. Principal results for the background of GWs for a seed object of $10^9 M_\odot$ that explodes at redshift $z_{c_i} = 120$. This is the maximum redshift, in which the explosive cycles originated from the $10^9 M_\odot$ seeds will be all consistent with the y_{Comp} distortion observed in COBE data. The cosmological model corresponds to an open Universe with $\Omega_0 = 0.1$ and $h_0 = 0.5$.

cycle	$z_{c_i} - z_{c_f}$	$\dot{\rho}_*$ ($M_\odot \, yr^{-1} \, Mpc^{-3}$)	$h_{BG_{max}}/(\varepsilon f_F)^{1/2}$	$[\nu_{min} - \nu_{max}]$ (Hz)	DC
1	120	1.0×10^9	2.6×10^{-21}	$7.8 - 42.9$	7.1×10^{10}
2	$119 - 69$	8.4×10^8	1.3×10^{-20}	$8.8 - 74.3$	1.9×10^{11}
3	$67 - 38$	6.6×10^7	5.6×10^{-21}	$15.4 - 133.3$	7.7×10^9
4	$34 - 19$	3.3×10^6	2.0×10^{-21}	$29.8 - 259.9$	1.4×10^8
5	$5.2 - 0$	MDP-1	2.4×10^{-24}	$172.7 - 5200$	5.8×10^{-2}
5	$5.2 - 0$	MDP-2	2.2×10^{-24}	$172.7 - 5200$	2.3×10^{-1}

RESULTS AND CONCLUSIONS

In Table 1 we present the main results for the model with seed mass $10^9 M_\odot$ that explodes at redshift $z_{c_i} = 120$. The maximum amplitude of the GWs is $h_{BG_{max}}/(\varepsilon f_F)^{1/2}$ (where f_F is the filling factor, that is the present fraction of the volume of the Universe occupied by shells), $[\nu_{min} - \nu_{max}]$ is the observed frequency band and DC is the duty cycle.

A complete description of the SFR density for each cycle is given in [7]. It is worth mentioning that the SFR density for the last cycle (MDP-1 and MDP-2) is obtained from Madau and collaborators ([8–10]) from the comoving luminosity density, so it does not depend on f_F (see [7]).

Since the star formation rate for all the cycles, except for the last one, is very high these cycles produce a significant amount of GWs. Also, these bursts of GWs produce high duty cycle values, which lead us to conclude that the stochastic GW background can be considered as continuous.

Figure 1 presents the GW background amplitude as a function of frequency for the cycles 1 to 4 presented in Table 1. Note that there is a partial superposition of the background produced by different explosive cycles enhancing, therefore, the amplitude of the GWs for most frequencies. Thus, the maximum amplitudes of the GWs present in Table 1 are lower limits to those frequency bands of that cycles.

Considering $\varepsilon \lesssim 7.0 \times 10^{-4}$ (see [11]), and $f_F = 0.1$ (see [1,2]) we then obtain $h_{BG_{max}} \sim 10^{-22}$ at $\nu_{obs} \simeq 50 \, Hz$. This result shows that the background of GWs produced in a putative explosive scenario is probable to be detected by VIRGO and LIGO. The detection in the future of a background of GWs could help to define the epoch when the first explosions related to the formation of the large structures in the Universe occurred.

Even if an explosive scenario did not take place in nature, our study shows that if a high star formation rate occurs at high redshifts a background of GWs could be produced and could in principle be detected. Thus, when the GW

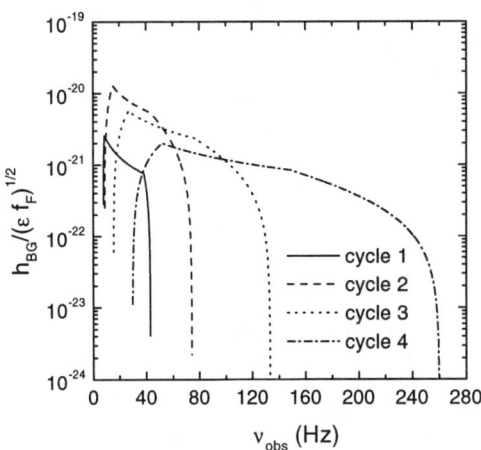

FIGURE 1. The background amplitude of the GWs as a function of ν_{obs} to the model presented in Table 1. The solid line represents $h_{BG}/(\varepsilon f_F)^{1/2}$ to the cycle 1, the dashed line to the cycle 2, the dotted line to the cycle 3 and the dashed-dotted line to the cycle 4.

observatories such as LIGO and VIRGO become operative, it will be possible to impose constraints on the models of structure formations of the Universe and, in particular, will be possible to put some light in the so called "dark age" obtaining some information on the epoch during which the first generation of stars were formed.

REFERENCES

1. Miranda, O.D., and Opher, R., *MNRAS* **283**, 912 (1996).
2. Miranda, O.D., and Opher, R., *ApJ* **482**, 573(1997).
3. Blair, D.G., and Ju, L., *MNRAS* **283**, 648 (1996).
4. Ferrari, V., Matarrese, S., and Schneider, R., *MNRAS* **303**, 247 (1999).
5. de Araujo, J.C.N., Miranda, O.D., and Aguiar, O.D., *Phys. Rev. D*, submitted for publication (1999).
6. Thorne, K.S., in *300 years of Gravitation*, eds. S.W. Hawking and W. Israel, Cambridge UP, Cambridge, 1987, ch.9, pp.331-458.
7. Miranda, O.D., de Araujo, J.C.N., and Aguiar, O.D., *MNRAS*, submitted for publication (1999).
8. Madau, P., Ferguson, H.C., Dickinson, M.E., Giavalisco, M., Steidel, C.C., and Fruchter, A., *MNRAS* **283**, 1388 (1996).
9. Madau, P., Della Valle, M, and Panagia, N., *MNRAS* **297**, L17 (1998a).
10. Madau, P., Pozzetti, L., and Dickison, M.E., *ApJ* **498**, 106 (1998b).
11. Stark, R.F., and Piran, T., in *Proceedings of the Fourth Marcel Grossmann Meeting on General Relativity*, ed. R. Ruffini, Elsevier Science Pub. B.V., 1986, pp. 327-364.

STATUS OF INTERFEROMETERS

The Status of LIGO

Mark W. Coles

LIGO Project
California Institute of Technology
Pasadena, CA 91125

Abstract. Construction activities at the LIGO Observatories near Hanford, Washington and Livingston, Louisiana are complete. Installation of detector components and initial commissioning of detector sub-systems is now under way. The scope of the overall project is reviewed. The current status of the commissioning effort and future plans are outlined.

INTRODUCTION

Albert Einstein predicted the existence of gravitational waves - ripples in the fabric of space and time - as part of the general theory of relativity. The Laser Interferometer Gravitational-Wave Observatory (LIGO) is one of a new generation of detectors [1-3] based on suspended mass laser interferometry which is designed to detect these waves directly, opening up a new vantage point from which to study the universe. LIGO is a national research facility designed by a team of scientists and engineers from the California Institute of Technology and the Massachusetts Institute of Technology through support from the National Science Foundation. LIGO consists of two facilities - near Livingston, Louisiana and 3000 km away near Hanford, Washington.

Each facility has an L-shaped laser interferometer with 4 km long arms. The two

FIGURE 1. Aerial photos of the LIGO Hanford Observatory, at left, and the LIGO Livingston Observatory, at right.

LIGO interferometers will operate in coincidence to reject noise sources that are local to a particular interferometer. To provide an additional confirmation of a genuine gravitational wave signal, a third interferometer, half as long as the other two, shares the vacuum space of the 4 km long interferometer at the Hanford site. The strain sensitivity of the interferometer increases linearly with length, consequently an important confirmation of direct detection of a gravitational wave will be the coincident observation of a half amplitude signal in the half length interferometer.

LIGO is designed to have two to three orders of magnitude improvement in detection sensitivity and band width relative to previous gravitational wave searches. The expected sensitivity of the initial LIGO interferometers is shown in figure 2. Sources of gravitational waves that can be plausibly detected with this sensitivity are: chirp signals from the coalescence of binary compact objects such as pairs of neutron stars and/or black holes, burst sources such as Type II supernova (provided that there is a sufficiently large quadrupole component to the collapse ejecta so that strong gravitational waves result), periodic signals resulting from the gravitational radiation given off by non-axisymmetric neutron stars with rotational frequencies f such that $2f$ lies within the LIGO detection band (as is the case with many known pulsars), and stochastic signals from the early universe which can perhaps be detected by measuring the correlations in the noise background between two or more detectors.

FIGURE 2. Strain sensitivity of the LIGO interferometers. Seismic noise, thermal noise, and shot noise limit the anticipated initial LIGO sensitivity to the red line indicated.

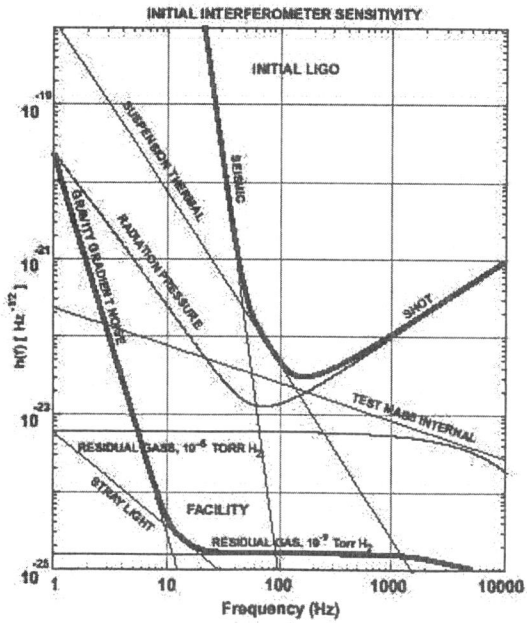

OVERVIEW

Each LIGO interferometer is basically a Michelson interferometer whose end mirrors and beam splitter are freely suspended. The arms of the interferometer have an important modification; input test masses, together with the end test masses, form 4 km long Fabry-Perot cavities which are resonant at the laser frequency, effectively increasing the number of round trips that the light makes back in forth in each arm by the quality factor of the cavity, which is around 50. A second enhancement to the optical topology of the interferometer is the inclusion of a "recycling mirror". In an ideal equal arm length Michelson interferometer, all of the light in the interferometer would destructively interfere in the direction of the interferometer output and would instead return from the beam splitter in the direction of the laser. The recycling mirror creates an optically resonant cavity between it and the two input test masses, causing the light reflecting towards the laser to remain in the interferometer and further boosting the total circulating laser power by about a factor of about 30. Principal parameters which characterize the overall design of LIGO are listed in Table 1.

TABLE 1. LIGO Design Parameters.	
Arm length	4000 meters
Light source	Nd3+:YAG, $\lambda=1.06\mu$
Input optical power to mode cleaner	6 watts
Power recycling factor	30
Fabry-Perot arm quality factor	50
Mass of suspended input and end mirrors	10.7 kg
Mirror diameter	25 cm
Mirror material	Fused silica
Mirror internal Q per mode	$>10^6$
Mirror internal adsorption at 1.06μ	~5 ppm
Mirror scattering loss (transmission + scattering)	50 ppm
Pendulum material	Steel wire, Q=200,000
Pendulum frequency	1 Hz
Seismic isolation at 100 Hz	> 110 dB

Each observatory site is laid out with a major building at the vertex of the interferometer and additional smaller buildings at the location of the end test masses 4 km away. At LHO, two additional buildings exist which surround the end test masses of the 2 km interferometer which shares vacuum space with the 4 km interferometer. Additional structures provide support infrastructure to the site such as storage, heating and cooling, fire protection, etc. The control room, support labs and shops, and meeting room space are located within the major building at the vertex. The high bay areas of the vertex and end stations at each observatory site have been constructed to accommodate the installation of an additional interferometer which could be added later without necessitating further civil construction.

The laser beam propagates in an evacuated beam tube running between the vertex and the mid and end stations. It is located above ground on a straight concrete slab. The beam tube is fabricated from 304L stainless steel having a wall thickness of only 3 mm. Prior to fabrication, the steel was specially processed to reduce the outgassing rate of hydrogen from the material to a level of less than 10^{-13} torr-liter/sec/cm^2. The tube is jacketed in thermal insulation so that it can be electrically heated for bake out to remove residual water from the clean inner surface of the tube. Bellows located every 130 feet accommodate thermal expansion of the tube. The beam tube is protected under pre-cast concrete arches placed in ten foot long segments over the entire length of the tube. Roughing pumps, turbo pumps, and liquid nitrogen cryo-pumps located at the mid and end stations (at LLO at the ends only) are used to maintain the vacuum within the beam tube. Additional pump-out ports located at 250 meter intervals along the beam tube provide additional access for diagnostic testing or the installation of additional capacity should it be required at some future point.

Large vacuum vessels house the suspended test masses of the interferometer, along with their associated vibration isolation suspension assemblies, as well as support optics for length and alignment sensing of the interferometer. Large gate valves are located between the vacuum vessels and the beam tube and between various chambers to allow them to be independently evacuated or vented without disturbing the beam tube vacuum. The vessels rest on thirty inches thick concrete floors which are isolated from the building footings to reduce ambient vibration levels in this area.

STATUS

Major construction activities, which began in 1995, are now complete at both observatory sites. At each location, laboratories and shops for support of site activities have been set up and are now functioning. Office space for resident staff and visitors and meeting areas also exist. High bay areas have been commissioned as "clean" work spaces. Control rooms have also been set up and are evolving as the detector installation and commissioning work progresses.

All vacuum vessels and beam tubes have been installed and accepted at both sites, along with associated pumps, vacuum measurement apparatus, and purge air systems. Bake out of the beam tube is complete at LHO and in progress at LLO. During bake out, an evacuated 2 km section of the beam tube is raised in temperature to 168 C for approximately 3 weeks by ohmic heating, as 2000 amps DC flows through the tube wall. Following cool down to ambient, internal pressures of approximately 1 nano-torr are obtained within the beam tube. The residual gas is dominated by hydrogen, while the partial pressure of water is approximately 10^{-13} torr, a million-fold reduction relative to the partial pressure of water observed prior to baking. It is expected that the bake out activity at LLO will be complete in May 2000.

The light source used by LIGO is a nominally 10 Watt single mode Nd:YAG laser, built by Light Wave Electronics. It is stabilized in frequency, intensity, and angle on an optical table before insertion into the interferometer vacuum. Collectively, the light source and associated optics and servo systems are called the "pre-stabilized laser" or PSL. Initial frequency stabilization is accomplished by locking the laser frequency to a thermally stabilized reference cavity. A triangular "pre-mode cleaner" located on the optical table acts as a spatial filter to reduce the frequency and angular noise of the light by about two orders of magnitude before it enters the mode cleaner. The pre-mode cleaner optical path length can be servo controlled via a PZT which displaces one of the three mirrors to allow locking of the pre-mode cleaner to the thermally stabilized frequency reference cavity. Before leaving the PSL table, the light is mode matched to the in-vacuum 15 meter triangular mode cleaner and modulated at three different frequencies used to lock the various interferometer cavities.

Initial construction and testing of the PSL was used to demonstrate the laser performance and to validate the design of the frequency and intensity stabilization servos and the pre-mode cleaner servo. At LHO, the PSL has been installed and operating for approximately one year. Frequency stabilization and intensity control have been demonstrated to be extremely reliable over this initial operating period, with locking periods of many hundreds of hours. At LLO, the PSL has been installed only recently. Similar acceptance and commissioning tests are now underway.

In-vacuum mirrors are each supported from a single loop wire pendulum clamped to a suspension "tower" (see figure 3). The wire contacts the mirror around its circumference at just above its mid-point, giving overall stability to the assembly. The optic is initially balanced and aligned normal to the design beam orientation within the vacuum chamber. Small magnets glued at four points around the circumference of the mirror and at one point on the circumferential surface can be independently pushed or pulled by small electromagnet coils mounted on the suspension tower adjacent to the optic, allowing the angular orientation and the displacement of the mirror to be externally controlled. Position sensing for control and damping is supplied by LED's and shadow sensors within the electromagnet coil assembly. Additional feedback from length and alignment sensing servos can be used to introduce frequency dependent actuation of the mirror.

FIGURE 3. Installation of a suspension tower assembly within the interferometer vacuum.

The mirrors are "super polished" fused silica with multi-layer dielectric coatings. LIGO optics are fabricated from specially selected glass to minimize absorption at 1.06 microns in order to reduce thermal distortions. The glass also exhibits high Q values (greater than one million) for body resonances in order to limit thermal noise contributions to the interferometer's performance. Precision measurements made at CSIRO and in LIGO's own metrology lab confirm that the surfaces of the mirrors are figured to an RMS tolerance of less than $\lambda/800$. LIGO's large optics have a approximately 10.7 kg mass, with the multi-layer dielectric coating extending over the entire 25 mm diameter surface. Large optics comprise the recycling mirror, end and input test masses, one of the mode matching mirrors used to match the recycling cavity to the output of the mode cleaner, and the beam splitter. Smaller optics used for guiding the input and output beams have approximately 130 mm diameters.

Mirror substrates for all three interferometers have been polished and coated. Procedures for vacuum preparation and assembly into suspension towers have been developed and are now being used to build up the suspension assemblies at each site. Installation of optics towers into the interferometer vacuum is underway at LHO. The mode cleaner components, recycling mirror, and end mirrors are now installed, with additional installation underway. At LLO, installation of suspended optics will begin in Fall, 1999.

The suspension towers are attached to seismic isolation platforms which passively isolate the suspended optics from ambient vibration present in the concrete slab supporting the vacuum chambers. The seismic support structures consist of three layers (for support optics) or four layers (for the test masses, beam splitter, and recycling mirror) of steel interleaved with layers of helical damped springs . The spring design utilizes an internal viscous layer, constrained between the the helical spring housing and a row of metal slugs along each spring's axis, to damp the spring's resonances. This limits the magnitude of the ambient vibration within the seismic isolation platform in the 1-10 Hz band   (the frequency region over which the lowest order structural resonances occur) while still providing good attenuation above 10 Hz.

Seismic installation work is underway at both sites and should complete for all three interferometers in 2000. Measurements of the in-vacuum vibration transfer functions of the installed large optic seismic assemblies have been made and agree well with engineering predictions of performance (see figure 4).

Input optics have been installed on the seismic isolation assemblies at LHO and commissioning of the 15 meter mode cleaner is now underway there. This is the first step in commissioning the interferometer which integrates the PSL controls with the suspension servo controls and the length and alignment sensing. Initial alignment and preliminary locking of the cavity were achieved in air, followed by more detailed studies that are now being conducted in vacuum.

Precision sensing of the alignment is accomplished using wave front sensing. Tilts of the mirrors cause a slight admixture of (0,1) or (1,0) modes to propagate in the mode cleaner along with the TEM00 mode. The amplitudes of these modes are used as error signals to control the orientation of the mirrors and maintain angular alignment. The commissioning of the mode cleaner is providing the opportunity to evaluate this system and validate that it is functioning properly. Similar work will be undertaken at LLO early in 2000.

FIGURE 4. In-vacuum measurement of the vertical displacement noise measured on a four layer seismic isolation system at LLO. The data above about 20 Hz is limited by the noise floor of the measurement instrumentation.

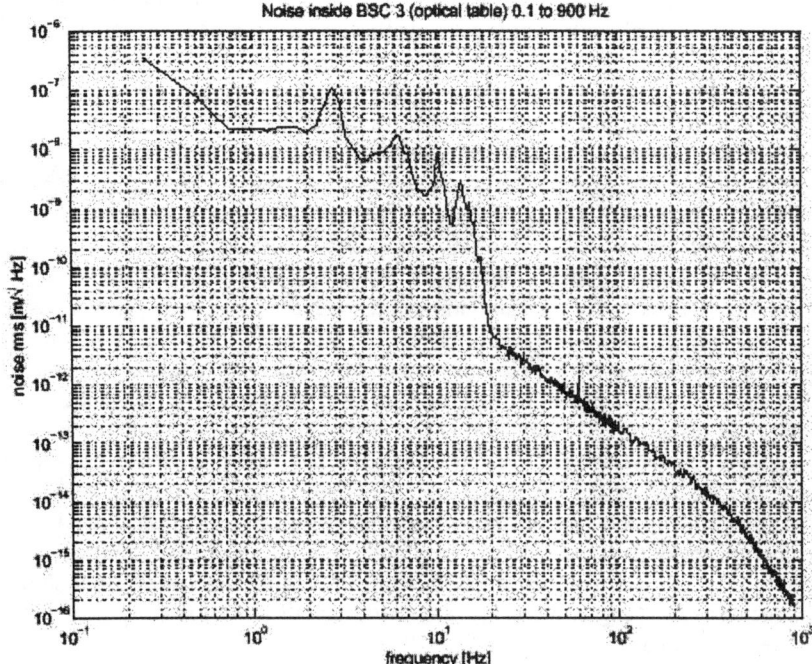

Following completion of the mode cleaner studies, a "long arm" test is planned that will optically couple the mode cleaner to each of the 2 km arms independently. This provides an additional level of system integration of the control system and provides an extremely sensitive test of the angular alignment control system over a 2 km baseline. This work will take place during the latter half of 1999.

In parallel with the installation of the optics and seismic systems, the data acquisition and control system installation work is also well under way. Local computing networks to handle interferometer data acquisition and control, general computing, on-site on-line data analysis, and global diagnostic monitoring have been designed. The hardware to implement these systems is now being installed at each site. The software which monitors and controls the vacuum system and PSL has been installed at both sites and environmental monitoring (of ground vibration, acoustics, the electromagnetic spectrum, and weather parameters) has also been installed. The suspension control hardware and software have been completely installed at LHO in preparation for the 2 km arm tests while installation work is now underway at LLO. Installation of the length and alignment sensing and control system for the mode cleaner and the end test mass suspensions is also complete for the 2 km interferometer at LHO. Installation activity is just beginning at LLO.

The LIGO Data Analysis System (LDAS) will provide the capability to perform initial analysis of interferometer data on-site. The design of the software and hardware configuration for this system is now complete. It makes use of a layered modular design that can easily accommodate extensions and revisions to the analysis flow based on operating experience. The system, when installed, will include the capability to distribute data to local and remote users for further analysis and will include an off-line system dedicated to archiving data and distributing it for computationally intensive re-analysis of the gravitational wave channel. The raw data (as much as 6 MBytes/sec, which includes all of the interferometer diagnostic data and environmental monitoring data as well as the gravitational wave channel) is ordered into "frames". These are a unified data structure which groups extensible sets of time series data into one second chunks for random access. The frame format will provide a standardized mechanism for communicating data within the LIGO scientific collaboration and between gravitational wave groups that adopt this format.

A major effort to construct a simulation and modelling software program is also underway. So far, an overall "end-to-end" model structure has been completed and released so that individual model elements can be built up and linked to the program. At present, a time domain model of the PSL has been completed and validation of the model using the installed PSL hardware at LHO is in progress. A simple model of the mechanical suspension system for the optics has also been developed and implemented. Detailed models of the seismic isolation system performance using the measured transfer functions and characterizations of the ambient seismic background are now being developed and will be available shortly. The servo control models which digitally represent the various LIGO control loops are now being documented and implemented. The long arm tests planned for this fall will provide a means to test and validate these models.

SCHEDULE

Tests of the LHO PSL and mode cleaner will continue through the fall of 1999 with "first light" down the interferometer arms planned for late in the year. Additional installation and commissioning activities at both sites will continue through 2000. Because of the limited staff available, many installation and commissioning tasks are performed sequentially from one interferometer to the next, making use of many of the same staff to repeat the installation and commissioning process. Thus the LHO 2km interferometer will be completed first, followed by the LLO 4km interferometer, and then the LHO 4km interferometer. On-going engineering studies of increasing complexity are planned as the installation, integration, and commissioning activities at LHO and LLO continue. The 2 km is being used to test essential features of the interferometer design while the LLO 4 km interferometer will follow these initial studies with detailed characterizations of the interferometer system. The LHO 4 km interferometer commissioning is expected to benefit significantly from this prior work.

Operation for engineering studies and evaluation will involve first one operating interferometer and then the simultaneous operation of two and then three interferometers for detailed studies of their individual and correlated performance.

The first scientific data runs involving all three interferometers operating in coincidence are planned in 2002.

ACKNOWLEDGMENTS

This work is supported under NSF Cooperative Agreement No, PHY-9210038 between the National Science Foundation, Washington, DC 20550 and the California Institute of Technology, Pasadena, CA 91125.

REFERENCES

1. F. Marion, VIRGO, this conference.
2. K. Tsubono, M. Ando, TAMA, this conference.
3. H. Lueck, GEO600, this conference.

Status of the VIRGO Experiment

Frédérique Marion
for the VIRGO Collaboration

Laboratoire de Physique des Particules (LAPP), Annecy-Le-Vieux, France
Dipartimento di Fisica dell'Universitá e INFN sezione di Firenze, Italy
Laboratori Nazionali dell'INFN, Frascati, Italy
Université Claude Bernard, IPNL, Villeurbanne, France
Dipartimento di Scienze Fisiche dell'Università e INFN Sezione di Napoli, Italy
Observatoire de la Côte d'Azur, Nice, France
Laboratoire de l'Accélérateur Linéaire, Université Paris-Sud, Orsay, France
Laboratoire de Spectroscopie en Lumière Polarisée, Ecole Supérieure de Physique et Chimie Industrielle, Paris, France
Dipartimento di Fisica dell'Università e INFN Sezione di Perugia, Italy
Dipartimento di Fisica dell'Università, INFN Sezione di Pisa e Scuola Normale Superiore, Italy
Dipartimento di Fisica dell'Università "La Sapienza" e INFN Sezione di Roma 1, Italy

Abstract. The status of the VIRGO experiment as of fall 1999 is presented here: progress in the construction is reported and next steps are outlined.

INTRODUCTION

Year 1999 has seen a lot of progress in the construction of VIRGO [1], the gravitational wave detector sponsored by CNRS and INFN. Not only the infrastructure housing the central part of the interferometer has been completed, but the works for the construction of the two 3 km long arms have started. The situation is similar for the vacuum chamber, since the central part has been completed and accepted, whereas the production of the modules for the long tubes has been initiated.

As far as the detector is concerned, the construction of the central interferometer (CITF) is well advanced, while all major contracts involved in the construction of the final interferometer are in place.

INFRASTRUCTURE

Buildings and Tunnels

The construction of the CITF buildings on the Cascina site is now over. They include the central building - housing the central part of the interferometer - which is now equipped with clean areas; the control building, housing the control and computer rooms as well as some offices and meeting rooms; the mode-cleaner and technical buildings.

The civil engineering works for the construction of the tunnels that are going to house the two 3 km vacuum tubes were initiated a few months ago. The first step in progress is to drive the piles supporting the tunnels.

Vacuum Chambers

Following the availability of the central building, we have proceeded with the assembly and integration of the central vacuum chamber. This has gone through putting into place the 7 lower vacuum chambers ("towers") intended for the mirrors and optical benches, the links between them, and the ovens and platforms surrounding them. All pumping elements have been installed, together with their control software, so as to make the pumping system operational.

The starting up of the pumping system has led to good results. All towers have been baked out. Figure 1 shows the residual gas spectrum measured after the bakeout of one of the towers, displaying partial pressures below 10^{-9} mbar for hydrogen, below 10^{-10} mbar for water and below 10^{-13} mbar for hydrocarbons.

The vacuum infrastructure now available includes the 144 m mode-cleaner tube, which has been mounted between the injection and mode-cleaner towers.

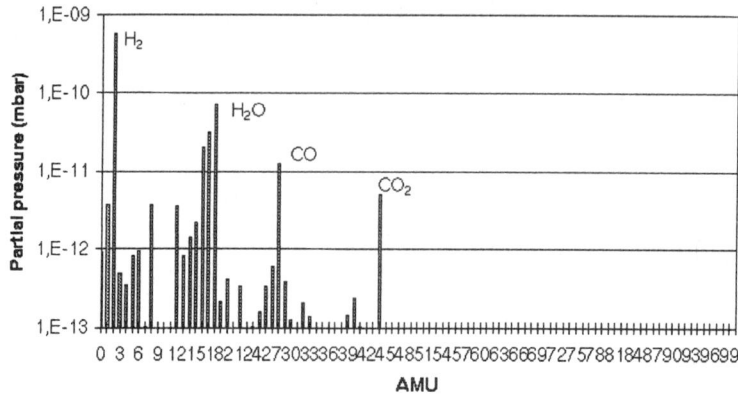

FIGURE 1. Residual gas spectrum measured in a tower after bakeout.

In parallel, the production of the modules for the long vacuum tubes has been initiated with tight control enforced at the contractor (**CNIM**) after tests on module 0 - though reasonably successful - have shown evidence for contamination. The results of those tests are summarized in table 1. The production, which all in all will last about two years, is now going on, and tube modules are progressively shipped to the **BELLELI** firm, in charge of the assembly on site, expected to start in October 2000. The production of other elements, like large valves and baffles, is also in progress.

DETECTOR

The construction of the detector is also advancing well. All the elements necessary for the CITF - suspensions, payloads, electronics, software - are progressively being brought to the site for assembly and integration. In parallel, R&D for those elements needed for the final configuration of the detector - mainly the large mirrors and monolithic suspensions - is going on.

TABLE 1. Summary of results obtained on tube module 0.

	Measured Value	Requirement
Air leak rate	$0.8\ 10^{-10}$ mbar.l.s^{-1}	$< 3\ 10^{-10}$
Hydrogen specific outgassing rate	$2.85\ 10^{-14}$ mbar.l.cm^{-2}.s^{-1}	$< 5\ 10^{-14}$
Hydrogen pressure extrapolated to 300m	$1.8\ 10^{-10}$ mbar	$< 10^{-9}$
Hydrocarbon pressure	$5\ 10^{-14}$ mbar	$< 10^{-13}$

Suspensions

Suspensions are key elements of the **VIRGO** design. They are of two kinds. *Long* suspensions - with five intermediate filters - are intended to hold the mirrors, while *short* suspensions - with only one intermediate filter - are intended to hold the optical benches (injection, mode-cleaner, detection).

Short Suspensions

The three short suspensions have been installed in their vacuum chambers. The mechanical performances of the three systems have been successfully tested. For each of them the normal modes have been identified and the transfer functions have been measured. Inertial damping on three degrees of freedom has been performed on one of the short suspensions, with good and robust results.

Long Suspensions

All components of the long suspensions being now available, the assembly of the filters is in progress. We are going to proceed with the installation of the first long suspension very soon, the introduction of the first payload itself being planned for March 2000.

Moreover, a prototype long suspension (called prototype SA) has been available in Pisa for some years, and R&D is still going on with this prototype, especially for the implementation of the inertial damping. Good attenuation factors have been obtained, with stable operation of three loops for two horizontal and one angular degrees of freedom [2]. Figure 2 shows the measured vertical transfer function performed on the prototype SA, compared to simulation results.

Last Stage

The last stage of a suspension is composed with a positioning device called "marionetta", to which are suspended a mirror and a reaction mass. The steel wires and clamps designed to hold the mirrors (or mirror holders for the CITF) in our reference solution are being produced. However, R&D is going on to try and make possible the use of monolithic suspensions from the very beginning of **VIRGO** in its final configuration [3].

Optics

The construction and installation of the optical elements is going on in parallel to that of the mechanical elements.

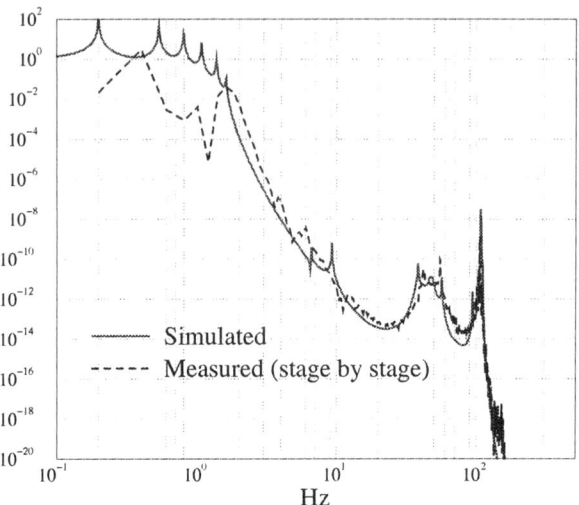

FIGURE 2. Suspension vertical transfer function as measured on prototype, compared to simulation results.

Injection System

The injection system is divided into three main parts: the laser itself, the input bench and the mode-cleaner.

The laser has been installed on site, with measured performances consistent with previous ones [4]. The delivered power is 10 W, of which 92% are in the TEM_{00} mode.

The input bench is ready for installation on site. Figure 3 shows pictures of it, suspended to the marionetta and holding the laser reference cavity.

Tests on the 30 m mode-cleaner prototype in Orsay have demonstrated the ability to lock the cavity [5]. The mirrors of the final mode-cleaner have been coated, and the complete laser system is expected to be operational in the first months of year 2000.

Mirrors

For the CITF, we are going to use small mirrors that will be replaced by larger ones to operate the interferometer in its final configuration.

For the small mirrors, the situation is as follows. All substrates have been received from **HERAEUS** and polished by **GENERAL OPTICS**. They have been characterized with results within the requirements: roughness below 0.5 Å RMS, flatness better than 25 nm RMS on active diameter, and curvature radius within ± 5 m around design value (93 m) for end mirrors. The substrates are being coated, and for

FIGURE 3. Pictures of the suspended input bench.

those that have already gone through this process, the obtained results are also compatible with the requirements: absorption lower than 5 ppm, scattering below 5 ppm and wavefront uniformity better than 25 nm RMS.

For the large mirrors, substrates are currently being manufactured by **HERAEUS** and a contract has been settled with **GENERAL OPTICS** for the polishing. The assembly of the coating facility dedicated to the large mirrors is in progress; the facility should be operational in January 2001.

Detection System

Figure 4 shows pictures of the detection system installed on the site. The system includes a suspended bench - supporting the output mode-cleaner - and an external bench - supporting the photodetectors. Tests on the output mode-cleaner have confirmed the expected performances, and have demonstrated that the cavity can be locked automatically. The electronics associated to the detection system is also ready and in place. Preliminary data acquisition tests carried through 33 hours have shown no evidence of excess noise from the electronics and have measured a noise spectral density a factor 5 below the **VIRGO** sensitivity [6].

Electronics and Software

The installation of the online electronics and software is going on in parallel to the other activities, and is done so as to support the assembly and integration of the other sub-systems. The network and the unix workstation cluster have been

FIGURE 4. Pictures of the detection system installed on the Cascina site: suspended bench (left) and external bench (right).

operational since summer 1998. Slow control and slow data acquisition (DAQ) are also available, and a permanent DAQ has been running since June 1999. The installation of the real time system is in progress, and local fast readouts can already be carried out. We are now installing first versions of the following elements: GPS timing, suspension control, photodiode readout, main frame builder, data distribution.

COMMISSIONING AND DATA ANALYSIS

The commissioning of the CITF is going to start with the first injection of laser light early in year 2000, and will be going on until the first 3 km arm is available (end of 2001). This phase will give us the opportunity to validate the technical choices made for VIRGO, and will train us operating a suspended interferometer over long periods of time. Hopefully, it will also allow us to evaluate the contribution of technical noises to the noise budget and to start studying their effects on data analysis.

The preparation of data analysis itself is in progress. The first step was to collect in a document [7] the suitable methods to be applied in the analysis of the different noises and signals. The second step is now to move towards implementation. Whereas the online data analysis system is progressively implemented, the needs still have to be evaluated as far as the offline analysis is concerned: software libraries, interactive environment, computing resources. In order to help in assessing those needs, we have undertaken performing prototype analyses based on simulated data.

CONCLUSION

The assembly and integration of the VIRGO central interferometer is in good progress and should be over in August 2000. The next milestones are as follows. The commissioning phase will start early in 2000 and will last till the end of 2001 when the first 3 km arm will be completed. The second arm will be ready mid 2002, soon after the large mirrors will have been made available. The commissioning of the interferometer in its final configuration will take place in the last term of 2002, so that data taking could start at the end of year 2002.

REFERENCES

1. VIRGO Coll., Final Design Report (1997)
2. Losurdo G. for Pisa and Florence VIRGO Groups, *these proceedings*.
3. Gammaitoni L., Kovalik J., Marchesoni F., Punturo M. and Cagnoli G., *these proceedings*.
   Cagnoli G., Gammaitoni L., Kovalik J., Marchesoni F. and Punturo M., *Phys. Lett.* **A213**, 245 (1996).
   Cagnoli G., Gammaitoni L., Kovalik J., Marchesoni F., Punturo M., Braccini S., De Salvo R., Fidecaro F. and Losurdo G., *Phys. Lett.* **A237**, 21 (1997).
4. Cleva F., Taubman M., Man C.N. and Brillet A., *Proceedings of the Second Edoardo Amaldi Conference on Gravitational Waves*, Singapore: World Scientific, 1998, pp. 321-327.
5. Barsuglia M., Bondu F., Cavalier F., Heitmann H., Man C.N. and Matone L., to appear in the *Proceedings of the Rencontres de Moriond on Gravitational Waves and Experimental Gravity*, 1999
6. Buskulic D., Caron B., Dcrome L., Flaminio R., Hermel R., Lacotte J.C., Marion F., Masserot A., Massonnet L., Morand R., Mours B., Puppo P., Verkindt D. and Yvert M., to appear in the *Proceedings of the Rencontres de Moriond on Gravitational Waves and Experimental Gravity*, 1999
7. VIRGO Coll., Data Analysis Document (1999)

The Status of GEO600

Harald Lück[1,2], P. Aufmuth[1], O.S. Brozek[2], K. Danzmann[1,2],
A. Freise[1], S. Goßler[1], A. Grado[1], H. Grote[1], K. Mossavi[2],
V. Quetschke[1], B. Willke[1,2], K. Kawabe[3], A. Rüdiger[3], R. Schilling[3],
W. Winkler[3], Ch. Zhao[3], K.A. Strain[4], G. Cagnoli[4], M. Casey[4],
J. Hough[4], M. Husman[4], P. McNamara[4], G.P. Newton[4], M.V. Plissi[4],
N.A. Robertson[4], S. Rowan[4], D.I. Robertson[4], K.D. Skeldon[4],
C.I. Torrie[4], H. Ward[4], B.F. Schutz[5], I. Taylor[5], B.S. Sathyaprakash[6]

Abstract. GEO600, the German/British gravitational wave detector currently being built in northern Germany, used advanced optical technologies to obtain a sensitivity comparable with the other, bigger detectors currently being built [1,2]. The installation of the ultra-high-vacuum system has almost been completed and the Mode Cleaners are operational.

INTRODUCTION

Five large scale gravitational-wave detectors are currently being built worldwide [1–3]. Despite the small size of GEO600 (600 m) in comparison to LIGO (4 km) and VIRGO (3 km), GEO600 will achieve a comparable peak sensitivity. This is possible by using advanced optical technologies such as dual-recycling [4,5]. This paper describes the current status of the GEO600 project and gives an introduction into the technologies used to fight the important noise sources.

---

[1] *Institut für Atom- und Molekülphysik, Universität Hannover, Callinstrasse 38, 30167 Hannover*
[2] *Max-Planck-Institut für Quantenoptik, Außenstelle Hannover, Callinstrasse 38, 30167 Hannover*
[3] *Max-Planck-Institut für Quantenoptik, Hans-Kopfermann-Strae 1, D-85748 Garching, Germany*
[4] *Physics & Astronomy, University of Glasgow, Glasgow G12 8QQ*
[5] *Max-Planck-Institut für Gravitationsphysik, Albert-Einstein-Institut, Haus 5, Am Mühlenberg, 14476 Golm, Germany*
[6] *Department of Physics and Astronomy, University of Wales Cardiff, P.O. Box 913, Cardiff, Wales, CF2 3YB.*

THE BASICS OF GEO600

Before the laser light enters the main interferometer it is sent through two successive mode cleaners to attenuate fluctuations of the laser beam.

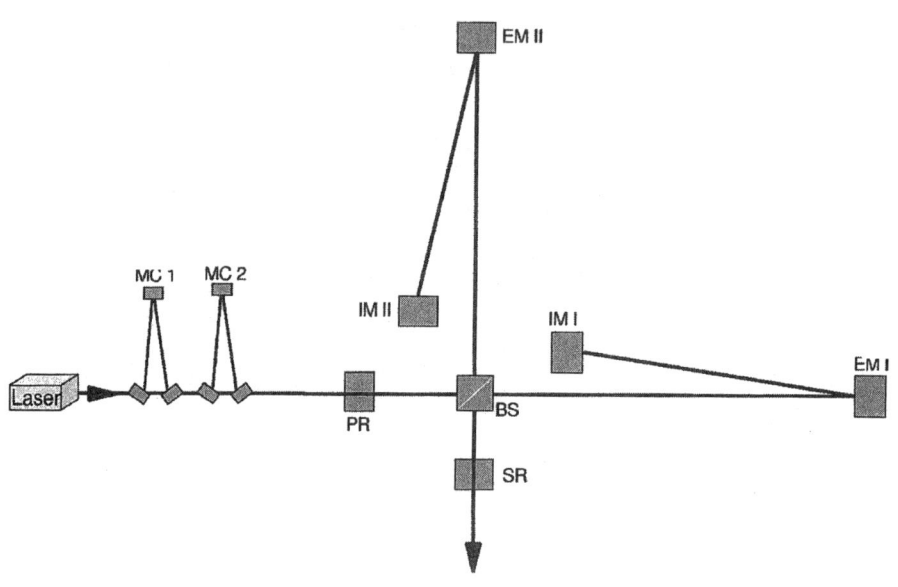

FIGURE 1. Optical layout of GEO600

The short arm length of GEO600 has to be compensated for by choosing a powerful optical layout. GEO600 uses a simple optical delay line which sends the light through the interferometer arms four times(DL4) to increase the optical round trip length to 2400 m (See Figure 1). The light is sent through the beam splitter (BS) to the end mirror (EM), is sent back to the inboard mirror (IM) and retraces it's incoming path back to the beam splitter. Figure 1 shows Power Recycling (PR) is employed to enhance the circulating light power by a factor of 2000. Hence with a laser power of 10 W and an assumed loss of 50% through the mode cleaner the laser power available inside the interferometer at the beam-splitter will be 10 kW. To further boost up the sensitivity Signal Recycling (SR) will be used. Signal Recycling reduces the shot noise in the vicinity of a tunable frequency and sacrifices sensitivity elsewhere. The spectral width of the sensitivity enhancement can be chosen by altering the reflectivity of the Signal Recycling mirror. Doing this online by using a thermally tunable etalon as a Signal Recycling mirror is an option currently being investigated. The sensitivity aimed at is shown in Figure 2. In addition to enhanced sensitivity SR also reduces the amount of non-TEM00 modes exiting through the output port due to imperfect mirror surfaces or imperfect alignment (Mode Healing) [6] and thus improves the signal to noise ratio.

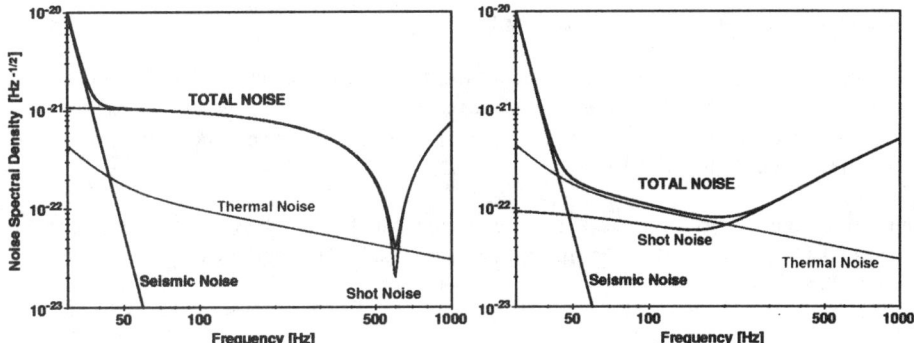

FIGURE 2. Sensitivity of GEO600 for broad- and narrow-band operation

The control signals for the differential Michelson length control and the Signal Recycling mirror will be obtained with Schnupp modulation from the output light and the light in one of the arms, respectively. This requires an arm length difference of about 10 cm. The position of the Power Recycling mirror will be controlled with the Pound-Drever-Hall (rf-reflection locking) technique.

STATUS OF THE INSTALLATION

Vacuum System

To reach the desired sensitivity in GEO600 the residual gas pressure has to be kept below 10^{-8} mbar. Hydrocarbons can degrade the performance of the highly reflective mirrors at even lower pressures. Hence GEO600 uses an all-metal vacuum system in an attempt to reduce the hydrocarbon content below the sensitivity limit of the mass spectrometers, i.e. 10^{-14} mbar. With a volume of about $400\,m^3$ the GEO600 vacuum system is currently the biggest UHV system in Europe.

The vacuum system of GEO600 is divided into 7 subsections:

- three tanks contain the two mode cleaner and injection optics. The mode cleaner tanks are separated from the central cluster by a 200 mm all metal gate valve.

- the central cluster of 4 tanks containing the Power Recycling mirror, beam splitter, compensation plate, and the inboard mirrors. The central cluster is set up, connected and pumped down (not yet baked) to a pressure in the upper 10^{-9} mbar region with a small air leak left to be tracked down.

- two beam tubes of 600 m length each, 60 cm diameter and a wall thickness of 0.8 mm. Corrugated walls give the stability against the air pressure. The tubes are separated from the other sections by 600 mm all metal gate valves. The pressure in the beam tubes is in the upper 10^{-9} mbar region. The tubes have received a bake at 200°C in air for two days and 250°C under vacuum for 5 days.

- two sections containing one end-tank each.

- two tanks containing the Signal Recycling mirror and output optics, separated from the central cluster by a 200 mm all metal gate valve to allow for a change of the Signal Recycling mirror without long down-times. These two tanks will be connected to the central cluster later on.

The whole vacuum system, except for the mode cleaner, is pumped by four magnetically levitated Turbo pumps with a pumping power of 1000 l/s (nitrogen) each backed by Scroll pumps (25 m^3/h).

The Laser

The laser for GEO600 is being developed in the Laser Zentrum Hannover. GEO600 uses an injection locked Nd-YAG master/slave system with the master being a stabilized diode-pumped 1 W NPRO (Non-Planar-Ring-Oscillator) and the slave a diode end-pumped bow-tie ring resonator giving an output power of 13 W. Currently the master laser is used on site for testing of the mode cleaner, whereas the master/slave system is still being optimized in the Hannover labs. At low frequencies the master laser is stabilized to a suspended rigid reference cavity made of ULE and it will be stabilized to the Power Recycling cavity in the measurement frequency band.

Mode Cleaner

GEO600 uses two mode cleaner in series to reduce fluctuations of the laser beam in shape, position and pointing. Each of the triangularly shaped mode cleaners has a round trip length of about 8 m and a Finesse of 1900. Two successive mode cleaner have been implemented to obtain high attenuation values in short cavities with moderate Finesse.

Vacuum The mode cleaner section is pumped by a magnetically levitated Turbo pump (170 l/s) backed by a scroll pump. Before the final bake the hydrocarbon content in the mode cleaner section was higher than expected (in the 10^{-11} mbar region). In order not to contaminate the other vacuum sections which contain the main optics a 1 cm aperture will be inserted which limits the conductance to about 10 l/s. Hence with a pumping power of 3000 l/s we can keep the hydrocarbon pressure in the central cluster in the 10^{-14} mbar region.

Optics The mirrors for the mode cleaners with a diameter of 100 mm and a thickness of 50 mm are made of fused silica (Suprasil1), polished by Rollei and ion-beam coated by the VIRGO group.

Auto-alignment Both mode cleaner are equipped with an auto-alignment system which adjusts the mirrors with respect to the incoming beam. A similar auto-alignment system is described in [8]. The bandwidth is currently restricted to 0.3 Hz, but will be increased to a few Hz.

Suspension All the optics for both mode cleaner are suspended as double pendulums in two vacuum tanks 1 m in diameter each. One mirror in each mode cleaner is hung as a double pendulum with a reaction mass which itself also is suspended as a double pendulum.

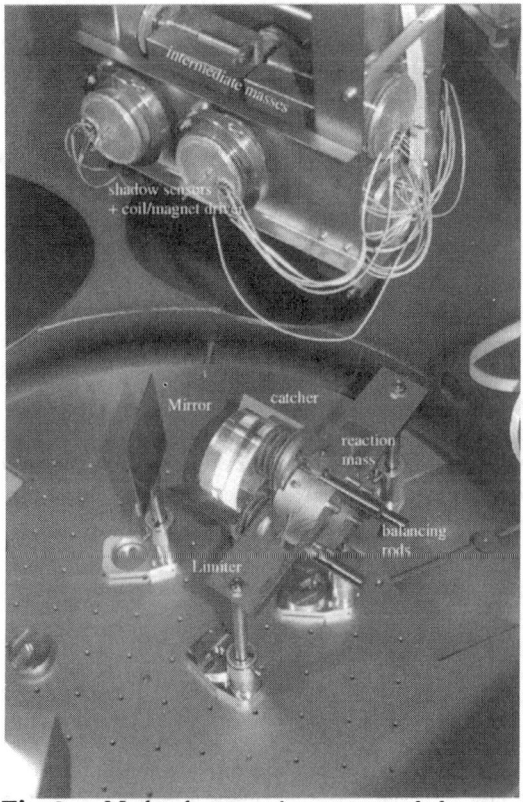

Fig. 3: *Mode cleaner mirror suspended as a double-double pendulum (see Text)*

Such a double-double pendulum can be seen in Figure 3. The pendulum resonances are electronically damped at the intermediate masses using coil/magnet actuators. The motion with respect to a rigid frame is sensed with shadow sensors. Control forces for fast movements are applied between the mirror itself and the reaction mass. The magnets used here are directly glued onto the mirror surface using an inorganic glue. The current is supplied through bare 150 μm copper wires. A 4 cm hole in the reaction mass allows to pass the laser beam pass through. The rods seen on the right hand side serve for balancing the reaction mass. Apropriately shaped stainless steel sheets limit the motion of the reaction mass such that is does not hit the mirror. Both the mirror and the reaction mass are suspended from stainless steel wires with a diameter of 60 μm (cannot be seen in this picture.)

Status Both mode cleaner are installed and have been locked in air to the laser wavelength. The locking was robust and could be maintained for a few hours. The contrast at the input of the first and second mode cleaner was about 85% and 95%, respectively. The mode matching into the first mode cleaner is not yet optimised. The throughput (about 60%) of the 'mounting unit' between the two mode cleaner carrying two Farady isolators and an elctro optical modulator (24 surfaces) is not yet satisfactory and remains to be improved.

Main Interferometer

Vacuum

The vacuum system, except for the end tanks, is finished. Small air leaks (internal or external) remain to be found but will not compromise the overall performance of GEO600.

Suspension

The suspension (see Figure 4) [9] of the mirrors has to ensure isolation of the mirrors against seismic disturbances and simultaneously allow microscopical movement and orientation. The first isolation stage in GEO600 is an active feedback stage consisting of a 3-dimensional piezoelectric actuator collocated with three geophones with a resonance frequency of 2 Hz. A feedforward loop using an external 1 Hz geophone will improve the isolation to about 35 dB. Atop of this active stage a passive rubber/stainless steel sandwich will give additional isolation above the resonance frequency of about 30 Hz. A hexagonal frame structure (stack stabilizer) is supported by three of these stacks and connected to them via flex pivots for rotational freedom. Another hexagonal frame (rotational stage) rests on the stack stabilizer and can be rotated against it to prealign the mirror. Polished Alumina/Sapphire disks serve as a bearing. From the rotational stage the mirror is suspended as a triple pendulum with two additional cantilever spring stages for vertical isolation (see Figure 4). Careful design of the wire lengths and inclination and all moments of inertia ensure optimum coupling of the motion of all pendulum stages to the upper mass where all six degrees of freedom are damped by collocated sensor/actuator pairs. The lowest stage of the mirror suspension, will be a monolithic suspension entirely made of fused silica. 'Hydroxide-catalysis bonding' [10] attaches the mirrors to fibres which in turn are bonded to the stage above. The fast feedback to the end-mirrors for longitudinal locking uses electro-static actuators positioned on reaction masses with an identical suspension. After a successful design and prototyping phase the first main suspension is being installed on site.

Optics

- The main mirrors of the interfeometer (i.e. the end- and inboard mirrors) are made of fused silica (Suprasil 1) with a diameter (D) of 18 cm and a thickness (d) of 10 cm. They are polished by General Optics and will be coated by the VIRGO group.

- The recycling mirrors are made of Suprasil 2 (a lesser grade for cost reasons) with D=15 cm and d=7.5 cm.

- The beamsplitter requires special attention because it is the only optical element which is transmitted by the 10 kW light inside the cavity. Suprasil 311SV is a fused silica with especially low absorbtion (in this case the absorbtion was unmeasurable, i.e. below 0.5 ppm/cm).

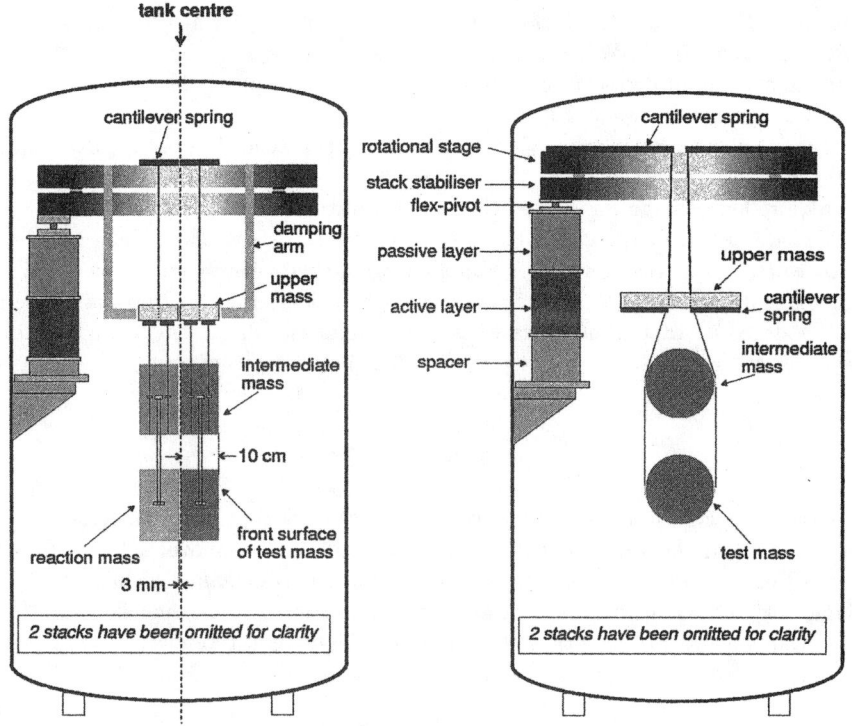

FIGURE 4. Side- and axial-view of the suspension for the main mirrors

- In the installation and debugging phase GEO600 will use test mirrors made from natural quarz hung in steel wire slings.
- The initial beam splitter will be made of Suprasil 312.

Detector control

Except for the active seismic isolation GEO600 uses analog feddback loops. Many of the loop parameters, e.g. gain or offset, are supervised by Labview software run on PCs. This allows remote control of mirror positions for prealignment purposes and gives short lock-aquisition times due to automatically setting optimum gains for each step.

Data Aquisition

The data aquisition system (DAQS) for GEO600 is based on the DAQS developed for LIGO but modified to the less demanding data rates of 0.5 MB/s in GEO600.

Each of the three buildings will contain one data collection unit (DCU). The DCU is a VME crate running VxWorks-RTOS and Tornado Software and contains an 24-bit ADC interface capable of digitising 32 (16 for the end buildings) channels at a rate of 8 kHz and 64 channels at a rate of 512 Hz.

Debugging and optimisation of GEO600 will make extended use of the data aquisition system and use TRIANA [11] as a software tool. First look analysis of the data that does not require high computing power will be performed on site.

The data of three days will be buffered on hard disk on site and simultaneously transferred to Hannover via a 34 Mbit/s radiolink, where they will be stored by a tape robot. The whole data-set will by transferred over night via Internet to Potsdam where it will also be stored on tape. The tapes written in Hannover will be sent to Cardiff via mail. Hannover will create a subset of data to be distributed to other groups.

Data Analysis

Data analysis will be distributed between the Albert-Einstein-Institut in Golm/Potsdam, the University of Wales in Cardiff and Hannover. The software will run on Beowulf clusters with a capacity of 12 GFlops (Potsdam/Cardiff) and 6 GFlops (Hannover). Potsdam will concentrate on the search for periodic signals, Cardiff will do burst searches and the time critical analysis will be done in Hannover.

PLANS

Within the next few months the optics in one interferometer arm will be installed and operated as a 1200,m Fabry-Perot cavity to gain experience with the long interferometer arms. The second interferometer arm, power- and signal recycling will be installed subsequently. During these phases GEO600 will be operated using test optics hung in wire slings to avoid the risk of damaging the high quality optics. Data taking is expected to commence towards the end of 2001.

REFERENCES

1. M. Coles, LIGO, this conference.
2. F. Marion, VIRGO, this conference.
3. K. Tsubono, M. Ando, TAMA, this conference.
4. B.J. Meers, *Phys. Rev. D* **38**, pp. 2317-2326, (1988)
5. K.A. Strain, B.J. Meers, *Phys. Rev. Lett* **66**, pp. 1391-1394, (1991)
6. G. Heinzel et al., Experimental demonstration of a Suspended Dual Recycling Interferometer for Gravitational Wave Detetction *Phys. Rev. Lett.* **81** pp. 5493-5496 (1998)
7. O.S Brozek et al., *Proc. 18th Moriond Workshop*, in press
8. G. Heinzel et al., *Opt. Comm.* **160**, pp. 321-334, (1999)

9. Design of Suspension System for GEO600, M.V. Plissi et al., *Proceedings of 8th Marcel Grossman Meeting on General Relativity and Gravitation*, Israel, ed. T Piran, World Scientific, pp. 1063-1065, (1999)
10. S. Rowan et al.
11. I. Taylor, Documentation for Triana OCL, http://www.astro.cf.ac.uk/pub/Ian.Taylor/Triana/index.html

TAMA Project: Design and Current Status

Masaki Ando, Kimio Tsubono,

Department of Physics, University of Tokyo, 7-3-1 Hongo, Bunkyo-ku, Tokyo 113-0033, Japan.

and the TAMA collaboration.

*Space-Time Astronomy Section, National Astronomical Observatory,
Institute for Cosmic Ray Research, University of Tokyo,
Department of Physics, University of Tokyo,
Department of Advanced Materials Science, University of Tokyo,
Earthquake Research Institute, University of Tokyo,
Miyagi University of Education,
Institute for Laser Science, University of Electro-Communications,
High Energy Accelerator Research Organization,
Yukawa Institute for Theoretical Physics, Kyoto University,
Graduate School of Science, Osaka University.*

Abstract. TAMA is a Japanese project to construct and operate an interferometric gravitational-wave detector with a baseline length of 300 m at the Mitaka campus of National Astronomical Observatory in Tokyo. The aim of this project is to develop techniques necessary for future large-scale interferometers and, moreover, to detect gravitational waves generated within our local group of galaxies. Now, stable operation of the TAMA300 interferometer has been realized under the final configuration, except for power recycling with a displacement noise level of 6×10^{-18} m/$\sqrt{\text{Hz}}$. The interferometer was operated continuously over 7 hours without unlock. In this article, the design of the TAMA300 interferometer is reviewed, and its current status of development is reported.

OVERVIEW

TAMA is a Japanese project to construct and operate an interferometric gravitational-wave detector at the Mitaka campus of National Astronomical Observatory in Tokyo [1-3]. The aim of this project is to develop techniques necessary for future large-scale interferometers and, moreover, to detect gravitational waves generated within our local group of galaxies. The target strain sensitivity of TAMA is $h_{\text{rms}} = 3 \times 10^{-21}$ at 300 Hz with a bandwidth of 300 Hz, which corresponds to

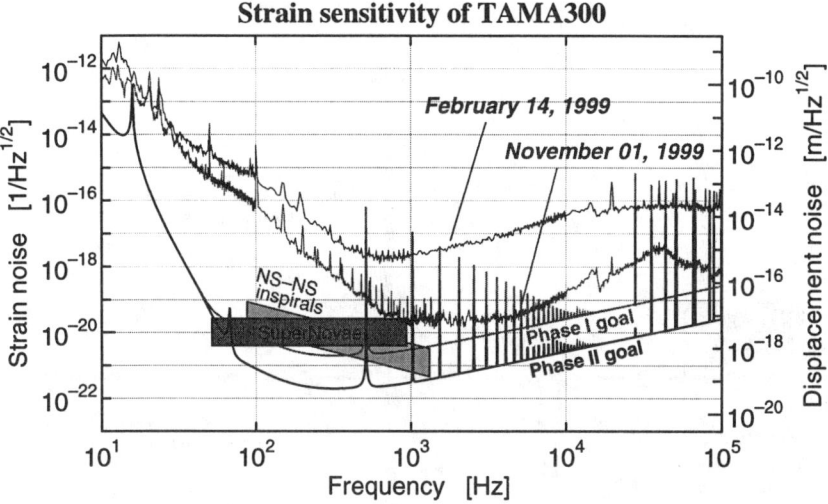

FIGURE 1. Strain sensitivity of the TAMA300 interferometer in the unit of $1/\sqrt{\text{Hz}}$. The target sensitivities in Phase I and Phase II, current sensitivity (November 01, 1999), and expected gravitational wave sources (in our galaxy) are shown together.

$h = 2 \times 10^{-22} /\sqrt{\text{Hz}}$ in the power spectrum density. With this sensitivity, the TAMA detector would be able to detect gravitational waves from supernova explosions, or the coalescence of compact binary systems within our local group of galaxies. Figure 1 shows the target sensitivities of the TAMA300 interferometer, the current sensitivity (November 01, 1999), and the expected gravitational-wave sources. In the first observation phase (Phase I), the TAMA interferometer is operated without power recycling, with a sensitivity of $h = 2 \times 10^{-21} /\sqrt{\text{Hz}}$. The final sensitivity of $h = 2 \times 10^{-22} /\sqrt{\text{Hz}}$ will be achieved in the second observation phase (Phase II) with power recycling and other improvements.

The interferometer, called TAMA300, is a Michelson interferometer with 300 m Fabry-Perot arm cavities (Fig. 2, Table 1). The arm cavities are designed to have a finesse of 516, and a cutoff frequency of 480 Hz. In the final configuration, power recycling will be implemented to enhance the effective laser power; the designed power recycling gain is 10, and the effective laser power will be around 30 W. An LD-pumped Nd:YAG laser with an output power of 10 W is used as the light source (developed by Sony Corp.) [4]. The mode cleaner of TAMA300 is an independently-suspended triangular ring cavity with a length of 9.75 m [5]. It has a finesse of 1700, and a cutoff frequency of 4.5 kHz. The electro-optic modulators (EOM) for phase modulation (12 MHz modulation for mode cleaner control and 15.235 MHz modulation for the main interferometer control) are placed in front of the mode cleaner. Thus, the wave-front distortion by the EOMs is rejected by the mode

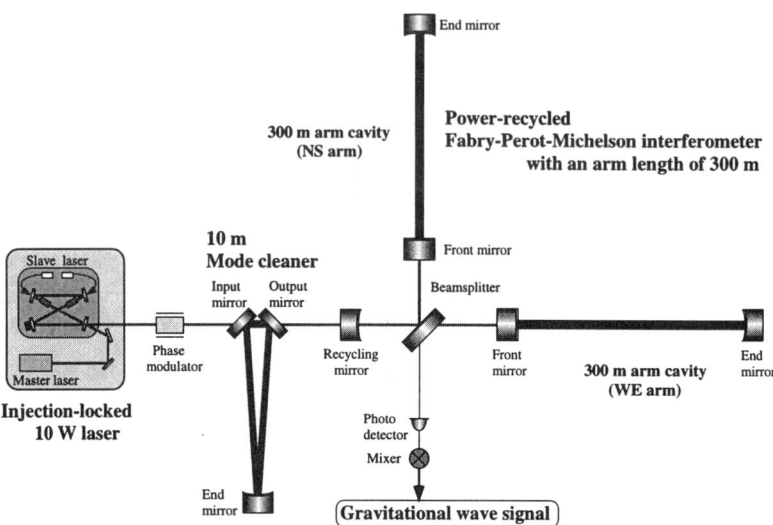

FIGURE 2. Conceptual design of the TAMA300 interferometer. The principal scheme is a power-recycled Fabry-Perot-Michelson interferometer with a baseline length of 300 m. A triangular ring cavity is inserted between the main interferometer and the laser source as a mode cleaner.

cleaner before entering into the main interferometers as well as the higher-modes from the laser source.

The mirrors of the main interferometer are made of fused silica (Suprasil-P10, produced by Shin-Etsu Quartz Products Co., Ltd.). Each mirror has a diameter of 100 mm, and a thickness of 60 mm; the lowest resonant frequency of axis-symmetric modes is around 28 kHz. The surface of the substrate is polished, and coated by an IBS (ion-beam sputtering) machine at JAE (Japan Aviation Electronics Industry Ltd.). The surface figure error is around $\lambda/40$, which satisfies the requirement of $\lambda/20$. In addition, the surface is super-polished so that the micro-roughness should be less than 10^{-10} m_{rms} [6]. The loss of the mirror is estimated to be 30 ppm from measurements with smaller sample mirrors [7,8].

The mirrors of the main interferometer and the mode cleaner are isolated from seismic motion with three-stage stacks and double pendulums (Fig. 3). The optical breadboard is supported by three legs of the stack [9]; each stack comprises three layers of rubber (Chloroprene) and disk blocks of stainless steel. Each rubber block is sealed in bellows so as to avoid outgassing. The suspension system [10] is a double pendulum; the mirror is suspended by two loops of tungsten wire from an

TABLE 1. Main parameters of the TAMA300 interferometer.

Features of the TAMA300 interferometer	
Sensitivity	$h_{\rm rms} = 3 \times 10^{-21}$ ($f_{\rm obs} = 300\,{\rm Hz}$, $\Delta f_{\rm obs} = 300\,{\rm Hz}$)
Site	National Astronomical Observatory, Mitaka, Japan
Main interferometer	Fabry-Perot-Michelson interferometer with power recycling
	Arm cavity length : 300 m
	Finesse of arm cavities : 516
	Power recycling gain : 10
	(Effective power in the interferometer: 30 W)
Light source	LD-pumped Nd:YAG laser
	Output power : 10 W
	Wavelength : 1064 nm
Mode cleaner	Triangle ring cavity
	Baseline length : 9.75 m
	Finesse : 1700
Vibration isolation	Three-stage stack + Double pendulum
	Vibration isolation ratio : < -165 dB
Data acquisition	16 bit, 20 kHz sample, 8 channels
Vacuum system	$< 10^{-6}\,{\rm Pa}$

upper mass, which is also suspended by four wires. The motion of the upper mass is damped by an eddy current with permanent magnets surrounding it. Since the magnets are also suspended with bending springs, they do not degrade the vibration isolation ratio around the observation frequency range; they damp the motion only around the resonant frequency of the pendulum. Vertical seismic motion is also isolated with bellows springs inserted between the suspension point and the upper mass. The suspension point is fixed to motorized stages, which are used for an initial adjustment of the mirror orientation. The fine position and orientation of each mirror is controlled with coil-magnet actuators; small permanent magnets are attached to the mirror.

The Q-factor of the suspension system is estimated to be 3×10^5 from the results of Q-factor measurement of the violin modes [11]. The thermal noise due to the mirror internal motion is one of the critical noise of the TAMA interferometer. The intrinsic Q-factor of the fused silica mirror is measured to be 3×10^6 [12].

The data-acquisition system comprises a high-frequency part for the main signals and a low-frequency part for detector diagnostics. The main output signals of the interferometer are recorded with high-frequency A/D converters (20 k samples/sec, 16 bit), after passing through whitening filters and 5 kHz anti-aliasing low-pass filters. Seven channel signals are recorded with a timing signal locked to UTC within an error of 110 ns. Along with the high-frequency system, environmental monitoring data (such as the pressure in the vacuum system, temperature, interferometer status, and so on) are collected with a low-frequency data-acquisition system, which is based on a EPICS (Experimental Physics and Industry Control System) and HP

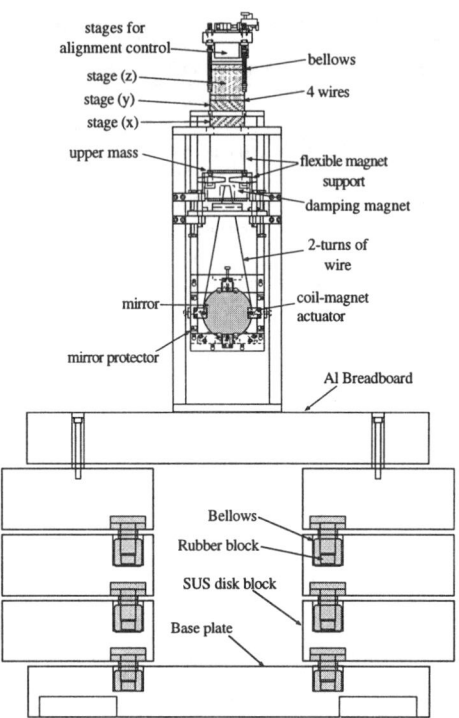

FIGURE 3. Design of the vibration isolation and suspension system. The optical breadboard is supported by three legs of the stack; each stack comprises three layers of rubber and disk blocks of stainless steel. The suspension system is a double pendulum; the mirror is suspended by two loops of tungsten wire from an upper mass, which is also suspended by four wires.

VEE (Visual Engineering Environment, Hewlett-Packard Co.) system.

The interferometer is housed in a vacuum system comprising eight chambers connected with vacuum ducts with a diameter of 400 mm (Fig. 4) [13]. Three vacuum chambers (MC1, BS, and RM) have a diameter of 1200 mm, while the others have that of 1000 mm. The inner surface of the vacuum system is polished by ECB (Electro-Chemical Buffing) so as to reduce the outgassing rate. With this surface processing, a vacuum pressure less than 1×10^{-6} Pa is achieved without baking [14]. The system is evacuated by rotary pumps and 8 turbo-molecular pumps, and kept in a vacuum with 16 ion pumps during operation of the interferometer.

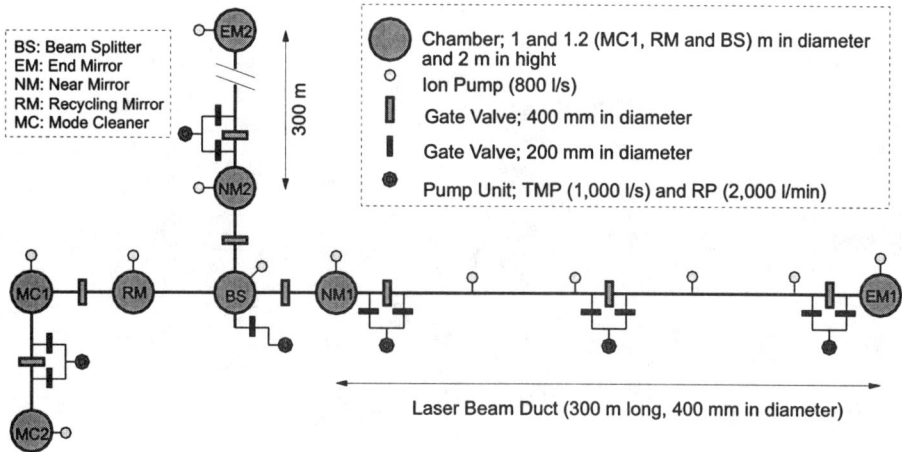

FIGURE 4. Vacuum system of TAMA300, which comprises eight chambers connected with vacuum ducts with a diameter of 400 mm. Three vacuum chambers (MC1, RM, and BS) have a diameter of 1200 mm, while others have that of 1000 mm.

OPTICAL AND CONTROL CONFIGURATION

The TAMA project started in 1995; at present, the interferometer is operated with the final configuration for the Phase I observation (without power recycling). Figure 5 shows the current optical and control configuration of TAMA300.

The laser source is an LD-pumped Nd:YAG laser with an output power of 10 W; a high-power slave laser is injection-locked to a 700 mW master laser. Two Nd:YAG rods in the slave laser are end-pumped by fiber-coupled laser diodes (SDL-3450-P6). The frequency of this laser system is controlled with a thermal actuator and the PZT on the master laser crystal, and an external EOM. In addition, another external EOM is inserted between the laser source and the mode cleaner for intensity stabilization. The mode cleaner is a triangular ring cavity with a length of 9.75 m. Its finesse is 1700, and the cutoff frequency is 4.5 kHz. The control signal to lock the cavity to the incident laser beam is extracted by the Pound-Drever-Hall scheme [15] with a modulation frequency of 12 MHz. The error signal is fed back to the 10 W laser; the frequency of the laser is stabilized with the mode cleaner. The EOMs for phase modulation (12 MHz modulation for mode cleaner control and

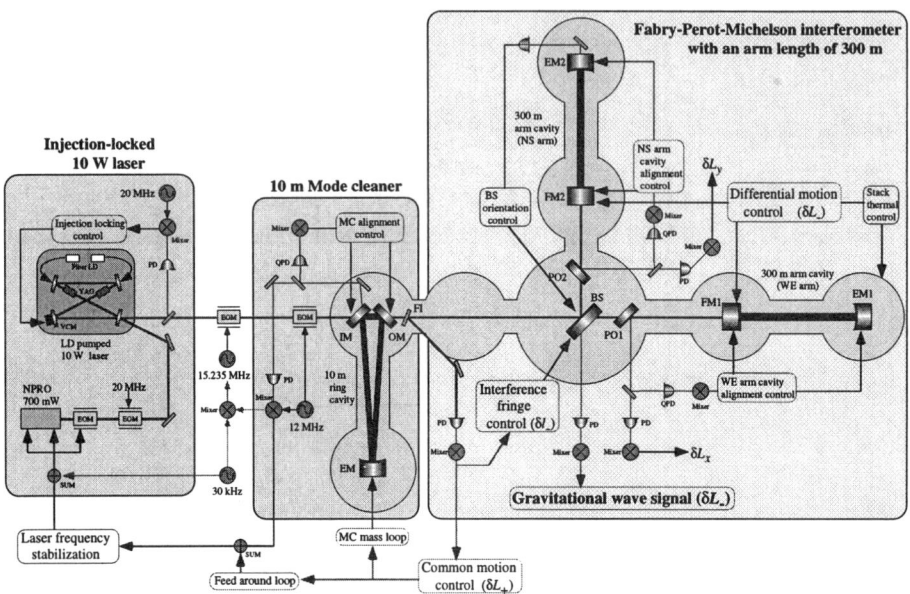

FIGURE 5. The optical and control configuration of TAMA300 for the first observation (without power recycling). An LD-pumped Nd:YAG laser with an output power of 10 W is used as the light source. The mode cleaner is a triangular ring cavity with a length of 9.75 m. The main interferometer is a Michelson interferometer with 300 m Fabry-Perot arm cavities.

15.235 MHz modulation for the main interferometer control) are placed in front of the mode cleaner. Thus, the wave-front distortion by the EOMs are rejected by the mode cleaner before entering into the main interferometer. The frequency of phase modulation for the main interferometer control is selected to be equal to the FSR (Free-Spectrum-Range) of the mode cleaner so that the RF sidebands should transmit through the mode cleaner. In addition, the difference of the modulation frequency and the FSR of the mode cleaner is sensed with additional modulation to control the modulation frequency.

In order to keep the main interferometer at the operational point (arm cavities are resonant with the incident beam, and interference fringe is dark at the detection port), three degrees of longitudinal freedom must be controlled: differential and common length change in the arm cavities (δL_- and δL_+), and the differential change in the Michelson arm length (δl_-). The control signals corresponding to these degrees of freedom are extracted with the frontal modulation scheme [16]. The differential signal of arm cavities (δL_-), which would contain gravitational wave signals, is extracted from the detection port and fed back to the coil-magnet actuators of the front mirrors differentially. The control bandwidth (the unity-

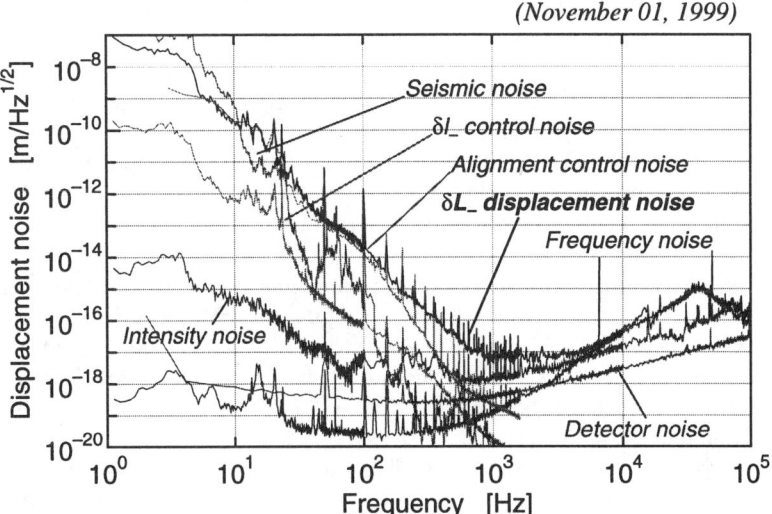

FIGURE 6. Sensitivity of TAMA300 and estimated noise sources. The floor level is 6×10^{-18} m/$\sqrt{\text{Hz}}$ (which corresponds to the strain sensitivity of 2×10^{-20} /$\sqrt{\text{Hz}}$). The sensitivity of the interferometer is limited by the alignment control noise from 10 Hz to 200 Hz. Above 5 kHz, the sensitivity is limited by the laser frequency noise; the CMRR (common mode noise rejection ratio) is about 1/60, which will be improved with fine alignment and cleaning of the mirrors. The noise source which limits the the floor level has not yet been identified. We are investigating the effect of scattered light.

gain frequency) of the δL_- loop is around 1 kHz, which is limited by the internal resonance of the mirror. The differential signal of the Michelson interferometer (δl_-) is extracted from the reflection port of the interferometer, and fed back to the coil-magnet actuators of the beamsplitter. The control bandwidth of the δl_- loop is about 30 Hz; this bandwidth is kept narrow in order to prevent the noise in the δl_- feedback signal from degrading the sensitivity of the interferometer. The common signal of arm cavities (δL_+) is also extracted from the reflection port, and fed back to the stabilized laser source part (10 W laser source and 10 m mode cleaner). In the frequency range lower than 1 kHz, the δL_+ signal is fed back to the end mirror of the mode cleaner to stabilize the mode cleaner length. On the other hand, the higher frequency region signal is fed back to the 10 W laser to expand the control bandwidth of the δL_+ loop; the unity-gain frequency is about 30 kHz for the δL_+ loop.

Besides the length control loops described above, alignment control loops are implemented to realize stable operation. The angular fluctuations of the suspended mirrors (three for the mode cleaner and four for the main interferometer) are controlled with bandwidths of around 10 Hz. Pick-off mirrors with 0.6% reflectivity are inserted between the beamsplitter and the front mirrors to sample the reflected beams from the arm cavities. The alignment control signals are extracted from these beams using a wave-front sensing scheme [17], and fed back to the corresponding degrees of freedom (in pitch and yaw directions of front and end mirrors). The alignment control signals for the mode cleaner are extracted from the reflected beam from the mode cleaner, and fed back to the input and output mirrors. With these alignment control loops, the interferometer is operated stably over 7 hours, while the interferometer goes out of lock within several minutes without an alignment control system.

In addition, two drift-control systems are implemented to realize long-term stability of the interferometer: a beamsplitter angular control and a δL_- drift control. Since the beamsplitter, which is also suspended, tilts in several tens of minutes, the alignment of the interferometer degrades because of finite control gains of the alignment control system; the contrast of the interferometer degrades by about 5% in several hours because of this drift in the beamsplitter orientation. Thus, the orientation of the beamsplitter is controlled with a 300 m-length optical lever. The beamsplitter tilt is sensed with a quadrant photo detector, which receives the transmitted beam through the north-south (perpendicular) arm cavity. The error signals in the pitch and yaw directions are fed back to the beamsplitter with unity-gain frequencies of about 0.3 Hz. Another drift-control system is a δL_- low-frequency control. Since the δL_- signal is fed back to the front mirrors, they are dragged by the large drifts of end mirrors during long-term operation. With this drift, the dark-fringe lock point is offset, and, moreover, the interferometer goes out of lock because of the finite range of the actuators. Thus, the drifts in the δL_- feedback signal is compensated with a thermal actuator attached at the bottom of the stack for the west (in-line) end mirror. The control unity-gain frequency is about 3 mHz.

CURRENT STATUS

Currently, the interferometer is operated under the configuration described above, and we are improving the sensitivity and stability of the interferometer. The displacement noise level of the interferometer is 6×10^{-18} m/$\sqrt{\text{Hz}}$ (Fig. 6). The interferometer has been operated stably for over 7 hours. During the three nights' data-taking run (September 17 to 20, 1999), the interferometer was in a locked state for about 94% of the total observation time.

The sensitivity of the interferometer is limited by the alignment control noise from 10 Hz to 700 Hz. The noise in the alignment control error signal causes displacement noise through coupling with miss-centering of the beam on the mirrors, and efficiency asymmetries of the coil-magnet actuators. This noise will be improved in three ways: improvement of the electric noise of the wave-front-sensor, fine adjustment of the beam centering on the mirrors, and a change in the align-

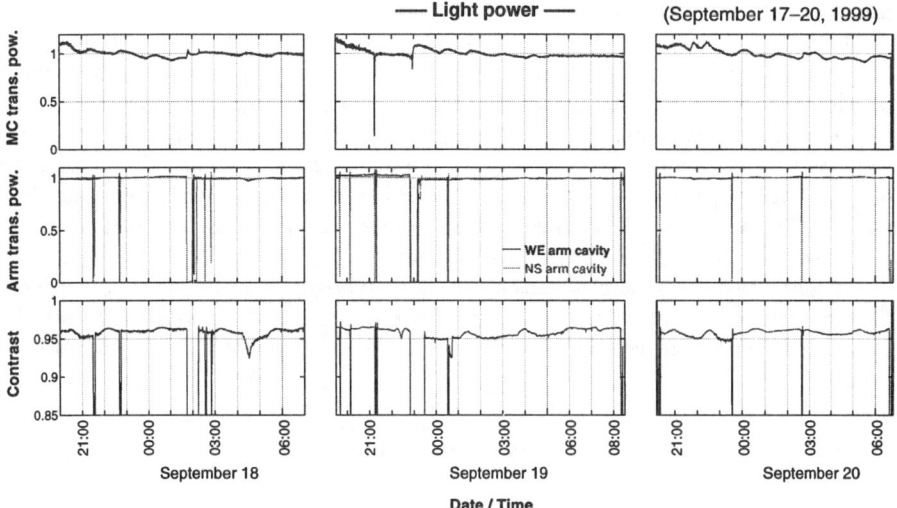

FIGURE 7. Power fluctuation of the TAMA300 interferometer during the data-taking run performed in September, 1999. The interferometer was operated stably for about 30 hours, 94% of the total time.

ment control filter. The alignment error signals are dominated by the electric noises of wave-front-sensors above 100 Hz, which will be reduced by a factor of about 10 by improving the quadrant photo detectors. Fine centering of the beam spot on the mirrors is also necessary to reduce the alignment control noise, and is also necessary to reduce the effect of thermal noise of the suspension system. Currently, the beam spot is estimated to be offset from the mirror center by around 10 mm, while the requirement is less than 1 mm. The effect of the alignment control noise is also reduced by a change in the design of the alignment control filter. We are installing new filters with a steep cutoff at 130 Hz using a 10th-order Chebyshev low-pass filter. With this filter, the alignment control noise will be dramatically reduced above 200 Hz.

The noise source limiting the floor level (around 800 Hz to 3 kHz) has not yet been identified. We are investigating the scattered light inside the beamsplitter tank as the origin of the floor noise level.

The operation of the interferometer is very stable. Figure 7 shows the power fluctuation of the interferometer during a data-taking run performed in September, 1999. The upper figures represents the transmitted light power through the mode cleaner. The vertical axis is normalized so that the averaged power should be unity. The middle figures show the transmissivity fluctuation through the arm cavities;

the vertical axis is normalized so that the averaged value should be unity. The bottom figures represent the fluctuation of the contrast, estimated from the dark fringe laser power.

The interferometer was operated for about 30 hours, 94% of the total time of this data-taking run. The transmitted light power through the mode cleaner fluctuated by about 10% because the alignment control system was not implemented to the mode cleaner at this time. On the other hand, the transmissivities of the arm cavities were very stable because of the alignment control system. The fluctuation of the transmissivity was less than 2%. The contrast was kept higher than 95%, and the fluctuation was 2%. The contrast will be kept at the highest level with the alignment control system of the mode cleaner.

The data taken in this observation run is now being processed with a matched-filtering scheme to find gravitational waves from compact binaries, or to set the upper-limit for the gravitational wave amplitude. The first results will be reported soon.

SUMMARY

The TAMA300 interferometer has been stably operated under the final configuration, except for power recycling. Though the target sensitivity with this configuration has not yet been attained, the best noise level has reached 6×10^{-18} m/$\sqrt{\text{Hz}}$ ($h = 2 \times 10^{-20}$ /$\sqrt{\text{Hz}}$). The interferometer is operated continuously for over 7 hours without unlock. A data-taking run was performed from September 17 to 20 in 1999. The first results of data analysis will be reported soon.

We will continue to improve the sensitivity and stability of the interferometer until spring of 2000. After that, installation of power recycling [18,19], an improvement of the isolation system, and a long-term observation run are planned. In addition, we have started a study of the next large-scale interferometric detector in Japan [20].

ACKNOWLEDGEMENTS

TAMA project is supported by a Grand-in-Aid for Creative Basic Research of the Ministry of Education.

REFERENCES

1. K. Tsubono, 300-m Laser Interferometer Gravitational Wave Detector (TAMA300) in Japan, in: *Gravitational Wave Experiments*, eds.: E. Coccia, G. Pizzella, and F. Ronga, World Scientific (1995) pp. 112-114.
2. K. Kuroda, Y. Kozai, M.-K. Fujimoto, M. Ohashi, R. Takahashi, T. Yamazaki, M. A. Barton, N. Kanda, Y. Saito, N. Kamikubota, Y. Ogawa, T. Suzuki,

N. Kawashima, E. Mizuno, K. Tsubono, K. Kawabe, N. Mio, S. Moriwaki, A. Araya, K. Ueda, K. Nakagawa, T. Nakamura, and Members of TAMA group, Status of TAMA, in: *Gravitational Waves*, eds.: I. Ciufolini and F. Fidecaro, World Scientific (1997) pp. 100-107.
3. Ed: K. Tsubono, *TAMA STATUS REPORT 99*, TAMA internal report (1999).
4. S. T. Yang, Y. Imai, M. Oka, N. Eguchi, and S. Kubota, Optics Letters 21 (1996) 1676.
5. A. Telada, *Development of a Mode Cleaner for a Laser Interferometer Gravitational Wave Detector*, Ph. D thesis, The Graduate University for Advanced Studies (1997).
6. M. Ohashi, Ouyou-butsuri 68 (1999) 663 (in Japanese).
7. N. Uehara, A. Ueda, K. Ueda, H. Sekiguchi, T. Mitake, K. Nakamura, N. Kitajima, and I. Kataoka, Optics Letters 20 (1995) 530.
8. S. Sato, S. Miyoki, M. Ohashi, M.-K. Fujimoto, T. Yamazaki, M. Fukushima, A. Ueda, K. Ueda, K. Watanabe, K. Nakamura, K. Etoh, N. Kitajima, K. Ito, and I. Kataoka, Applied Optics 38 (1999) 2880.
9. R. Takahashi, F. Kuwahara, and K. Kuroda, Vibration isolation stack for TAMA300, in: *Gravitational Wave detection*, eds.: K. Tsubono, M.-K. Fujimoto, K. Kuroda, Universal Academy Press (1997) pp. 95-102.
10. A. Araya, K. Arai, Y. Naito, A. Takamori, N. Ohishi, K. Yamamoto, M. Ando, K. Tochikubo, K. Kawabe, and K. Tsubono, Design of a mirror suspension system used in the TAMA 300m interferometer, in: *Gravitational Wave detection*, eds.: K. Tsubono, M.-K. Fujimoto, K. Kuroda, Universal Academy Press (1997) pp. 55-62.
11. N. Ohishi, in: *TAMA Status Report 99*, ed.: K. Tsubono (1999) pp. 22-35 (in Japanese).
12. K. Numata, talk in the TAMA project meeting (1999).
13. Y. Saito, N. Matsuda, Y. Ogawa, G. Horikoshi, Vacuum 47 (1996) 609.
14. Y. Saito, Y. Ogawa, G. Horikoshi, N. Matuda, R. Takahashi, M. Fukushima, Vacuum 53 (1999) 353.
15. R.W.P. Drever, J. L. Hall, F. V. Kowalski, J. Hough, G. M. Ford, A. J. Munley, and H. Ward, Applied Physics B 31 (1983) 97.
16. M. W. Regehr, F. J. Raab, and S. E. Whitcomb, Optics Letters 20 (1995) 1507.
17. E. Morrison, B. J. Meers, D. I. Robertson, and H. Ward, Applied Optics 33 (1994) 5037.
18. M. Ando, K. Arai, K. Kawabe, and K. Tsubono, Physics Letters A 248 (1998) 145.
19. S. Sato, Ph. D thesis, The Graduate University of Advanced Studies (1998).
20. K. Kuroda, M. Ohashi, S. Miyoki, D. Tatsumi, S. Sato, H. Ishizuka, M.-K. Fujimoto, S. Kawamura, R. Takahashi, T. Yamazaki, K. Arai, M. Fukushima, K. Waseda, S. Telada, A. Ueda, T. Shintomi, A. Yamamoto, T. Suzuki, Y. Saito, T. Haruyama, N. Sato, K. Tsubono, K. Kawabe, M. Ando, K. Ueda, H. Yoneda, M. Musha, N. Mio, S. Moriwaki, A. Araya, N. Kanda, M. E. Tobar, *Large-scale Cryogenic Gravitational wave Telescope*, International Journal of Modern Physics D (in press).

Status of the Australian Consortium for Interferometric Gravitational Astronomy

D.E. McClelland, M.B. Gray, D.A. Shaddock, B.J. Slagmolen, S.M. Scott, P. Charlton, B.J. Whiting[1], R.J. Sandeman

Department of Physics, The Faculties,
The Australian National University, Canberra, 0200, Australia

D.G. Blair, L.Ju, J. Winterflood, D. Greenwood, F. Benabid, M.Baker, Z. Zhou,

Department of Physics,
The University of Western Australia, Nedlands, 6009, Australia

D. Mudge, D. Ottaway, M. Ostermeyer, P.J. Veitch, J. Munch, M.W. Hamilton, C. Hollitt,

Department of Physics and Mathematical Physics,
The University of Adelaide, Adelaide, Australia

Abstract. We report progress on the development of gravitational wave research facilities by the Australian Consortium for Interferometric Gravitational Astronomy (ACIGA) and significant R&D advances across the four major subsystems related to an interferometric gravitational wave detector: configurations; lasers and optics; isolation, suspension and thermal noise; and data analysis.

INTRODUCTION

The Australian Consortium for Interferometric Gravitational Astronomy (ACIGA) (1) is primarily a collaboration between three main institutions: The Australian National University, The University of Adelaide and The University of Western Australia. We summarize here progress on detector subsystem R&D, with extended articles on some areas appearing elsewhere in this proceedings. We begin with a report on the status of new facilities being built in Australia to allow high sensitivity suspended mass experimental tests to be performed.

FACILITIES DEVELOPMENT

Over the last year ACIGA has overseen the construction of two new research facilities: the Advanced Research Interferometer (ARI) Facility at Gingin, 60km north

---

[1] Permanent Address: Department of Physics, University of Florida, Florida, U.S.A.

of Perth in Western Australia, and the Gravitational Wave Research Facility on The Australian National University campus. Fig.1 shows an aerial view of The ARI development, featuring a central 20mx20mx10m laboratory area and two end stations located 80m from the central building. This will allow the construction of a test interferometer of up to 80m in arm length. The facility is located on a site which could eventually accommodate a 4 km long detector.

The 15mx15mx6m ANU facility will be used for R&D on suspended mass systems, with the focus in the first instance being on measuring off resonance thermal noise.

FIGURE 1 Aerial view of the ARI site in Gingin, W.A.

CONFIGURATIONS

As part of the Advanced interferometer Configurations LIGO Science Collaboration (LSC) we are developing a control system for a Resonant Sideband Extraction (RSE) interferometer (2). The RSE layout is shown in Figure 2(a). The rationale behind our system is to provide the ability to be able to tune the interferometer to any desired frequency. This leads to a more complicated modulation system than is needed for a fixed frequency instrument. Figure 2(b) indicates the various resonance conditions on the 4 modulations we propose to use. Currently we are testing this system on a bench top, non suspended interferometer. Further details can be found in the article by Gray et al in this volume.

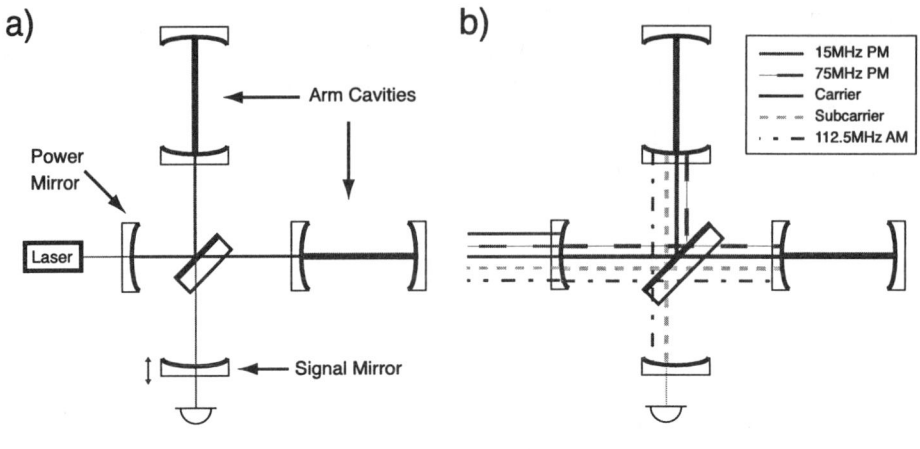

FIGURE 2 (a) RSE Layout   (b) Modulation frequencies

LASERS AND OPTICS

Laser Development

The high power Nd:YAG laser being developed at The University of Adelaide consists of a chain of injection-locked lasers (Figure 3). Injection locking offers several advantages compared to MOPA systems. The gain of the slave lasers is fully saturated by the injected power and thus the system will be more efficient and will produce less amplified spontaneous emission (ASE). Further, the slave resonators provide spatial mode discrimination and thus the output can efficiently be mode-matched into the interferometer. Our architecture reduces the perceived risk of injection locking by lowering the power gain per stage, which increases the injection-locking range.

MASTER LASER	MEDIUM POWER SLAVE LASER	HIGH POWER SLAVE LASER
100mW	5W	100W

FIGURE 3. A three-stage injection-locked laser. Power from the master laser is injected into the medium power slave laser, the output of which is injected into the high power slave laser. The indicated laser powers are nominal.

We have already demonstrated a compact and efficient, injection-locked 5W Nd:YAG laser that reliably produces a diffraction-limited TEM_{00} mode which has low frequency and intensity noise (3). The frequency noise is just that of the NPRO master laser, as expected, and the contribution by the slave agrees with theoretical predictions.

We have also shown that the intensity noise of the injection-locked laser can be reduced to less than $10^{-6}/\sqrt{Hz}$ in the gravitational wave (GW) band by feedback to the multiple-emitter pump diode of the slave laser (4). The high frequency intensity noise is also very low: it is estimated (4) that the intensity noise at 24.5 MHz will be only 0.6% above the shot noise limit for 600mA of detected photo-current.

The high power slave laser, shown schematically in Figure 4, will use a resonator that is stable in the horizontal plane and unstable in the vertical plane (5). The gain medium is side pumped by fiber-coupled diode lasers and side cooled using water. Since the pumping and cooling are coplanar, this architecture can be scaled to arbitrarily high powers if the vertical width of the lowest-order laser mode can be increased. The resonator is therefore unstable in the vertical direction. The pump diodes are fiber coupled to provide a simple means for tailoring the vertical gain profile. This will ensure that the unstable plane can have simultaneously a magnification that is sufficient to provide good mode discrimination, and an output that can efficiently be mode-matched into the interferometer. Note that there will be no excess spontaneous emission as the unstable resonator will be injection-locked (6,7).

We have performed design verification experiments using a laser that is side-pumped by 50W per side in a standing wave, stable resonator configuration (8). The laser produced a multi-mode output power of 32W with an optical efficiency of 32%, and an almost-diffraction-limited TEM_{00} power of 20W. The reflectivity of the optimum output coupler was 70%, in agreement with theoretical predictions.

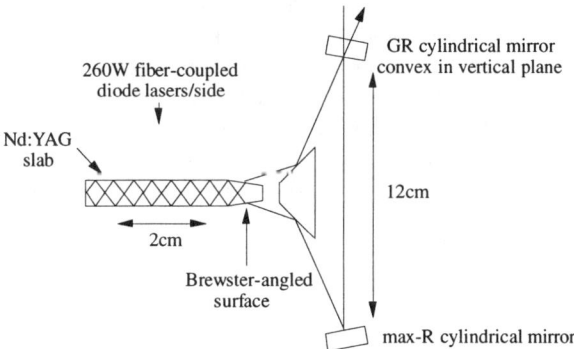

FIGURE 4. A schematic of the high power slave laser resonator. The resonator is stable in the (horizontal) plane of the page, and unstable in the (vertical) plane defined by the normal to the page and the direction of propagation. The curvature of the max-R mirror in the vertical plane is dictated by the strength of the thermal lens in the unstable direction - it is convex (concave) if the thermal lens is strong (weak).

We have also demonstrated that the thermal lensing in the unstable plane can be adjusted simply from a focal length of 47mm to several meters, using Peltier-effect heaters on the top and bottom surfaces of the slab. This flexibility will simplify the design of the unstable plane of the resonator, relax the design specifications for the graded reflectivity (GR) mirror, and ensure that the lowest order mode efficiently couples to the pumped gain volume.

The thermal lens in the horizontal direction has a focal length of order meters and is unaffected by the action of the Peltier-effect heaters. Thus, the horizontal aperture defined by the Brewster-angled entrance and exit faces will provide mode discrimination sufficient to restrict the horizontal mode to its lowest-order.

We are presently assembling the high power laser in a standing-wave configuration. The pump diodes (26 × 20W fiber-coupled) and diode drivers will be delivered soon. The temperature controllers for the diodes are nearing completion. The laser head has been constructed, and the GR mirror is due to be delivered in mid December. We therefore expect to begin testing the stable/unstable laser in early January, 2000.

Sapphire Test Masses

The excellent thermal and mechanical properties of sapphire make it a promising material for use in test masses. Compared with silica material, the high thermal conductivity means the thermal lensing problem is minimized, while the very high Young's modulus means the internal resonant frequencies will be high. The highest reported Q-factor of sapphire at room temperature is 4×10^8 by Braginsky (9) and recent measurements reported a Q-factor of 3×10^8 in small samples (10). We have observed a reproducible Q-factor of 5×10^7 in a $\phi 50 \times 100$mm sapphire sample using a Nb cantilever suspension (11), and a Q-factor of 1.5×10^8 using a wire suspension (12). Studies (13) on the optical properties of sapphire samples showed a level of Rayleigh scattering comparable with fused silica. Tests (14) on several sapphire samples also show that the inhomogeneous birefringence in sapphire is less than 0.1 degree/cm which is within the loss requirement of interferometers, although it places a strict requirement on the alignment of the interferometer. Sapphire has been reported to have a very low optical absorption level (~3ppm/cm). However a recent study suggests that X-ray radiation is responsible for the degradation in optical absorption in sapphire, and re-annealing of the sapphire sample may not be able to fully recover the excellent absorption performance. Further details can be found in the paper by Benabid et al in this volume.

ISOLATION, SUSPENSIONS AND THERMAL NOISE

Our recent suspension research has been directed at reducing the residual low frequency (0.2 to 2Hz) motion of the suspended test mass to as low a value as can be obtained with reasonable effort. This is motivated by the difficulty of achieving and holding lock in a high finesse cavity in the presence of even micron level low-frequency motion, and also the difficulty of preventing injected noise from electronics and seismic coupling when these low-frequency locking forces have to be applied near to the test mass. We have focussed on passive techniques as far as possible for their simplicity of implementation and operation. Figure 5 illustrates the conceptual design of our proposed isolation system.

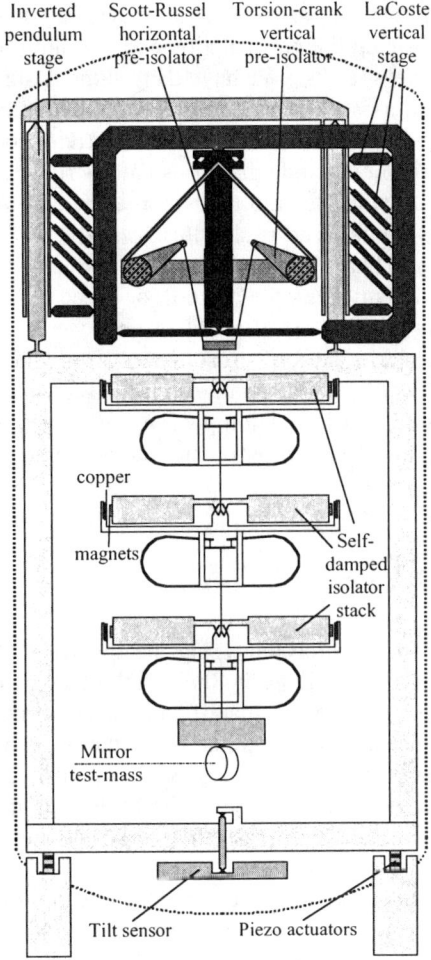

FIGURE 5 Conceptual design of the isolation system, including 2 stages of pre-isolation and 'self damping'.

One thrust has been to find a good method of strongly damping the internal resonant modes of our all-metal isolator stack. A new method we have termed "self-damping" viscously cross-couples modes of different motion and frequency (rocking and swinging) on each single isolator stage to each other. Sufficient viscous force can be obtained from magnetic eddy current coupling - for high vacuum compatibility.

In order to reduce low-frequency residual motion below the seismic level, we have continued our effort in the ultra low frequency pre-isolator area (15,16,17). We have built a tilt-rigid 3-D pre-isolator stage which is intended to be followed in cascade by higher performance horizontal and vertical pre-isolator stages. This extra stage allows shake testing and percussion adjustment of the following high performance pre-isolators in situ, centimeter translation of the isolation stack without compromising the operating point of the high performance stages, and provides passive operation of the units in cascade.

Analysis indicates that with adequate damping and pre-isolation in place, residual motion becomes dominated by seismic tilt bypassing the pre-isolation and coupling in as horizontal acceleration at the top of the isolator stack. We have done some design work and experimental measurements on a dual axis tilt sensor in order to eventually actively servo against seismic tilt. With reasonably attainable tilt sensor sensitivity and control loop gain bandwidth, in addition to our damping and dual pre-isolator structure, we expect the low-frequency residual motion to be reduced to the nanometer level. A fuller description can be found in the article by Winterflood et al in this volume.

Predictions for the noise thermal floor of long baseline interferometers are usually based on measuring the Qs of the suspensions and inferring the noise floor based on an assumed damping mechanism. In collaboration with the VIRGO Project we have embarked on an experiment to directly measure the 'off resonance' frequency distribution of thermal noise. Our set up employs two unique features: the insertion of the test cavity into a feedback loop to suppress classical radiation pressure fluctuations (18) and the use of 'tilt locking' for cavity control (19). Further details can be found in the article by Gray et al in this volume.

DATA ANALYSIS

The ACIGA Data Analysis group (20) has focussed on analyzing the spectrum of data produced by two experimental interferometers - the Caltech 40m and the Glasgow 10m instruments. The aim has been to develop algorithms to identify and characterize the stationary non-Gaussian noise sources present in the detectors. These include, for example, interference from mains power supply and resonances in the mirror suspension wires. As many signal-detection techniques, such as matched filtering, are optimal only when the underlying noise is Gaussian, it is essential that any non-Gaussian noise be accounted for and removed whenever possible.

Our primary source of data has been the experimental run taken on the Caltech 40m instrument in November 1994. This data is stored as integers in the range -2048 to +2047, with typical values in the range of ±100 when the instrument is in lock. For our analysis, we divide the data into blocks of 1000 points. Since the data was sampled at 9868 Hz, each block corresponds to approximately 0.1 seconds.

As we are mainly concerned with stationary properties, our initial work was aimed at distinguishing the useful blocks of data from those which contain transient bursts of noise far above normal levels, as it was found that these introduced spurious features into the frequency spectrum. Another source of such features is the behavior of the instrument immediately after it comes into lock, since there is then usually a period of about a minute when the interferometer output level is effectively zero (apart from a small DC offset), due to saturation of the output amplifiers. It was found that blocks of good data typically had a standard deviation of between 15 and 200. Accordingly, any block whose standard deviation was outside this range was discarded, as were all blocks originating less than 60 seconds after lock was attained.

We compiled frequency histograms from the remaining data, our aim being to determine the statistical properties of these histograms - an important step in characterizing the stationary non-Gaussian noise sources for this particular detector. In order to obtain higher frequency resolution, the blocks of 1000 data points were padded with zeroes to form blocks of $8,192 = 2^{13}$ points. A power of 2 was chosen to maximize the speed of the Fast Fourier Transform (FFT), which was then performed on each block.

Using the first 550 blocks of data, a histogram was compiled for the real part of the FFT for each of the 4,097 frequencies, and another histogram for the imaginary part of the FFT. Histograms formed from each group of 550 blocks of data were combined to form final histograms representing the entire day of data taking. The mean μ and standard deviation σ were calculated for each histogram. Figure 6 depicts the histogram for the real part of the FFT for the frequency 599.91 Hz.

A measure of the Gaussianity for each histogram was performed by calculating the likelihood ratio l (actually $l^{1/N}$) that the distribution was obtained from a Gaussian distribution with the same mean and standard deviation. Figure 7 depicts the histogram for the real part of the FFT for the frequency 299.96 Hz, together with the Gaussian distribution with the same mean and standard deviation. The value of $l^{1/N}$ is 0.7724 – the distribution is clearly far from Gaussian, which is not surprising since the frequency 299.96 Hz is a harmonic of the mains frequency. A more detailed explanation of these and other properties of the frequency histograms will be given in an article to appear in the Proceedings of the Second Tama Workshop on Gravitational Wave Detection (21).

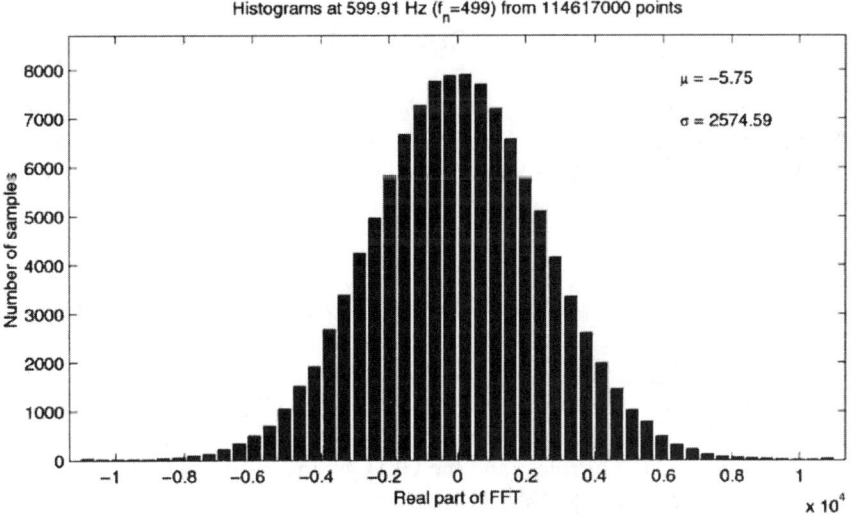

FIGURE 6: Histogram of real part of FFT at 599.91 Hz showing Gaussian structure

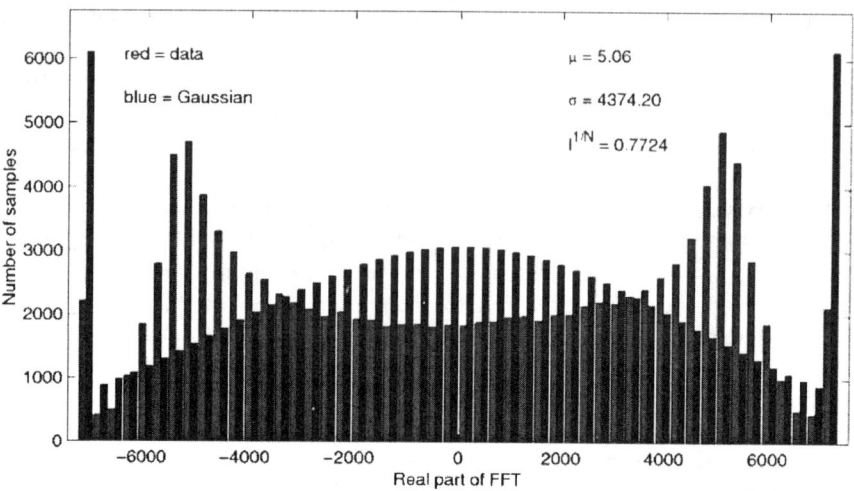

FIGURE 7: Histogram of real part of FFT at 299.96 Hz showing non-Gaussian structure. Gaussian fit is also shown.

SUMMARY

The development of the 80 m test interferometer in Western Australia along with an R&D program which is tightly linked to the international gravitational wave detector projects will ensure that ACIGA continues to make a strong contribution to the world wide effort to detect gravitational waves.

ACKNOWLEDGEMENTS

We acknowledge financial support from the Australian Research Council, the Western Australian Government, the Shire of Gingin, and ACIGA member universities. We thank the LIGO Project and the VIRGO Project for in kind support. We also thank the LIGO Project and the GEO group for access to data from the 40 m prototype interferometer at CALTECH and the 10 m prototype interferometer at Glasgow (respectively).

REFERENCES

(1) ACIGA Web Page: http//:www.anu.edu.au/Physics/ACIGA.
(2) Meers, B.J., *Phys.Rev.***D38**, 2317-2326 (1988); Mizuno, J., Strain, K.A., Nelson, P.G., Chen, J.M., Schilling R., Rüdiger, A., Winkler, W., and Danzmann, K., *Phys.Lett.***A175**, 273-276 (1993).
(3) Ottaway, D.J., Veitch, P.J., Hamilton, M.W., Hollitt, C., Mudge, D., and Munch, J., *IEEE J of QE* **34**, 2006-2009 (1989).
(4) Ottaway, D.J., Veitch, P.J., Hollitt, C., Mudge, D., Hamilton, M.W., and Munch, J., *IEEE J of QE*, submitted October, (1999).
(5) Mudge, D., Veitch, P.J., Munch, J, Ottaway, D., and Hamilton, M.W., *IEEE J of Selected Topics in Quantum Electronics* **3**, 19-25 (1997).
(6) Siegman, A.E., *Phys. Rev. A* **39**, 1253-1263 (1989).
(7) New, G.H.C., *J. Mod. Opt.* **42**, 799-810 (1995).
(8) Mudge, D., Ostermeyer, M., Veitch P.J., Munch, J., Middlemiss B., Ottaway D., and Hamilton, M.W., *IEEE J of Selected Topics in Quantum Electronics*, submitted December 1999.
(9) Braginsky, V.B., Mitrofanov, V.P. and Panov ,V.I., *System with Small Dissipation*, The University of Chicago Press, Chicago (1985).
(10) Rowan, S., et. al, "Mechanical loss factors of material and suspension systems for advanced gravitational wave detectors", this proceeding
(11) Taniwaki, M., Ju, L., Blair, D.G. and Tobar, M.E., *Phys Lett. A*, **246**, 37-42 (1998).
(12) Locke, C., University of Western Australia, Private Communications
(13) Benabid, F., Notcutt, M., Ju, L. and Blair, D.G., *Optics Communic.***167**, 7-13 (1999)
(14) Benabid, F., Notcutt, M., Ju, L. and Blair, D.G., *Phys. Lett. A***237**. 337-342 (1998)
(15) Winterflood, J., Blair, D.G, *Physics Letters***A222**, 141-147 (1996)
(16) Winterflood, J., Blair, D.G., *Physics Letters* **A243**, 1-6 (1998).
(17) Winterflood, J., Losurdo, G., Blair D.G., *Physics Letters* **A** (1999) (in press).
(18) Buchler, B.C., Gray, M.B., Shaddock, D.A., Ralph, T.C. and McClelland, D.E., *Opt.Lett.***24**, 259-261 (1999)
(19) Shaddock, D.A., Gray, M.B. and McClelland, D.E., *Opt.Lett.***24**, 1499-1501 (1999).
(20) Whiting, B.F., Coldwell, C., Scott, S.M., Evans, B., and McClelland D.E., *Gen.Rel.&Grav.*(in press).
(21) Scott, S.M., Charlton, P., Whiting, B.F., Sandeman, R.J., and McClelland, D.E., "Detector Characterization", *Proceedings of the Second TAMA International Workshop on Gravitational Wave Detection*, Tsubono, K., Fujimoto, M.-K., and Kuroda, K., (eds), Univerrsal Academy Press (in press).

OVERVIEWS

Suspensions

Stefano Braccini

*VIRGO Project, Istituto Nazionale di Fisica Nucleare,
Sezione di Pisa, Italy*

Abstract. Special suspension systems are used in gravitational wave detectors to reduce the transmission of seismic vibrations to test masses by many orders of magnitude. In ground-based interferometric antennas, this allows to detect gravitational signals even below a few tens of Hz, where seismic vibrations are very strong. The state of the art on this topic is presented.

INTRODUCTION

The detection of gravitational waves in the low frequency range (below a few hundreds of Hz) by means of ground-based interferometric antennas is hindered by seismic noise, namely by continuous vibrations of the soil induced by geophysical phenomena and by human activities. These vibrations cause large spurious motions of interferometer test masses, masking by many orders of magnitude small displacements induced by gravitational waves.

Since many astrophysical sources, such as pulsars and coalescing binaries, are expected to emit mainly low frequency gravitational waves (from a fraction of a Hz to several hundreds of Hz), it is of great importance to lower the frequency detection threshold as much as possible. This requires to develop suspension systems for the interferometer optical components able to reduce strongly the transmission of seismic vibrations. In general, a vibration isolation suspension has to fulfill the following requirements:

- To isolate the test mass from seismic noise in all the detection band.
- Not to induce drift of the test mass position on the long period.
- To allow a noiseless remote control of the test mass position.
- Not to have internal noise mechanisms able to affect the antenna sensitivity.
- To be compatible with the high vacuum level of the antenna.

Precise specifications and technical solutions will be discussed in the next sections.

SEISMIC ISOLATION

Seismic vibrations of a point on the ground are usually of the same order of magnitude in all directions. Their linear spectral displacement, between 1 Hz and a few hundreds of Hz, has been measured to be about A/f^2, where A ranges from 10^{-6} to 10^{-9} m·Hz$^{3/2}$, depending on the location and on the time of the measurement. These vibrations are many orders of magnitude larger than the spurious displacements of the interferometer test masses induced by thermal noise (linear spectral displacements in the low frequency range of a few 10^{-18} m·Hz$^{-1/2}$). Once seismic noise is suppressed, thermal noise limits the antenna sensitivity between a few Hz and a few hundreds of Hz. In order to make seismic noise negligible it is thus necessary to attenuate the seismic vibrations transmitted to the test masses well below the thermal noise floor, starting from a few Hz.[1]

The general approach to provide an attenuation of seismic vibrations is the following. In an N-stage pendulum, well above the highest resonant frequency, the ratio between the linear spectral horizontal displacement of the last mass (test mass) and that of the suspension point (where the excitation acts) decreases with frequency as C/f^{2N}, where C is the product of the squares of the N resonant frequencies. This means that a few Hz above the highest resonance, the noise induced by horizontal seismic vibrations can be strongly suppressed. It is important to stress that as the pendulum resonant frequencies decrease, the attenuation at a given frequency increases.

Vertical vibrations are not attenuated by a chain of simple pendula. These vibrations would be partially transferred to the horizontal direction, along the interferometer beam, because of unavoidable mechanical couplings between different degrees of freedom.[2] Conceptually, vertical attenuation of seismic noise can be achieved by using the same principle, namely replacing each suspension wire with a spring, forming a cascade of vertical harmonic oscillators.

Passive Vibration Isolation Stacks

A simple and reliable application of the principles described above are the "*Passive vibration isolation stacks*", used in the two interferometric antennas of the American LIGO project [2]. A circular table, from which the test mass hangs, is supported by three isolation stacks. Each stack is formed by three stiff masses connected by elastomer springs (Fig.1). The basic design of a stack can be thought of as a series of four harmonic oscillators both in vertical and horizontal. Above the resonant frequencies a strong attenuation of seismic vibrations is thus expected in all directions.

---

[1] Below a few Hz, the detection is prevented by "gravity gradient noise", namely by the large spurious displacements of the test masses induced by fluctuations of the static gravitational field [1]. Isolation of the antenna from seismic vibrations below a few Hz is thus useless.

[2] Transmission coefficients of the test mass vertical displacements to the beam direction up to a few % have been evaluated, depending on the design of the suspension.

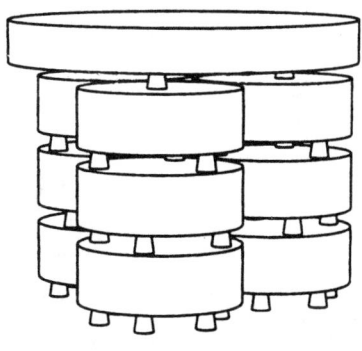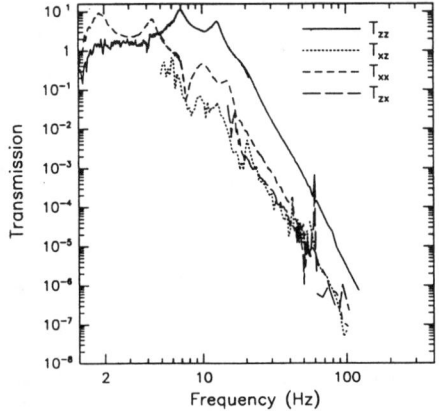

FIGURE 1. Perspective drawing of the prototype passive vibration isolation stack suspension used in the LIGO project (from reference [2]). The overall height is 68 cm and the width 78 cm. On the right side the transmission matrix elements of base motion to top motion for a single stack prototype are plotted as a function of the frequency (Z and X denote the vertical and horizontal direction respectively). See reference [2] for details.

The measurement of the modulus of the transmission matrix elements for a single stack, connecting the horizontal and vertical linear spectral displacements of the base and of the top, is reported in Fig.1. One can notice the resonant peaks (suppressed by viscoelastic damping of elastomer springs), followed by rapid roll-off with frequency. At 100 Hz, horizontal and vertical transmission (T_{xx} and T_{zz}) are 10^{-7} and $3 \cdot 10^{-6}$, respectively, while the cross-coupling terms (T_{xz} and T_{zx}) are between these values.

Low Frequency Chains

Due to the high stiffness of the elastomer springs, the resonant frequencies of Stacks are very high. As a consequence, the attenuation starts to be strong only above a few tens of Hz. The use of *"Low Frequency Chains"* has been proposed in order to achieve a strong attenuation starting from a few Hz [3,4,5]. A typical Low Frequency Chain is a cascade of mechanical filters connected each to the other by steel wires and designed to attenuate seismic vibrations in all degrees of freedom, starting from a few Hz. While strong attenuation in the low frequency range can be achieved in the horizontal directions by using long suspension wires (i.e. low frequency pendula), it is difficult to reduce the vertical resonant frequencies. Vertical springs able to support large loads with very soft spring constants have to be designed for the purpose. Using vertical cantilever springs with dimensions fitting the typical sizes of suspension systems, chains with the highest vertical mode around 10 Hz can be constructed.[3]

---

[3] The use of cantilever springs is favored by their large stability with temperature. Moreover, they can be designed to have only high frequency internal modes (above 100 Hz, where seismic noise is small).

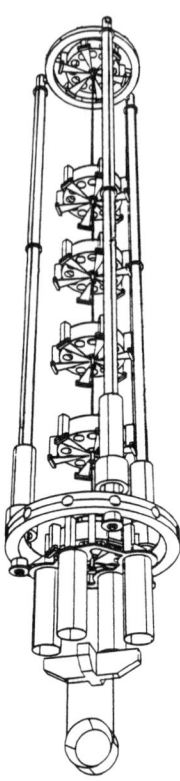

FIGURE 2. A typical Low Frequency Chain is the VIRGO Superattenuator [5]. It is essentially a cascade of pendula where each mass is replaced by a special vibration isolation mechanical filter. All the transmission curves have been measured to be below 10^{-10} starting from about 3 Hz.

Special systems of antisprings (both mechanical [6] and magnetic [7]) are used to reduce the stiffness of the springs in order to displace all the chain vertical modes below a few Hz. In the interferometric gravitational antenna VIRGO a Low Frequency Chain (named "Superattenuator") has been assembled and tested (Fig.2). It is a cascade of five mechanical filters (about 6 meters long) from which the optical payload hangs. Each filter is a rigid cylindrical tank (about 100 kg mass), supporting a set of converging cantilever blade springs and a system of magnetic antisprings. As shown in [5], all the fundamental modes of the chain are smaller than 2.5 Hz and all the transmission curves are below 10^{-10}, starting from about 3 Hz. This guarantees that residual seismic noise does not affect the antenna sensitivity above this frequency.

STABILITY OF THE SUSPENSIONS

The stability of the position of the suspended test mass on the very long period (i.e. at frequencies well below the detection band) has been widely studied. In Low

Frequency Chains, the main effect is the test mass vertical swing due to changes of suspension length induced by temperature variations. Daily vertical swings of the test mass of a few hundreds of microns occur in the vacuum tank containing the suspension (where temperature variations are less than 0.1 °C peak to peak). Long period vertical swings of test masses up to a few mm can be compensated by the control system used for the interferometer locking.

The thermal swing of the Stacks, which become shorter with increasing temperature, has been measured to be much smaller (tens of microns per °C) [2]. In the Stacks, the dominant effect is a lowering of the test mass vertical position of a few mm per year due to the plastic deformation of the elastomer springs. This very long term compression of the springs can be compensated with periodic adjustments to the height of the isolation system support structure.

SUSPENSION CONTROL

At frequencies below the attenuation range, seismic noise is transmitted to the test mass amplified by suspension mechanical resonances. The amplitude of the low frequency swing increases with the quality factor of the resonant peaks in the transmission curve. While the viscoelastic damping of the elastomer springs strongly reduces the quality factor of the resonant peaks in the Stacks (see Fig.1), in Low Frequency Chains the amplification of seismic noise can be very high. Test mass displacements along the beam direction of a few tens of microns occur in the frequency range of resonances (typically between a few tens of mHz and a few Hz). The active control system of suspension has to compensate for this large swing in order to keep the interferometer in the working position. The specification is to chill the root mean-squared displacement of the test mass along the beam direction down to 10^{-14} m when the interferometer is locked. The error displacement signal to be compensated by the active system is provided by the interferometer.

It is impossible to perform a so large reduction of the swing (from tens of microns to 10^{-14} m) acting directly on the test mass and thus with a plain control strategy. The finite dynamics of actuation systems implies that too large compensation forces, even if in the low frequency range or in DC, increase the noise floor in the detection band. The maximum displacement which can be compensated by acting directly on the test mass without affecting the interferometer sensitivity is only of a few nanometers. The techniques to reduce the test mass horizontal swing down to nanometers are illustrated in this section.

Ultra Low Frequency Pre-Isolator Stages

A reduction of horizontal seismic noise of test mass can be achieved even in the frequency range of resonances by suspending the entire chain from a pre-isolator stage, conceived to have an horizontal resonant frequency of a few tens of mHz or lower.

A typical horizontal pre-isolator stage is the inverted pendulum. An ideal inverted pendulum is made of a massless vertical beam of length l, connected to ground by means of an elastic joint of stiffness k, and supporting a mass M on its top. In such a pendulum, gravity acts as an antispring and the resonant frequency

$$f_0 = \frac{1}{2\pi}\sqrt{\frac{k}{M} - \frac{g}{l}}$$

can be lowered down to mechanical instability by adding load. In the VIRGO Superattenuators, a 6 m-high system made of three inverted pendula supporting a top table, at which the entire filter chain is attached, will be used (Fig.2). The system, described in [8], operates at a frequency of 30 mHz, providing an attenuation of horizontal seismic vibrations by about 40 dB at frequencies around 300 mHz.

A long period horizontal pendulum based on Scott-Russel linkage has been developed at University of Western Australia. In this system, described in [9], gravity, mechanical restoring forces and mechanical reactions counterbalance so that the total horizontal stiffness is very small. A period in excess of 20 seconds has been measured.

On similar principles, researchers of the TAMA interferometric antenna in Japan, have developed a pre-isolator stage, named "X pendulum". This system, described in [10], operates with a period of about 10 s. It has the advantage to be very compact, even if it exhibits many low frequency internal resonances.

Thanks to their very low horizontal stiffness, pre-isolator stages can be moved by small forces. This means that they are suitable for the horizontal remote adjustment of the filter chain suspension point.

Active Inertial Platforms

The use of active methods to reduce input seismic noise at the chain suspension point has been proposed. The idea is to measure the acceleration of the suspension point by accelerometers. A feed-back force is provided by coil-magnet actuators to compensate the acceleration measured. This method (limited by the noise of the sensors) can be applied both in the low frequency range (to reduce the large swing of the test mass) and in the detection band.

The main problem is due to the couplings between different degrees of freedom. Closing a loop in a given degree of freedom affects the mechanical response (and thus the loop design) along the other ones. For this reason it is very important to minimize mechanical couplings in suspension design.[4]

In the advanced LIGO configuration, an inertial top platform resting on three isolation stacks will be used for each test mass [12]. At frequencies smaller than 10 Hz seismic noise on the top table will be lowered by more than one order of magnitude by

---

[4] In order to minimize the interference between different loops, special control techniques to "diagonalize" the response of a mechanical system are used [11].

using seismographs and coil-magnet actuators. At higher frequency, the active control system will attenuate linear spectral displacements of inertial platform down to about 10^{-14} m·Hz$^{-1/2}$. This strong active reduction of input seismic noise will allow to extend the LIGO detection band down to about 10 Hz just by a small passive attenuation achieved by a few stage Low Frequency Chain.

In the VIRGO antenna, use of coil-magnet actuators and very sensitive accelerometers is made to reduce low frequency displacements of the top table of the inverted pendulum [11]. With this technique, the horizontal low frequency swing of the test mass, on time scale of tens of seconds, is reduced from a few tens of microns to about one micron.

Hierarchical Active Control

A fruitful strategy, used for instance in the British-German antenna GEO [13], is to compensate the test mass displacements larger than one nanometer by acting on the upper stages of the suspension. As mentioned above, the forces required to compensate these large displacements induce a strong electro-mechanical noise in the detection band. But this noise is injected to the actuation stage and thus its transmission to the test mass is suppressed by the mechanical filters below.

Seismic noise increases at low frequencies. Very large seismic displacements (above a few tens of microns) are expected in the very low frequency range (below a few tens of mHz). They can be compensated by acting on the suspension point by means of large dynamics actuators. The transmission of the induced mechanical vibrations to the test mass is strongly reduced by the entire filter chain.

In the resonance frequency range (between a few tens of mHz and a few Hz) seismic displacements with amplitude up to a few tens of microns are expected. As mentioned above, a preliminary reduction down to about one micron can be performed in this frequency range by means of pre-isolator stages and (or) active systems on the suspension point. Once the residual swing has reached this level, it can be compensated by a not too difficult control strategy, acting on the stage immediately above the test mass. The mechanical noise induced by the compensation force is filtered below the antenna displacement sensitivity by the test mass pendulum suspended from the actuation stage.

Displacements smaller than one nanometer, that can be easily compensated in a noiseless way by acting directly on the test mass, are expected at frequencies larger than a few Hz (i.e. in the detection band).

INTERNAL NOISE SOURCES

Noise processes occur inside suspensions, inducing spurious vibrations of test masses. For instance, stress from load acting on the vertical springs and on suspension wires induces plastic strains. This effect, named micro-creep, is due to a series of microscopic yielding events that induce a mechanical shot-noise on the test mass. The use of special precipitation hardened steels allows the construction of springs and wires showing an acceptable stability under high stress [14]. Elastomer springs also exhibit microcreep under load. For this reason the Stacks are usually designed to keep the strain in the springs well below the elastic limit (a few tens %). Very sensitive acoustic and optical techniques [15,16] are used to detect and study the small pulses induced by single creep events on the test mass.

Motor controlled mechanisms with large dynamics (several mm) are necessary to adjust the position of the suspension items on the very long period. These systems induce strong mechanical vibrations. They can be employed only in the upper part of the suspensions so that the induced vibrations are filtered by the stages below.

For the fine adjustment of test mass position (displacements below a few tens of microns) coil-magnet actuators are usually employed. It is important to assemble the coils on recoil masses hanged from the suspension and thus isolated from seismic vibrations [13]. This allows to avoid the injection of seismic noise during the actuation.

In order to reduce mechanical vibrations induced by couplings to varying external magnetic fields, magnets used in suspensions (in antisprings or coil-magnet actuators) have to be arranged so as to minimize the dipole moment [17]. For the same reason, use of shielded and twisted pair cables for the control system is recommended.

VACUUM COMPATIBILITY

In order to reduce to negligible values the noise due to fluctuations of the light optical path induced by variations of the gas density, interferometric antennas will operate at a total pressure less than 10^{-8} mbar. The hydrocarbon partial pressure has to be smaller than 10^{-13} mbar in order to avoid the long term pollution of the interferometer optical components.[5]

Steels and other standard materials used for mechanical constructions exhibit acceptable outgassing rates. Trouble arises from glues (to attach magnets or other components), resins and rubbers (to build self-damped springs), magnets (for coil-magnet actuators or antisprings), motors (for remote adjustment mechanisms) and varnishes (used for sealing or attaching coil winding). In order to preserve the vacuum quality, all components have to be selected by severe vacuum tests. The outgassing rates of several components used in suspensions can be found in internal reports of the main gravitational wave detector collaborations.

CONCLUSIONS

The attenuation performances, the stability and the vacuum compatibility of prototypes of Stacks and Low Frequency Chains have been successfully tested. These

---

[5] This stringent specification is very conservative and probably could be relaxed.

systems will be used in the ground-based interferometric antennas presently under construction in order to isolate test masses from seismic vibrations. The design for the noiseless control of the test mass position has been defined in all these projects. However, only the interferometers will be sensitive enough to check if the noise induced by the control system and by spurious mechanisms occurring inside the suspensions will not affect the sensitivity of antennas.

REFERENCES

1. Beccaria, M., et al., *Classical and Quantum Gravity* **15**, 3339-3362 (1998).
2. Giaime, J., et al., *Rev. Sci. Instr.* **67**, 1, 208-214 (1996).
3. Ju, L., and Blair, D.G., *Rev. Sci. Instr.* **65**, 11, 3482 - 3488 (1994).
4. De Salvo, R., in this volume.
5. Frasconi, F., et al., "Performances of the R&D superattenuator chain of the VIRGO experiment", to be published in *Proceedings of the XXXIVth Rencontres de Moriond*, 1999.
6. Winterflood, J., and Blair, D.G., *Phys. Lett. A* **243**, 1-6 (1998).
7. Beccaria, M., et al., *Nucl. Instr. and Meth. in Phys. Res. A* **394**, 397-408 (1997).
8. Losurdo, G., et al., *Rev. Sci. Instr.* **70**, 5, 2507-2515 (1999).
9. Winterflood, J., and Blair, D.G., *Phys. Lett. A* **222**, 141-147 (1996).
10. Barton, M.A., Kanda, N., Kuroda, K., *Rev. Sci. Instr.* **67**, 11, 3994-3999 (1996).
11. Losurdo, G., in this volume.
12. Richman, S.J., et al., *Rev. Sci. Instr.* **69**, 6, 2531-2538 (1998).
13. Plissi, M.V., et al., *Rev. Sci. Instr.* **69**, 8, 3055-3061 (1998).
14. Beccaria, M., et al., *Nucl. Instr. and Meth. in Phys. Res. A* **404**, 455-469 (1998).
15. Braccini, S., et al., "Preliminary measurements for the characterization of the Virgo filters by acoustic technique", submitted to *Measurement Science & Technology*, 1999.
16. Ageev, A.Yu., et al., *Phys. Lett. A* **246**, 479-484 (1998).
17. Drever, R.W.P., "Techniques for Extending Interferometer Performance Using Magnetic Levitation and Other Methods" in *Proceedings of the Conference on Gravitational Waves Sources and Detectors*, Cascina (Pisa), Italy, 19-23 March 1996, pp. 316-320.

VIRGO SUSPENSION R&D: FUSED SILICA AND CREEP

L. Gammaitoni[a,b], J. Kovalik[a], F. Marchesoni[a,c], M. Punturo[a] and G. Cagnoli[a,b]

(a) Istituto Nazionale di Fisica Nucleare, Sezione di Perugia, VIRGO Project, Perugia (Italy)

(b) Dipartimento di Fisica, Universita' di Perugia, Perugia (Italy)

(c) Istituto Nazionale di Fisica della Materia, Universita' di Camerino, Camerino (Italy)

Abstract. The sensitivity of the VIRGO interferometer is limited by thermal noise between 10 Hz and 50 Hz which can be reduced by using low loss fused silica suspensions. The excess noise due to creep in steel wire suspensions has been examined and the upper limits indicate that it is not a significant noise source. The anelastic elongation of fused silica fibres does not seem to pose a problem for long term drift of a suspensions made of this material.

I FUSED SILICA SUSPENSIONS

The main limit to the VIRGO sensitivity in the 10–50 Hz range is given by the thermal noise of the last stage suspension pendulum mode. There are two dissipation mechanisms that must be reduced: frictional losses in the upper and lower wire clamps and the material internal loss angle.

Clamp losses have been reduced in VIRGO using special high pressure clamps for the current C85 steel suspension wires, but a monolithic solution (wires with large heads) has been demonstrated to be the best solution.

The dependence of the material internal loss angle ϕ_w and the material breaking strength T_b on the thermally induced fluctuations $X(\omega)$ of the VIRGO test mass enters through the formula:

$$X^2(\omega) \approx 4k_B T \cdot \frac{g}{\omega^5} \cdot \frac{1}{L^2} \sqrt{\frac{Eg}{4\pi Nm}} \cdot \frac{\phi_w}{T_b C_S} \tag{1}$$

where only the pendulum horizontal mode has been considered. The current VIRGO reference solution is to use C85 steel wire (with a breaking strength of ~ 2.9 GPa for 0.2 mm diameter) which shows the best ratio of ϕ_w/T_b [2].

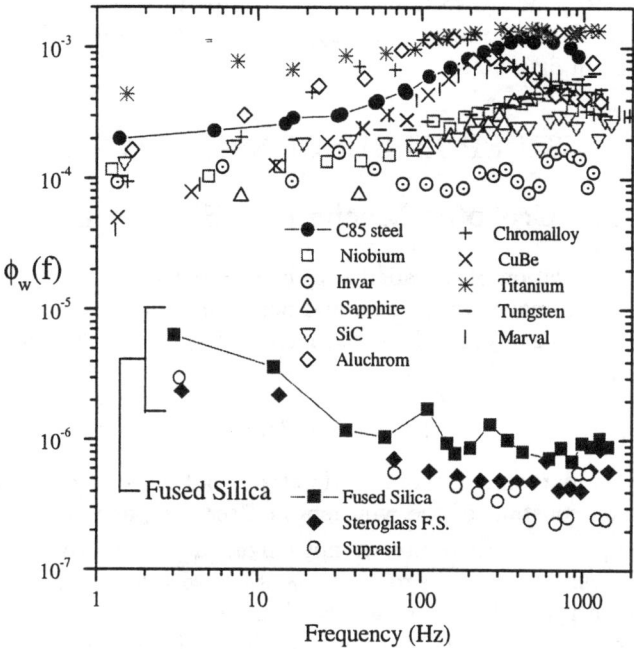

FIGURE 1. Loss angle vs frequency for several materials.

From figure 1, it is clear that fused silica is the best candidate for the suspension of the VIRGO last stage since at low frequencies it has a $\phi_w \sim$ 50–100 times lower than C85 which implies a factor 7–10 improvement in terms of thermal noise. Furthermore, it is relatively easy to produce monolithic suspensions with fused silica, while it has been impossible to do so using C85 steel. The low ϕ_w of fused silica has been demonstrated in a medium scale prototype in Perugia (in collaboration with the GEO600-Glasgow group), by suspending a 2.8 kg pendulum with two fused silica wires 0.4 mm in diameter and 300 mm in length. With this system, a $Q \approx 2.3 \cdot 10^7$ (possibly limited by clamp losses) has been measured.

The breaking strength of fused silica is really a puzzle. It strongly depends upon how the wire has been produced, how it has been stored and how it has been handled. In fact, by producing wires with a graphite susceptor in a RF oven, we obtained a wire breaking strength of only 300 MPa due to carbon impurities released by the oven in an argon atmosphere. Wires produced by pulling fused silica rods with an $O_2 - H_2$ flame, show a very high breaking strength: $T_b \geq 3$ GPa. With four 0.2 mm diameter wires, we suspended a 21 kg VIRGO dummy mirror. The positive result of this test is promising and the VIRGO collaboration will modify the mirror design so that it can be compatible with a monolithic fused silica suspension. The Virgo sensitivity using either C85 or fused silica suspension wires is compared in figure 9.

II CREEP IN SUSPENSION WIRES

A Creep as Mechanical Shot Noise

The anelastic elongation of a loaded wire can be modeled as a series of instantaneous glitches or slippages with size q and occurring at a rate λ (see Marchesoni et al. [4] for a detailed analysis). The static length $l(t)$ of a loaded wire is modeled to grow with time as

$$\tau_l \ddot{l} = -\dot{l} + \eta_S(t), \qquad (2)$$

where $\beta = 1/\tau_l$ is a damping constant and $\eta_S(t)$ a shot noise source. The constant β is the relaxation constant of the wire length after a slippage has occurred: the wire glitch takes place in a time interval not longer than l_0/c_l, where $c_l = \sqrt{E/\rho}$ is the longitudinal sound velocity in rods and l_0 is the wire rest length. The average τ_l will be $l_0/2c_l$ with $\beta = 10^4$ s^{-1} for a typical metallic wire.

The shot noise $\eta_S(t)$ can be represented by a Poissonian sequence of equal pulses $h(t - t_i)$, each associated with a glitch of the wire. For simplicity, the Poisson rate is λ and the η_S pulse at time t_i has the following shape

$$h(t - t_i) = \frac{q_S}{\tau_S} \exp[-(t - t_i)/\tau_S] \qquad (t > t_i) \qquad (3)$$

with integrated intensity q_S (glitch size). Moreover, we assume that $\tau_l \gg \tau_S$, namely the grain slippage duration τ_S is much shorter than the propagation time τ_l, so that $\lim_{\tau_S \to 0} h(t-t_i) = q_S \delta(t-t_i)$. Note that Eq. (2) can be easily integrated: due to a single pulse, l increases by

$$\Delta l = \int_{t_i}^{\infty} h(s - t_i) ds = q_S, \qquad (4)$$

whence the step-wise growth of $l(t)$.

The power spectral density (p.s.d.) $S_S(\omega)$ of the stochastic force $\eta_S(t)$ is

$$S_S(\omega) = 2\pi q_S^2 \lambda^2 \delta(\omega) + \frac{q_S^2 \lambda}{1 + (\omega \tau_S)^2}. \qquad (5)$$

so that $\langle \eta_S \rangle = q_S \lambda$ and $\sigma_S^2 = \langle (\eta_S - \langle \eta_S \rangle)^2 \rangle = q_S^2 \lambda / 2\tau_S$. The δ-like term of $S_S(\omega)$ corresponds to a linear increase $l_{dc}(t) = q_S \lambda t$ of the wire length $l(t)$, while the Lorentzian curve on the r.h.s. of Eq. (5) characterizes its *zero-mean* noisy component $\delta_l(t)$.

The VIRGO suspension control system compensates for the stationary creep which corresponds to subtracting the d.c. component of $l(t)$. The p.s.d. $S_\delta(\omega)$ of the subtracted wire length $\delta_l(t) = l(t) - l_{dc}(t)$ can be written as

$$S_\delta(\omega) = \frac{1}{1+(\omega\tau_l)^2} \frac{S_S(\omega)}{\omega^2}. \tag{6}$$

At low frequencies, $\omega\tau_l \ll 1$ and $\omega\tau_S \ll 1$,

$$S_\delta(\omega) \simeq q_S^2 \frac{\lambda}{\omega^2}. \tag{7}$$

The vertical displacement $z(t)$ of a mass suspended by an elastic wire with constant static length, is given by the differential equation

$$\ddot{z} = -\gamma\dot{z} - \omega_0^2 z + \omega_0^2 \delta_l(t), \tag{8}$$

where $\nu_0 = \omega_0/2\pi$ is the resonant frequency of the vertical mode and $\gamma = \gamma(\omega)$ is the damping constant due to the internal friction, $\gamma(\omega) \propto \phi(\omega)/\omega$.

The p.s.d. for the $z(t)$ becomes

$$S_z(\omega) = |\chi(\omega)|^2 \, S_\delta(\omega), \tag{9}$$

where

$$\chi(\omega) = \omega_0^2/[(\omega_0^2 - \omega^2) + i\gamma\omega]. \tag{10}$$

The corresponding vertical fluctuation amplitude $|z(\omega)|$ is:

$$|z(\omega)| \simeq \frac{\omega_0^2}{\sqrt{(\omega^2 - \omega_0^2)^2 + \gamma^2\omega^2}} \frac{q_S\sqrt{\lambda}}{\omega}, \tag{11}$$

or, for $\nu_0 \ll \nu \ll \tau_l^{-1}, \tau_S^{-1}$,

$$|z(\nu)| \simeq \left(\frac{\nu_0}{\nu}\right)^2 \frac{q_S\sqrt{\lambda}}{2\pi\nu}. \tag{12}$$

which is Eq. (7) filtered by the vertical transfer function of the wire.

The minimum vertical-horizontal coupling factor is due to the earth's curvature over the distance between two VIRGO mirrors, $L = 3$ km, and is of the order of $\theta = 3 \cdot 10^{-4}$rad (a larger coupling factor ($10^{-2} - 10^{-3}$) could come from the mechanical imperfections on the suspension chain):

$$|x(\nu)| = \theta \, |z(\nu)| \tag{13}$$

By adding in quadrature the fluctuations created by the mechanical shot noise on each mirror of the interferometer and introducing the geometric factor $\kappa_m \geq 2$ which accounts for the two wire loops of in the suspension, the the contribution of creep noise to the VIRGO sensitivity curve becomes

$$\tilde{h}_C(\nu) = \theta \frac{\kappa_C}{\kappa_m} \left(\frac{\nu_0}{\nu}\right)^2 \frac{q_S}{L} \frac{\sqrt{\lambda}}{2\pi\nu}. \quad (14)$$

The wire elongation or $q\lambda$ can be measured by loading a wire and recording its change in length. Typically, this creep velocity depends on the wire load, temperature and their respective histories. By heating the wire under load, one can artificially age the wire and discharge most of the creep. The only problem that remains is how to estimate the residual creep noise. From Eq. (11), one must now either measure q or λ individually and not just $q\lambda$.

In this case, a mechanical resonance which is driven by the creep noise was constructed. The excitation of a resonance driven by creep noise produces a spectrum

$$x(f) = \frac{q\sqrt{\lambda}}{2\pi f \sqrt{(1 - (f/f_0)^2)^2 + 1/Q^2}} \quad (15)$$

where f_0 is the frequency and Q is the mechanical quality factor of the resonance. The resonant frequency and Q is easily found by performing a transfer function measurement of the system. The displacement noise spectrum is measured interferometrically as is the wire elongation rate. With these parameters known, one can separate the glitch size q from the data.

The mechanical system consists of a light mass attached to a wire that is loaded using a spring (see fig. 2). This allows the wire to have a high load without using a heavy mass which in turn gives the system a high resonant frequency. The high resonant frequency allows good isolation from seismic and acoustic noise. The wire in this case was C85 tool steel approximately 3 cm in length and loaded with 7 kg.

A mirror was placed on the mass of the resonant system and it formed the corner of a Mach Zender interferometer (see fig. 3). An optical fibre was used to introduce light from a 100 mW Nd:YAG laser at 1.06 μm into a vacuum system. This was phase modulated at 12 MHz and the noise floor of the interferometer was approximately $5 \times 10^{-15} m/\sqrt{Hz}$. The interferometer was built on an aluminum plate 30cm by 30 cm in area and was suspended by 4 viton cords about 1 m in length. The vacuum system into which the interferometer was placed sat on an optical table.

The transfer function of the resonance was found by exciting the mass electrostatically and is shown in figure 4. The measured Q was 762 ± 5 and the frequency 780.78 ± 0.02 Hz. The elongation of the wire was measured by turning off the servo locking the interferometer and recording the drift in the interferometer signal as shown in figure 5. The measured elongation rate was $2.1 \pm 0.2 \times 10^{-10} m/s$.

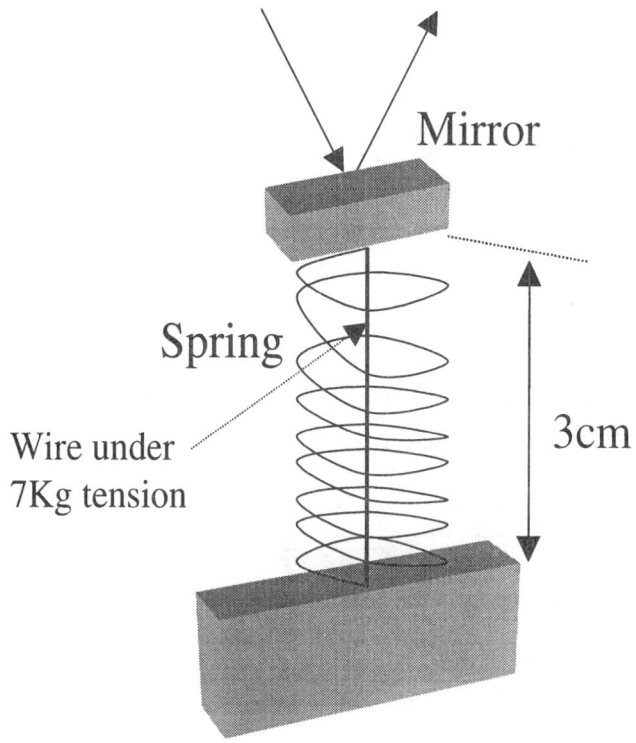

FIGURE 2. Sketch of resonant system where a mass is attached to a wire that is loaded by a spring.

The displacement noise spectrum is shown in figure 6. The noise is fitted using equation. The resulting value for q is $1.2 \pm 0.1 \times 10^{-15} m$ and λ is $1.8 \pm 0.1 \times 10^5 s^{-1}$. The measured noise is consistent with the thermal noise for that system. Thus, the resulting value for q is only an upper limit since the actual creep noise is probably much less than the thermal noise.

The very small value for q is much less than a microscopic process like a grain slippage that would actually cause a creep glitch. It is likely that the glitch is somewhere in the wire and only affects a small part of the wire cross section. When the slippage occurs, the rest of undamaged wire cross section elastically deforms slightly and compensates for the defect that is produced. The high rate of creep glitch events indicated by this experiment suggests that the shot noise model is probably close to describing the statistics of this process. The foreseen creep noise contribution to the VIRGO sensitivity curve is shown in figure 9.

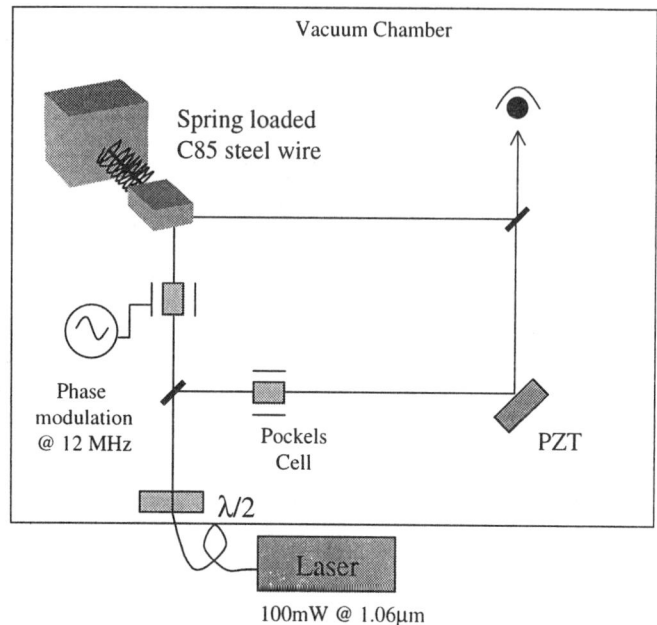

FIGURE 3. Sketch of Mach Zender interferometer used to measured the motion of the resonant system.

The only remaining question to answer is whether any infrequent, large creep events take place which could create an impulse like signal in a gravity wave interferometer.

B Creep in Fused Silica Fibres

The creep of fused silica fibres while under load is not a well measured phenomena. In this experiment, a fused silica fibre was produced from a 5 mm rod by heating it with an oxygen-hydrogen torch and then pulling it by hand. The resulting fibre was approximately 61 cm in length and was approximately 200μm at its thinnest point in the middle of the fibre and slowly tapering to about 900μm at the point which it rapidly expands back into the 5 mm rod. The majority of the wire length was between 200μm and 400μm in diameter.

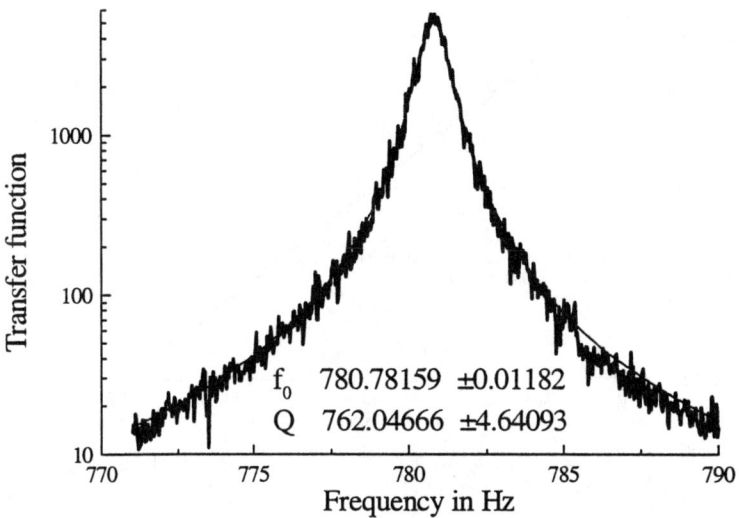

FIGURE 4. Plot of mechanical transfer function obtained by exciting resonance electrostatically.

The wire was then hung from a three prong 8 mm chuck. The chuck was attached to a 1 m long INVAR vertical bar. A 5.5 kg mass was attached to the bottom of the fibre also using a three prong 8 mm chuck. The position of the mass or wire length was measured with a stabilized shadow meter. The system was covered with a steel tube and evacuated. The temperature of the tube was then heated to 34.5 degC and stabilized.

Figure 7 shows the temperature and fibre length as a function of time. Figure 8 shows the fibre length as a function of time on a larger scale. The majority of change in the wire length is coming from temperature fluctuations. In fact, it is not clear why the long term effect is a shortening of the wire. In absolute terms, the wire length did not change by more than 0.5 μm in over twenty days which indicates that the fused silica fibre is creeping at an insignificant rate to cause any long term stability problems for suspending masses.

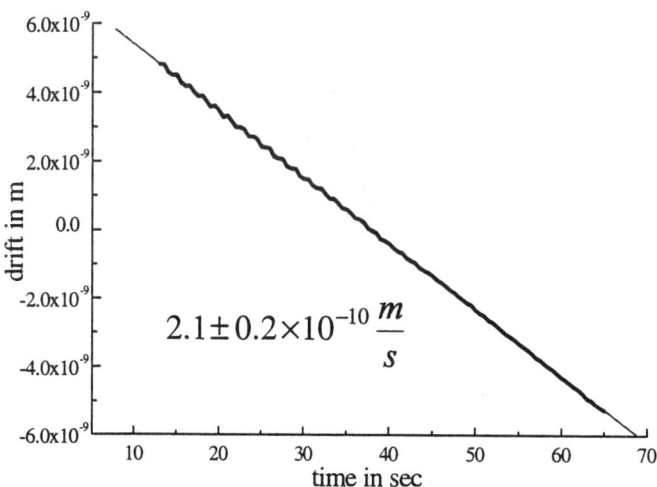

FIGURE 5. Plot of wire elongation vs. time.

REFERENCES

1. G. Cagnoli, L. Gammaitoni, J. Kovalik, F. Marchesoni and M. Punturo, Phys. Lett. **A213**, 245 (1996)
2. G. Cagnoli, L. Gammaitoni, J. Kovalik, F. Marchesoni and M. Punturo, "Low Frequency Internal Friction in Clamped–Free Thin Wires", Physics Letters A **255**, 230 (1999).
3. P. R. Saulson, Phys. Rev. **D42** 2437 (1990).
4. G. Cagnoli, L. Gammaitoni, J. Kovalik, F. Marchesoni, M. Punturo, S. Braccini, R. De Salvo, F. Fidecaro and G. Losurdo, Physics Letters A **237**, 21 (1997).

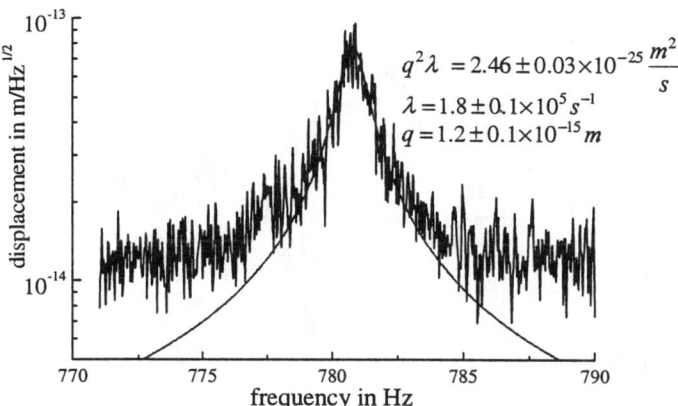

FIGURE 6. Plot of displacement noise of spring-wire mass-resonance. The noise level corresponds to the thermal excitation of that resonance.

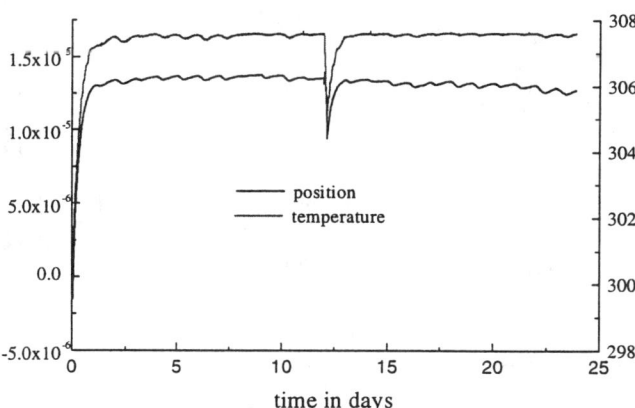

FIGURE 7. Plot of temperature of column supporting fused silica fibre and length of fused silica fibre vs. time.

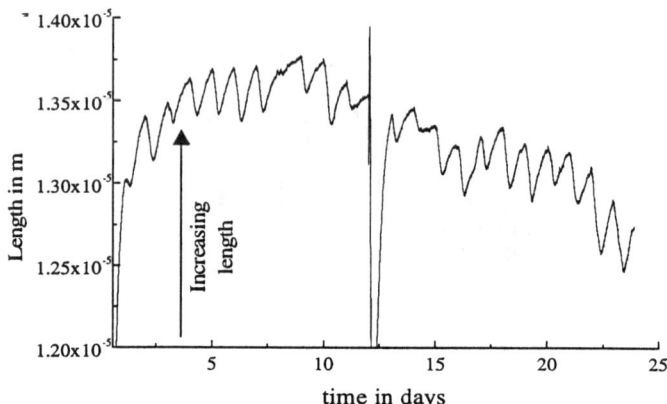

FIGURE 8. Creep of a 60 cm long 200 μm diameter fused silica wire loaded with 5.5 kg.

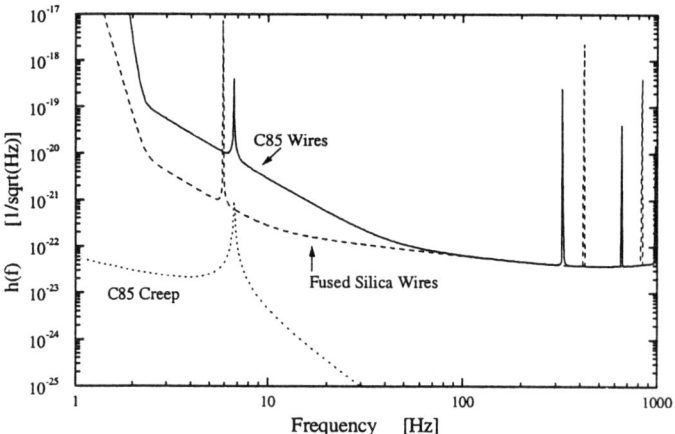

FIGURE 9. Virgo sensitivity curve: the total sensitivity and the creep contribution are plotted for C85 suspension wires; the total sensitivity for fused silica suspension wires is also plotted.

Interferometer Configurations – An Overview

Kenneth A. Strain

University of Glasgow, Glasgow G12 8QQ, Scotland.
email: k.strain@physics.gla.ac.uk

Abstract. The configuration of an interferometer, i.e. the disposition of the optical components which compose it, is chosen to provide sensitivity to a wide range of signals. Two basic configurations – Michelson and Sagnac interferometers – are described, as well as a range of advanced 'extensions' which improve performance. The main limitations to sensitivity are discussed and methods of pushing the limits in the future are introduced.

WHAT DRIVES INTERFEROMETER CONFIGURATION DESIGN?

The strains produced by gravitational waves are expected to be small [1] and as a result interferometer technology has been driven strongly toward more and more fundamental limits. At the same time it is desired to target a range of signal types: narrow or broad band with known or unknown spectra. For example, neutron-star – neutron-star binary inspirals would be broad-band signals with a known spectrum, whereas pulsars would be narrow-band and have either known or unknown frequencies. This suggests that either a versatile optical system, or several interferometers with different responses would be beneficial.

The main limits to sensitivity can be divided into those which cause movements of the mirrors (or 'test-masses'), independent of any sensing method, and those which limit the performance of the interferometric measurement. The most significant limitations being, in the first category: thermal noise, and in the second category: the Standard Quantum Limit.

Thermal noise in the observation band arises since the mirrors and their pendulum suspensions have finite mechanical dissipation [2]. The resulting motion, away from mechanical resonances which should be placed outside the desired observation band, is minimised by careful choice of materials and by employing construction techniques which minimise the dissipation. In many materials the effect can be reduced by cooling the system, but this has not yet been implemented in long baseline interferometer systems. The use of low-loss materials and techniques

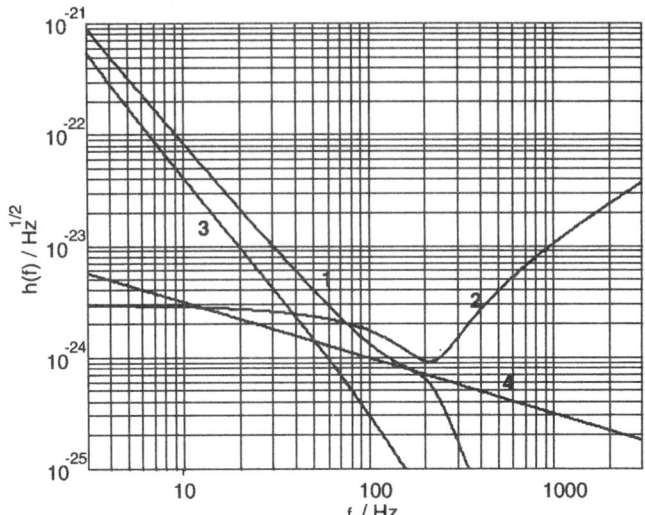

FIGURE 1. *The main contributions to the sensing noise – shot noise (curve 1) and radiation pressure noise (curve 2) – are plotted. The graph shows equivalent strain noise spectral density as a function of frequency. The sensing limit may be compared to the test mass motion caused by dissipation in the suspension (curve 3) or in the mechanical structure of the mirror (curve 4). The parameters used were similar to those proposed for LIGO II (30 kg sapphire mirrors with a material loss of $\phi = 5 \times 10^{-9}$, suspended from fused-silica ribbon fibres with a material loss of $\phi = 3.3 \times 10^{-8}$. The interferometer arms were taken to be 4 km long. Signal recycling is employed to give a broad response peaked at $\sim 200\,Hz$*

places a practical limit on the allowed mirror mass, with present technology this limit is of order 30 kg.

The sensing limit for the interferometric measurement is set by the Standard Quantum Limit, if classical detection techniques are used. For a given mirror mass, the relationship between *shot noise* and *quantum radiation pressure noise* is fixed. To push down this limit requires either heavier mirrors or Quantum Non-Demolition methods [3].

A sensitivity example. The performance that may be expected using classical, room temperature techniques and 30 kg mirrors has been estimated and is shown in figure 1. It can be seen that the sensing limit approaches to the level of thermal motion expected, for the typical parameters chosen. These parameters reflect the state of the art for planned 'second generation' detector systems [4].

BASIC INTERFEROMETER CONFIGURATIONS

There are two fundamental arrangements that have been considered for use in interferometric detectors, Michelson and Sagnac interferometers.

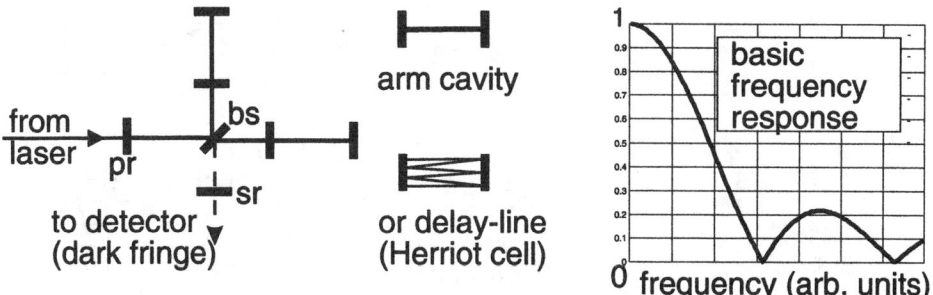

FIGURE 2. *A Michelson interferometer is shown with Fabry-Perot cavities in the arms, and both power recycling (pr) and signal recycling (sr) mirrors added (left). Delay-lines could be used instead of cavities (centre). The frequency response of a basic Michelson interferometer is peaked at zero frequency (right).*

Michelson Interferometers are the conventional choice, used in all existing interferometric detectors. They have symmetry between the two arms, but have separate components defining the two paths taken by the light (are not common path) and therefore are susceptible to differences between the components in the two arms (e.g. mirror figure) that lead to imperfect interference. The performance of Michelson interferometers can be enhanced by incorporating delay-lines or Fabry-Perot cavities in the arms, and by adding power and signal recycling (see below). The Michelson interferometer tends to give its peak of response at low (near zero) frequency. The basic Michelson interferometer is shown in figure 2. See below for a description of the extensions also shown in the figure.

The Sagnac interferometer has been proposed as an alternative that offers a few potential advantages due to its 'common path' nature. (Each of the separated beams travels around the same paths and visits the same components, but in counter-rotating directions.) This makes this system relatively insensitive to figure errors and similar defects. It also eliminates the need to actively control the state of the interference – one can obtain a dark-fringe condition by design. Unfortunately these 'common-path' benefits are mostly lost when the enhancements (arm-cavities, recycling techniques) needed to achieve the required level of performance are added. Another difficulty with Sagnac systems is that the peak of response tends to be close to the reciprocal of the light travel time τ for a single round trip of the arms. To obtain peak response in the interesting ~ 100 Hz frequency range requires very long paths in the arms. See figure 3.

EXTENDING THE BASIC CONFIGURATIONS

One of the earliest and most obvious extensions to the basic interferometer systems was the idea of putting optical delay-lines in the arms to increase their effective length. This is a simple way to extend storage time in arms towards the optimum

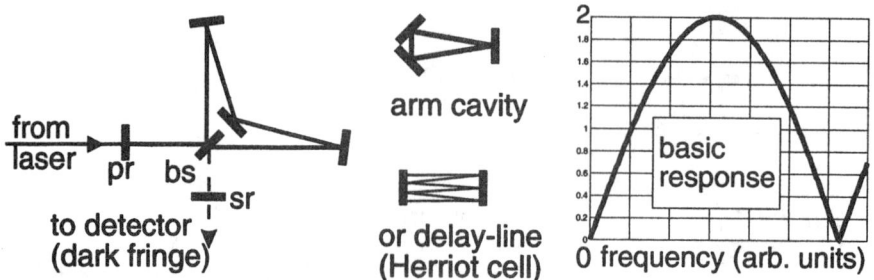

FIGURE 3. *A Sagnac interferometer is shown with both power recycling (**pr**) and signal recycling (**sr**) mirrors added (left). Delay-lines or ring cavities can be used in the arms (centre). The frequency response of a basic Sagnac interferometer is peaked at non-zero frequency (right).*

semi-period of the gravitational waves. This technique works well for small number of reflections (3 ∼ 7), but with increased risk of multiple scattering and trapped paths for stray light for larger numbers of reflections.

A slightly less obvious approach is to put Fabry-Perot cavities in the arms of the interferometer. A single optical mode is formed and this minimises problems due to scattering, while achieving similarly long storage times for the light. The disadvantage is that the cavities must be kept on resonance with the light and so control is more complicated than with delay lines. Also the simplest solution requires light to pass through substrates and this leads to thermal problems (see below).

The development of ultra-low-loss mirrors in the last two decades has allowed the technique called 'power recycling' to be usefully applied [5]. The interferometer is operated with the output at a dark fringe (interference minimum) with a modulation about that point to extract the signal. A partially transmitting 'power recycling mirror' is added to the highly reflective interferometer to form a resonant cavity in which the light power can be enhanced. This maximises the light energy in the system for a given laser power and mirror losses.

The final 'extension' considered here is the technique called signal recycling [6]. A 'signal recycling' mirror is placed at the output of the interferometer. Thus another cavity is formed in which the light that would otherwise reach the photodiode can resonate. This light is composed of (GW) signal components and 'waste' light due to imperfect interference. Since the signal-related light components are in a resonant system the frequency response of the interferometer is modified. Tuning the resonance, by microscopic adjustments of the position of the signal recycling mirror, allows the interferometer response to be targeted at expected signal frequencies, not tied to peak response at zero frequency (or $1/\tau$ for Sagnacs). It is also possible to vary the system bandwidth by changing the reflectivity of the signal recycling mirror. This can be used to compensate either for too short storage time in the arms (in which case the technique is called dual recycling) or too long storage time (when the technique is called resonant side-band extraction) [7].

The signal recycling mirror can also stabilise the spatial mode of the system, and reduces the amount of waste light reaching the photodetector [8].

SENSING LIMITS IN CLASSICAL INTERFEROMETERS

Two 'quantum' processes determine the sensing limit for interferometers using classical detection techniques. The limit is obtained for a given test mass, interferometer configuration and response. The first of these processes has played a major role at all phases of the development of the field. It is the shot noise, seen as the Poissonian fluctuations in the detected light. The power spectral density of this noise can be expressed in several ways, for example as the single-sided noise-equivalent dimensionless strain power spectral density:

$$\tilde{h}^2_{\text{shot}}(f) \geq \frac{2}{\pi^2\,\tau\,\nu^2 \mathcal{N}_\text{p}\,\mathcal{G}^2(f)}, \tag{1}$$

where $\mathcal{N}_\text{p}$ is number of photons of frequency ν stored for time τ in the arms of the interferometer and $\mathcal{G}(f)$ describes the normalised frequency response of the system (unity for a simple Michelson interferometer at low frequency). This expression, applicable to any classical interferometer configuration, is discussed more fully elsewhere [9].

Power recycling, especially in combination with cavities in the arms of the interferometer, enables a large 'carrier' storage time τ to be obtained for the photons. This allows one to use the minimum optical power to obtain a desired number of photons $\mathcal{N}_\text{p}$.

Adding signal recycling provides flexible and 'independent' choice of $\mathcal{G}(f)$ – in the sense that the frequency response of the system can be quite different for carrier photons and for the signal.

The optical power cannot, however, be usefully increased without limit, due to the existence of quantum radiation pressure noise and also thermal distortion effects in the optics.

There are several 'pictures' that explain the existence of quantum radiation pressure noise in the interferometer. The simplest one is like the Heisenberg microscope experiment. Here one can picture randomly fluctuating photon number in the two arms. This leads to a time-varying light pressure on the mirrors and a corresponding acceleration. More sophisticated models have been presented [10].

Again there are several ways to express the limit set by the quantum radiation pressure noise, but following the style chosen to express the shot noise one can obtain:

$$\tilde{h}^2_{\text{qrpn}}(f) \geq \frac{\alpha\,h_\text{p}^2\,\nu^2}{2\,\pi^4\,\mathcal{M}^2\,f^4\,c^4\,\tau^3}\,\mathcal{N}_\text{p}\,\mathcal{G}^2(f), \tag{2}$$

where h_p is Planck's constant, $\mathcal{M}$ is the mass of a mirror and α is a factor allowing for correlations in the measurement: $\alpha = 4$ for the common interferometer configurations. This result applies at frequencies between the mirror suspension mode and the free spectral range of the arms.

The balance between shot noise and radiation pressure is an 'uncertainty relation' and is called the standard quantum limit.

The remaining power limits are of a technical nature and arise due to imperfect materials and engineering. Absorption in substrates and at coatings of mirrors, together with finite thermal conductivity, leads to non-uniform heating. Since all materials have a temperature dependent refractive index (the photo-refractive effect) the mirrors acquire a non-uniform phase transmission. This can both reduce the interferometer contrast or even produce an unstable power recycling cavity. In either case it sets a limit on the internal light power. Better materials (lower absorption, higher conductivity, lower photo-refractive coefficients) and adaptive correction of the effect by application of heat to give uniform temperature distribution produces systems that can tolerate as much distortion as possible.

Low-noise lasers have finite power (~ 100 W, at present, more soon) and input optics (modulator/isolators/mode-cleaners) have finite power handling, but these are seen as less significant limits on overall performance.

SOLUTIONS TO THE SENSING LIMITS

All reflective (diffractive) interferometers have been proposed to avoid transmission of high power through optical substrates, removing one of the technical limits to interferometer power handling. The first challenge is to remove the need to send light through the coupling mirrors of Fabry-Perot arm cavities. This can possibly be done by using weak diffractive coupling into the cavity. The remaining challenge is to replace the transmissive beamsplitter with an all-reflective one. This may be more difficult since diffractive optics tend to have losses (higher orders etc.) that are related to the coupling strength.

A promising idea is to use weak coupling into high finesse arm cavities in an interferometer with strong resonant sideband extraction. The power at the beamsplitter would be quite low, and a conventional beamsplitter may be adequate [11].

Squeezing was the first 'non-classical' measurement technique suggested for use in interferometric detectors. The idea here is to replace the vacuum fluctuations which enter the system (and lead in the end to the shot noise and quantum radiation pressure noise) with an appropriate squeezed state. This works to change the balance between shot noise and quantum radiation pressure noise according to the applied squeezing. This method allows the standard quantum limit to be reached with lower power [10].

Quantum non-demolition measurements would provide one way of pushing down the sensing limits in the event of heaver mirrors being technologically infeasible. Such systems are discussed in a subsequent talk at this conference [3].

CONCLUSION

Modern extensions of conventional systems provide sensitivity close to the standard quantum limit with modest mirror masses (of order 30 kg). Further advances are needed to push up the mirror mass and light power together unless quantum non-demolition techniques can provide a loop-hole through the standard quantum limit.

ACKNOWLEDGEMENTS

The author would like to thank all who have made contributions to the exciting field of configurations design, especially colleagues in the GEO and LIGO projects, who have provided many interesting ideas over the years. The author's work in configurations has been supported by The University of Glasgow and by PPARC, and this support is gratefully acknowledged.

REFERENCES

1. See for example: S. Phinney, *Astrophysical GW Sources* in the present proceedings.
2. P.R. Saulson, Phys.Rev **D 42** (1990) 2437 – 2443
3. See: *Energetic Quantum Limitations in the Antennae on Free Masses*, F. Khalili, in the present proceedings.
4. E. Gustafson, D. Shoemaker, K. Strain and R. Weiss, *LIGO Science Collaboration White Paper on Detector Research and Development*, http://www.ligo.caltech.edu/docs/T/T990080-00.pdf
5. R.W.P. Drever in *Gravitational Radiation* Eds. N. Daruelle and T. Piran (North Holland, Amsterdam 1983).
6. B.J. Meers, Phys.Rev. **D 38** (1988) 2317 – 2326
7. J. Mizuno *et al.* Phys. Letts. **A 175**(1993) 273 – 276
8. B.J. Meers and K.A. Strain, Phys. Rev. **D 43** (1991) 3117 – 3130.
9. K.A. Strain and J.Hough *paper in preparation*
10. C.M. Caves, Phys. Rev. Lett. **45** (1980) 75 – 78
11. M. Zucker *personal communication* (1999)

Energetic Quantum Limit in Large-Scale Interferometers

Vladimir B. Braginsky, Mikhail L. Gorodetsky,*
Farid Ya. Khalili,* and Kip S. Thorne[†]

Physics Faculty, Moscow University, Moscow Russia
[†] *Theoretical Astrophysics, California Institute of Technology, Pasadena, CA 91125*

Abstract. For each optical topology of an interferometric gravitational wave detector, quantum mechanics dictates a minimum optical power (the "energetic quantum limit") to achieve a given sensitivity. For standard topologies, when one seeks to beat the standard quantum limit by a substantial factor, the energetic quantum limit becomes impossibly large. Intracavity readout schemes may do so with manageable optical powers.

I THE ENERGETIC QUANTUM LIMIT

It is well known that quantum mechanics limits the sensitivity of traditional position measurements by the Standard Quantum Limit (SQL). Several methods of overcoming the SQL have been proposed. There is almost no doubt that in the next several years large-scale gravitational wave antennae will reach the level of the SQL and possibly will beat it by factor $2 \div 3$. Are there any other quantum limits beyond the SQL?

One possible answer is: the next serious limitation is the Energetic Quantum Limit. A gravitational wave in an interferometric antenna changes the phase of the optical field. In order to detect this phase shift, the uncertainty of the phase $\Delta\phi$ must be sufficiently small. In particular, due to the uncertainty relation

$$\Delta\mathcal{E}\Delta\phi \geq \frac{\hbar\omega_0}{2} \quad (1)$$

(where ω_0 is the optical frequency), a large uncertainty of the optical energy $\mathcal{E}$ is required.

This is not a peculiar property of interferometric meters only, but a consequence of a more general principle: In order to detect an external action on a quantum object, the uncertainty of the interaction Hamiltonian $\hat{\mathcal{H}}_I$ must be sufficiently large [1–3]:

$$\left\langle \left(\int_{-\infty}^{\infty} \Delta\mathcal{H}_I(t)dt\right)^2 \right\rangle \geq \frac{\hbar^2}{4}. \quad (2)$$

The uncertainty of $\mathcal{H}_I$ is related to the signal-to-noise ratio by

$$\frac{S}{N} = \frac{4}{\hbar^2}\left\langle \left(\int_{-\infty}^{\infty}\Delta\mathcal{H}_I(t)dt\right)^2\right\rangle \tag{3}$$

For laser interferometer gravitational-wave antennas this is equivalent to

$$\frac{S}{N} = \frac{4}{\hbar^2 L^2}\int_{-\infty}^{\infty} x_{signal}(t)x_{signal}(t')B_{\mathcal{E}}(t,t')dtdt', \tag{4}$$

where L is the length of the arms of the antenna,

$$x_{signal}(t) = \frac{Lh(t)}{2} \tag{5}$$

is the effective change of L caused by a gravitational wave, $h(t)$ is the variation of the wave's metric, and $B_{\mathcal{E}}(t,t')$ is the correlation function of the optical energy in the antenna, or the correlation function of the difference of energies in the two arms of the antenna if two-arm-topology is used.

The origin of the limitation (2) is the Heisenberg uncertainty relation. To detect a small displacement of the mirrors, it is necessary to apply to them a sufficiently strong random kick. The only source of this kick is the uncertainty of the optical energy in the antenna or of the difference of energies in the two arms.

We shall limit ourselves here to the stationary regime, for which the quantum state of the electromagnetic field in the interferometer does not depend explicitly on time. In this case formula (4) can be rewritten in spectral form as

$$\frac{S}{N} = \frac{4}{\hbar^2 L^2}\int_{-\infty}^{\infty} |X_{signal}(\omega)|^2 S_{\mathcal{E}}(\omega)\frac{d\omega}{2\pi}, \tag{6}$$

where $X_{signal}(\omega)$ is the Fourier transform of $x_{signal}(t)$, and $S_{\mathcal{E}}(\omega)$ is the spectral density of the fluctuations of the optical energy. It is important to note here that formula (4) is the ultimate limit on the sensitivity for any measurement technique, and formula (6) describes the ultimate sensitivity for all stationary procedures.

II COMPARISON WITH THE SQL

In all estimates below we will use the value of the Standard Quantum Limit (SQL) as a convenient measure of sensitivity. The SQL, as it was defined more than thirty years ago [4], is the sensitivity of an ordinary position meter, i.e. a position meter which does not use any non-stationary or correlation methods to increase the sensitivity. The forms of the SQL as usually given in the literature, are not convenient since they are based on some assumed shape of the force's time dependence (most commonly a single-cycle sinusoid or a long, monochromatic wave train). Here we prefer a more general form of the SQL expressed in terms of the spectral density $S(\omega)$ for the net noise of a measurement device. This (double-sided) spectral density is defined in such a way that for optimal signal processing the signal to noise ratio is equal to

$$\frac{S}{N} = \int_{-\infty}^{\infty}\frac{|X_{signal}(\omega)|^2}{S(\omega)}\frac{d\omega}{2\pi}. \tag{7}$$

In the case of an ordinary position meter the spectral density of the net noise is equal to

$$S_{SQL}(\omega) = S_x + \frac{S_F}{M^2\omega^4}, \qquad (8)$$

where M is the test mass, $S_x(\omega)$ is the spectral density of the noise that the meter superimposes on the output position signal, $S_F(\omega)$ is the spectral density of the fluctuating back-action force that the meter exerts on the test mass, and these noises satisfy the uncertainty relation [5],

$$S_x S_F \geq \hbar^2/4 . \qquad (9)$$

We assume that the position meter is a perfect one, corresponding to equality in this uncertainty relation, and the spectrum of the signal is concentrated near the frequency ω_{signal}. In this case the meter can be optimally tuned to minimize the net noise at this frequency:

$$S_{SQL}(\omega_{signal}) = \frac{\hbar}{M\omega_{signal}^2}. \qquad (10)$$

As follows from formulas (6), (7) and (10), in order to beat this SQL for the spectral density of x_{signal} by a factor $\xi^2 < 1$, the spectral density of the fluctuations of the optical energy must obey the condition

$$S_{\mathcal{E}}(\omega_{signal}) \geq \frac{\hbar^2}{S_{SQL}(\omega_{signal})\xi^2} = \frac{\hbar M\omega_{signal}^2 L^2}{4\xi^2}. \qquad (11)$$

Let us calculate now the values of $S_{\mathcal{E}}$ for different possible topologies of gravitational-wave antennae.

III ONE-ARM SCHEME

The simplest of possible topologies, shown in Fig. 1, consists of a single Fabry-Perot resonator excited by a pumping laser. The optimally chosen variable of the reflected beam is registered. A spectral or time-domain variational measurement [6], or any other advanced technique can be used to increase the sensitivity. Simple calculations yield that in this case

$$S_{\mathcal{E}}(\omega) = \frac{2\zeta^2 \hbar \omega_0 \overline{\mathcal{E}} \delta}{\delta^2 + \omega^2}, \qquad (12)$$

where ω_0 is the eigenfrequency of the resonator, δ is the half-bandwidth of the resonator, $\overline{\mathcal{E}}$ is the mean value of the optical energy in the resonator, and ζ is the squeeze factor of the quantum state of the input wave ($\zeta = 1$ for a coherent quantum state and $\zeta > 1$ for a squeezed state). As was noted by Caves almost twenty years ago [7], the use of a squeezed state allows one to reduce the value of $\overline{\mathcal{E}}$ by the factor ζ^2. Unfortunately,

FIGURE 1. One-arm topology.

experimental achievements in squeezing are still quite modest, so we will keep ζ in our formulas but use $\zeta = 1$ (i.e. coherent state) in all numerical estimates.

Substituting formula (12) into (11) we obtain that in this case

$$\overline{\mathcal{E}} = \frac{M\omega_{signal}^2(\delta^2 + \omega_{signal}^2)L^2}{8\xi^2\zeta^2\omega_0\delta}, \tag{13}$$

and, correspondingly,

$$\overline{W} = \frac{M\omega_{signal}^2(\delta^2 + \omega_{signal}^2)Lc}{16\xi^2\zeta^2\omega_0\delta}, \tag{14}$$

$$\overline{w} = \frac{M\omega_{signal}^2(\delta^2 + \omega_{signal}^2)L^2}{16\xi^2\zeta^2\omega_0}, \tag{15}$$

where $\overline{W}$ is the mean value of the optical power circulating in the resonator and $\overline{w}$ is the mean value of the input pump power.

The values of $\overline{\mathcal{E}}$ and $\overline{W}$ may be minimized by choosing for the resonator's half-bandwidth

$$\delta = \omega_{signal}. \tag{16}$$

In this case

$$\overline{\mathcal{E}} = \frac{M\omega_{signal}^3 L^2}{4\xi^2\zeta^2\omega_0}$$
$$= \frac{2 \cdot 10^8 \text{erg}}{\xi^2\zeta^2} \left(\frac{M}{10^4\text{g}}\right)\left(\frac{\omega_{signal}}{10^3\text{s}^{-1}}\right)^3 \left(\frac{L}{4\cdot 10^5\text{cm}}\right)^2 \left(\frac{\omega_0}{2\cdot 10^{15}\text{s}^{-1}}\right)^{-1}, \tag{17}$$

$$\overline{W} = \frac{M\omega_{signal}^3 Lc}{8\xi^2\zeta^2\omega_0}$$
$$= \frac{0.75 \cdot 10^{13}\text{erg/s}}{\xi^2\zeta^2} \left(\frac{M}{10^4\text{g}}\right)\left(\frac{\omega_{signal}}{10^3\text{s}^{-1}}\right)^3 \left(\frac{L}{4\cdot 10^5\text{cm}}\right) \left(\frac{\omega_0}{2\cdot 10^{15}\text{s}^{-1}}\right)^{-1}, \tag{18}$$

$$\overline{w} = \frac{M\omega_{signal}^4 L^2}{8\xi^2\zeta^2\omega_0}$$
$$= \frac{1 \cdot 10^{11}\text{erg/s}}{\xi^2\zeta^2} \left(\frac{M}{10^4\text{g}}\right)\left(\frac{\omega_{signal}}{10^3\text{s}^{-1}}\right)^4 \left(\frac{L}{4\cdot 10^5\text{cm}}\right)^2 \left(\frac{\omega_0}{2\cdot 10^{15}\text{s}^{-1}}\right)^{-1}. \tag{19}$$

All known methods of overcoming the SQL for position measurements of free test masses [6,8,9] apply an additional restriction on the optical energy and pumping power. This restriction arises as follows. All such methods rely on a correlation between two meter noises: the (radiation-pressure-induced) back action noise on the test masses, and the output light's shot noise. This correlation is damaged by electromagnetic fluctuations which enter the resonator wherever there is optical dissipation. Such fluctuations produce radiation pressure noise that cannot correlate with the shot noise, so all these schemes are limited by the level of intrinsic losses (dissipation) in the electromagnetic resonator. It can be shown that, due to this, the amount by which these methods beat the SQL cannot exceed the limit

$$\xi = \left(\frac{\delta_{intr}}{\delta_{load}}\right)^{1/4}, \tag{20}$$

where δ_{intr} is the part of the electromagnetic resonator's half-bandwidth $\delta = \delta_{intr} + \delta_{load}$ due to intrinsic losses (i.e. absorption in the mirrors and beam-splitters and nonzero transmittance of end mirrors) and δ_{load} is the part which characterizes the coupling of the resonator to the output light beam. Hence the optimization (16) is possible only if

$$\xi > \left(\frac{\delta_{intr}}{\omega_{signal}}\right)^{1/4} \tag{21}$$

Using the best known mirrors and the LIGO value of $L = 4 \times 10^5$ cm one can achieve the value of $\delta_{intr} \sim 10^{-1} \text{s}^{-1}$. Hence, if $\omega_{signal} \simeq 10^3 \text{s}^{-3}$ then condition (20) may be satisfied only if $\xi \geq 0.1$.

Any further increase of sensitivity (decrease of ξ) makes the optimization (16) impossible. Instead, the resonator's half-bandwidth will be forced to increase, with decreasing ξ, as

$$\delta \simeq \delta_{load} = \frac{\delta_{intr}}{\xi^4} > \omega_{signal}, \tag{22}$$

which leads to the necessity of increasing the pumping power:

$$\overline{\mathcal{E}} = \frac{M\omega_{signal}^2 \delta_{intr} L^2}{8\xi^6 \zeta^2 \omega_0}, \tag{23}$$

$$\overline{W} = \frac{M\omega_{signal}^2 \delta_{intr} Lc}{16\xi^6 \zeta^2 \omega_0 \delta}, \tag{24}$$

$$\overline{w} = \frac{M\omega_{signal}^2 \delta_{intr}^2 L^2}{16\xi^{10} \zeta^2 \omega_0}. \tag{25}$$

Numerical values of $\mathcal{E}$, $\overline{W}$ and $\overline{w}$ calculated using formulas (17)–(19) and (23)–(25) are plotted in Figs. 5, 6 and 7 correspondingly, as a functions of ξ^{-1} (see the upper solid curves at the end of this paper). In these figures, LIGO parameter values are used: $M = 10^4$g, $L = 4 \times 10^5$cm, $\omega_0 = 2 \times 10^{15} \text{s}^{-1}$, and $\omega_{signal} = 10^3 \text{s}^{-1}$. These figures show a very strong dependence of $\mathcal{E}$, $\overline{W}$ and $\overline{w}$ on ξ in the region to the left of the inflection point $\xi = (\delta_{intr}/\omega_{signal})^{1/4}$.

This strong ξ dependence ($\overline{\mathcal{E}} \propto \xi^{-6}$, $\overline{W} \propto \xi^{-6}$, $\overline{w} \propto \xi^{-10}$) makes it practically impossible to achieve sensitivities $\xi < 0.1$ using conventional interferometer topology: For $\xi = 0.1$ the circulating power in the resonator arms is $\overline{W} \sim 100$MW, corresponding to a power density ~ 300kW/cm^2 in the ~ 10cm-diameter light spots on each mirror's surface. This power density is already so high that it is questionable whether future mirrors will be able to handle it; and the further increase as $\overline{W} \propto \xi^{-6}$ makes it highly implausible that future mirrors will allow sensitivites better than $\xi \simeq 0.1$.

IV RECYCLING SCHEMES

In the next generation of large-scale gravitational-wave antennae, recycling [10] will be used for several purposes, including reduction of the optical pump power $\overline{w}$. Meers [11] has discussed several different variants of recycling, which can be divided into two main groups, depending on the structure of the electromagnetic modes in the antenna: wide-band recyling, and narrow-band or resonant recycling. We shall illustrate these two types of recycling schemes by two concrete topologies, but the main properties in each group are independent of the concrete topology.

A Wide-band recycling

In wide-band recycling schemes (Fig. 2) the optical field in the interferometer may be regarded as a superposition of two electromagnetic modes – the input mode (also called "symmetric mode"; solid lines in Fig. 2) and the output mode (also called "antisymmetric mode"; dashed lines). The Bandwidths of these modes depend on the transmittances of the mirrors D and D', correspondingly, and may be substantially different. Mirror D' may be completely absent, which corresponds to the case of simple power recycling.

All previous expressions for the mean resonator energy $\overline{\mathcal{E}}$ and circulating power $\overline{W}$ [formulas (13), (14), (17), (18), (23)] remain valid for wide-band recycling schemes, but with replacement of δ by the half-bandwidth of the output mode δ_{out}. On the other hand, the value of the input (pump) power is defined by the bandwidth of the input mode:

$$\overline{w} = 2\overline{\mathcal{E}}\delta_{in}. \qquad (26)$$

This allows one to decrease the input power by holding $\delta_{out} \sim \omega_{signal}$ and choosing δ_{in} as small as possible. For example, using the best known mirrors, the value of $\delta_{in} \sim 10^{-1} s^{-1}$ can be obtained, which allows one to decrease the input power by several orders of magnitude (see the dotted curve in Fig. 7).

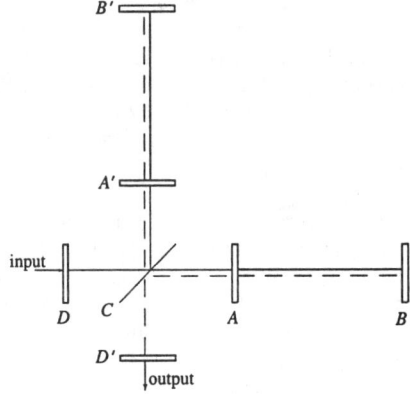

FIGURE 2. Wide-band recycling toplogy.

B Narrow-band recycling

Narrow-band recycling schemes differ from wide-band ones only by the values of the distances between mirrors and the mirror transmittances. However, narrow-band schemes lead to a substantially different behavior of the system. Its eigenfrequencies form doublets. The frequencies in each doublet are separated from each other by a frequency difference Ω which depends on the the mirror configuration and which should be chosen close to ω_{signal}. The spectral density of stored energy $S_{\mathcal{E}}$ in this case is maximimal at the frequency Ω instead of at zero frequency:

$$S_{\mathcal{E}} = \frac{4\hbar\omega_0 \overline{\mathcal{E}} \omega^2 \delta}{(\Omega^2 - \omega^2)^2 + \Omega^2 \delta^2}. \tag{27}$$

This allows one to obtain an especially high value of $S_{\mathcal{E}}$ in a narrow band in the vicinity of Ω by choosing $\Omega \gg \delta$. It can be shown that in this case

$$\overline{\mathcal{E}} = \frac{M\omega_{signal}^2 (\delta^2 + \Delta^2) L^2}{8\xi^2 \zeta^2 \omega_0 \delta} \tag{28}$$

where Δ is the bandwidth over which ξ is close to its desired minimum value. This stored energy can be further minimized by choosing $\delta = \Delta$, so that

$$\overline{\mathcal{E}} = \frac{M\omega_{signal}^2 \Delta L^2}{4\xi^2 \zeta^2 \omega_0}. \tag{29}$$

It is evident from comparison of formulas (17) and (29) that this regime allows one to reduce the necessary value of $\mathcal{E}$, but the price is a proportional reduction of the bandwidth of the signal.

V INTRACAVITY SCHEMES

The above considerations show that the Energetic Quantum Limit does not permit one to obtain sensitivities substantially better than the SQL using the traditional optical topologies of gravitational-wave antennae, even if recycling is employed.

A new method of extracting information from a gravitational-wave antenna was proposed in the article [12]. Instead of measuring the time phase shift of an output optical wave, it was suggested to detect directly the spatial phase shift of the optical field inside the antenna using some QND-type method. In this case the necessary level of uncertainty of the optical energy (2) is provided by a fluctuational redistribution of the energy between the two arms of the antenna instead of by the shot noise of the resonator energy. In principle this allows one to increase the sensitivity without increasing the mean stored energy.

Two practical realizations of this idea were considered in the articles [13,14]. From our point of view, the most promising of them is the "optical bar" scheme [13]. The structure of the electromagnetic modes in this case is similar to that of narrow-band recycling topologies: It has frequency doublets separated by a frequency Ω, which depends on the transmittance of the central mirror. One of the possible topologies for the "optical bar" scheme is shown in Fig. 3.

In this scheme (we omit now all intermediate details) the optical fields in the two arms of the antenna work (via their radiation pressure) as mechanical springs with rigidity proportional to the optical energies $\overline{\mathcal{E}}$ stored in the two arms. The optical fields' rigidities move the internal mirror C when a gravitational wave moves the end mirrors. This displacement may be measured relative to an additional reference mass which is not affected by the optical field in the antenna, using one of the measurement methods developed for resonant-mass gravitational-wave antennae.

The value of the displacement of mirror C may be close to $Lh(t)/2$ if the stored energy $\overline{\mathcal{E}}$ is sufficiently high:

$$\overline{\mathcal{E}} \geq \frac{m\omega_{signal}^3 L^2}{2\omega_0}, \tag{30}$$

where m is the mass of the internal mirror. The structure of this expression is similar to expression (17) for traditional schemes, but it contains the mass m of the internal mirror C instead of the masses of the end mirrors M, and, which is more important, it does not depend on ξ. Hence, in this case it is possible to increase the sensitivity beyond the level of the SQL without any necessity to increase $\overline{\mathcal{E}}$. If, for example, $m = 10^3$g and the values of all other parameters are the same as above, then $\overline{\mathcal{E}} \simeq 4 \cdot 10^7$erg.

It should be noted that internal losses in the interferometer limit the sensitivity of the intracavity schemes at the level,

$$\xi \simeq \sqrt{\frac{\delta_{intr}}{\omega_{signal}}} \tag{31}$$

[square root, by contrast with the 1/4 power for conventional interferometers, Eq. (21)]; so if $\delta_{intr} = 10^{-1}\text{s}^{-1}$, then the value $\xi \simeq 10^{-2}$ may be reached in principle.

VI METHODS OF MEASUREMENT

It is evident that for intracavity schemes the focus of the problem shifts to the local meter that measures the position of the central mirror. Intracavity schemes require high precision in this meter's measurement, exceeding the SQL. During the last twenty five years several possible methods of such measurement were proposed. We consider here briefly only one of them – the so-called " dual-resonator speedmeter" [9,15]. It is based on two microwave resonators, one of which couples evanescently to the position of the test mass (see Fig. 4). The sloshing of the resulting signal between the resonators,

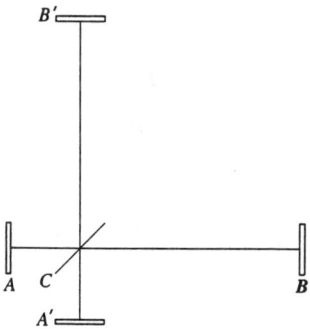

FIGURE 3. Optical bars.

and a wise choice of where to place the resonators' output waveguide, produce a signal in the waveguide that (for sufficiently low frequencies) is proportional to the test-mass velocity, but not its position. This permits the speed meter to achieve sensitivities better than the SQL.

Unfortunately, this scheme has two major disadvantages that are typical for all similar schemes. First, when using this scheme to monitor the central mass, the level of optical energy in the antenna required to overcome the SQL has to be substantially higher than the theoretical limit (30):

$$\overline{\mathcal{E}} \geq \frac{M\omega_{signal}^3 L^2}{2\xi^2 \omega_0} \tag{32}$$

and even a bit higher than the energy (17) (see the dotted curves in Figs. 5, 6, 7. This is because the speedmeter satisfies the criterion for QND measurements only when monitoring a free mass, and the internal mirror C coupled to the optical field and through this field to the end mirrors is a more sophisticated object than a free mass; it has several degrees of freedom. This mirror behaves like a single solid free mass only if the optical energy is sufficiently large, which leads to the requirement (32).

The second disadvantage is an inherent property of all known devices for overcoming the SQL for position measurements of free test masses. As mentioned above, all such measurement devices are very sensitive to the level of intrinsic losses in their resonators [see formula (20) and the discussion above]. In the particular case of the speedmeter, this disadvantage limits its sensitivity to $\xi \gtrsim 1/3$ even if the best known microwave resonators are used [15]. An additional fee for this scheme is the necessity to use cryogenic temperatures.

The first disadvantage (abnormally high optical power) may be evaded by using a local "variational meter" [6] based on a microwave parametric transducer or on a relatively short Fabry-Perot sensor. As follows from the formula (18), the necessary circulating optical power is proportional to the length of the resonator. But for such a variational meter, the second disadvantage (a severe limitation on ξ due to electromagnetic dissipation) still remains.

Two co-authors of this report think that the most promising way to evade this obstacle is to use an oscillator instead of free test mass [16]. In the case of a harmonic oscillator it is possible to overcome the SQL without using a correlation of meter noises, and thus the oscillator is free from the limitation (20).

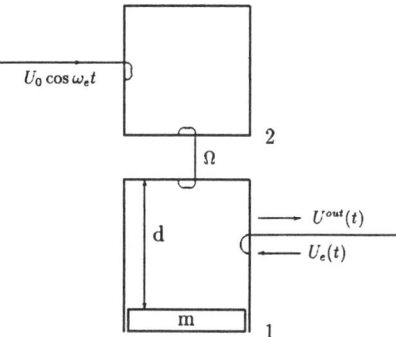

FIGURE 4. Quantum speed meter.

REFERENCES

1. F. Ya. Khalili, Vestnik Moskovskovo Universiteta, series 3, no. 3 (1983), p. 17.
2. Yu. I. Vorontsov, F. Ya. Khalili, Vestnik Moskovskovo Universiteta, series 3, number 3 (1985).
3. V.B.Braginsky, F.Ya.Khalili, "Quantum Measurement", Chap. IX, ed.by K.S.Thorne, Cambridge University Press, 1992.
4. V. B. Braginsky, Zhur. Eksp. Teor. Fiz., **53**, 1436 (1967).
5. V.B.Braginsky, F.Ya.Khalili, "Quantum Measurement", Chap. VI, ed.by K.S.Thorne, Cambridge University Press, 1992.
6. S.P.Vyatchanin, E.A.Zubova, Physics letters A201 (1995) 269.
7. C.M.Caves, Phys.Rev.D23 (1981) 1693.
8. W. G. Unruh, in *Quantum Optics, Experimental Gravitation, and Measurement Theory*, eds. P. Meystre and M. O. Scully, (Plenum, 1982), p. 647.
9. V.B.Braginsky, F.Ya.Khalili, Phys.Lett. A147 (1990) 251.
10. R.W.P.Drever, in "Gravitational Radiation", ads. N.Deruelle and T.Piran, NATO Adv. Inst., page 321, 1982.
11. B.J.Meers, Phys.Rev. D38 (1988) 2317.
12. V.B.Braginsky, F.Ya.Khalili, Phys. Letters A218 (1996) 167.
13. V.B.Braginsky, M.L.Gorodetsky, F.Ya.Khalili, Phys. Letters A232 (1997) 480.
14. B.Braginsky, M.L.Gorodetsky, F.Ya.Khalili, Phys. Letters A246 (1998) 485.
15. V.B.Braginsky, M.L.Gorodetsky, F.Y.Khalili, K.S.Thorne, Phys. Rev D, in press.
16. V.B.Braginsky, F.Ya.Khalili, Phys. Letters A257 (1999) 241.

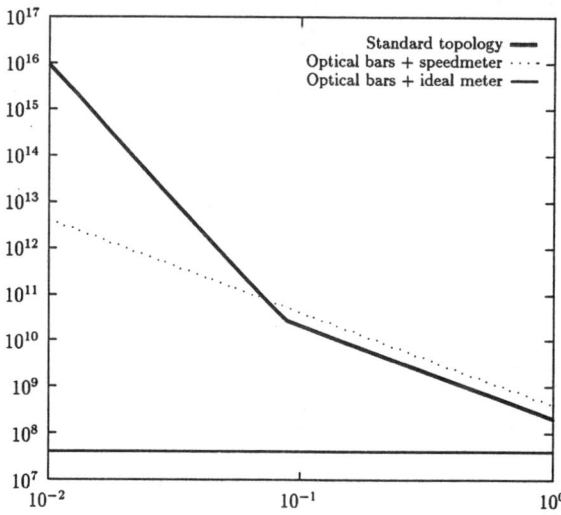

FIGURE 5. Energy as a function of ξ, erg.

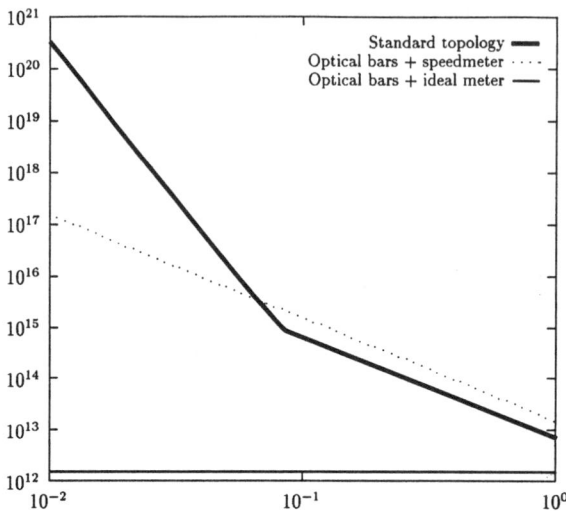

FIGURE 6. Circulating power as a function of ξ, erg/s.

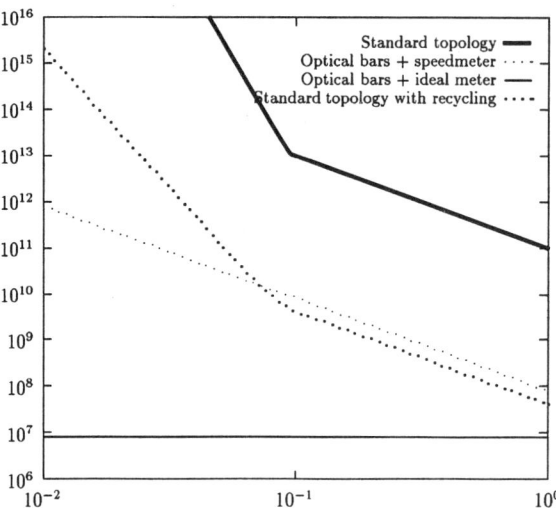

FIGURE 7. Pump power as a function of ξ, erg/s.

ADVANCED CONFIGURATIONS

A power recycled Michelson interferometer with resonant sideband extraction

M B Gray, D A Shaddock and D E McClelland

*Department of Physics, Faculty of Science,
The Australian National University, A.C.T. 0200, Australia.*

Abstract. We present a system for the control and signal extraction of a power recycled Michelson interferometer with Resonant Sideband Extraction. This system maintains excellent orthogonality between error signals for the control of all five degrees of freedom. In addition this orthogonality is maintained for arbitrary detunings of the signal cavity.

INTRODUCTION

Due to its robustness to optical distortions and tunable frequency response, Resonant Sideband Extraction (RSE) [1,2], see Fig. 1a, has been selected as the reference design for the next stage of the LIGO Project [3]. It is a configuration which can be used to maximize the power build up in the arm cavities without reducing the signal bandwidth of the instrument or increasing the power transmitted through substrates such as the beam splitter or input test masses.

However, control of an RSE interferometer presents a serious technical problem requiring the locking of 5 degrees of freedom: arm cavity common and differential modes, Michelson (differential) mode, the power recycling cavity and the signal recycling cavity. Here we present one possible solution to this complex control problem which allows continuous tuning of the signal mirror. We begin by briefly summarizing the principle of operation of RSE before presenting the control system and summarizing progress to date on testing it on a bench top interferometer.

The system, as depicted in Fig. 1a, consists of high finesse arm cavities and a signal mirror (a power mirror is optional). The Michelson interferometer is locked in such a way that all unused power is reflected back to the laser. If this is the case, the carrier never senses the signal mirror and sees only the high finesse arm cavities as shown in Fig. 1b. This results in a large power build up in the arms, which improves the shot noise limited sensitivity. A consequence of increasing the arm cavity finesse is an increase in the storage time of the arms which would drastically reduce the signal bandwidth of the instrument in a non-RSE device.

The gravitationally induced signal sidebands however, interfere constructively towards the dark fringe and so do experience the signal mirror. The sidebands effectively see a three mirror compound cavity, as shown in Fig. 1c. This coupled cavity can be thought of as a single two mirror cavity where one mirror has a tunable reflectivity (and phase shift). By moving the signal mirror by a fraction of a wavelength the reflectivity of this "tunable mirror" can be reduced to near zero. In this way the storage time of the signal sidebands can be reduced, thereby increasing the signal bandwidth of such a device. Moreover, by tuning the signal mirror position carefully, the signal response of the instrument can be changed continuously from narrow band (dual recycling) to broad band (RSE).

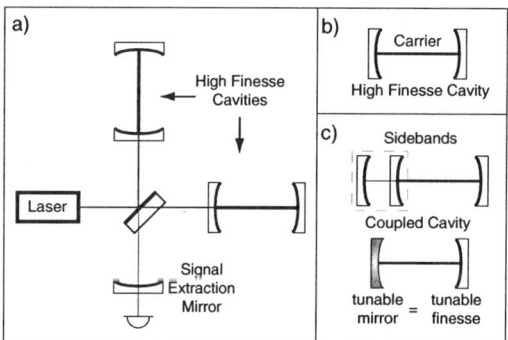

FIGURE 1. a) schematic of a Resonant Sideband Extraction Michelson interferometer, b) effective arm cavity configuration as experienced by the carrier and c) by the signal sidebands.

CONTROL SYSTEM

The aim of the control system design is to lock the arm cavities, power cavity and Michelson interferometer to a fixed position (on resonance or to a dark fringe) with the capability of detuning the signal cavity to an arbitrary position. Two strategies can be employed to ensure the control signals for the other degrees of freedom are not influenced by the detuning of the signal cavity. The first is to calculate the offsets added due to the detuning of the signal mirror on all other control signals and then subtract this offset to produce the error signal. This technique has the advantage that only a few modulation frequencies are required however it relies on extremely accurate knowledge of interferometer parameters and is difficult to implement for arbitrary signal cavity detunings. The second technique is to use a modulation scheme which gives the maximum optical isolation against the signal cavity response for all other degrees of freedom. This can be achieved by careful selection of interferometer lengths and modulation frequencies. For example, a modulation frequency that is completely reflected by the Michelson is immune to changes in the signal cavity as it never senses the signal mirror. One disadvantage is that this technique requires many modulation frequencies leading to a complex input beam preparation problem. We have chosen such a system, sacrificing some of the optical and modulation simplicity for flexibility of the detuning of the signal cavity with maximum orthogonality between error signals.

The control system presented here requires three modulation frequencies and a tunable subcarrier (single sideband). For the bench top system we use two lasers (high power carrier and low power tunable sub-carrier), two resonant phase modulators (at 75 MHz and 15 MHz) and one amplitude modulator (at 112.5 MHz). All error signals are derived by demodulation of the outputs of the three photodetectors at the reflected port, transmitted port and power cavity port. Below we discuss how each error signal is derived and what type of modulation/demodulation is required.

Arm cavity common mode

The arm cavities are the highest finesse cavities and thus will dominate many of the error signals. This important error signal is obtained using the field detected at the reflected port. Phase modulation sidebands are imposed on the carrier at a frequency of 75 MHz. In order to isolate the error signal from changes due to the tuning of the signal mirror we set the Michelson arm length mismatch so that the interferometer is completely reflective at this frequency. The reflectivity of the Michelson is proportional to $\cos^2(4\pi\nu_m \Delta L/c)$ where c is the speed of light, ΔL is half the total Michelson arm length mismatch and ν_m is the modulation frequency. For zero transmission to the signal mirror a Michelson arm length mismatch of $\Delta L = c/(4\nu_m)$ is required. Using 75 MHz modulation a Michelson arm length mismatch of ± 1 m ensures any error signals derived from demodulating at 75 MHz are well isolated against detuning of the signal cavity. The predicted error signal derived from modulation of the reflected port is shown in Fig. 2a.

Power mirror

This error signal is obtained using the same 75 MHz PM sidebands except from the output of the photodetector at the power cavity port, see Fig. 2d. As we are using the 75 MHz modulation this error signal also has excellent isolation from the detuning of the signal cavity. This signal is sensitive to both the arm cavity common mode and the power mirror however the slope of the power mirror error signal has a sign change from reflected port, see Fig. 2c, to the power cavity port, Fig. 2d. This sign change ensures that the power cavity and arm cavity error signals are linearly independent and allows extraction of the power cavity error signal. This can be achieved either with a hierarchical locking structure or by electronic processing of the error signals from both ports.

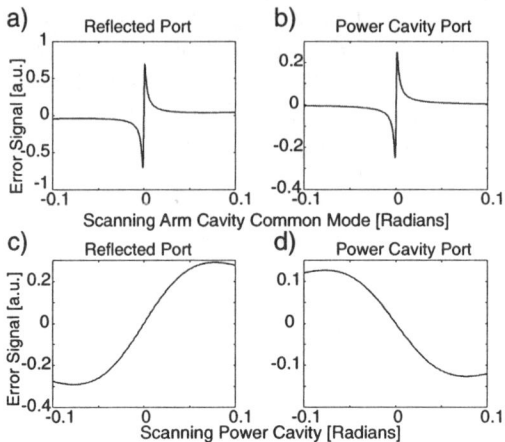

FIGURE 2. Error signal for the arm cavity common mode taken from a) reflected port, b) power cavity port and error signal for the power mirror from c) reflected port and d) power cavity port.

Signal Mirror

The signal mirror is locked with the use of a tunable sub-carrier. For our bench top experiment this sub-carrier will be provided by a (tunable) offset phase-locked laser. This offset sub-carrier is phase modulated at 15 MHz. The sub-carrier is resonant in the combined power-signal cavity whilst the 15 MHz sidebands are not. This gives a Pound-Drever-Hall [4] type error signal when the reflection port photocurrent is demodulated at 15 MHz. The sub-carrier is nearly anti-resonant in the arm cavities making the signal cavity locking immune to the arm cavity common and differential modes. The signal cavity is tuned by tuning the offset frequency of the sub-carrier. The signal mirror changes the power signal cavity resonance to track the sub-carrier and keep it on resonance.

Arm Cavity Differential Mode

By observing the beat between the carrier and sub-carrier on the transmission port, a strong error signal for the arm cavity differential mode is obtained. As the sub-carrier is locked to the resonance of the power-signal cavity it receives no phase shift on transmission, regardless of the tuning of the signal mirror. The largest cross talk term in this error signal is from the Michelson differential motion and can be ignored, see Fig. 3.

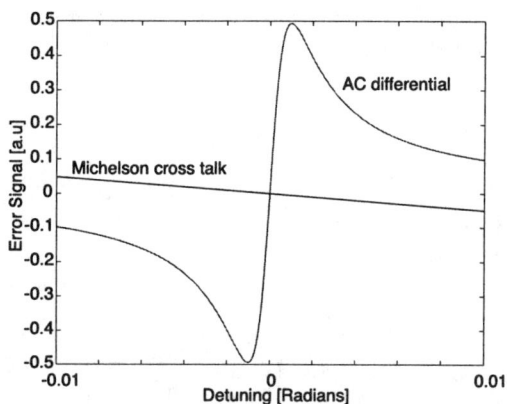

FIGURE 3. Error signal for arm cavity differential mode and the most significant cross talk term; the Michelson (differential) mode.

Michelson (differential)

This is by far the most difficult error signal to isolate from the effects of tuning the signal mirror. This is because in order to obtain an error signal for the Michelson we need some component of the field to be partially transmitted through the beam splitter hence we cannot use the isolation technique of sections and . We also cannot easily use a sub-carrier which is always resonant with the power-signal cavity as the zero crossing point of the Michelson error signal is a function of signal cavity detuning.

There are two ways that an error signal can be corrupted by the tuning of the signal mirror. The first is a change in the slope around the desired lock point, and the second is a voltage offset from zero. We have chosen a system which gives a change in slope, but not a change in offset. The result of this is the gain of the locking loop changes as the signal mirror is tuned whilst the lock position does not, as shown in Fig. 4. The gain change can be calculated fairly accurately and compensated by an electronic gain change.

We use modulation sidebands placed at an integral number of power-signal cavity free spectral ranges from the carrier. This ensures that as the power-signal cavity resonance moves positive and negative sidebands experience the same phase shift and attenuation. The Michelson is locked using the beat between these sidebands and the 75MHz PM sidebands used in and . We use this intermodulation beat in order to make the error signal insensitive to the arm cavities. The sidebands must have amplitude modulation symmetry to obtain a zero crossing error signal. For maximum transmission through the Michelson, and thus maximum sensitivity, the sideband frequency must be restricted to 37.5 MHz plus multiples of 75 MHz and must also be resonant in the power-signal cavity. We choose a frequency of

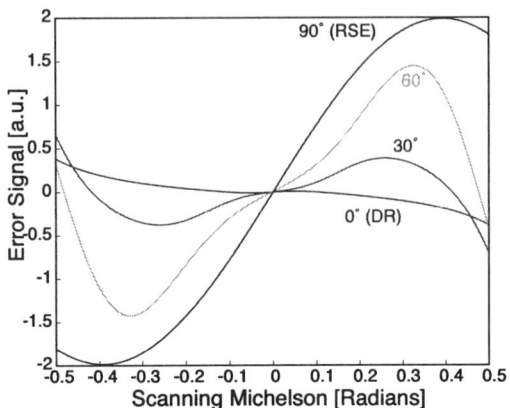

FIGURE 4. Error signal for Michelson (differential) mode illustrating the gain change as the signal mirror is detuned. Note the zero crossing point remains unchanged. RSE: Resonant Sideband Extraction, DR: Dual Recycling.

112.5 MHz for the amplitude modulation allowing the signal and power cavities to maintain a reasonable length.

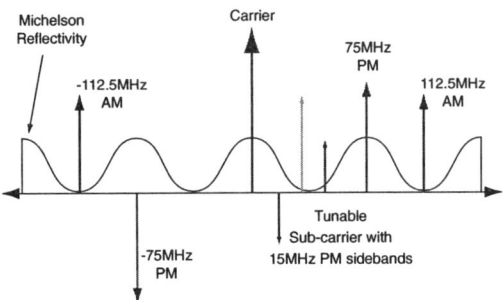

FIGURE 5. Schematic showing frequency sidebands indicating the type (sub-carrier, PM, AM) and position (MHz). The sub-carrier offset from main carrier is tunable.

INPUT FIELD GENERATION

Figure 5 shows the frequency of the various sidebands required for this locking scheme superimposed on the Michelson reflectivity curve.

We use a 700mW Nd:YAG laser (Lightwave model 126) to provide the main carrier beam. This laser is frequency stabilized to a modecleaner cavity [5] to ease the gain requirements on all further control subsystems. We cannot tolerate

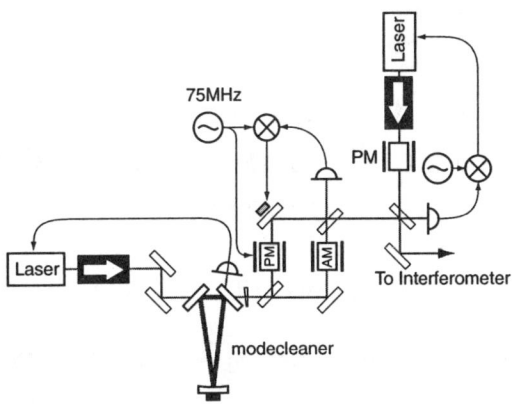

FIGURE 6. Experimental layout used for input field generation.

intermodulation products on the 75 MHz and 112.5 MHz modulations as these destroy the orthogonality of the Michelson and arm cavity locking signals. Hence these sidebands must be added in parallel using a Mach-Zehnder interferometer. The Mach-Zehnder interferomter is locked using the 75 MHz phase modulation.

The subcarrier is provided by a second Nd:YAG laser (50 mW Lightwave model 120). This subcarrier is offset phase locked using a tunable signal generator and a broadband photodetector [6]. This enables us to continuously tune the offset frequency across the full range of required values. The sub-carrier itself is phase modulated at 15 MHz. Figure 6 shows the experimental progress to date. The input beam has all of the necessary modulation components including the phase locked sub-carrier. The construction of the main interferometer is currently underway.

This research was supported by the Australian Research Council through the Australian Consortium for Interferometric Gravitational Astronomy.

REFERENCES

1. Mizuno, J., Strain, K., Nelson, P., Chen, J., Schilling, R., Rüdiger, A., Winkler, W. and Danzmann, K. *Phys. Lett. A* **175**, 273 (1993).
2. Heinzel, G., Mizuno, J., Schilling, R., Winkler, W., Rüdiger, A. and Danzmann, K. *Phys. Lett. A* **217**, 305 (1996).
3. *LIGO II conceptual project book.* http://www.ligo.caltech.edu/docs/M/M990288-A1.pdf
4. Drever, R., Hall, J., Kowalski, F., Hough, J., Ford, G., Munley, A. and Ward, H. *Appl. Phys. B* **31**, 97 (1983).
5. Shaddock, D., Gray, M. and McClelland, D. *Opt. Lett.* **24**, 1499 (1999).
6. Gray, M., Shaddock, D., Harb, C. and Bachor H. *Rev. Sci. Instrum.* **69**, 3755 (1998).

The Polarization Sagnac Interferometer as a Candidate Configuration for an Advanced Detector

Peter T. Beyersdorf, R.L. Byer, M.M. Fejer

Ginzton Laboratory, Stanford University, Stanford, CA 94305

Abstract. We present a delay-line polarization Sagnac (1) interferometer that utilizes the increased laser power which will likely be available to future advanced detectors (2) to greatly simplify many laser and interferometer control requirement. We describe how the signal sidebands and the local oscillator are made common path by using a polarizing beamsplitter and controlling the circulating light's polarization state. The common path nature of the interfering beams simplify alignment, reduce the interferometer control effort, and lower the laser amplitude and frequency stability requirements by 6 orders of magnitude over a frontally modulated Michelson interferometer. We present experimental verifications of the robustness of this configuration, which was implemented on a 2m tabletop interferometer

The advantage of common mode noise rejection of the Sagnac interferometer due to the symmetry of the counter-propagating beam's paths is extended by utilizing the beamsplitter in a symmetric fashion. Detection of the dark fringe of interference on the symmetric port of the beamsplitter is achieved in a polarization Sagnac interferometer by using a polarization beamsplitter to split the input light into two polarization components which counter-propagate around the interferometer loop. An in-loop polarization-changing element, such as a half wave plate oriented at 45° with respect to the polarizing beamsplitter axis, swaps the polarization states of the counter-propagating beams. The reflected polarization is then transmitted upon reencountering the beamsplitter, and the transmitted polarization is then reflected.

The polarization state of the output is effected by any relative phase delay between the two polarization components inside the interferometer loop. In this way the measurement of differential phase is transformed into a measurement of polarization ellipticity. The polarization state of the output beam is given by the Jones vector representing transmission through the entire optical system

$$\vec{E}_{out} = \left(P_2^t W(\theta) P_2^r e^{+i\frac{\Delta\phi}{2}} + P_2^r W(-\theta) P_2^t e^{-i\frac{\Delta\phi}{2}} \right) \vec{E}_{in} \tag{1}$$

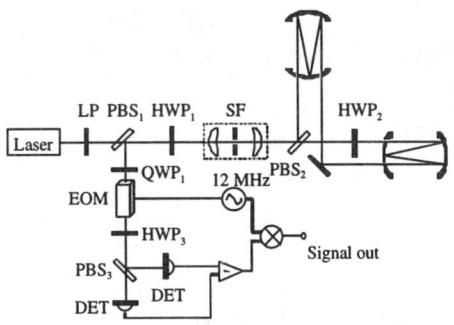

FIGURE 1. The layout of the polarization Sagnac interferometer. The polarization of the input light is set by HWP_1 to have equal components of s and p polarization. A phase delay between these components as they propagate in opposite directions around the interferometer loop produces a polarization component that is separated from the output light by PBS_1.

where P is the Jones matrix representing PBS_2 and W is the Jones matrix representing the HWP_2. The superscript t represents transmission through, and r represents reflection from, the polarizing beamsplitter. When the input polarization is linear at 45° the output polarization is elliptical

$$\vec{E}_{out} = \begin{bmatrix} e^{+i\phi/2} \\ e^{-i\phi/2} \end{bmatrix} \cdot \vec{E}_{in} \qquad (2)$$

with the minor axis of the ellipse the difference between the interfering beams and the major axis of the ellipse the sum of the interfering beams.

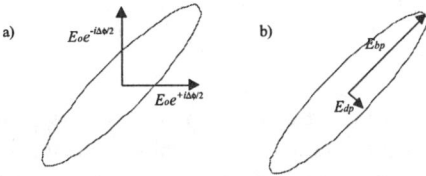

FIGURE 2. Elliptical polarization state of interferometer output. a) The relative phase shift of $\Delta\phi$ between the orthogonally components of the beam produces an elliptical polarization state. b) The output is resolved into polarization components along the principle axes of the ellipse. The dark polarization, the minor axis of the ellipse, contains the signal field and is a measure of the phase shift. The bright polarization, the major axis of the ellipse, is the carrier field and contains most of the power.

The polarization ellipticity is measured by using a polarizing beamsplitter with imperfect contrast to separate the dark polarization from the output light. The beamsplitter leaks a bit of the bright polarization for use as a local oscillator in a heterodyne detection scheme. The orthogonallity of the local oscillator and the signal polarization allow waveplates to be used to set the relative phase of the local oscillator. Heterodyne sidebands are added by modulating only the local oscillator polarization. The polarization components are mixed by a properly oriented polarizing beamsplitter to produce the sum and difference of the dark fringe and the local oscillator for balanced detection.

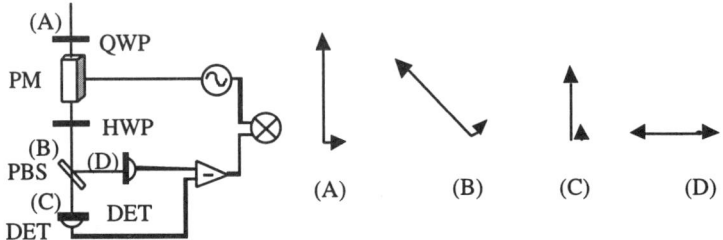

FIGURE 3. Heterodyne detection scheme. The polarization state of the signal and the local oscillator is shown at (A) pick-off from interferometer (B) after rotation by the half wave plate HWP and (C) and (D) on each of the balanced detectors. The phase modulator PM acts only on the local oscillator polarization.

The sensitivity of an interferometer to fluctuations of the input laser amplitude and frequency is directly related to the asymmetric path length between the interfering beams. For an asymmetric path length of ΔL, a laser frequency n, and intensity I_0, the conversion of laser frequency (3) and laser intensity noise to output phase noise is

$$\tilde{\phi}(\omega) = \Delta L \frac{\delta\tilde{v}(\omega)}{v}$$

$$\tilde{\phi}_{eq}(\omega) \approx \frac{\omega \Delta L}{2c} \frac{\tilde{I}_{in}(\omega)}{I_0}. \qquad (3)$$

Both are directly proportional to the asymmetry in the path length between the interfering beams. For a polarization Sagnac interferometer this asymmetry can be made on the order of a wavelength, 6 orders of magnitude smaller than the 1m asymmetry length considered for generating a length-sensing signal in a Michelson interferometer (4). This results in an interferometer virtually insensitive to input amplitude and frequency fluctuations.

The common optical path of the local oscillator with the signal sidebands eliminates the need to actively control the phase or the alignment of the local oscillator. The phase relation between the interfering beams is constant and does not need to be controlled, nor does the longitudinal position of the core optics outside of the measurement band.

Another advantage offered by the polarization Sagnac interferometer is that the collinearity of the input and output beams allows a spatial filter to be placed between the laser and the interferometer to select a single spatial mode for input into the interferometer and output to the detector. By using a single spatial filter in such a reciprocal fashion the maximum overlap between the input and output spatial modes is achieved.

We have demonstrated a 2m arm length polarization Sagnac interferometer with 75-bounce delay lines in each arm. The input frequency was modulated by an electrooptic modulator placed between the laser and the interferometer. The output of the interferometer was seen to be insensitive to the input frequency modulation.

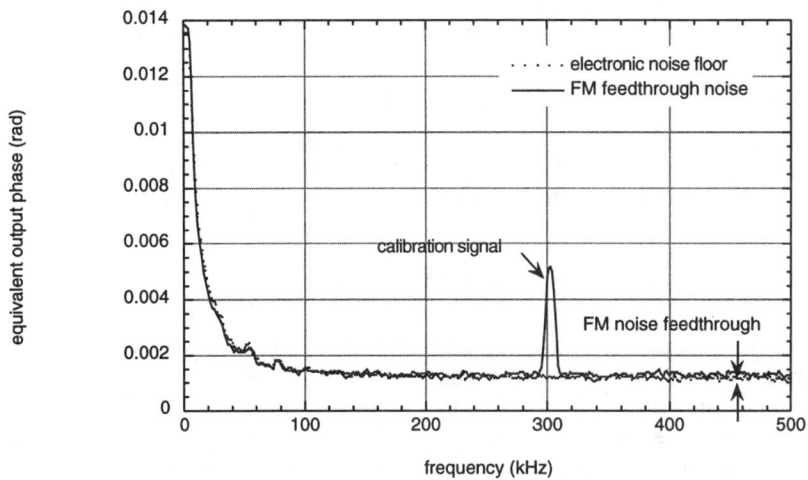

FIGURE 4. Measurement of the frequency noise suppression of the polarization Sagnac interferometer. A broadband phase modulation is imposed on the input light to the interferometer with a modulation depth of 2.5 rad. The output power levels are calibrated to an amplitude modulated signal with a modulation depth of 0.005. The measurement bandwidth is 9.1 kHz. The dotted line is the electronic noise floor of the interferometer. The difference between the two traces represents the conversion of the 2.5 rad of input frequency noise to amplitude noise on the detector.

ACKNOWLEDGMENTS

This research was supported by the National Science Foundation (NSF PHY 9630172). We thank Daniel Shaddock for useful discussions relating to this work.

REFERENCES

1 Beyersdorf, P. T., Fejer M. M., and Byer R. L., JOSA B 16, 1354-8 (1999)
2. Tulloch, W. M., Rutherford, T., Gustafson, E. K., Byer, R. L., Journal of Quantum Electronics (Accepted 1999)
3 Takahashi, R., Mizuno, J., Miyoki, S., et al., Physics Letters A 187, 157-62 (1994).
4 Flaminio, R., and Heitmann, H., Physics Letters A 214, 112-22 (1996).

Transfer Functions for Fields in a 3-mirror Nested Cavity

Malik Rakhmanov

LIGO Project, California Institute of Technology, Pasadena, CA 91125

Abstract. Transfer functions for the fields in a 3-mirror nested cavity are obtained. Explicit formulae for their poles and zeros are found. These results are used for simple analysis of the response of power recycling and signal recycling interferometers.

INTRODUCTION

Fabry-Perot cavities are an essential part of the optical topology of emerging interferometric gravitational wave detectors. It is well-known that the sensitivity of these detectors can be improved if one or two external cavities are formed by adding more mirrors to the interferometer. The result is equivalent to having a cavity inside a cavity or a nested cavity. Optical topologies with equivalent nested cavities are power recycling [1], signal recycling [2] and resonant sideband extraction [3]. In the first generation of the interferometric gravitational wave detectors the only equivalent nested cavity appears in the dynamics of the common mode of the arm-cavity motion and the laser frequency. The next generation of detectors may have two or more nested cavities. In this paper we obtain the transfer functions for fields in an arbitrary nested cavity and describe how to calculate the most important parameter of these transfer functions: the lowest order pole.

FABRY-PEROT CAVITY

We begin with a reminder of how the transfer function of a single Fabry-Perot cavity can be obtained. Let the length of the Fabry-Perot cavity be L and the delay time T ($T = L/c$). If the amplitude of the incident field is $E_{\text{in}}(t)$ then the amplitude of the field in the cavity, $E(t)$, is defined by the self-consistent equation

$$E(t) = t_a E_{\text{in}}(t) + r_a r_b E(t - 2T), \tag{1}$$

where $r_{a,b}$ is the reflectivity and $t_{a,b}$ is the transmissivity of the cavity mirrors. Figure 1 shows how the self-consistent field is formed.

FIGURE 1. Self-consistent field in the Fabry-Perot cavity.

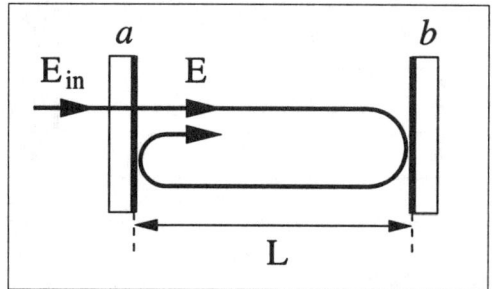

By applying the Laplace transformation to both sides of equation (1) we obtain the relation between the fields in s-domain:

$$\tilde{E}(s) = t_a \tilde{E}_{in}(s) + r_a r_b \tilde{E}(s) e^{-2sT}, \quad (2)$$

where tilde stands for the Laplace transform. Therefore, the cavity acts as a linear operator and its transfer function is

$$H(s) \equiv \frac{\tilde{E}(s)}{\tilde{E}_{in}(s)} = \frac{t_a}{1 - r_a r_b e^{-2sT}}. \quad (3)$$

This transfer function has an infinite set of poles:

$$z_n = z_0 + i\frac{\pi}{T}n, \quad \text{where} \quad z_0 = \frac{\ln(r_a r_b)}{2T}. \quad (4)$$

Here z_0 is the lowest order pole and n is an integer.

NESTED CAVITY

A nested cavity can be formed by adding a mirror to the existing Fabry-Perot cavity, as shown in Figure 2. Then there will be two fields and their amplitudes, E and E', satisfy the self-consistent equations:

$$E(t) = t_a E'(t - T') + r_a r_b E(t - 2T), \quad (5)$$
$$E'(t) = t_c E_{in}(t) - \eta r_a r_c E'(t - 2T') + \eta t_a r_b r_c E(t - T' - 2T), \quad (6)$$

where $\eta = e^{-2i\phi}$ and ϕ is the detuning phase of the outer cavity. Let us apply the Laplace transformation to these equations. The result can be written as a matrix equation for complex 2-vectors:

$$\mathbf{M} \begin{bmatrix} \tilde{E} \\ \tilde{E}' \end{bmatrix} = t_c \begin{bmatrix} 0 \\ \tilde{E}_{in} \end{bmatrix}, \quad (7)$$

FIGURE 2. Self-consistent fields in a 3-mirror nested cavity with partial length L and L'.

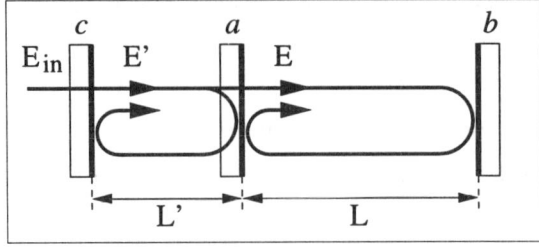

where **M** is the matrix function of s:

$$\mathbf{M} = \begin{bmatrix} 1 - r_a r_b \, e^{-2sT} & -t_a \, e^{-sT'} \\ -\eta t_a r_b r_c \, e^{-sT'-2sT} & 1 + \eta r_a r_c \, e^{-2sT'} \end{bmatrix}. \tag{8}$$

By inverting the matrix we obtain the two transfer functions:

$$G_1(s) \equiv \frac{\tilde{E}(s)}{\tilde{E}_{in}(s)} = \frac{-t_c M_{12}}{\det \mathbf{M}}, \tag{9}$$

$$G_2(s) \equiv \frac{\tilde{E}'(s)}{\tilde{E}_{in}(s)} = \frac{t_c M_{11}}{\det \mathbf{M}}. \tag{10}$$

These transfer functions share the same poles, which are defined by

$$\det \mathbf{M} \equiv 1 - r_a r_b e^{-2sT} + \eta r_a r_c e^{-2sT'} - \eta(t_a^2 + r_a^2) r_b r_c e^{-2s(T'+T)} = 0. \tag{11}$$

In addition, the transfer function $G_2(s)$ has zeros, which are defined by $M_{11} = 0$. Note that the zeros are $s = z_n$, where z_n are given by the equation (4). In other words, the zeros of the outer cavity appear at the frequencies where the poles of the inner cavity were before the nested cavity was formed. This is a general result.

The equation for the poles (eq. (11)), cannot be solved analytically, unless some approximations are made. In the low frequency approximation we replace the exponents by linear functions and obtain the solution for the lowest order pole:

$$s_0 \approx \frac{1 + \eta r_a r_c - r_a r_b - \eta(t_a^2 + r_a^2) r_b r_c}{2[\eta r_a r_c T' - r_a r_b T - \eta(t_a^2 + r_a^2) r_b r_c(T' + T)]}. \tag{12}$$

This approximation is valid as long as $|s_0|T \ll \pi$, that is the frequency of the pole must be much less than the free spectral range of the longer cavity.

Another formula for the lowest order pole can be obtained if one of the cavities is much shorter than the other. Assume that $T' \ll T$ then we can neglect T' in equation (11) and obtain the solution

$$s_0 \approx \frac{1}{2T} \ln \left[\frac{1 + \eta r_a r_c}{r_a r_b + \eta(t_a^2 + r_a^2) r_b r_c} \right], \tag{13}$$

which is valid as long as $|s_0|T' \ll \pi$, that is the frequency of the pole must be much less than the free spectral range of the shorter cavity.

At low frequencies the transfer functions can be approximated up to a constant by a low-pass and a lead-lag filter respectively:

$$G_1(s) \approx \frac{1}{s - s_0}, \quad G_2(s) \approx \frac{s - z_0}{s - s_0}. \tag{14}$$

The formulae above can be used for a simple analysis of optical response of gravitational wave detectors. First consider a power recycling interferometer, for example, the 4km LIGO interferometer. Its equivalent nested cavity appears in the dynamics of the common mode of the arm motion. The target bandwidth of the detector sets the lowest order pole of the arm cavity: $|z_0|/2\pi = 90$ Hz. This condition defines the transmission of the input mirror ($t_a^2 = 0.03$) provided the transmission of the end mirror is fixed ($t_b^2 = 20$ ppm.) The condition for optimal coupling of the laser power to the interferometer defines the transmission of the recycling mirror ($t_c^2 \approx 0.024$). The result of the recycling is that the circulating power reaches 16 kW for an input power of only 6 W. From either equation (12) or (13) with $\eta = 1$ we find that the lowest order pole for the common mode is $|s_0|/2\pi = 1.1$ Hz. Thus the transfer function of the power recycling cavity has a pole at 1.1 Hz and a zero at 90 Hz.

Now consider an interferometer with LIGO parameters, which has a signal recycling mirror instead of a power recycling mirror. Then an equivalent nested cavity appears in the dynamics of the differential mode of the arm motion. Such an interferometer can be analyzed as follows. The optimal coupling of the laser power to the arm cavities defines the transmission of the input mirror $t_a^2 = 120$ ppm under the assumption that the losses are 50 ppm per mirror. This sets the arm cavity pole $|z_0|/2\pi = 0.7$ Hz, and also results in a high circulating power (25 kW for the same 6 W of input power). The reflectivity of the signal recycling mirror will define the detection bandwidth according to the equation (12) or (13) with $\eta = -1$. These equations show that in order to obtain the same 90 Hz-bandwidth the signal recycling mirror must have transmission $t_c^2 = 0.011$. Note that the transfer function of the signal recycling cavity will have a pole at 90 Hz and a zero at 0.7 Hz. This means that in general the response of the cavity will be poor at low frequencies, and, therefore, it will be hard to control the signal recycling mirror. These two examples show how the formulae for the lowest order pole of a nested cavity can be used in a simple analysis of the interferometer response.

REFERENCES

1. Drever, R.W.P, in *Gravitational Radiation*, eds. N. Deruelle and T. Piran, North Holland, Amsterdam, 1983.
2. Meers B.J., *Phys. Lett.* **A142**, 465, (1989).
3. Mizuno, J., Comparison of optical configurations for laser-interferometric gravitational-wave detectors, *MPQ-Report* **203**, (1995).

Signal Extraction and Length Sensing for LIGO II RSE

James Mason and Phil Willems*

*California Institute of Technology, Pasadena, California 91125

Abstract. In anticipation of an upgrade of the LIGO detector to an advanced configuration in 2004, a tabletop prototype of resonant sideband extraction is being designed and constructed at Caltech. We present here two frontally modulated length sensing and control schemes, one in which the signal extraction/recycling mirror is a simple mirror, and one in which it is a Fabry-Perot cavity. Issues regarding the controllability, RF sideband transmission, shot noise, and noise couplings are discussed.

Advanced detectors for the LIGO interferometer are planned which use a configuration in which the position of a mirror at the output is used to tune the frequency response, such as resonant sideband extraction (RSE) [1] or dual recycling [2]. RSE can also allow for the storage time of carrier light in the Fabry-Perot arms to be increased, hence the power recycling in the interferometer can be made smaller, reducing thermal lensing in transmissive optics such as the beamsplitter or input test masses without reducing the bandwidth of the detector to gravitational waves.

The sensitivity of RSE is well understood [3], but issues of control, lock acquisition and signal extraction are still open, and are the subject of our research. We are considering schemes that, like LIGO I, employ frontal modulation, and which use a set of fixed sidebands on the optical carrier so as to guarantee transmission of all sidebands through an input mode cleaner before entering the interferometer, regardless of the tuning of the output mirror.

In LIGO, the signal extraction and control are sensitive to the sideband transmission through the interferometer. The RF sideband transmission to the dark port is well approximated by a three-mirror coupled cavity consisting of the input power recycling mirror, the compound mirror formed by the Michelson interferometer (i.e., the beamsplitter and arm mirrors), and the output signal extraction mirror. In the broadband mode of RSE, where the signal extraction cavity (SEC) is resonant for the optical carrier, both RF sidebands are resonant in this three-mirror cavity and are efficiently transmitted to the output. However, in detuned RSE, when the signal extraction cavity is not resonant for the optical carrier, the

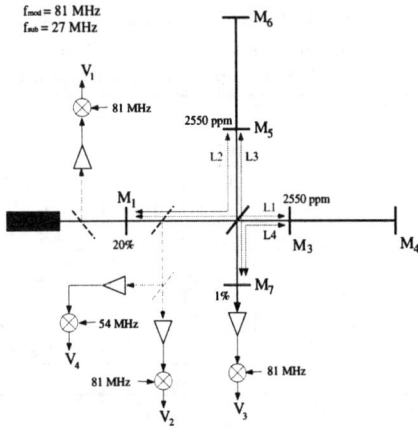

FIGURE 1. Optical configuration for RSE with a single output mirror.

two RF sidebands cannot both be transmitted with high efficiency, and they will resonate unequally in the interferometer.

We have identified two schemes for implementing RSE which address the problem of RF sideband transmission while detuned. The first, or single-sideband, scheme (SSB) accepts the loss of one RF sideband and detunes by making a macroscopic change in the signal extraction mirror position, thus allowing one RF sideband to resonate in the SEC. Once this macroscopic position is chosen there is a limited range of additional detuning that is gotten by microscopic length changes to the SEC. Compared to having both RF sidebands at the dark port, this scheme gives one-half the signal but also $\sqrt{3}$ less shot noise, so the sensitivity is degraded only slightly. However, the unequal power buildup in the two RF sidebands introduces cross couplings into many of the degrees of freedom of the interferometer as seen by the length sensing pickoffs.

The second scheme is to replace the signal extraction mirror with an output coupling cavity (OCC) that is optimally coupled for the RF sidebands and antiresonant for the carrier. This makes the OCC 'invisible' to the RF sidebands so that they resonate in the interferometer exactly as in the well-understood power-recycled case. The RF sidebands transmit with high efficiency for all detunings, which can be obtained through only microscopic changes in the SEC length, and the standard shot-noise-limited sensitivity is restored. All this comes at the price of additional complexity at having to control another suspended mirror.

The SSB scheme (displayed in Figure 1) requires an independent local oscillator to measure the RF sideband phase variation in the SEC. We add a frequency-shifted subcarrier at one-third the RF modulation frequency f_{mod}. This frequency can pass through an input mode cleaner and resonate in a power recycling cavity of suitable length. We then use the beatnote at $f_{\mathrm{mod}} - f_{\mathrm{mod}}/3$ at the PRC pickoff (labelled

TABLE 1. Matrix of discriminants for the single-sideband RSE scheme. Initial values are for no detuning (broadband case), and values in parentheses are for high detuning (narrowband case). Boldface values represent signals used for length sensing.

PD	Φ_+	Φ_-	$I_{81\text{MHz}}$ ϕ_+	ϕ_-	ϕ_s
1	130(-1400)	0(0)	**-0.26(.50)**	0(-.53)	-3.7(-1.6)
2	**-1.6e4(-2.3e4)**	0(-1.7e-2)	1.1(-4.3)	0(-26)	-180(-80)
3	0(0)	**73(43)**	0(0)	.046(.027)	0(0)
			$Q_{81\text{MHz}}$		
1	0(-730)	-1.1e-3(0)	0(.32)	-1.6(-1.2)	0(.27)
2	0(-3600)	-5.1e-2(-3.9e-2)	0(-4.5)	**-81(-61)**	0(13)
3	0(0)	0(-14)	0(0)	0(-9.0e-3)	0(0)
			$I_{54\text{MHz}}$		
1	-1.2e-3(-1.2e-3)	0(0)	-.11(-.11)	0(0)	-2.0(-2.0)
2	-1.3e-2(-1.3e-2)	0(0)	.88(.88)	-7.4e-3(-7.3e-3)	**-19(-19)**
3	0(0)	0(0)	0(0)	5.3e-3(5.3e-3)	0(0)

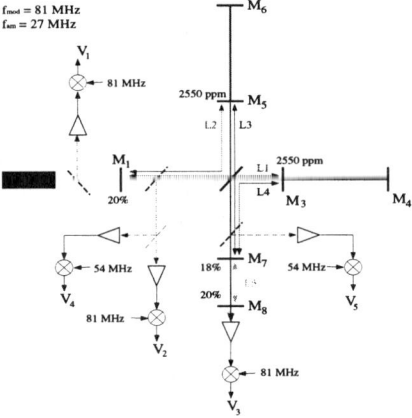

FIGURE 2. Optical configuration for RSE with an output-coupling cavity.

V_4) to control the SEC length. The control signal for the SEC is independent of detuning; however, those for the other degrees of freedom of the interferometer are not. As mentioned earlier, cross-couplings occur and are strong when the SEC is slightly detuned. For highly detuned RSE, the cross couplings are weaker, but the signal gains tend to change by as much as an order of magnitude for reasonable ranges of detuning. Sample control matrices for the broadband and detuned cases are shown in Table 1. Note that the cavity signals are taken from the demodulation of the optical signals at the pickoffs both in phase with (I) and in quadrature to (Q) the local oscillator.

The OCC scheme uses AM sidebands at $f_{\text{mod}}/3$, as well as another pickoff signal

TABLE 2. Matrix of discriminants for the output-coupling cavity scheme.

PD	Φ_+	Φ_-	$I_{81\text{MHz}}$ ϕ_+	ϕ_-	ϕ_s	ϕ_{occ}
1	73(72)	0(0)	**.15(.15)**	0(0)	0(0)	-.22(-.22)
2	**1.7e4(1.7e4)**	0(0)	-.71(-.71)	0(0)	-2.5e-2(-2.5e-2)	-11(-11)
3	0(0)	9.2(9.2)	0(0)	5.8e-3(5.8e-3)	0(0)	0(0)
4	0(0)	0(0)	0(0)	0(0)	0(0)	0(0)
			$Q_{81\text{MHz}}$			
1	0(0)	0(0)	0(0)	9.7e-2(9.7e-2)	0(0)	0(0)
2	0(0)	3e-3(3e-3)	0(0)	**4.7(4.7)**	0(0)	0(0)
3	0(0)	0(0)	0(0)	0(0)	0(0)	0(0)
4	0(0)	**-77(-77)**	0(0)	-4.9e-2(-4.9e-2)	0(0)	0(0)
			$I_{54\text{MHz}}$			
1	0(0)	0(0)	-.18(-.23)	0(.12)	0(0)	.19(.17)
2	0(0)	0(0)	.30(.34)	-1.3(-.87)	**-1.5(-2.1)**	-.21(-.30)
3	0(0)	0(0)	0(.32)	0(.86)	0(.18)	**-.65(-.62)**
4	0(0)	0(0)	0(0)	6.3e-2(6.6e-2)	0(0)	0(0)

(shown as V_5 in Figure 2). The beat note at this pickoff provides signal for length control of the OCC, while the pickoff at the PRC provides signal for the SEC length. As seen in Table 2, the control matrix for this scheme is more diagonal for both the broadband and the detuned case. The degrees of freedom for the power-recycled Michelson portion of the interferometer are barely dependent upon detuning. The degrees of freedom for the signal extraction cavity and output coupling cavity are also only weakly dependent on detuning for many designs. The layout for the OCC scheme is shown in Figure 2.

We have begun experiments to test these LSC schemes using a fixed-mirror tabletop interferometer before trying them on a more representative and more complicated interferometer using suspended mirrors. These experiments will help to evaluate the advantages and disadvantages of the two schemes and to investigate the process of interferometer lock acquisition. Our power recycled arm cavity has been tested and characterized (our arm finesse is $\sim$1000), and we have characterized our power recycled Michelson and signal recycled Michelson interferometers without the arm cavities and found them to be in excellent agreement with our models. We have also locked the dual-recycled Michelson interferometer, and are beginning to introduce the $f_{\text{mod}}/3$ subcarrier into the interferometer and test detuned RSE schemes.

REFERENCES

1. Jun Mizuno, Ph.D. thesis, Max-Planck-Institut für Quantenoptik, 1995.
2. Strain, K. A. and B. J. Meers, *Physical Review Letters* **66**, 1391 (1991).
3. Heinzel, G. *et al.*, *Physics Letters A* **217**, 305 (1996).

LASERS AND OPTICS

The GEO 600 Stabilized Laser System and the Current-Lock Technique

B. Willke[1,2], O. S. Brozek[1,2], K. Danzmann[1,2], C. Fallnich[3],
S. Goßler[1], H. Lück[1,2], K. Mossavi[1], V. Quetschke[1], H. Welling[3]
I. Zawischa[3],

[1] *Institut für Atom- und Molekülphysik, Universität Hannover, Callinstr. 38, D-30167 Hannover, Germany*
[2] *Max-Planck-Institut für Quantenoptik, Hans-Kopfermann-Str. 1, 85748 Garching, Germany*
[3] *Laser Zenrum Hannover e.V., Hollerithalle 8, D-30419 Hannover, Germany*

Abstract.
This talk will give an overview over the GEO 600 laser-diode pumped Nd:YAG laser system. After an introduction which defines the requirements, we describe the laser design and the laser frequency stabilization scheme.

Due to its low power noise in the radio frequency region and due to its good spatial beam quality, an injection-locked master-slave design is used for the GEO 600 laser. This laser system has an output power of 12 W and a good spatial profile ($M^2 \leq 1.05$).

A monolithic non-planar ring-oscillator (NPRO) with an optical power of 0.8 Watt is used as the master laser. By stabilizing the NPRO to a reference cavity we achieve an frequency noise spectral density of less than $1\text{mHz}/\sqrt{\text{Hz}}$ for fourier frequencies between 10 Hz and 1 kHz at the feed-back loop error point.

Finally we present a new scheme to stabilize the frequency of a NPRO by feeding back to the power of its laser-diode pump-source. First experiments with this so-called current-lock technique led to a robust lock and simultaneously to a reduction in power-noise of the NPRO without an additional power stabilization control system.

INTRODUCTION

All gravitational wave detectors currently under construction will use laser diode pumped Nd:YAG laser systems as their light source. The reasons for the selection of Nd:YAG systems are their high efficiency and the availability of a very stable master-oscillator, the so-called NPRO (Non-Planar Ring-Oscillator). Especially the low free-running frequency-noise of the NPRO makes it possible to reach the demanding frequency stability requirements of gravitational wave detectors. Although the GEO 600 detector use a dual-recycled interferometer with a folded optical path in the interferometer arms, the stability requirements for the laser system are very similar to the needs of the LIGO [2], VIRGO [7], and TAMA [10] detectors, which use different optical layouts.

To reach shot noise limited performance of the GEO 600 detector for fourier frequencies above 200 Hz a light power of 5 W needs to be injected into the power recycling cavity. Due to losses in the optical chain before the power recycling mirror an output power of the pre-stabilized laser system of more than 10W is anticipated. The power-stability requirement on the GEO 600 laser is set by analyzing two different paths on which power noise can couple into the interferometer output: a) Deviations from the dark fringe locking point of the interferometers will transfer low frequency power noise of the laser into an artificial gravitational wave signal. The power stability requirements in the low fourier frequency region are plotted in Figure1. b) In the radio-frequency region amplitude modulation of the laser light will change the amplitude of the modulation sidebands and therefore couple directly into the heterodyne readout of the interferometer. As in the GEO 600 detector the light passes sequentially two suspended modecleaners we decided to phase modulate the laser beam with the heterodyne rf-frequencies ω_{rf} behind the first modecleaner. This allows the first modecleaner to act as a passive power noise filter at ω_{rf} and relaxes the power noise requirements of the pre-stabilized laser system in the rf-region.

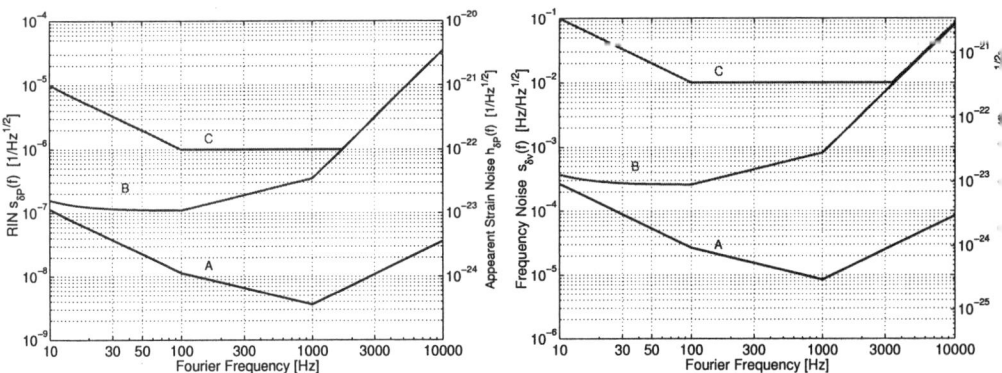

FIGURE 1. Requirements for the power-stability (left) and frequency-stability (right) of the GEO 600 laser system. The upper curves labeled with C are the demanded requirements for the pre-stabilized laser, the curves labeled with B reflect a limit for the fluctuations allowed for light entering the power recycling cavity and the lower curves show the stability requirements at the beam splitter.

The frequency stability requirements for the pre-stabilized laser system are also plotted in Figure1. GEO 600 will use a method based on the Schnupp locking-scheme to control the interferometer longitudinal degree of freedom. As this method requires a small difference in the interferometer armlength, $\Delta L \leq 10$cm, frequency noise of the laser system will be transfered into a signal at the gravitational wave output port of the detector.

Both graphs in Figure 1 show three different curves: the upper curves labeled with C are the demanded requirements for the pre-stabilized laser, the curves labeled with B reflect a limit for the fluctuations allowed for the light entering the power recycling cavity and the lower curves show the stability requirements at the beam splitter. The difference between the curves B and C defines the noise reduction that needs to be achieved by the so called second-loop control systems which measure the fluctuations of the light before the power recycling cavity and feed back to the laser system to reduce these fluctuations. Therefore the pre-stabilized laser system has to provide low-noise actuators for these feed-back control systems.

In this contribution we introduce the layout and performance of the GEO 600 high-power laser-system and the frequency stabilization scheme employed. The free-running noise spectral density as well as the achieved noise reduction will be presented and the source of the remaining fluctuations will be discussed. Finally we present a new scheme to stabilize the frequency of a NPRO by feeding back to the current of its laser-diode pump-source and the advantages of this scheme will be discussed.

THE INJECTION-LOCKED LASER-SYSTEM

The GEO 600 high-power Nd:YAG laser system uses the injection-locking method [9] to transfer the high frequency stability of a low-power master laser to a high-power oscillator, called the slave laser. A schematic drawing of the GEO 600 laser is shown in Figure 2. The master laser used in the GEO 600 laser system is a laser-diode pumped monolithic Nd:YAG ring-laser (NPRO) with an output power of 0.8 W (Innolight, Modell Mephisto 800). The advantage of the monolithic design invented by Kane and Byer [6] is the high intrinsic frequency and power stability which is orders of magnitude better than the free-running stability of Ar^+ lasers. After passing an optical diode to isolate the NPRO against light coming from the high-power slave and an electro-optical phase modulator used in the injection-locking control scheme, the light of the NPRO is injected into the slave cavity. Two Nd:YAG rods each of which is end pumped by a fiber-coupled laser-diode-module with 17 W optical power are used as the gain elements in a four mirror slave-oscillator ring-cavity. One of the mirrors is mounted on a piezo-electric transducer (PZT) to control the length of the cavity in order to keep the difference between the frequency of the slave laser and the NPRO frequency well within the injection locking range of 1.6 MHz. The control loop bandwidth of approximately 10 kHz was limited by mechanical resonances in the PZT. To increase the mechanical stability of the slave laser cavity the mirrors and the PZT were mounted on a rigid copper spacer. This reduced the vibrations of the slave cavity in the acoustic frequency region by an order of magnitude.

Two brewster plates in the slave laser cavity compensate for the astigmatism introduced by the non-normal reflection of the curved mirrors, reduce the depolar-

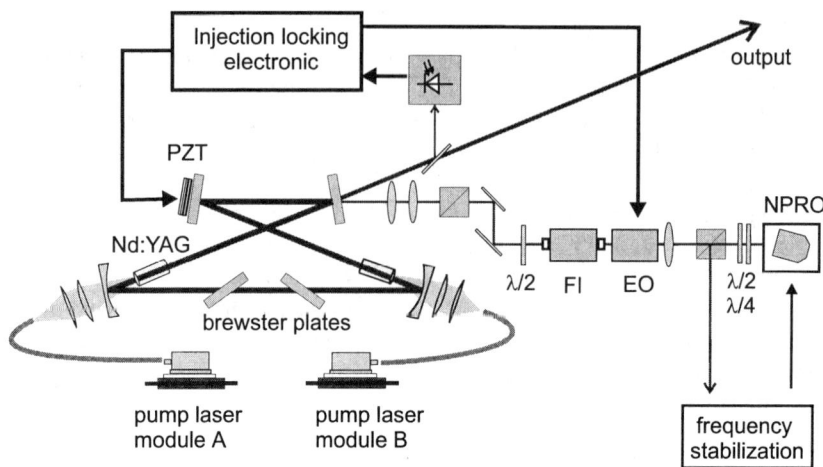

FIGURE 2. Setup of the GEO 600 high power laser system. A frequency-stabilized NPRO with 0.8 W output power is used as a master laser. The injection-locked slave laser with 12 W output power and a $M^2 \leq 1.05$ is pumped by two fiber-coupled diode-laser-modules each with 17 W optical power.

ization losses and define the polarization of the 12 W output beam. The spatial profile of the high-power laser beam was measured to have a M^2 value of less than 1.05 which agrees well with the maximum light power of 95 that could be coupled into the TEM_{00} mode of a cavity.

FREQUENCY STABILIZATION

With a robust and high-bandwidth injection-locking control loop in place, the frequency-stability of an injection-locked laser system is determined by the stability of its master laser [1], [4]. Therefore a frequency stabilization of the NPRO to the resonance frequency of a high-finesse ring-cavity could be used to reduce the frequency fluctuations of the high-power system. The ring-cavity used for GEO 600 is formed by three mirrors which are optically contacted to a rigid spacer made from ultra-low expansion material (ULE). To avoid contaminations and length fluctuations of the cavity introduced by acoustics the reference cavity was placed inside a high-vacuum-system ($p \leq 1 * 10^{-8}$ mbar). We used the Pound-Drever-Hall method to get an error-signal for the frequency stabilization loop. By feeding back to the PZT frequency actuator of the NPRO the laser was first locked with a low bandwidth loop to the cavity to perform a measurement of the free-running frequency noise of the NPRO (see Figure 3, upper curve). According to the work of Day et al. [3] this free-running frequency noise is mainly due to fluctuations of the index of refraction in the NPRO crystal caused by fluctuations of the pump-diode power.

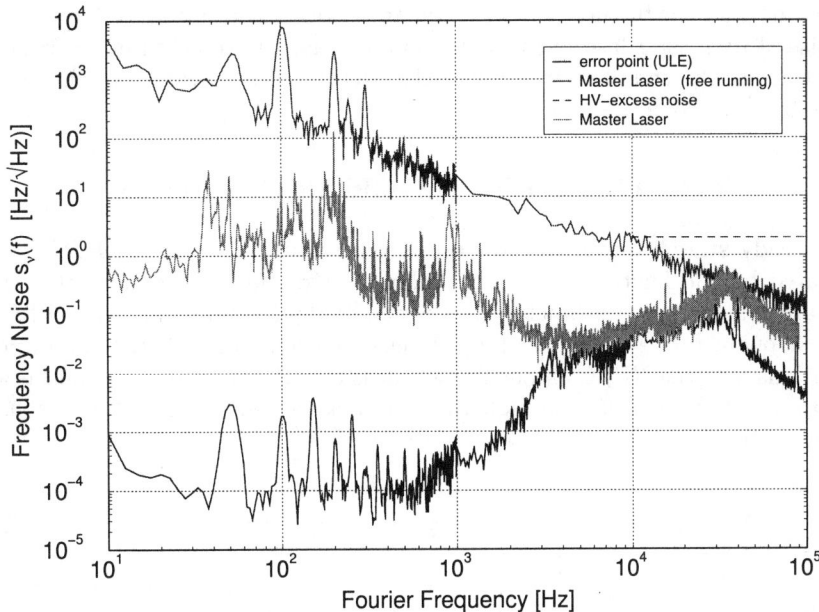

FIGURE 3. Frequency noise of the free-running NPRO (upper curve), of the stabilized NPRO at the error point of the stabilization loop (lower curve) and of the stabilized NPRO with respect to an analyzer cavity (middle curve).

By adding a fast electro-optical phase shifter to the control system we increased the bandwidth to 1 MHz and achieved high low-frequency gain. The lower curve of Figure 3 shows the spectral density of the error point noise of this feed-back loop. The high noise reduction at low fourier frequencies reflects the high gain of the control system in this frequency region.

By splitting the NPRO beam and shifting the frequency of one part of it with an acousto-optical modulator (AOM) we were able to lock the NPRO to the reference cavity and simultaneously couple a fraction of the light into an analyzer cavity. A second Pound-Drever-Hall scheme was used to measure the frequency fluctuations of the frequency-shifted NPRO beam with respect to the analyzer cavity. The middle curve in Figure 3 shows the measured noise spectral density which is significantly higher than the error-point noise (lower curve.) This additional noise is probably due to mechanical resonances of the optical table as well as vibrations coupled into the reference and analyzer cavity. We expect, that a pendulum suspension of the cavities will reduce the out-of-loop noise significantly as it was demonstrated by Nakagawa et al. [8].

By feeding back to the AOM we were able to lock the frequency-shifted NPRO to the analyzer cavity. The measured drift between the reference and the analyzer cavities was 300 Hz/s.

To assure that the frequency noise of the injection-locked system is dominated by the noise of its master laser we used the analyzer cavity to measure the frequency fluctuations of the complete injection locked system. No difference to the noise of the master laser could be found.

THE CURRENT LOCK TECHNIQUE

As already stated above, the free-running frequency noise of a NPRO laser is dominated by the fluctuations of the index of refraction in the Nd:YAG crystal which are caused by power-fluctuations of the diode-laser pumping the NPRO [3]. To get a quantitative understanding of the cross-coupling between pump-power fluctuations and the NPRO frequency we measured the transfer function between fluctuations of the current through the pump-laser-diodes and NPRO frequency (see Figure4).

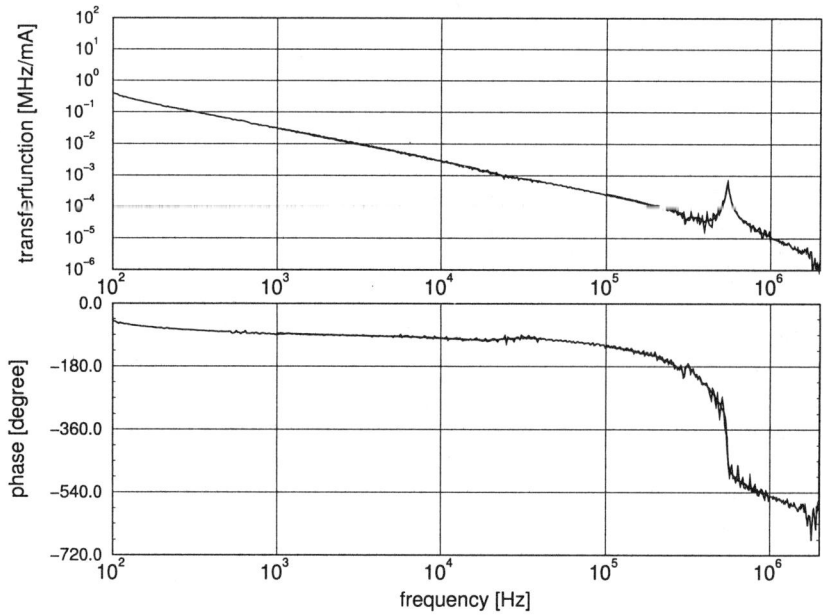

FIGURE 4. Transferfunction between fluctuations of the current through the pump-laser-diodes and NPRO frequency. The structure around 500 kHz corresponds to the relaxation oscillation of the used NPRO.

Based on this transfer function we designed a controller to lock the laser frequency to a reference cavity by feeding back to the current of the NPRO pump laser-diodes.

This so-called current-lock [11] has several advantages:

-To design a high-bandwidth frequency-stabilization control system only one actuator is required (instead of the PZT and the phase-correcting pockels-cell needed in the conventional scheme).

-The pump-diode-current actuator needs only a low voltage signal (compared to the PZT and the pockels cell which both need high voltage amplifiers).

-No beam-pointing is introduced by changing the pump-diode-current (instead of the pointing introduced by the PZT actuator).

-No PZT frequency-actuator is needed on the NPRO which makes the fabrication easier and allows a better thermal engineering and a different design of the magnetic-field that is needed to assure single direction operation of the NPRO.

In addition to the demonstration of a stable lock one very interesting result was obtained: Instead of increasing the NPRO-power-fluctuations by feeding a signal to the pump-diode-current, the NPRO power fluctuations were reduced by about 6dB. The reason is, that the current lock decreases the pump-diode-power fluctuations which are also driving the power fluctuations of the NPRO. Earlier experiments [5] to reduce the NPRO intensity noise were limited by the spatial variation of the pump-laser-fluctuations and the lack of information on which spatial part of the pump beam is sensed by the NPRO. The NPRO frequency deviations used as an error signal in the current-lock technique, however are only sensitive to pump-diode-fluctuations in the active path of the NPRO crystal.

Further current-lock experiments are planned to show how much NPRO power-noise-reduction is achievable and if the current lock technique can be employed in the stabilization scheme of the GEO 600 laser.

ACKNOWLEDGMENTS

This work was supported by the Volkswagenstiftung of the stare Niedersachsen/Germany.

REFERENCES

1. R. Barillet et al. *Meas. Sci. Technol.*, **7**, p. 162 (1996)
2. M. Coles, this conference
3. T. Day, Ph.D. thesis, Stanford University, USA (1990)
4. A.D. Farinas et al. *J. Opt. Soc. Am.*, **12**, p. 328, (1995)
5. C. C. Harb et al. *J. Opt. Soc. Am.*, **11**, p. 2936 (1997)
6. T. J. Kane et al. *Opt. Lett.*, **10**, p. 65 (1985)
7. F. Marion, this conference
8. K. Nakagawa et al. *Appl. Opt.*, **33**, p. 6383 (1994)
9. A. E. Siegman,"Lasers", University Science Books (1986)
10. K. Tsubono, this conference
11. B. Willke et al., Opt. Letters, in preparation

The influence of X-ray damage on high purity sapphire optical absorption and investigation on the origin of the residual absorption @1064 nm

F. Benabid, M. Notcutt, L. Ju, D.G. Blair

Physics Department, The University of Western Australia
Nedlands, W.A. 6907, Australia

Abstract. We show that the optical absorption of high purity sapphire is degraded by X-radiation. We suggest that the level of absorption at 1064 nm is due to the presence of complex clusters of Ti^{3+}, Ti^{4+}, Fe^{2+}, Fe^{3+} and oxygen vacancies, responsible for the well known 800 nm peak, called the infrared residual absorption, as well as for a 920 nm and a 1100 nm absorption bands. We show that the annealing can reduce the effect of X-ray on the absorption level from an average level of ~80 ppm/cm to ~24 ppm/cm. These results indicate that a further reduction of Ti and Fe in sapphire (<0.1 ppm) as well as oxygen vacancies, which play the role of precursors in the formation of the mentioned clusters, would reduce the absorption level.

INTRODUCTION

The results of the measurements on sapphire mechanical Q-factor [1] and optical properties [2-4] show that sapphire is an excellent candidate material for Laser interferometric gravitational-wave (LIGW) detectors. An absorption level of 3 ppm/cm has even been observed [3] on a high purity heat exchange grown sapphire sample. However, upon X-radiation of the samples for XRF spectroscopy purposes, this level has not been reproduced and the different absorption measurements on the same material gave a level 10 times higher with inconsistent results and show a wide range of absorption coefficient within one sample (from ~10 ppm/cm to >100 ppm/cm).

In this paper, we explain the rise of the absorption at 1064 nm of the samples mentioned in reference 3 and a way to decrease it using a set of experiments, including XRF spectroscopy, photothermal absorption measurement, UV-Vis photospectroscopy and annealing. Moreover, in terms of improving the transmission of sapphire material at 1064 nm it is important to identify the nature of the parasitic absorbers in sapphire crystals. Among those absorbers there are impurities (extrinsic defects) such as Ti, Cr and Fe, and vacancies (intrinsic defects) such as Oxygen vacancies which have been identified in several reports in heat exchange grown sapphire. Using the results of recent works [5-7] we suggest that the level of absorption at 1064 nm is due to the presence of clusters containing Ti^{3+}, Ti^{4+}, Fe^{2+}, Fe^{3+} and oxygen vacancies. Those defects have been assigned as responsible for the well known 800 nm peak, called the infrared residual absorption, and of the 2 wide peaks centered at ~ 920 nm and ~1100 nm. We show that the annealing can reduce the effect of X-ray on the absorption level. Moreover, with the 20 ppm/cm of absorption reached here, sapphire is a very promising test mass material for the next generation interferometers.

EXPERIMENT

The 2 sapphire samples used in this work are of a high purity grade from Crystal Systems [8] and grown by heat exchange method. They are both of cuboidal shape (Typically 20 x 10 x 10 mm^3). One is CSI-White grade and the other is Hemex-Ultra grade, which has a lower content of impurities than the Csi-White grade. Precedent photothermal absorption measurements on the as-grown samples gave 55 ppm/cm for the Hemex-Ultra sample and 3.4 ppm/cm for the CSI-White sample [3].

The samples have followed the following sequence: XRF spectroscopy, UV-Vis photospectroscopy and photothermal absorption measurements at 1064 nm, and then annealing. Finally, a photothermal absorption measurement and UV-Vis absorption spectrum were taken on the samples.

XRF spectroscopy is done using a tube with Sc and Mo sources with 90 kV and 40 mA of accelerating voltage and current respectively, exposing one of the sample surfaces under X-ray beam of a few millimeters diameter, for 1 hour for the CSI-white sample and 1 and a half hours for the Hemex-Ultra sample.

The absorption coefficient measurement at 1064 nm is taken using the photothermal beam deflection method [9,10]. A 30 W cw Nd:Yag laser-pump beam, with a modulated power with a mechanical chopper at a frequency of 128 Hz. is directed on the sample where a fraction of the power is absorbed creating a modulated gradient index. A He-Ne laser-probe beam is deflected by the thermally induced index changes. The deflection signal is then detected by a quadrant photodetector and sent to a Lock-in amplifier and a voltmeter. With the sample being on a translation stage, the measurements were carried out over several points covering a whole surface parallel to the side of the sample facing the pump laser.

The UV-Vis absorption spectra were obtained using a Cray spectrophotometer, and taken in the range 200-1200 nm before annealing and 200-1000 nm after annealing.

RESULTS

The XRF measurement has not given a quantitative result on the amount of impurities present in the samples but rather traces (<10 ppm) of the presence of Ca, Mo, Fe, Ti, which agree with the results of McGuire [11]

Figure 1 shows the map of the photothermal absorption measurement results on the Hemex-Ultra sample and represents the typical procedure for all the photothermal measurements carried out on the samples before and after annealing. The results are summarized in table 1 where a statistic of the different measured points over the scanned surface is shown. Figure 1 clearly shows the effect of the X-ray beam on the exposed surface. A similar pattern was present in the Csi-white sample before annealing also. The assumption that the pattern is the result of X-ray exposure is corroborated by the same absorption measurements done on another Csi-White sapphire which has been exposed to X-ray on a known area of the sample and which clearly shows an enhancement of the absorption level at the mentioned area compared with the rest of the scanned surface of the sample.

The results show that for the X-irradiated sample and before annealing (see Fig. 1) the absorption coefficient achieves a maximum of 109 ppm/cm, a minimum of 53 ppm/cm for the Hemex-Ultra sample and a maximum 76 ppm/cm and a minimum of 46 ppm/cm for the Csi-White sample.

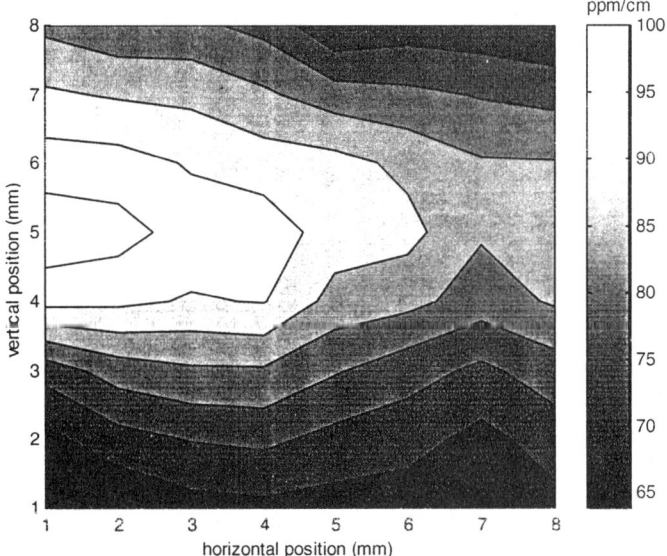

FIGURE 1. Map of absorption coefficient in ppm/cm over a determined surface of Hemex-Ultra sample before annealing.

TABLE 1. Results of the absorption measurements scan

Statistics of the absorption coefficient (ppm/cm) over the scanned surface	Hemex-Ultra		Csi-white	
	Before Annealing	After Annealing	Before Annealing	After Annealing
Maximum absorption coefficient	109	25	76	68
Minimum absorption coefficient	53	23	46	30
Average absorption coefficient	81	24	64	40
Standard deviation	12	0.6	7	9

After annealing, the absorption decreased in both samples. The Hemex-Ultra sample shows a much higher decrease in the absorption level which was brought back to an average of 24 ppm/cm and very uniform pattern, whereas the difference in absorption coefficient between two random point of the scanned surface is less than 2 ppm/cm which is very close to the 1 ppm/cm uncertainty of our measurement set-up (see table 1). However, although the Csi-White sample shows a decrease in absorption coefficient level and a more uniform pattern compared with the one before annealing, the average absorption coefficient is 40 ppm/cm which is very close to the varied results on different as-grown samples of Csi-White grade.

UV-Visible absorption measurements also show obvious changes in the spectrum of the X-irradiated samples compared with the as-grown one (see Fig. 2 and Fig. 3). At the UV region, the F-band (201 nm) and the narrow F^+-band [12] (254 nm) increased upon X-radiation and two new peaks centered at ~255 nm, ~321 nm appeared. In the near infrared range, we notice an enhancement of the hump at ~800 nm and a clear band centered at ~921 nm which we have not been able to see with as-grown sapphire because of the 900 nm upper limit of the apparatus.

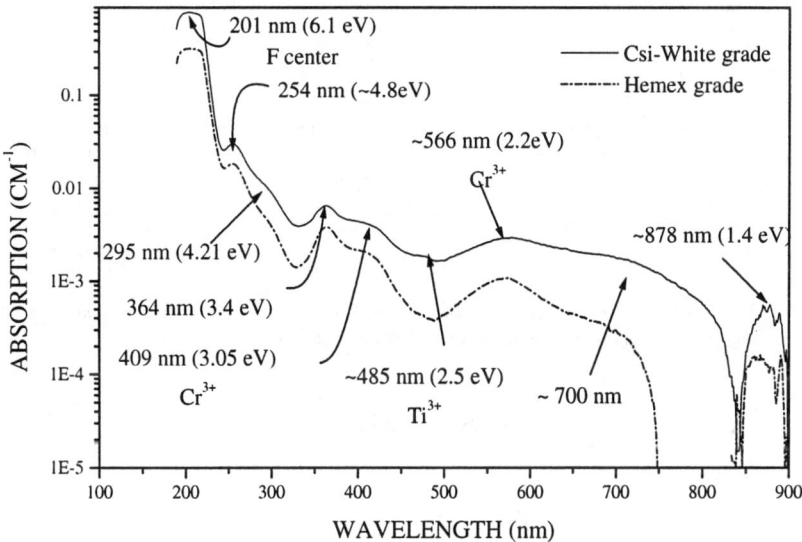

FIGURE 2. Typical UV-Vis spectrum of as-grown high purity sapphire from Crystal Systems. Continuous line for Csi-White grade sample and dot-dashed line for Hemex grade sample

The spectra of the annealed samples also show changes with respect to one of the same samples before annealing (Fig. 3). In both samples the F-band decreased, the 321 nm band vanished and the 255 nm increased. Moreover, the 921 nm decreased in both samples, the 800 nm hump witnessed a slight increase in the Csi-White sample while it diminished significantly in the Hemex-Ultra sample (Fig. 3b). The 486 nm band, which is due to Ti^{3+} decreased in the Csi-White sample after annealing.

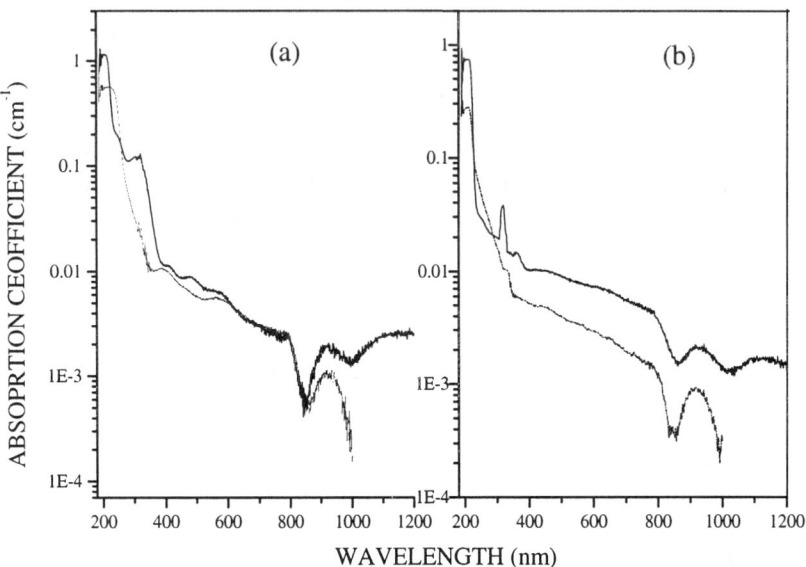

FIGURE 3. UV-Vis-IR absorption spectrum of (a) Csi-White sample and (b) Hemex-Ultra sample before annealing (continuous lines) and after annealing (dashed lines)

DISCUSSION

Absorption bands assignment

The rise of the dominant absorption occurring at ~201 nm after X-radiation, due to F centers (2 electrons trapped in oxygen vacancy) [6,13,14] and known to be present in sapphire from Crystal Systems because the crystals are grown in a vacuum, which results in a stoichiometric imbalance of cations and anions, is explained by the fact that electrons and holes are constantly trapped at and released from defect centers throughout irradiation which results in the change in the defect concentrations and thus on the overall absorption spectrum after irradiation. However, the X-irradiated Csi-White sample shows a wider peak due to ~255 nm band (Fig. 3a). The assignment of this band is controversial. Indeed, Zha et al [15] gave a theoretical calculation based on a cluster model of Ti^{3+}-Ti^{4+} pairs, where he explains the ~800 nm large band (known residual IR absorption) and the UV hump at ~255 nm observed in Ti doped sapphire. The model also predicts an absorption band at around 294 nm (14 $A_1 \rightarrow$ 17E transition) mainly due to Ti^{3+} instead of the Ti^{3+}-Ti^{4+} pair. Moreover, Evans corroborated the latter assignment by assigning the 255 nm hump to Ti^{4+} which should not be confused with the narrow absorption band centered at ~254 nm (see Fig. 2) and associated with F^+ [16]. On the other hand, optical absorption measurements on Iron doped sapphire done by Ferguson et al. and Blum et al. [17,18] show that the 255 nm hump is linked with the iron. The stronger presence of this shoulder in Csi-white sample than in that of Hemex-Ultra clearly shows that the 255 nm hump is impurities related, and its enhancement in Csi-White after annealing along with the decrease of the Ti^{3+} absorption band (489 nm) (see Fig. 3a), due to the oxidizing atmosphere of the annealing, leading to an increase of the Ti^{4+} ions which is corroborated in the very slight increase of the hump around ~800 nm, pleads that the 255 nm shoulder is due to Ti^{4+}.

The ~321 nm band, present in both X-irradiated samples (Fig. 3) is assigned by Moon et al [5] to exchanged- coupled pairs of Fe^{3+}-Fe^{2+} provoked by a complex defect cluster such as oxygen vacancies. Rumyantsev [19] demonstrated that this band corresponds to tetracoordinated trivalent iron which agrees with our oxidizing annealing, resulting in a decrease of Fe^{2+} and increase of Fe^{3+}.

Moon et al. [5] studied the polarization, isothermal, and the isochronal behavior of the optical bands in Ti and Fe doped sapphire. He concluded that the 571 nm and 700 nm bands (Fig. 2) are related to an electronic transition involving Fe^{2+} and Ti^{4+}, while the 878 nm band is associated with Fe^{2+}. Accordingly, the 700 nm has been assigned to an intervalence charge transfer between edge-sharing Fe^{2+}-Ti^{4+}. More precisely the 700 nm band has been assigned to an intervalence charge transfer in a complex cluster containing Fe^{2+}, Ti^{4+} and oxygen vacancies. In the same way the 878 nm band is also the result of a complex cluster including Ti^{3+}, Ti^{4+}, Fe^{2+}, Fe^{3+} and oxygen vacancies.

Fig. 3. shows clearly a peak at ~800 nm which has been reported and identified to be due to a complex centre formed by Ti^{3+} and Ti^{4+} [20,21].

The ~920 and 1040 nm band are associated with Fe^{2+}/Fe^{3+} [19]

Taking into account the above assumptions, it is clear that further improvement in sapphire transparency at 1064 nm is possible and requires decrease of the Ti and Fe impurities level to less than 0.1 ppm/cm because of the high absorption cross section of Ti^{4+} at 800 nm band, and the Fe^{2+}/Fe^{3+} cluster responsible of the ~920 nm band. Moreover, a decrease of the oxygen vacancies in heat exchange grown sapphire would diminish the likelihood of clusters and electronic state exchanges to occur.

Sapphire absorption induced noise in LIGW detectors

An absorption level of 20 ppm/cm in post-grown annealed Hemex-Ultra sapphire which we have shown that is reachable is within the requirements of the next generation of the laser interferometric gravitational-wave detector (e.g. LIGO II requires 40 ppm/cm). Moreover, we have shown [23] that the high thermal conductivity of sapphire sets higher tolerable absorption level in interferometers with sapphire test mass than with fused silica ones. For instance, with a level of 20 ppm/cm of absorption in sapphire a circulating power of higher than 40 kW is achievable, while with high transparent fused silica (1 ppm/cm of absorption), the circulating power is limited to 27 kW by optical distortions such as thermal lensing and thermal induced birefringence.

CONCLUSION

The effect of X-radiation on the rise of the sapphire absorption is shown. The annealing brought the level back to ~23 ppm/cm for the Hemex-Ultra sample and 45 ppm/cm for the Csi-White sample. The level of absorption at 1064 nm is mainly set by the contribution of three elements, Ti, Fe and Oxygen vacancies. The latter are responsible for the absorption bands centered at ~800, 900 and 1100 nm. Decreasing the absorption to the level of 1 ppm/cm or below is then possible with a reduction of the titanium and iron content below the ppm level and a post-growth annealing in oxidizing conditions in order to diminish the amount of oxygen vacancies. Although an absorption level of 20 ppm/cm in sapphire is already promising for implementing sapphire material in next generation LIGW detectors, a further improvement by reducing the amount of impurities and a better control of the post growth oxidizing annealing would make sapphire a more attractive material to fulfill the challenging requirements of the next generation laser interferometric gravitational-wave detectors.

REFERENCES

1. V. B Braginskii and et al, *System with small dissipation* (Chicago, 1985).
2. F. Benabid, M. Notcutt, L. Ju *et al.*, Phys. Lett. A (237), 337-342 (1998).
3. D.G. Blair, Frederick Cleva, and C. Nary Man, Optical materials (8), 233-236 (1997).
4. F. Benabid, M. Notcutt, L. Ju *et al.*, Opt. Commun. **167**, 7-13 (1999).
5. A. R. Moon and M. R. Phillips, J. Am. Ceram. Soc **77** (2), 357-367 (1994).
6. B. D. Evans, J. Nucl. Mater. **219**, 202-223 (1995).
7. W. C. Wong and D. S. McClure, Phys. Rev. B **51** (9), 5682-5692 (1995).
8. Crystal Systems and Inc., "Salem, USA," .
9. V. Loriette, Ann. Phys. **23** (2), 1-146 (1998).
10. W. B. Jackson, N. M. Amer, A. C. Boccara *et al.*, Appl. Opt. **20** (8), 1333-1344 (1981).
11. S. C. McGuire, "Trace transition metals in sapphire for high Q microwave resonator applications," presented at the LSC Meetings 5, Stanford University, July 19-22, 1999 (unpublished).
12. B. D. Evans and M. Stapelbroek, Phys. Rev. B **18** (12), 7089-7098 (1978).
13. B. G. Draeger and G. P. Summers, Phys. Rev. B **19** (2), 1172-1177 (1979).
14. B. D. Evans, Gerald J. Pogatshnik, and Yok Chen, Nuclear Instruments and Methods in Physics Research B **91**, 258-262 (1994).
15. F. X. Zha, J. H. Zhang, and S. D. Xia, J. Phys-Cond. mat. **6** (32), 6497-6505 (1994).
16. B. D. Evans, Journal of Luminescence **60-1**, 620-626 (1994).
17. J. B. Blum, H. L. Tuller, and R. L. Coble, J. Am. Ceram. Soc **65** (8), 379-382 (1982).
18. J. Ferguson and P.E. Fielding, Chem. Phys. Let. **10** (3), 262-265 (1971).
19. V. N. Rumyantsev, S. V. Grum-Grzhimailo, and O. N. Boksha, Soviet Physics-Crystallography **16** (2), 373-375 (1971).
20. R. L. Aggarwal, A. Sanchez, R. E. Fahey *et al.*, Appl. Phys. Lett. **48** (20), 1345-1347 (1986).
21. A. Sanchez, A. J. Strauss, R. L. Aggarwal *et al.*, IEEE J. Quantum Electron. **24** (6), 995-1002 (1988).
22. W. C. Wong and D. S. McClure, Phys. Rev. B **51** (9), 5693-5698 (1995).
23. F. Benabid, M. Notcutt, L. Ju *et al.*, Opt. Commun. **170** (1-3), 9-14 (1999).

LISA

Technology of Free Fall for LISA

Stefano Vitale and Rita Dolesi

Department of Physics, University of Trento, Povo, Trento, I-38050, Italy

Abstract. In this paper we discuss the status of technology for achieving free fall of test-masses for LISA. The main part of this effort consists of developing an inertial sensor with parasitic force isolation better than 10^{-15} N/√Hz at 0.1 mHz, and residual coupling of test-mass to spacecraft with stiffness less than 10^{-7} N/m. We also briefly report on experiments to test sensor performances on ground and in space.

INTRODUCTION

LISA (1) test-masses have to be in free fall within $10^{-15} \left[(m/s^2)/\sqrt{Hz}\right]\left[1+(f/3\,mHz)^2\right]$ along the direction of the laser beam coming from the distant spacecraft. In order to achieve this, the spacecraft follows the test-mass acting as a shield against external disturbances like, for instance, solar radiation pressure. As it has been repeatedly shown (2), the major sources of disturbance consist of internal random forces f_{int} acting between the spacecraft and the test-mass, and of the residual coupling of the test-mass to the spacecraft through any parasitic stiffness k_p. This last mechanism perturbs the test-mass free fall as it couples back the motion of the spacecraft $x_{s/c}$ to the test-mass by creating a random force $k_p x_{s/c}$.

In order to reduce spacecraft motion, this is sensed relative to the test-mass, and this information is fed back to an active control loop. This control loop fires a proper set of micro-Newton authority thrusters that force the spacecraft to follow the free-falling test-mass (the so-called drag-free control).

A closer inspection reveals that, within some simplifying assumptions, the test-mass is subject to an overall disturbance force:

$$f_{tot} \equiv m(a-g) \approx \vec{f}_{int} + k_p\left(x_n + \frac{F_{ext}}{k_{fb}}\right) \quad (1)$$

where x_n is the noise affecting the position measurement, F_{ext} represents any external force acting on the spacecraft, and k_{fb} is the displacement-to-force gain of the control loop. Notice that F_{ext} also includes thrusters force noise.

Thus, in essence, the key features to be achieved in this strategy of free fall are a high degree of dynamical isolation, a low motion-sensor noise, a low thrust noise and high force gain for the control loop.

These features are clearly conflicting with each other. Indeed the capacitive, parametric bridge sensor that is currently base-lined for sensing test-mass motion, for instance, introduces both stiffness and parasitic forces. It is easy to calculate (2) that, as the sensitivity of the sensor depends on the voltage bias, a better sensitivity implies larger electrical forces applied onto the test-mass. Actually electric forces are proportional to voltage square, thus a simplistic analysis would show that the lower the bias the smaller would be the overall disturbance to the test-mass. Unfortunately there is an obvious lower limit to this procedure, when other forces begin to dominate over the electrical one.

There are at least three experimental studies on the subject (2)(3)(4). Despite some differences in approach, they all aim at getting inertial sensors with residual stiffness around 10^{-7} N/m and parasitic forces, between the test-mass and the rest of the sensor, below 10^{-15}N/√Hz, while keeping displacement sensitivity better than ≈1÷10 nm/√Hz.

EXPERIMENTAL DEVELOPMENT OF SENSORS OF FREE FALL

A sensor of free fall share many features with an accelerometer or in general with any other motion sensor, differences being related to the numerical values of some key parameters. It is of no surprise then that the experimental groups working to this kind of technology have been tempted by the development of a sensor of free fall. However a few main features have to be drastically changed with respect to a standard accelerometer to fit the specific needs of LISA.

An example of one of the sensors under development (2)(5) is shown in figure 1.

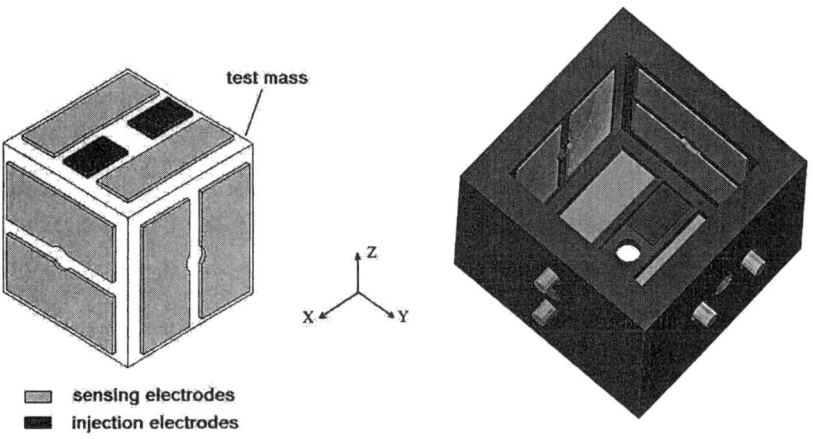

FIGURE 1. View of the inertial sensor.

A cubic metallic test mass is surrounded by electrodes. Each opposite electrodes pairs is part of one of 6 independent parametric capacitive bridges whose purpose is to sense the motion of the test mass along is 6 degrees of freedom.

Contrary to many standard accelerometers, there is no mechanical contact to the test mass, and ac-voltage bias is achieved by auxiliary injection electrodes. This is due to the need to lower any residual dissipation, the source of thermal noise, below a level equivalent to an imaginary stiffness of $\approx 5\ 10^{-14}$ N/m. This figure cannot be achieved, within present day material technology, with any wire of reasonable diameter contacting the test mass.

The most relevant difference with a standard accelerometer is the low level of coupling and displacement sensitivity these instruments aim at. In a standard capacitive accelerometer, gaps between electrodes and the test-mass are small, $d \approx 30\text{-}300$ µm, and ac-voltage bias is high ($V_{ac} \approx$ 1-10 V). This way displacement sensitivities better than 10^{-13} m/√Hz have been achieved.

This however brings about a huge electrical negative stiffness, $k_e \approx -C_o V_{ac}^2/d^2$, where C_o is the capacitance of each electrode toward the test-mass. This can be tolerated for an accelerometer because, for this instrument, the equivalent force noise is, within the same language of eq.(1):

$$f_{noise} \approx \vec{f}_{int} + (k_p - m\omega^2)x_n \qquad (2)$$

where ω is the signal angular frequency. Thus it is not worth reducing the intrinsic stiffness below the value $m\omega^2$ and, in order to reduce the contribution of the sensor noise x_n, the only way is to make it as low as possible. This brings about an increase of k_p that can be obviously tolerated until (6) $k_p \approx m\omega^2$.

The need to lower this stiffness has pushed experimentalist to work to configurations with large gaps and low voltage bias. In the design shown in figure 1, for instance, electrodes are separated form the test-mass by 2 mm gaps and voltage bias of ≈ 300 mV are used. This remembers of what has been used in the pioneering experiment performed on board TRIAD (7), where a sensor with three degrees of freedom was flown to perform the spacecraft drag-free control.

As an alternative, one of the groups (3) is developing a sensor with only one preferred axis with very low stiffness. Along this axis, using electrodes parallel to mass motion reduces the stiffness. One can show that within this arrangement the stiffness can be effectively reduced for a given displacement-to-voltage gain, provided that a good level of cross-talk suppression is achieved with other degrees of freedom.

In standard accelerometer, as the electrostatic stiffness is negative, an electrostatic feedback loop is used to stabilize the test-mass by providing a positive active stiffness. For LISA an active spring has the same damaging effects as any other spring. This is the reason why, even if such a control loop is still needed to stabilize degrees of freedom that are not used to control the spacecraft, it has to provide a low stiffness too. With an intrinsic stiffness of order -10^{-7} N/m, this can be achieved by a positive stiffness of same order; indeed a very weak control loop compared to the $\approx$ N/m typical values of accelerometers that include electrostatic control.

In the most advanced accelerometers, as thermo-mechanical distortion is basically equivalent to displacement noise, materials with very low thermal expansion are usually selected. Glass ceramics with coefficient of linear thermal expansion $< 10^{-7}$ K^{-1} are then preferred choices. For LISA's inertial sensor parasitic forces are more important. Among those, fluctuations in a variety of radiative effects are one of the limiting sources of disturbance. For instance radiometer effect exerts a net force on the test mass $S\dfrac{P}{4}\dfrac{\Delta T}{T}$, with S the area of the lateral surface, P the pressure, T the temperature and ΔT the difference of temperature across the test mass. Similarly, the difference of thermal radiation pressure gives a force $\approx 10 S\dfrac{\sigma}{c}T^3\Delta T$, with σ the Stefan constant and c the speed of light. Thus limiting thermal gradients and their fluctuations becomes an important target and materials with higher thermal conductivity than glass ceramics can be employed.

The device illustrated in figure 1 is machined in Molybdenum with insulating parts made of high thermal conductance ceramics. A linear thermal expansion coefficient of order 5×10^{-6} K^{-1}, like that of Molybdenum, converts into effective displacements of $\approx$ 1 nm/K for standard machining accuracy. As thermal stability of better than 10 μK/√Hz is foreseen for the sensor, the effect of thermo-mechanical distortion is clearly negligible.

The design shown in figure 1, with the proper implementation of the readout electronics, is calculated to achieve a displacement sensitivity of $x_n\approx0.5$ nm/√Hz, a stiffness of order 10^{-7} N/m and a total force disturbance of $\sim 10^{-16}$ N/√Hz at 0.1 mHz. Unfortunately this performance can only be calculated as a full test on ground is prevented by the presence of the gravitational force due to Earth. Devising a strategy to at least partially test those performances is the subject of the following section.

TESTING DYNAMICAL ISOLATION ON GROUND

A standard way to test accelerometers on ground is to lift the test-mass by means of an intense electric field E applied orthogonal to its upper face (with area A). The electrostatic force $(1/2)A\,\varepsilon_o E^2$ overcomes the gravitational one mg when the mass "surface density" m/S> $(1/2)\varepsilon_0 E^2/g$, a number that cannot in practice exceed 45 kg/m^2. LISA test-mass has instead m/A$\approx$800 kg/m^2. This prevents levitation on ground.

Levitation of a hollow test-mass still would allow in principle a very significant test of parasitic forces applied to its surface. However such a test, though crucial from a functional point of view, has a few important limitations.

First the high vertical electric field used for levitation, $\approx 10^7$ V/m, being much larger than the $\approx 10^2$ V/m used for displacement sensing along horizontal axes, is likely to contribute an appreciable stiffness also along horizontal axes. This would alter the result of any stiffness measurement along the horizontal axes.

In addition the high electric field enhances dielectric losses, and thus thermal noise (2), in a hardly controllable way. Extrapolating this information to flight operating conditions becomes unreliable.

Finally, the need to control the angular degrees of freedom makes the 1g gravity to cross talk into horizontal directions, a limitation to force noise measurements. To be specific, as the mass is unstable along all axes, an electrostatic control loop has to be used to control it. In particular, rotation around horizontal axes has to be controlled, and the control loop noise for these degrees of freedom converts into a random tilt $\delta\theta_n$ in the 1g gravitational field of the laboratory. This in turn produces a force noise $mg\delta\theta_n$ along the horizontal axes. This has been, up to now, the main limitation to tests in ground-based laboratories, where measurements have been limited to $\approx 10^{-9}$ ms^{-2}Hz$^{-1/2}$ and to frequencies larger than ≈ 10 mHz due to this phenomenon.

To circumvent this problem we are developing a test bench where the test-mass, though moving among its sensing electrodes, is made part of the inertial member of a torsion pendulum. Torsion pendulums of very high sensitivity have been repeatedly used for the measurements of very low forces at low frequency. Just as an example, the pendulum used to test for violation of the Equivalence Principle at the University of Washington (8), has shown to be dominated by its own thermal torque noise of $\approx$ 1.5×10^{-14} J/$\sqrt{\text{Hz}}$ in the frequency range 0.5-1.5 mHz that includes the resonance frequency. By using an effective arm length for its inertial member, this figure converts into a force sensitivity of $\approx 3\times10^{-13}$ N/$\sqrt{\text{Hz}}$.

A schematic of the pendulum is shown in figure 2.

4 hollow test masses hang from a supporting structure at ≈ 10 cm from the pendulum axis. Hollow, and then lighter test-masses are used to reduce the load on the fibre in order to be able to use thinner torsion fibres. The mass is surrounded by its casing carrying the sensing electrodes. A 5-axis micro-positioner allows adjusting the electrodes position around the equilibrium location of the test mass. A further micro-positioner allows adjusting the vertical position and the equilibrium angular deflection of the inertial member. An autocollimator allows an independent measurement of the pendulum angular deflection. The pendulum is enclosed in a high vacuum chamber evacuated by means of an ion pump.

One can calculate that 4, 0.25 kg each test-masses can be supported by a Silica torsion fibre of 60 µm diameter resulting in a resonant frequency of 0.65 mHz. With the electrodes geometry of the sensor in figure 1, an ac-voltage bias of 1 V applied to the electrodes, would decrease the resonance frequency of the pendulum by 20%. Even the foreseen operational bias of 0.3 V changes the resonant frequency by $\approx$2%, a quantity that should be measurable with reasonable accuracy. In addition any asymmetry in the resonance frequency vs. dc-voltage bias, would allow measuring the presence of dc stray fields like those due to charge patches (2) with a resolution in the mV range, i.e. 0.1 % of the level of stray fields that are expected to induce unbearable stray forces in LISA.

With the fibre above, the pendulum readout would show a deflection-to-force gain of $\Gamma \approx 2.5\times10^6$ rad/N. Expressing the total pendulum noise as an equivalent force noise at input f_e, one gets:

$$f_e \approx f_n + \Gamma \theta_n \left[1 - \frac{\omega^2}{\omega_{res}^2} + i\frac{1}{Q}\right] \quad (3)$$

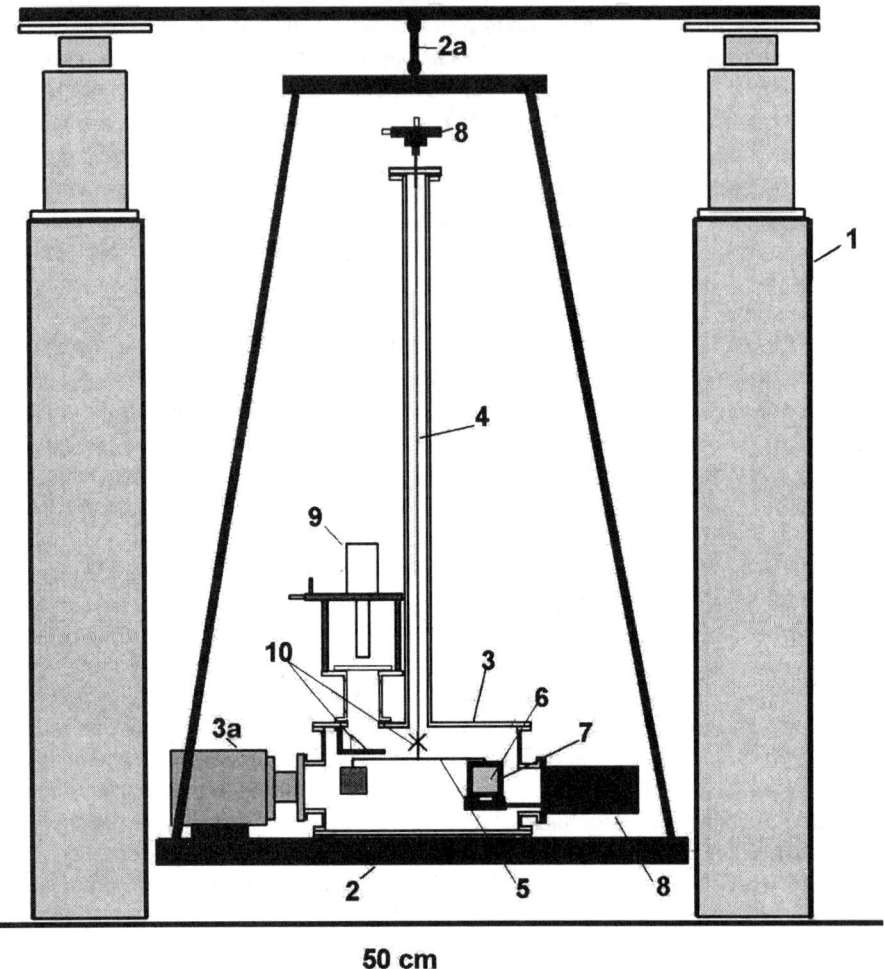

FIGURE 2. Schematic of the torsion pendulum apparatus. 1. Vibration isolation. 2. Pendulum support. 3. High vacuum chamber. 3a Ion pump. 4. Torsion fiber. 5. Supporting element. 6. Hollow test-mass. 7. Instrument case with sensing electrodes. 8. Micro-positioners. 9. Autocollimator. 10. Mirrors

where ω_{res} is the pendulum angular frequency, Q its quality factor and f_n is the real force noise acting on the test body like, for instance, thermal noise. Eq. (3) shows that, below resonance, readout with a resolution of 1 μrad/√Hz allows for a measurement of the stray forces with a resolution of 4×10^{-13} N/√Hz. At resonance this figure is improved by a factor Q, a number surely in excess of 10.

Silica, in addition, can be fabricated as a very high Q material. Q factors in excess of 10^6 have been reported though at much higher frequencies. Would one be able to reproduce that figure also at around 1 mHz, the pendulum would show imaginary torsion stiffness equivalent to a linear value of 2×10^{-12} N/m. In principle then one could measure an imaginary stiffness of order $\approx 2\times10^{-13}$ N/m as a 10% correction to the Q factor of the pendulum. Any dissipation related to the electrical field is expected to scale as the electric field squared. In order to detect a dissipation of order of that which would be armful for LISA, 5×10^{-14} N/m at 0.1 V of dc bias, one could then just increase the bias up to 0.2 V. It should be borne in mind however that measurements of such high Q's at such low frequency are challenging and can hardly be found in the literature.

DISCUSSION

Testing dynamical isolation with accuracy of $\approx 10^{-13}$ N/$\sqrt{Hz}$ looks like a logical step toward a space test of the inertial sensor with accuracy around $\approx 10^{-14}$ N/$\sqrt{Hz}$ i.e. within one order of magnitude from the accuracy foreseen for LISA. Such a test is actively studied worldwide (9). The basic idea of such a test is to fly on board of one spacecraft two identical sensors like those described above. Laser interferometer readout is then used to measure the relative displacement of the two test-masses. The ac-bias voltage can be regulated so that the coupling of the two test-masses to the spacecraft is different. Specifically if the stiffness of the coupling of one of the two test bodies to the spacecraft is substantially reduced in respect to that of the other one, the first body can be in practice used as an inertial reference. In order to be able to perform the measurement with enough accuracy at ≈ 1 mHz, one needs a relative displacement resolution of 0.25 nm/$\sqrt{Hz}$, a figure that is easily reached by laser interferometers working at higher frequency and at appears at hand also at these low frequencies.

REFERENCES

1. Bender, P. et al., LISA Pre-Phase-A Report, Max Plank Institute for Quantum Optics, MPQ233-1998.
2. Vitale S. and Speake C.C., Proc 2nd International LISA Symposium, JPL 1998. W. Folkner Ed. AIP Conf. Proc., **456,** 172-177 (1998)
3. Touboul P., Rodrigues M., and Le Clerc, G.M. *Classical and Quantum Gravity,* 13, A259-A270 (1996).
4 G.M. Keiser and S. Buchman, Stanford University (Private communication)
5. The sensor is developed at the University of Trento, Italy, in collaboration with the Rutherford Appleton Laboratory (M.C.W. Sandford), the University of Birmingham (C.C. Speake) and the Imperial College (T. J. Sumner), partially under *ESTEC contract #13691/99/NL/FM (SC)*
6. Notice that, if the stiffness is of electrical origin, it is a negative number, so that there is no zero at the resonance $|k_p|=m\omega^2$. In addition in this simplified notation, k_p also include any damping contribution in the form of a non zero imaginary part.
7. DeBra, D. B., Proc 2nd International LISA Symposium, JPL 1998. W. Folkner Ed. AIP Conf. Proc., **456,** 199-206 (1998)
8. Y. Su et al., Phys. Rev. **D50**, 3614 (1994)
9 S. Buchmann et al. Disturbance Reduction System for DS5, Stanford 1998 (unpublished). Y. Jafry ELITE proposal, ESA/ESTEC 1998 (unpublished)

Deep Surveys of Massive Black Holes with LISA

Alberto Vecchio

Max Planck Institut für Gravitationsphysik, Albert-Einstein-Institut Am Mühlenberg 1, D-14476 Golm, Germany

Abstract. Massive black hole binary systems – with mass in the range $\sim 10^5\,M_\odot$ - $10^8\,M_\odot$ – are among the most interesting sources for the Laser Interferometer Space Antenna (LISA); gravitational radiation emitted during the last year of in-spiral could be detectable with a very large ($\sim 10^3$) signal-to-noise ratio for sources at cosmological distance. Here we discuss the impact of LISA for astronomy and cosmology; we review our present understanding of the relevant issues, and highlight open problems that deserve further investigations.

INTRODUCTION

The Laser Interferometer Space Antenna (LISA) [1] is a gravitational wave (GW) observatory in the low-frequency band which is currently accessible only through non-dedicated (and low sensitivity) experiments based on the technique of Doppler tracking of interplanetary spacecraft [2,3]. As of this writing, LISA is identified as an ESA Cornerstone mission in the Horizon 2000-plus program, but is presently studied by both ESA and NASA with the view of a joint mission with launching date 2008-2010. The instrument has an optimal sensitivity in the milli-Hz frequency range, $h_{\rm rms} \approx 3 \times 10^{-22}$ for $f \sim 1$ mHz, covering the band $\sim 10^{-5}\,{\rm Hz} - 30\,{\rm mHz}$. It consists of a constellation of tree drag-free spacecraft placed at the vertices of an ideal equilateral triangle with sides of $\simeq 5 \times 10^6$ km, forming a three-arms interferometer [1,4].

The low frequency band is populated by a *plethora* of GW sources, that are out of reach for Earth-based detectors, and could be easily detectable by LISA [1]: they include *guaranteed* sources, such as *known* galactic short-period binary stars; neutron stars (NS's) and/or low-to-intermediate mass black holes ($\sim 10\,M_\odot - 10^3\,M_\odot$) falling into a massive companion ($\sim 10^5\,M_\odot - 10^8\,M_\odot$); massive black hole binary systems (MBHB's), with mass in the range $\sim 10^5\,M_\odot - 10^8\,M_\odot$; stochastic backgrounds of primordial origin, and generated by the incoherent superposition of unresolved binary systems in the Universe.

The purpose of this contribution is to discuss the impact of LISA for astronomy. Being impossible to cover all aspects, we will concentrate on one specific class of sources: massive black hole binary systems. We will describe how LISA works as GW telescope – we are ultimately dealing with a new branch of observational astronomy – summarize our present understanding of the main issues, and highlight open questions that deserve further investigations.

MBHB's are possibly the strongest sources of GW's that LISA will be able to detect; for typical objects of mass $\sim 10^6 \, M_\odot$ at redshift $z \sim 1$, the signal-to-noise ratio (SNR) is $\sim 10^3$, as show in Fig.1. The instrument is able to detect the

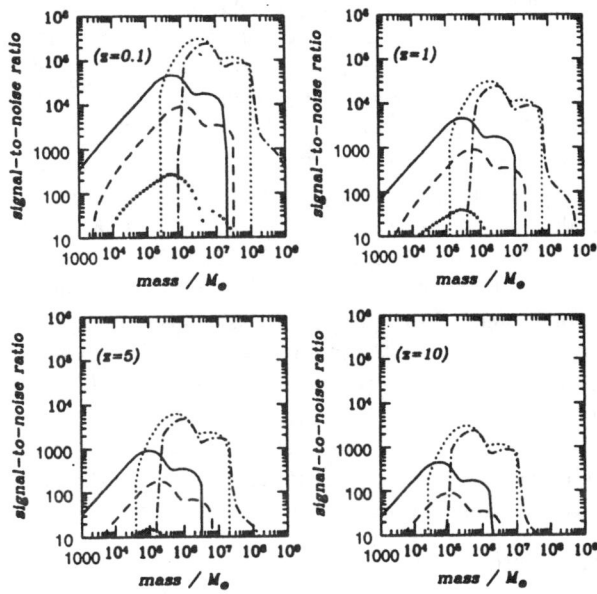

FIGURE 1. The sensitivity of LISA to coalescing black hole binary systems. The plots show the angle-averaged signal-to-noise ratio which characterizes LISA observations of the three phases of BH coalescence – in-spiral, merger, and ring-down – as a function of the mass m_1 of the primary object. The solid, dotted and dotted-dashed lines refer to the in-spiral, merger and ring-down signal, respectively, of two BH's with $m_1 = m_2$; the dashed line and the bold dots describe the in-spiral signal from two BH's with $m_1 = 100 \, m_2$, and a secondary BH of $10 \, M_\odot$ orbiting m_1, respectively. The SNR of the in-spiral signal refers to the final year of life, with cut-off frequency $f_{\mathrm{isco}} \simeq 4.4 \times 10^{-3} \, [m(1 + z)/10^6 \, M_\odot]^{-1}$; the quasi-normal ringing is assumed to occur at $f_{\mathrm{qnr}} \simeq 6.5 \times 10^{-2} \, [m(1 + z)/10^6 \, M_\odot]^{-1}$, and the energy radiated during the merger and the ring-down phases are computed according to [5]. The four panels refer to different distances of the fiducial source, where we adopt, for simplicity, a luminosity distance given by $D = z/75 \, \mathrm{km \, sec^{-1} \, Mpc^{-1}}$. The instrument low and high frequency cut-offs are (conservatively) 10^{-4} Hz and 3×10^{-2} Hz, respectively. The noise spectral density takes into account both the instrumental noise and the so-called confusion noise [6,7].

radiation emitted during one (or more) of the three phases of black hole coalescence (in the GW jargon: in-spiral, merger, and ring-down) for a very wide range of masses – in principle from $\sim 1\,M_\odot$ to $\sim 10^9\,M_\odot$, depending on the mass m_1 and m_2, and the source distance – see Fig. 1, possibly beyond redshift $z \sim 5$, if BH's do already exist, and are involved in catastrophic events with copious release of energy through GW's.

LISA will be able to carry out a deep and extensive census of black hole populations in the Universe, providing an accurate demography of these objects and their environment. Compelling arguments suggest the presence of MBH's in the nuclei of most galaxies, and they are invoked to explain a number of phenomena, in particular the activity of quasars and active galactic nuclei [8,9]. However, the observational evidences of MBH existence come mainly from observations of relatively nearby galaxies, whose nuclei do not show significant activity [10–12]. Massive black holes seem to be clustered in the mass-range $10^6\,M_\odot - 10^9\,M_\odot$ [13]; at the lower edge of the BH mass-spectrum, we find evidences for solar-mass BH candidates [14]. No information is presently available regarding BH's with mass between $\sim 10\,M_\odot$ and $\sim 10^6\,M_\odot$, although some recent X-ray observations are interpreted as possible (but not compelling) indications of "middleweight" BH's [15,16]. LISA – and Earth-based laser interferometers – will definitely show whether this gap is simply due to a "selection" effect of present electro-magnetic observations, or indeed Nature does not provide intermediate mass black holes: an important feature of LISA is its capability of detecting BH's with mass $\sim 10^3\,M_\odot - 10^4\,M_\odot$, still far from coalescence at high redshift, see Fig. 1. LISA is also likely to monitor binary systems with a wide spectrum of BH spins and orbital eccentricities, which will enable us to carry out high precision tests of general relativity [17–19], and to derive a map of the distribution of these physical parameters in astrophysical objects.

One of the most interesting observations would be the detection of GW's *and* electro-magnetic radiation from the merger of two BH's. We do not know as yet, whether a burst of electro-magnetic radiation is emitted during MBH collisions [21]. Determining where and when a MBH merger takes place, and possibly alerting in advance the astronomers is of paramount importance; this issue is directly linked to the identification of the source host galaxy: it would allow us to establish correlations between MBH's and their environment, and use LISA observations to estimate the fundamental cosmological parameters [1,20].

We have not discussed so far the rate at which we expect to detect such signals. A fair statement would probably be that, essentially, we do not know it. However, we can summarize our present knowledge as follows. For MBHB systems, the event rate depends strongly on theoretical prejudices and model assumptions; the "canonical" value is $\sim 1\,\mathrm{yr}^{-1}$, but rates as high as $\sim 10^3\,\mathrm{yr}^{-1}$ or as low as $\sim 10^{-2}\,\mathrm{yr}^{-1}$ are consistent with theoretical models [21–24]. For low-mass black holes captured by a massive one in galactic cores, we believe to have a better understanding, and current astrophysical estimates yield a rate of a few events per year up to $z \simeq 1$ [25,26].

THE LISA TELESCOPE

We are dealing with a new generation of telescopes, both regarding the kind of radiation they observe (gravitational waves) and the frequency window in which they operate ($\sim$ mHz). It is therefore instructive to analyze the features that enable LISA to extract accurate information about GW sources.

We consider here only the in-spiral portion of the whole coalescence waveform, neglecting the merger and ring-down, both easily detectable, cfr. Fig. 1. The merger waveform is still poorly understood from the theoretical point of view; significant progresses have been made using either full numerical schemes or semi-analytical approximations, but both approaches are still far from returning a satisfactory answer for GW observations (see [27,28] and references therein). We do however expect to gain key information by detecting GW's emitted during the final plunge, for instance how energy and angular momentum are radiated during this extreme strong-gravity phase. The ring-down signal, on the contrary, is theoretically well know; in order to limit the level of complexity of our analysis, we do not include it into the signal that we consider here; however, future investigations should keep it (as well as the final plunge, if/when available) into account, as it might change (conceivably improve) LISA performances in a number of astrophysical situations.

There are two main features that distinguish the in-spiral signals recorded by LISA from the ones that we expect to detect with Earth-based interferometers: (i) they last for months-to-centuries (depending on the masses) in the instrument observational band, and therefore are not burst-signals; in fact, the (Newtonian) time to coalescence is $\tau \simeq 1.2 \times 10^7 \left(f_0/10^{-4}\,\text{Hz}\right)^{-8/3} \left[m(1+z)/10^6\,M_\odot\right]^{-5/3} (\eta/0.25)^{-1}$ sec; here $m = m_1 + m_2$ is the total mass, and $\eta = \mu/m$ is the symmetric mass ratio, where $\mu = m_1 m_2/m$ is the reduced mass; (ii) the structure of the waveform is in general much more complex; in fact, we can expect to detect black holes that are fast spinning *and* live on highly elliptical orbits, in particular for the extreme mass ratio case, $\eta \ll 1$ [29]. As an example, in LIGO observations one will likely monitor no more than 10 cycles of precession of the orbital plane and the spins, whereas in the LISA band, for a typical observation time of 1 year, they could be as many as ~ 1000, see Table 1. An useful figure, for both detection and parameter estimation, is also the number of wave cycles recorded by LISA: during the final year of in-spiral, they range from

TABLE 1. The number of precession cycles observed by LISA. The table shows the number of cycles ($\mathcal{N}_{\text{prec}}$) of $\hat{\mathbf{L}}$ and $\hat{\mathbf{S}}$ around the constant direction of the total angular momentum $\mathbf{J} = \mathbf{L} + \mathbf{S}$ during the final year of in-spiral for BH binary systems with selected masses (in units of $M_\odot$) and spins.

S/m^2	$m_1\ m_2$	$\mathcal{N}_{\text{prec}}$	$m_1\ m_2$	$\mathcal{N}_{\text{prec}}$	$m_1\ m_2$	$\mathcal{N}_{\text{prec}}$	$m_1\ m_2$	$\mathcal{N}_{\text{prec}}$	$m_1\ m_2$	$\mathcal{N}_{\text{prec}}$	$m_1\ m_2$	$\mathcal{N}_{\text{prec}}$
0.95	$10^7\ 10^6$	11	$10^6\ 10^6$	25	$10^7\ 10^5$	34	$10^6\ 10^5$	23	$10^7\ 10^2$	404	$10^6\ 10^2$	1262
0.50	$10^7\ 10^6$	7	$10^6\ 10^6$	20	$10^7\ 10^5$	19	$10^6\ 10^5$	16	$10^7\ 10^2$	276	$10^6\ 10^2$	708
0.10	$10^7\ 10^6$	4	$10^6\ 10^6$	16	$10^7\ 10^5$	5	$10^6\ 10^5$	9	$10^7\ 10^2$	74	$10^6\ 10^2$	150
0.01	$10^7\ 10^6$	3	$10^6\ 10^6$	16	$10^7\ 10^5$	2	$10^6\ 10^5$	8	$10^7\ 10^2$	8	$10^6\ 10^2$	16

$\sim 10^3$ (for $m_1 \sim m_2$) to $\sim 10^5$ (for $\eta \ll 1$).

In general, 17 parameters describe the waveform. No analysis has been carried out so far dealing with such general situation. Here, we will introduce some simplifying assumption, while retaining most of the key physical ingredients. The main limitation of our approach derives from considering circular orbits; this is probably quite realistic for binary systems of two MBH's which have undergone a common evolution inside a galactic core, but is almost for sure violated for solar mass compact objects and/or low mass BH's orbiting a massive one [29]. We do, however, take into account spins; in this case we assume that either the masses of the BH's are roughly equal, or one of the BH's has a negligible spin (which still describe a wide range of astrophysical situations): the binary system undergoes the so-called *simple precession* [30], where the orbital angular momentum **L** and the total spin $\mathbf{S} = \mathbf{S_1} + \mathbf{S_2}$ are locked together, and precess around the (almost) constant direction of the total angular angular momentum $\mathbf{J} = \mathbf{S} + \mathbf{L}$. We also use the post$^{1.5}$-Newtonian approximation of the GW phase [31]. As a consequence of this chain of approximations, the number of parameters describing the signal drastically reduces, from 17 to 11.

It is useful now to review some of the instrumental features, in order to understand how LISA works as GW observatory:

(i) LISA is an *all-sky monitor*, and one gets for free all-sky surveys. During the observation time, however, LISA changes location and orientation. The LISA orbital motion is rather peculiar – the baricenter of the instrument is inserted in a heliocentric orbit, following by 20° the Earth; the detector plane is tilted by 60° with respect to the Ecliptic and the instrument counter-rotates around the normal to the detector plane with the same 1-yr period – and is conceived in order to keep the configuration as stable as possible during the mission, as well as to give optimal coverage of the sky. It also turns out to be a key factor in reconstructing the source location in the sky.

(ii) The sources are distinguished in the data stream by the different structure and time evolution of the signals at the detector output; the recorded in-spiral signal reads:

$$h_\alpha(t) = A_{\rm gw}(t)\, A_{\rm p}^\alpha(t)\, \cos[\phi_{\rm gw}(t) + \varphi_{\rm p}^\alpha(t) + \phi_{\rm D}(t)] \tag{1}$$

where $A_{\rm p}(t)$ and $\varphi_{\rm p}(t)$ are the time-varying polarization amplitude and phase, respectively, and $\phi_{\rm D}(t)$ is the Doppler phase shift induced by the motion of the detector around the Sun; an example of in-spiral signal at the output of LISA is given in Fig. 2. The signal is therefore amplitude and phase modulated by the motion of the LISA centre-of-mass around the Sun, the change of orientation of the detector arms, and of the binary orbital plane. All these effects encode information about some of the source parameters.

(iii) There is only one LISA detector currently planed; correlations and/or time-of-flight measurements are not possible; they would be highly desirable in order to improve the estimation of the source parameters, in particular the source location and distance; however, as the gravitational wavelength is $\lambda_{\rm gw} \simeq 2\,(f/1\,{\rm mHz})^{-1}\,{\rm AU}$, a second detector would have to be placed at several AU from the first one in order to provide useful information on the position of a source in the sky; however LISA is a three-arms instrument; Cutler [32] has shown that the outputs from each arm can be combined in such a way to form a pair of data sets, $\alpha = 1,2$ in Eq. (1), whose noise is uncorrelated at all frequencies, that are equivalent to the data streams recorded by two co-located interferometers, rotated by $\pi/4$ one with respect to the other.

Indeed, there will be two data streams available to extract all source parameters. Correlations between the parameters are inevitable, and conspire to degrade the accuracy of the parameter measurements. It should also be clear that for LISA the measurement errors depend crucially on the actual value of the source parameters, and one therefore needs to explore a very large parameter space to give a fair description of the instrument performances.

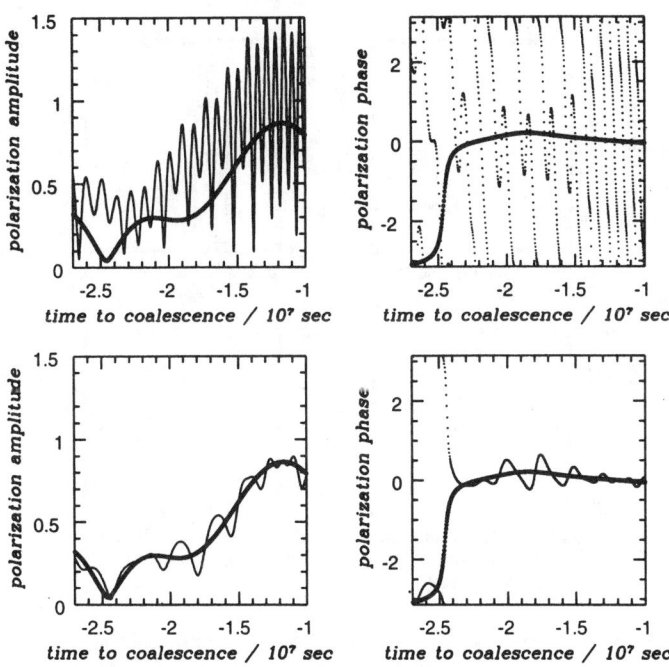

FIGURE 2. In-spiral signals at the LISA output. The plots show the evolution of the polarization amplitude $A_{\rm p}(t)$ (on the left) and phase $\varphi_{\rm p}(t)$ (on the right) as a function of time, cfr. Eq. 1, for black holes with $S = 0$ (bold solid line) and $S \neq 0$ (thin solid line, for $A_{\rm p}(t)$, and dotted line, for $\varphi_{\rm p}$). The two plots at the top refer to a source with masses $m_1 = 10^7\,M_\odot$ and $m_2 = 10^5\,M_\odot$; in the case of spinning black holes the parameters are: $S/m^2 = 0.95$, $\hat{\bf S}\cdot\hat{\bf L} = 0.5$. The plots at the bottom refer to a MBHB with $m_1 = m_2 = 10^6\,M_\odot$: when spins are present the choice of parameters is according to: $S/m^2 = 0.3$, and $\hat{\bf S}\cdot\hat{\bf L} = 0.9$.

SURVEYS OF MASSIVE BACK HOLES

We have discussed in the Introduction the sensitivity of LISA: there is little doubt that such interferometer will be able to survey a fairly large fraction BH populations in the Universe. We would like to stress that in the present discussion, we assume to be able to monitor the whole final year of in-spiral. This is a key and delicate point which affects the capability of surveying sources at increasingly higher z and/or with larger m, and measuring precisely the parameters: in fact, at some frequency (between 10^{-4} Hz and 10^{-5} Hz) the instrumental noise will completely dominate the signal, allowing to pick up only the very final portion of the in-spiral (say a few days), or even preventing the detection; the redshifted radiation simply falls outside the observational band, cfr. Fig. 1. It is clear that the higher the redshift, the lower the typical mass for which LISA reaches the optimal sensitivity. Super-massive black holes of mass $\sim 10^9\,M_\odot$ might be observable, by detecting ring-down

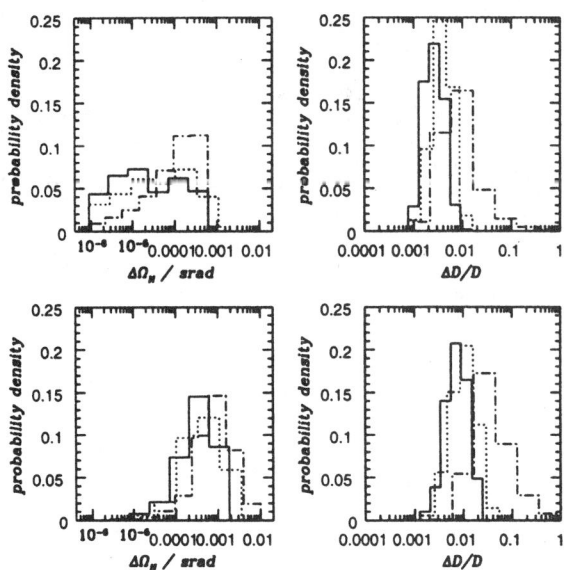

FIGURE 3. The probability distribution of the angular resolution $\Delta\Omega_N$ and the relative error of the distance determination $\Delta D/D$, with which LISA can identify a MBHB by observing the final year of in-spiral. The histograms show the result of a Monte-Carlo simulation, where 1000 sources, with masses $m_1 = m_2 = 10^6\,M_\odot$ at redshift $z = 1$, have been randomly located and oriented in the sky. The top panels refer to measurements carried out by both LISA detectors, whereas the bottom panels report the results obtained by using only a single interferometer. The plots compare the estimated errors in the measurement of the parameters assuming three different values of the spin: $S/m^2 = 0.9$ (solid line), 0.3 (dotted line), and 0 (dotted-dashed line). The total noise is given by the sum of the instrumental noise and the confusion noise.

signals at low redshifts ($z \lesssim 0.1$), if the sensitivity window extends to $\sim 10^{-5}$ Hz.

Several analysis have been carried out so far dealing with the accuracy of the parameter measurements with LISA [32–36]; however, they have been mainly focussed on investigations of the instrument angular resolution; moreover, spin effects have been either ignored or explored for a very limited portion of the total parameter range. Here we will try to give a more comprehensive description of the performances of LISA as GW observatory. The accuracy of the parameter measurements is very sensitive to the actual source parameter values; it is therefore almost impossible to give *typical figures* for LISA as GW telescope, that can be applied to a wide range of binary systems. We discuss in some detail the case of an equal-mass MBHB, with $m_1 = m_2 = 10^6\, M_\odot$, and give some general criteria to extend these results to other parameter values. It turns out that the source location and orientation with respect to the detector play a key role. We have therefore performed Monte-Carlo simulations, where we fix the source distance and the physical parameters, and vary randomly the "geometrical" parameters, $\hat{N}$, $\hat{J}$ and $\hat{S}$. We compute the estimated mean squared errors associated to the parameter measurements by means of the so-called *variance-covariance* matrix [37,38].

The main results are presented in Figs. 3 and 4, and can be summarized as follows. The angular resolution is $\Delta \Omega_N \sim 10^{-5}$ srad; however, depending on

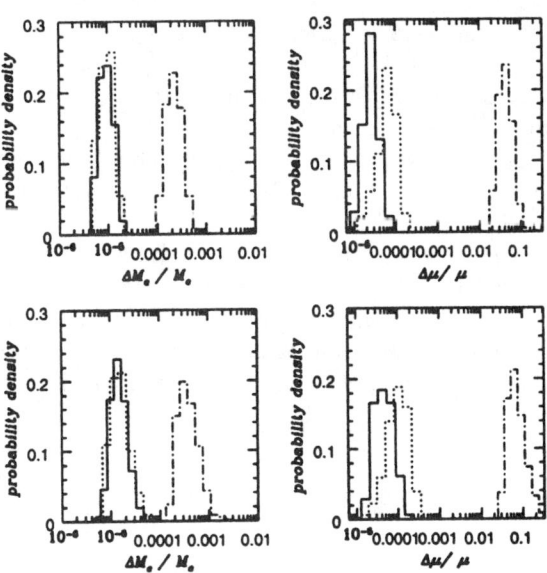

FIGURE 4. The probability distribution of the errors with which the source masses can be measured by LISA in one year of observation. Same as Fig. 3, but now the histograms show the distribution of the errors regarding the mass parameters, where our choice corresponds to the chirp mass, $\Delta M_c/M_c$ (panels on the left), and the reduced mass, $\Delta \mu/\mu$ (panels on the right).

the location and orientation of the source it varies over a wide range of values, $10\,\mathrm{arcmin}^2 \lesssim \Delta\Omega_N \lesssim 3\,\mathrm{deg}^2$. Typically, large spins and misalignment angles – the angle between $\hat{\mathbf{L}}$ and $\hat{\mathbf{S}}$ – allow us to measure more precisely the source location; for a small region of these parameters, the "error-box" in the sky could possibly be only a fraction of arcmin^2. The distance is usually measured with an error $0.1\% \lesssim \Delta D/D \lesssim 1\%$. The timing accuracy is very high, and the instance of coalescence can be identified within ~ 10 sec. Masses and spins can be measured very precisely; typically, the errors affecting the determination of the chirp and reduced mass are $\Delta\mathcal{M}/\mathcal{M} \sim 10^{-5}$ and $\Delta\mu/\mu \sim 10^{-4}$, respectively; the so-called spin-orbit parameter β can be determined with an error $\Delta\beta \sim 10^{-3}$. There is one general rule that can be derived from this analysis: if BH's are highly spinning and the misalignment angle is large, the parameter determination improves. This is due to the fact that the parameters leave peculiar finger prints on the recorded signal, cfr. Fig. 2: in particular, A_p and φ_p undergo strong modulations, which carry information not only on the position of the source and the orientation of the angular momenta, but also on the physical parameters, such as the masses. This is an effect which is similar – although the physics behind it is different – to the one that takes place when spins are not present, but one considers not only radiation emitted at twice the orbital frequency, but also at other harmonics [35] (notice that in Fig. 3 and 4, for the case $S = 0$, we report results obtained considering only the dominant harmonic; we refer the reader to [35] for more details).

We can now ask how these results change by selecting different source parameters. MBHB's with $m_1 \sim m_2 \sim 10^7 M_\odot$ would be typically observed with larger errors, by a factor ≈ 10, than the ones reported here. If we fix m_1 and vary m_2, the measurement accuracy is fairly constant – within, say, a factor ≈ 2 – as long as $m_2/m_1 \gtrsim 0.1$, then is starts degrading: this is due to a rather complex competition between several effects, in particular the SNR and the number of wave/precession cycles [39,40].

MBHB's will be visible several months before the final coalescence. This will allow us to pick up the signal when the binary system is still far from merging, and refine the source parameter measurements as the source proceeds toward the deadly plunge [39]: for a limited region of the parameter space, it could be possible to determine the source location in the sky with enough precision to have a realistic chance of observing the same field with other telescopes.

REFERENCES

1. P. Bender et al., *LISA Pre-Phase A Report; Second Edition*, MPQ 233 (1998).
2. F. B. Estabrook and H. R. Wahlquist, Gen. Rel. Grav. **6**, 439 (1975).
3. B. Bertotti, A. Vecchio, and L. Iess, Phys. Rev. D **59**, 082001 (1999).
4. K. Danzmann, theses proceedings.
5. E. Flanagan and S. Hughes, Phys. Rev. D **57**, 4535 (1998).
6. D. Hils, P. L. Bender, and R.F. Webbink, Astrophys. J. **360**, 75 (1990);

7. P. L. Bender and D. Hils, Class. and Quantum Grav. **14**, 1439 (1997).
8. Y. B. Zel'dovich and I. D. Novikov, Sov. Phys. Dokl. **158**, 811 (1964).
9. E. E. Salpeter, Astrophys. J. **204**, L1 (1964).
10. M. Miyoshi, J. Moran, J. Herrnstein, L. Greenhill, N. Nakai, P. Diamond and M. Inoue, Nature **373**, 127 (1995).
11. A. Eckart and R. Genzel, Nature **383**, 415 (1996).
12. E. Maoz, Astrophys. J. **494**, L181 (1998).
13. D. Richstone et al., Nature **395**, A14 (1998).
14. M. J. Rees, in *Black holes and relativistic stars*, edited by R. Wald (University of Chicago Press, Chicago, 1998), pp. 79-101;
15. E. J. M. Colbert and R. F. Mushotzky, pre-print astro-ph/9901023
16. A. Ptak and R. Griffith, astro-ph/9903372
17. S. Hughes, these proceedings.
18. E. Poisson, Phys. Rev. D **54**, 5939 (1996).
19. F. D. Ryan, Phys. Rev. D **56**, 1845 (1997).
20. B.F. Schutz, Nature **232**, 675 (1986).
21. M.C. Begelman, R. D. Blandford, and M. J. Rees, Nature **287**, 307 (1980).
22. O. Blaes, these proceedings.
23. M.G. Haehnelt, Mont. Not. Roy. Astron. Soc. **269**, 199 (1994);
24. A. Vecchio, Class. and Quantum Grav. **14**, 1431 (1997).
25. S. Sigurdsson and M. J. Rees, Mont. Not. Royal Astron. Soc. **284**, 318 (1996).
26. S. Sigurdsson, Class. and Quantum Grav. **14**, 1425 (1997);
27. See the Binary Black Hole Grand Challenge Alliance web site for recent results and publications: http://jean-luc.ncsa.uiuc.edu/
28. J. Pullin, in *Gravitation and Relativity: At the turn of the Millennium*, eds. N. Dadhich and J. Narlikar (IUCAA, India), pp. 87-106 (1998).
29. D. Hils and P. L. Bender, Astrophys. J. **447**, L7 (1995).
30. Apostolatos, T. A., Cutler, C., Sussman, G. S., and Thorne, K. S., Phys. Rev. D **49**, 6274 (1994).
31. L. Blanchet, T. Damour, B.R. Iyer, C.M. Will and A.G. Wiseman, Phys. Rev. Lett. **74**, 3515 (1995).
32. C. Cutler, Phys. Rev. D **57**, 7089 (1998).
33. A. Vecchio and C. Cutler, in *Laser Interferometer Space Antenna*, ed. W. M. Falkner (AIP Conference Proceedings 456; 1998), pp. 101-109.
34. A. Vecchio, in *Recent Development in General Relativity*, ed. F. Francaviglia, (Springer Verlag), pp. 221-238), 1999.
35. A. Sintes and A. Vecchio, these proceedings.
36. T. A. Moore and R. W. Hellings, gr-qc/9910116.
37. C. Cutler and E. E. Flanagan, Phys. Rev. D **49**, 2658 (1994).
38. D. Nicholson, and A. Vecchio, Phys. Rev. D **57**, 4588, (1998).
39. A. Vecchio, C. Cutler and A. Sintes, in preparation.
40. A. Vecchio, in preparation.

Supermassive Black Holes As Gravitational Wave Sources for LISA

Omer Blaes

Department of Physics, University of California at Santa Barbara, Santa Barbara, CA 93106

Abstract. I briefly review the astrophysics of supermassive black holes that is relevant to their role as sources of gravitational waves for the proposed *Laser Interferometer Space Antenna* mission.

INTRODUCTION

Probably the most exciting of the realistic gravitational wave sources for LISA are supermassive black holes (SMBH's) [1]. By astrophysical standards, the case for the existence of SMBH's is very strong. Stellar and gas dynamics can be used to infer the presence of massive dark objects at the centers of most bright galaxies. The most spectacular extragalactic case is the galaxy NGC 4258, where water masers have been observed in an edge-on, warped disk of gas in the nucleus. The masing gas has a Keplerian velocity profile, consistent with a central compact mass of 3.6×10^7 $M_\odot$ inside a radius less than 0.13 pc, implying a density greater than 4×10^9 $M_\odot$ pc^{-3} [2]. Stellar proper motions in our own Galactic center imply the existence of a compact mass of 2.6×10^6 $M_\odot$ inside a radius of $\sim 10^{-2}$ pc, giving a density greater than 10^{12} $M_\odot$ [3,4]. Such high densities can be used to rule out alternatives such as dense clusters of stars or compact objects [5], leaving SMBH's or perhaps SMBH binaries as the only plausible explanation. Massive dark objects appear to be ubiquitous in the nuclei of bright galaxies, and there is even a correlation between the dark mass and the galaxy bulge mass: $M_\bullet \simeq 0.006 M_{\text{bulge}}$ [9]. This is roughly consistent with the integrated quasar light in the universe, albeit with an average radiative efficiency of around a percent [10], somewhat less than the ten percent usually assumed in accretion.

It is the existence of active galactic nuclei and quasars that is perhaps the most compelling argument for the existence of SMBH's. These sources can exhibit compact relativistic outflows as well as relativistically broadened X-ray lines [6]. Rapid variability in the X-rays and gravitational microlensing in the optical [7] can constrain the size of the central source to be very small, while variability in the broad optical emission lines can be used to infer high central masses [8].

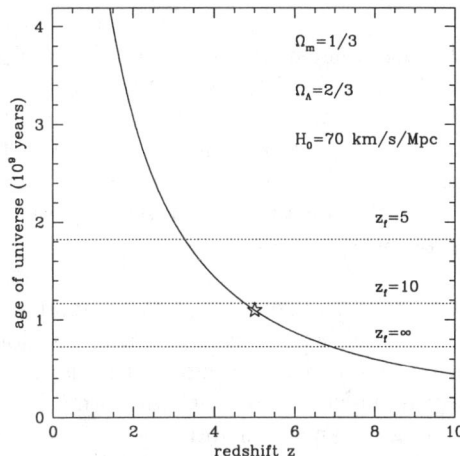

FIGURE 1. Age of the universe as a function of redshift, for the currently popular flat universe model with nonzero cosmological constant (adapted from [14]). Dashed horizontal lines indicate the time when a 10^8 $M_\odot$ SMBH could have formed by Eddington limited accretion with ten percent radiative efficiency, assuming an initial mass of 10 $M_\odot$ at the indicated redshift z_f. The star indicates the redshift of the most distant quasars known.

SMBH FORMATION

The SMBH mass/bulge mass correlation strongly suggests that SMBH formation is an integral part of the overall process of galaxy formation. The peak of quasar activity occurred at redshifts between two and three, comparable to the epoch of peak star formation in the universe (e.g. [10]). Quasars have been detected out to higher redshifts, the current record holders being at $z = 5.00$ and 5.03 [11,12]. These high redshift objects indicate that structure formation was well on its way early on in the universe, and that SMBH's formed fast (Figure 1, [13]).

There are a variety of formation pathways for SMBH's which are summarized in a famous flowchart by Rees [15], some of which might be copious sources of gravitational waves for LISA. An oversimplified viewpoint is that there are two classes of mechanisms to form SMBH's: stellar dynamical and gas dynamical.

The most detailed work has been done with stellar dynamics. An exciting scenario from LISA's point of view is the dynamical collapse and merger of a dense cluster of neutron stars and/or black holes. Quinlan & Shapiro [16] explored this using Fokker-Planck calculations and found that mass segregation in the merging remnants of the cluster resulted in several 100 $M_\odot$ black holes at the center. They could not follow the subsequent evolution, but if ~ 500 $M_\odot$ seeds could form and merge, this would make a very strong gravitational wave source for LISA. Lee [17] followed up on this work by performing a full N-body calculation of 1000 compact objects, and found runaway growth of *one* object. This would therefore not be as spectacular a source of gravitational waves in the LISA bandpass.

Both of these calculations suffer from the problem of artificial initial conditions. It is hard to see how such a cluster of compact objects could form in the first place without also having more ordinary stars which have much larger collision cross-sections. Fokker-Planck calculations of the evolution of a dense cluster of main sequence stars have been performed [18]. Bearing in mind that the physics of the collisions is very uncertain, it appears that a $> 10^3$ $M_\odot$ massive star forms rapidly within the cluster relaxation time.

It therefore seems reasonable to assume that gas dynamics dominates the SMBH formation process, involving either the collapse of a gas cloud, or the formation of supermassive stars or disks. Unfortunately, this is a much harder problem (cf. our poor theoretical understanding of star formation!). Rees has argued for a long time that the gas dynamical formation of a SMBH would be a very weak source of gravitational waves (e.g. [19]). In the collapse of a supermassive star by the post-Newtonian instability, for example, the gravitational wave energy could be spread out over a long time because not all the matter would reach the center at once (i.e. within a time interval comparable to GM/c^3, where M is the mass of the star or resulting black hole). This is because the free-fall time $\sim (G\rho)^{-1/2}$ depends only on the average density ρ inside a given radius, and supermassive stars do not have a uniform density distribution. However, early numerical spherical collapse calculations (e.g. [20]) found that, in spite of the initial density nonuniformity, the collapse proceeds homologously, so that it could be a strong source of gravitational waves provided the necessary asphericities do not affect this conclusion. Supermassive stars are essentially $n = 3$ polytropes, and such models are known to have (Newtonian) homologous collapse solutions [21]. (The inner parts of the collapsing stars have substantial pressure support, and are therefore *not* in complete free fall and need not run away from the outer parts.) Even so, given that the energy released in gravitational waves is very sensitive to the collapse time scale, this scenario is asking for a lot of optimism. Even small departures from a homologous collapse would greatly reduce the luminosity. More recent post-Newtonian spherical collapse calculations by Fuller et al. [22] find in fact that electron-positron pair production and neutrino cooling destroy the homologous nature of the collapse over all but the inner ten percent of the mass of the star.

Loeb and Rasio [23] have simulated the isothermal collapse of rotating, optically thin gas clouds. Assuming the gas can continue to cool, much of it fragments into clumps, with a small fraction forming a rotating supermassive disk in the center, leading to the possibility that a newborn SMBH grows by accretion from such a disk. These disks may themselves be subject to instabilities, and this begs the obvious question: can galaxies be born with binary or multiple SMBH's in their nuclei? After all, we know this is common in star formation, although making quantitative theoretical predictions is currently not possible.

SMBH BINARIES

Various pieces of observational evidence for the actual existence of SMBH binaries have been presented over the years, but none are as yet compelling. Tantalizing evidence that time variations in the double peaked Hβ emission line profile of the quasar 3C 390.3 could be interpreted as being due to a binary SMBH [24] were ruled out by later observations when the line did not behave as predicted by the model [25]. Periodic "wiggles" in VLBI radio jets have been interpreted as being due to the orbital motion in a SMBH binary (e.g. the quasar 1928+738, [26]), but this is not the only explanation for this phenomenon. The optical light curve of the BL Lac OJ 287 exhibits flares which regularly recur every twelve years, and a binary SMBH has been proposed to explain this periodicity [27,28]. SMBH binary evolution can scour out the density profile of a surrounding star cluster [29–31], which provides a plausible explanation for the formation of observed inner optical surface brightness "core" profiles of bright elliptical galaxies. However, adiabatic growth of a single SMBH can also produce the same profiles, and can even explain some of the observed correlations in those profiles [32].

Even if SMBH's are not born as binaries, binaries are likely to form when galaxies merge, a common process in the currently popular hierarchical scenarios for structure formation. We even see such mergers in the local universe - the most interesting example being the radio source 3C 75, which actually consists of two radio galaxies in the process of merging, with radio jets that appear to interact and merge together on one side [33]. At least in this case we *know* that the two merging galaxies each contain a SMBH.

The evolution of the two SMBH's in the galaxy merger is thought to proceed in four stages [34]: (1) Dynamical friction brings the SMBH's to the center of the merged stellar core on a characteristic time scale $t_{fric} \sim 6 \times 10^7$ yr$(M_\bullet/10^6 M_\odot)^{-1}$, assuming the stellar core has a density $\sim 10^3$ $M_\odot$ pc^{-3}, velocity dispersion ~ 300 km s^{-1}, core radius ~ 100 pc, and core mass $\sim 8 \times 10^9$ $M_\odot$. (2) The SMBH's form a bound binary when they approach within ~ 5 pc$(M_\bullet/10^6 M_\odot)^{1/3}$ of each other. Dynamical friction causes the semimajor axis of the binary to continue to shrink. (3) The binary becomes "hard", meaning that its orbital velocity exceeds the velocity dispersion of the surrounding stars. Stellar encounters continue to shrink the binary, causing it to harden further. (4) Eventually, gravitational radiation takes over the job of driving the binary coalescence when the semimajor axis shrinks to $\sim 2 \times 10^{-3}$ pc$(M_\bullet/10^6 M_\odot)^{3/5}$, the time scale for shrinking being a rapidly decreasing function of semimajor axis. The longest and most uncertain stages of this process are stages (2) and (3). Ejection of stars on radial orbits ("loss cone depletion"), which are the ones which have close encounters with the binary and therefore dominate the hardening, is a major uncertainty. The eccentricity of the merging binary can evolve, either increasing (reducing the time when gravitational radiation starts to take over) or decreasing.

Many authors have explored aspects of this problem, but the most detailed work so far is by Quinlan [35] and Quinlan & Hernquist [30]. The first paper explored the evolution of SMBH binaries in a fixed stellar background. Among the interesting results was the fact that the eccentricity of the binary could increase while it was hardening, but if the SMBH's first became bound on low eccentricity (i.e. circular) orbits, then the eccentricity stayed low. The initial eccentricity of the bound binary is therefore an important quantity to determine. More worrying for LISA, Quinlan could calculate the rate of mass ejection of stars by the binary and use it to estimate when the loss cone would be entirely depleted, even if he was not calculating the evolution of the star cluster itself. He found that *all* low eccentricity binaries stalled before they were able to get close enough for gravitational radiation to take over. Quinlan & Hernquist addressed this problem in their second paper by calculating the back reaction of the binary on the stars and found a new, important effect. Because the star cluster is approximately spherical, the net force on the SMBH binary is very small, and so the center of mass of the binary *wanders* within the star cluster. This allows the loss cone to be sufficiently replenished to continue to shrink the binary. They were unable to make large explorations of parameter space, but they estimated that wandering implies that the overall hardening rate would occur 10 to 50 times slower than if the binary was in an absolutely fixed stellar background. This means that most high mass SMBH binaries will merge within a Hubble time. Unfortunately, SMBH binaries with masses less than 10^7 $M_\odot$ may still take too long to merge, and it is these which are of primary interest for the LISA bandpass. There are still considerable uncertainties, e.g. the effects of gas dynamics which would presumably accelerate mergers. Surprisingly, there are even uncertainties as to whether a SMBH binary can form at all in galaxy mergers, because some simulations find that the SMBH's end up on wide orbits where dynamical friction is slow in bringing them into the nucleus [36], although these results may have been due to unrealistic initial conditions [29]. Clearly much more work needs to be done on the dynamical evolution of SMBH's in galaxy mergers.

All the SMBH binary merger rate estimates for LISA ignore these details and assume that every galaxy merger produces a merged SMBH binary. Currently popular hierarchical structure formation scenarios (e.g. ΛCDM) lead to many mergers over time, and can successfully model both the historical star formation rate and quasar luminosity function, albeit with assumptions! (See e.g. [37] for recent work along these lines.) A very crude estimate based on multiplying the local number density of bright galaxies by a Hubble volume and assuming that every one underwent one merger in its history leads to a merger rate of roughly 1 per year [38], i.e. it only takes every galaxy to do one exciting thing in its lifetime in order to get interesting event rates for LISA!

The observed black hole mass/galaxy bulge mass relationship combined with the quasar luminosity function can be used to argue that black hole mergers do in fact occur [10,38]. Quasar activity peaked in the universe at redshifts between two and three, and the brightest quasars we observe then have luminosities corresponding to black hole masses of at least 4×10^8 $M_\odot$, assuming that their luminosity is at most the Eddington luminosity. The comoving number density at that time was 10^{-6} Mpc^{-3}, but if the black hole masses remained unchanged since these quasars faded, then the corresponding galaxy bulges today have number densities $\sim 10^{-3}$ Mpc^{-3}! One way to explain this discrepancy is to assume that the time over which quasars are bright is only $\sim 10^6$ yr, i.e. 10^{-3} times shorter than the duration of the bright quasar epoch ($\sim 10^9$ yr between redshifts two and three, cf. Figure 1), giving a much higher quasar number density. But then why is the black hole mass/galaxy bulge mass relationship roughly consistent with the total integrated quasar light, implying that black hole mass growth was around 1 percent radiatively efficient (mass e-folding time of 5×10^6 yr)? An alternative possibility is that these black holes later merged - a factor 4 increase in black hole mass (two equal mass mergers) would mean that the corresponding bulges would be several hundred times rarer. On the other hand, perhaps the bright quasars are not quite at the Eddington limit, and black holes are already four times larger at those epochs. Such (ad hoc) sub-Eddington growth to 2×10^9 $M_\odot$ from an initial mass of 10 $M_\odot$ would require $\sim 3 \times 10^9$ yr at ten percent radiative efficiency, which is way too long (cf. Figure 1). However, the 1 percent radiative efficiency inferred from the black hole mass/galaxy bulge mass relationship would allow this to be possible.

Finally, I should briefly mention that SMBH's could be lit up today in gravitational waves by orbiting stellar mass compact objects. Such sources would be extremely interesting from the physics point of view, as they make ideal high precision tests of strong field general relativity. This is mainly for the simple reason that orbits in a black hole spacetime are relatively easy to compute compared to, say, the gravitational wave signal from two merging equal mass black holes. Once again, the event rate is extremely uncertain: Sigurdsson [39] has recently estimated ~ 1 yr^{-1} to ~ 1 day^{-1}!

CONCLUSIONS

The predicted event rates for LISA are terribly uncertain when it comes to SMBH's. It is perhaps sobering to compare this situation with the predictions of neutron star binary mergers for ground-based gravitational wave observatories. The predicted event rates are based on only *three* known binaries which will merge within a Hubble time, of which only one (PSR 1534+12) dominates the merger rate extrapolation [40,41]. So while the event rates are clearly highly uncertain, at least one can take some consolation in knowing that these sources really exist! In contrast, we do not know, but strongly suspect, that SMBH binaries exist and merge. On the other hand, there is astronomical (never mind physics!) justification enough for LISA, if only because it is guaranteed to see Galactic binary stars. The fact that we do not know the SMBH merger rate, or how SMBH's formed, actually implies that LISA will be a unique astronomical tool in probing the origin and evolution of SMBH's, which, as is becoming increasingly clear, are intimately tied to the growth of structure in the universe.

REFERENCES

1. Thorne, K. S., & Braginsky, V. B., *ApJ*, **204**, L1-L6 (1976).
2. Miyoshi, M., et al., *Nature*, **373**, 127-129 (1995).
3. Genzel, R., Eckart, A., Ott, T., & Eisenhauer, F., *MNRAS*, **291**, 219-234 (1997).
4. Ghez, A. M., Klein, B. L., Morris, M., & Becklin, E. E., *ApJ*, **509**, 678-686 (1998).
5. Maoz, E., *ApJ*, **494**, L181-L184 (1998).
6. Tanaka, Y., et al., *Nature*, **375**, 659-661 (1995).
7. Wambsganss, J., Paczyński, B., & Schneider, P., *ApJ*, **358**, L33-L36 (1990).
8. Peterson, B. M., & Wandel, A., *ApJ*, **521**, L95-L98 (1999).
9. Magorrian, J., et al., *AJ*, **115**, 2285-2305 (1998).
10. Richstone, D., et al., *Nature*, **395**, A14-A19 (1998).
11. Fan, X., et al., *AJ*, **118**, 1-13 (1999).
12. Fan, X., et al., *AJ*, submitted (1999); astro-ph/9909169.
13. Turner, E. L., *AJ*, **101**, 5-17 (1991).
14. Krolik, J. H., *Active Galactic Nuclei*, Princeton: Princeton University Press, 1999.
15. Rees, M. J., *ARA&A*, **22**, 471-506 (1984).
16. Quinlan, G. D., & Shapiro, S. L., *ApJ*, **343**, 725-749 (1989).
17. Lee, M. H., *ApJ*, **418**, 147-162 (1993).
18. Quinlan, G. D., & Shapiro, S. L., *ApJ*, **356**, 483-500 (1990).
19. Rees, M. J., *Class. Quantum Grav.*, **14**, 1411-1415 (1997).
20. Shapiro, S. L., & Teukolsky, S. A., *ApJ*, **234**, L177-L181 (1979).
21. Goldreich, P., & Weber, S. V., *ApJ*, **238**, 991-997 (1980).
22. Fuller, G. M., Woosley, S. E., & Weaver, T. A., *ApJ*, **307**, 675-686 (1986).
23. Loeb, A., & Rasio, F. A., *ApJ*, **432**, 52-61 (1994).
24. Gaskell, C. M., *ApJ*, **464**, L107-L110 (1996).
25. Eracleous, M., et al., *ApJ*, **490**, 216-226 (1997).
26. Roos, N., Kaastra, J. S., & Hummel, C. A., *ApJ*, **409**, 130-133 (1993).
27. Sillanpää, A., et al., *ApJ*, **325**, 628-634 (1988).
28. Lehto, H. J., & Valtonen, M. J., *ApJ*, **460**, 207-213 (1996).
29. Makino, J., & Ebisuzaki, T., *ApJ*, **465**, 527-533 (1996).
30. Quinlan, G. D., & Hernquist, L., *New Astronomy*, **2**, 533-554 (1997).
31. Faber, S. M., et al., *AJ*, **114**, 1771-1796 (1997).
32. van der Marel, R. P., *ApJ*, **117**, 744-763 (1999).
33. Owen, F. N., O'Dea, C. P., Inoue, M., & Eilek, J. A., *ApJ*, **294**, L85-L88 (1985).
34. Begelman, M. C., Blandford, R. D., & Rees, M. J., *Nature*, **287**, 307-309 (1980).
35. Quinlan, G. D., *New Astronomy*, **1**, 35-56 (1996).
36. Governato, F., Colpi, M., & Maraschi, L., *MNRAS*, **271**, 317-322 (1994).
37. Kauffmann, G., & Haehnelt, M., *MNRAS*, submitted (1999); astro-ph/9906493.
38. Richstone, D., *Laser Interferometer Space Antenna*, ed. W. M. Folkner, Woodbury: American Institute of Physics, 41-44 (1998).
39. Sigurdsson, S., *Laser Interferometer Space Antenna*, ed. W. M. Folkner, Woodbury: American Institute of Physics, 53-56 (1998).
40. Narayan, R., Piran, T., & Shemi, A., *ApJ*, **379**, L17-L20 (1991).
41. Phinney, E. S., *ApJ*, **380**, L17-L21 (1991).

The Angular Resolution of Space-Based Gravitational Wave Detectors

Thomas A. Moore* and Ronald W. Hellings†

Physics Department, Pomona College, Claremont, CA 91711
†*Jet Propulsion Laboratory, Pasadena, CA 91109*

Abstract. Currently proposed space-based gravitational wave detectors consist of satellite arrays deployed in an equilateral triangle either in the ecliptic plane or in a precessing plane tilted at 60° with respect to the ecliptic. This paper summarizes recent calculations of how well these two types of detectors can resolve the angular positions of two kinds of sources: low-mass compact binary sources and massive-black-hole binaries just before coalescence. Our results show that including higher-order post-Newtonian terms in the gravitational wavefunctions is important when predicting the angular resolution of ecliptic-plane detectors.

This paper summarizes the results of some new calculations that we have undertaken to predict the angular resolution of two types of space-based gravitational wave detectors. In the "ecliptic-plane" option (the configuration of the OMEGA project, shown in Fig. 1a) where the detector satellites orbit the earth in an equilateral triangle in the ecliptic plane. In "precessing-plane" option, (the currently proposed LISA configuration shown in Fig. 1b), the detector satellites orbit the sun in an equilateral triangle in a plane whose normal is tilted 60° toward the sun away from the ecliptic normal. Though we used OMEGA's and LISA's noise curves respectively for these cases, the basic nature of the results should apply to any kind of detectors having these configurations.

The sources we considered were spinless compact binaries in inspiralling quasicircular orbits. We considered both low-mass, essentially monochromatic sources as well as massive-black-hole mergers. Our most important goals for this particular study, extending solid previous work by Curt Cutler and Alberto Vecchio (1), were to model both the waveform phase and amplitude to 4/2 post-Newtonian (PN) order using equations described by Blanchet, et. al. (2), to systematically explore the dependence on various parameters, and to carefully compare results for both detector configurations.

One can extract information about the angular position of the source from its observed waveform as follows. In the ecliptic-plane case, the wave is doppler-shifted as the detector array orbits the sun. The phase of the dopper shift determines the azimuth angle Φ; the magnitude of the dopper shift determines the polar angle Θ. In the precessing-plane case, this doppler shift information is supplemented by an amplitude variation of the signal as the detector array plane precesses.

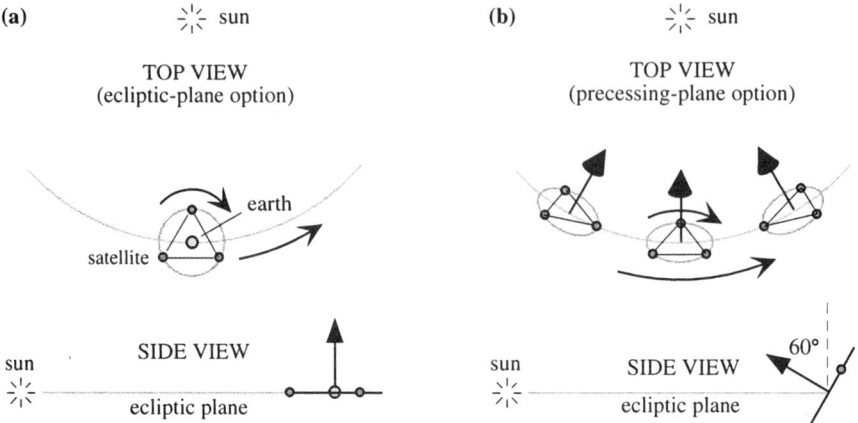

FIGURE 1. Spacecraft configurations for the ecliptic-plane and precessing-plane deployment options. In the ecliptic-plane case (Fig. 1a), the satellites orbit the earth in the ecliptic plane. In the precessing-plane case (Fig. 1b), the satellites orbit the sun in such a way that they form an apparently rigid equilateral triangle whose plane is tilted from the plane of the ecliptic 60° toward the sun. This triangular array also rotates clockwise around its center once a year. In both cases, the arrow with the conical head represents the normal to the plane of the detector array.

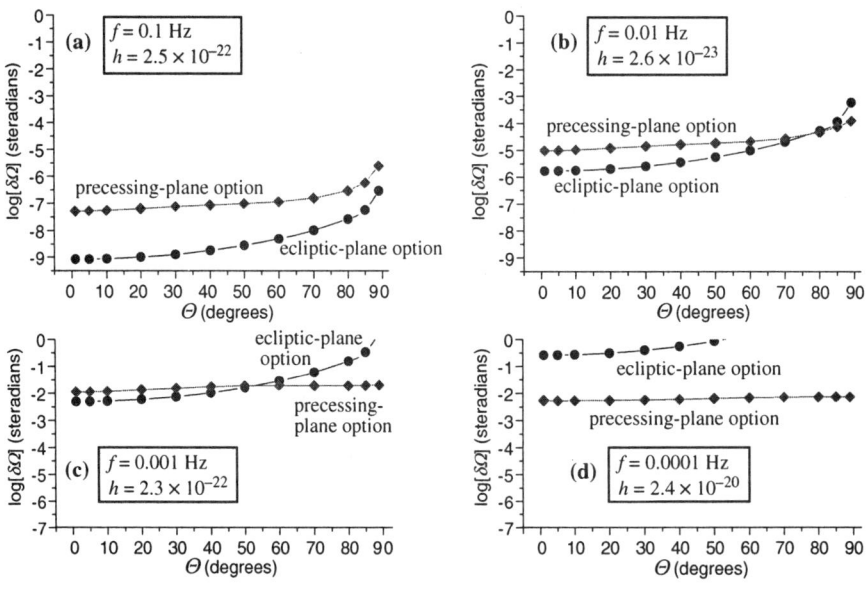

FIGURE 2. Angular uncertainty as a function of Θ for compact binaries with various different (but essentially constant) frequencies. Masses were chosen so that the times to coalescence were long compared to one year, and the distance was chosen so that the signal-to-noise ratio was about 10 in the ecliptic plane case (the amplitude h for each graph is shown). The inclination of the binary orbit to the line of sight was chosen to be almost face-on (the cosine of the inclination angle was 0.8).

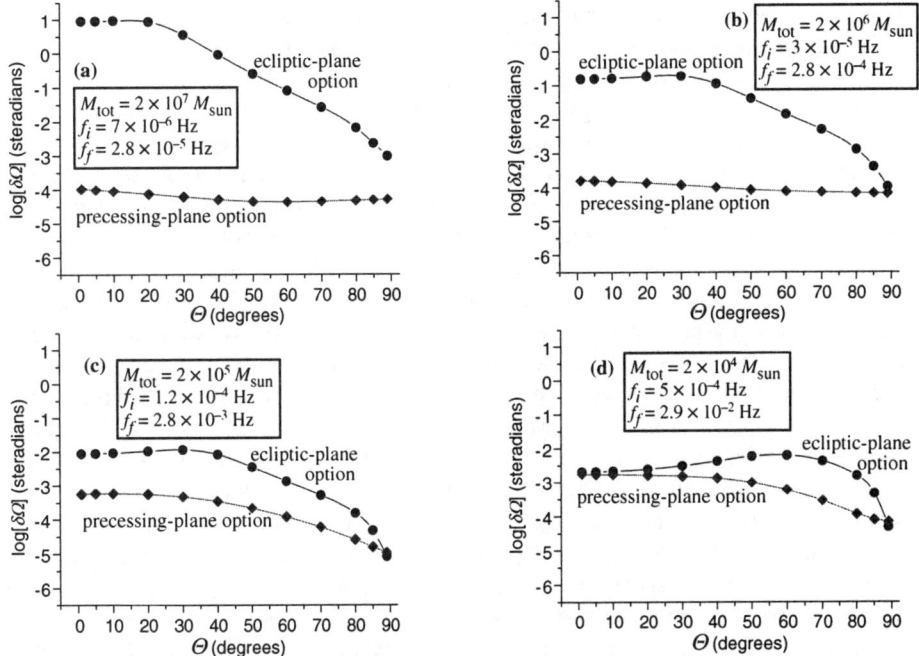

FIGURE 3. Angular uncertainty as a function of Θ for various equal-mass pairs of supermassive black holes undergoing coalescence. In each case the pair was assumed to be at redshift $z = 1$ and the time to coalescence was chosen to be one year. The calculation was terminated roughly 10 orbits before coalescence. The inclination of the binary orbit to the line of sight was chosen to be almost face-on (the cosine of the inclination angle was 0.8).

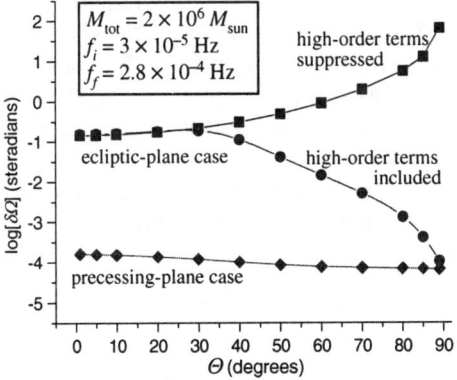

FIGURE 4. Angular uncertainty as a function of Θ for the situation shown in Fig. 3b. This illustrates that the higher-order terms in the post-newtonian approximation for the gravitational waveform have a large effect on the angular resolution in the ecliptic-plane case for values of Θ near the ecliptic plane.

One can indeed extract information about no less than nine independent parameters (that describe the total mass and mass ratio of the binary pair, the orientation, initial size, and phase of its orbit, and its distance as well as Θ and Φ) from the observed waveform. We wrote a computer program that modeled the waveform from a source with specified values for the parameters, added realistic noise to the waveform, and then calculated the consequent uncertainty in all nine parameters at the same time using a linear least-squares covariance analysis for a 1-yr observation period. In this paper, however, we will focus on the angular position uncertainty as a function of polar angle Θ.

Figure 2 shows the base-ten logarithm of the angular position uncertainty as a function of Θ for essentially monochromatic binaries at frequencies between 0.1 mHz and 0.1 Hz. In the ecliptic-plane case, note that position sensitivity improves with increasing frequency, as the doppler shift becomes easier to track. There is also reduced sensitivity as one approaches $\Theta = 90°$ (that is, the plane of the ecliptic) because the doppler shift becomes relatively insensitive to changes in Θ for sources near the plane of the ecliptic. In the precessing-plane case, there is moderate sensitivity at low frequency because of the information provided by the precession of the array plane, and this sensitivity improves as frequency increases because of the additional doppler-shift information. The sensitivity in this case is relatively independent of Θ because the precession means that the ecliptic plane is not as special as it is in the ecliptic case.

Figure 3 shows angular uncertainty as a function of Θ for large-mass coalescing sources. All of these sources have equal mass partners (at 10^7, 10^6, 10^5, and 10^4 solar masses respectively) are at $z = 1$, and are observed for the last year before coalescence. Note that in the precessing plane case sensitivity is good at the lowest frequencies, and stays relatively the same as the mass decreases, because even as the signal strength drops, the noise curve also drops. In the ecliptic plane case, again the sensitivity improves as the frequency increases (as expected), but curiously, and in sharp contrast to the monochromatic case, the sensitivity *improves* sharply as theta approaches 90°.

This is because the higher-order post-newtonian corrections to the waveform amplitude are significant in the massive-black-hole case, while in the monochromatic case, they are not. Figure 4 clearly shows that when we artificially turn off the higher-order terms, the angular resolution now decreases with Θ for the massive black-hole case, just as it does for the monochromatic case. The lowest-order term simply does not provide enough information to resolve uncertainty-producing correlations that develop as $\Theta \to 90°$ between a number of angular parameters and the source distance, information that the higher-order terms provide. In short, we see that in the ecliptic plane case, keeping track of the higher-order terms in the wavefunction is crucially important.

REFERENCES

1. A. Vecchio and C. Cutler, "LISA's Angular Resolution for Monochromatic Sources, and "LISA: Parameter Estimation for Massive Black Hole Binaries," in CP456, *Laser Interferometer Space Antenna*, edited by W. M. Folkner, AIP, 1998, pp. 95-109.
2. L. Blanchet, B. R. Iyer, C. M. Will, and A. G. Wiseman, *Class. Quantum Grav.* **13**, 575-584 (1996).

BAR ANTENNAE

An optical transduction chain for the AURIGA detector

L. Conti*, F. Marin[†], M. De Rosa[×], G. A. Prodi*, L. Taffarello[•],
J. P. Zendri[•], M. Cerdonio[°], S. Vitale*

* Dipartimento di Fisica, Università di Trento, and INFN Gruppo Coll. Trento,
I-38050 Povo (Trento), Italy
[†] Dipartimento di Fisica, Università di Firenze, and INFN Sez. Firenze,
L.go E.Fermi 2, I-50125 Arcetri (Firenze), Italy
[×] European Laboratory for Nonlinear Spectroscopy (LENS),
L.go E.Fermi 2, I-50125 Arcetri (Firenze), Italy
[•] INFN Sez. Padova, Via F.Marzolo 8, I-35100 Padova, Italy.
[°] Dipartimento di Fisica, Università di Padova, and INFN Sez. Padova,
Via F.Marzolo 8, I-35100 Padova, Italy.

Abstract. We describe the principle of operation of an opto-mechanical readout for resonant mass gravitational wave detectors; with such a device the AURIGA detector is expected to reach a sensitivity at the level of $\sqrt{S_{hh}} = 10^{-22}/\sqrt{Hz}$ over a bandwidth of about $40Hz$. Recent developments in the implementation of this transduction chain are also reported. In particular we acheive quantum limited laser power noise in the frequency range of $200Hz$ around the bar fundamental frequency (about $1kHz$) by means of active stabilization. We also set up a reference cavity of finesse 40000 with optically contacted mirrors on a $0.2m$ long Zerodur spacer. The cavity can be heated from room temperature to about $100°C$ and temperature stabilized with fluctuations within $1mK$ over a period of several days. The cavity is under vacuum and isolated from mechanical disturbancies by means of a double stage cantilever system.

INTRODUCTION

The sensitivity of presently working g.w. detectors, which are of the resonant type, is limited by read-out noise [1]. As a result, in the past several kinds of devices have beeen proposed to detect the small vibration of a resonant bar determined by the passage of a gravitational wave. One of the most appealing one is that proposed by J.P.Richard [2], who suggested to use laser interferometric techniques for this purpose. Actually Richard not only suggested the new idea and calculated the expected sensitivity, but also begun the needed experimental work [3].

A more recent version of the Richard's idea is the object of the work described in this paper. The transduction chain is designed in order to take fully advantage from the fast growing technical improvements in the optics industry and takes into account the many problems that an integration into a real detector may arise.

PRINCIPLE OF OPERATION

The fundamental idea for signal optical transduction is to have a resonant optical cavity of lenght L formed by a mirror attached to a bar end face and a second fixed mirror. Then a relative motion ΔL of the two mirrors, due to bar vibration induced eventually by a g.w., is converted into a change $\Delta \nu$ of the optical resonant frequency ν according to:

$$\frac{\Delta L}{L} = \frac{\Delta \nu}{\nu} \qquad (1)$$

If the second mirror is not fixed but is attached to a resonator having the same frequency as the bar (i.e. to a resonant transducer) then the bar vibration amplitude can be amplified by a factor equal to the square root of the oscillators mass ratio. The system formed by a bar and a resonant transducer can be viewed, indeed, as that of two harmonic oscillators coupled together.

The idea expressed in eq.(1) has been developed in the scheme shown in Figure 1 [4]. The optical resonant cavity cited above is here named *transducer cavity*. The phase modulated beam produced by a Nd:YAG laser source is split in two

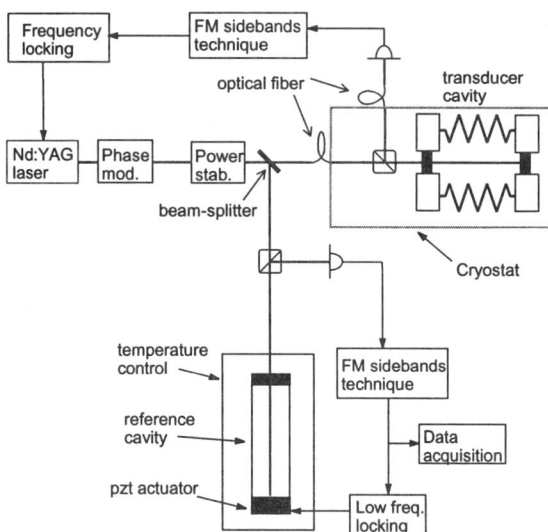

FIGURE 1. Scheme of the optomechanical readout for resonant bar g.w. detectors.

parts, one of which is sent to the transducer cavity inside the cryostat that houses the bar by means of an optical fiber. The reflected beam is conveyed into another fiber cable and is then detected by a photodiode outside the cryostat. The error signal obtained with the Pound-Drever-Hall technique [5] is used by a servo system to frequency-lock the laser source to the transducer cavity; thus the light frequency carries information on the bar motion.

In order to extract this piece of information, the second beam is sent to a second Fabry-Perot cavity, called *sensing cavity*, which acts as frequency reference. Again, the reflected beam is detected by a photodiode and its current used to obtain a signal which is proportional to the frequency difference between the sensing cavity and the laser. In this way, this error signal carries information on the bar and transducer motion and is to be analysed for g.w. detection. Anyway, at this purpose, one must guarantee that the sensing cavity is in resonance condition with the light: therefore one has to correct its length at low frequency (i.e. in a frequency range far from the kHz range, e.g. up to some tens of Hz) in order to follow any slow drift of the transducer cavity (due for instance to temperature variations). The error signal to be used for such controls is the same that, at higher frequencies, is acquired for g.w. detection.

An important point that should be made is that laser power noise acts as a *back-action noise*, in the sense that it ends up exciting and heating the system of bar and transducer. This occurs because the laser intensity noise induces fluctuations in the radiation pressure on transducer cavity mirrors, thus generating a noise force. For this reason it is important that the laser power noise is reduced before entering the cryostat.

With the figures given in Table 1, the expected total strain noise referred at detector input reaches minimum level of $\sqrt{S_{hh}} = 10^{-22}/\sqrt{Hz}$ and the useful bandwidth (defined at $+3dB$ for S_{hh}) is about $40Hz$ in a frequency range centered on detector resonance. Assuming that the signal is a $1ms$ burst of g.w. with central frequency equal to bar resonance, the wave amplitude that can be detected with unitary signal-to-noise ratio is: $h_{min} = 3 \cdot 10^{-20}$.

TABLE 1. Parameters used for calculating the sensitivity of a bar detector equipped with the optical readout.

bar temperature	$0.1K$	transducer effective mass	$6.5kg$
bar res. freq.	$920Hz$	Q_{bar}	$5 \cdot 10^6$
transd. res. freq.	$920Hz$	Q_{transd}	$5 \cdot 10^6$
trasd.-cav. length	$0.01m$	sens.-cav. length	$0.2m$
laser wavelength	$1064nm$	mirror losses	$1ppm$
sens.-cav. input mirror T	$150ppm$	sens.-cav. output mirror T	$2ppm$
transd.-cav. input mirror T	$15ppm$	transd.-cav. output mirror T	$2ppm$
photodiodes quantum efficiency	0.85	laser phase mod. amplitude	$0.06rad$

LASER POWER NOISE

The first task was to reduce the laser intensity noise in the $1kHz$ region of the laser source: the experimental setup is shown in Figure 2 and follows ref. [6]. The linearly polarized laser beam enters an electro-optic modulator (EOM) with the crystal axis sligtly rotated with respect to the laser polarization direction; the EOM is followed by a polarizer parallel to the input light polarization. The output beam passes through a beam-splitter with reflectivity R: the reflected beam is detected by a photodiode whose output, properly amplified, is fed back to the EOM. The outgoing beam is that to be used in the experiment: for measureament purpuse, this beam was split by a 50% beam-splitter and the beams detected by two balanced photodiodes, whose outputs were summed and/or subtracted. The subtraction gives the shot noise level of the beam impinging the beam-splitter.

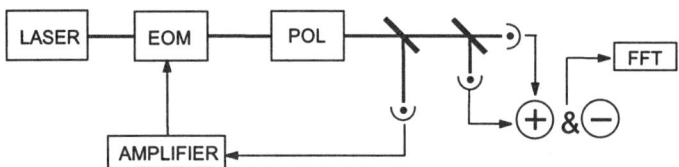

FIGURE 2. Experimental setup for the laser intensity noise reduction and measurement.

The free-running laser intensity noise was thus measured and compared to shot noise level: the calibration was checked using a halogen lamp, whose intensity power noise around $1kHz$ was found to scale linearly with intensity, as expected from shot noise limited light fluctuations. The measureaments in the frequency range $(900 \div 1100)Hz$ are shown in Figure 3 (left): the reflectivity R of the first beam-splitter was varied by replacing it with a half-wave plate followed by a polarizing beam-splitter. The measureaments were repeated with the intensity noise reduction loop switched on and are shown in Figure 3 (right): experimental points are fitted by a theoretical curve with constant total loop gain (dashed line). Best achievable performance can be obtained with infinite total loop gain [7], as shown in Figure 3 (right; solid line).

The intensity noise was also measured at the output of a single-mode polarization maintaining optical fiber, coupled to the outgoing beam: the fiber was found to deteriorate the acheived laser intensity noise reduction, maximally at low frequencies as evident from Figure 4. This figure refers to beam-spitter $R = 50\%$ and $15mW$ transmitted by the fiber: the measured noise power is $8dB$ above the shot level, in the frequency range of interest. Scaling this to the $2mW$ fed to the transducer-cavity, the acheived intensity stabilization will allow the power reaching the resonators to be close to the shot-noise limit.

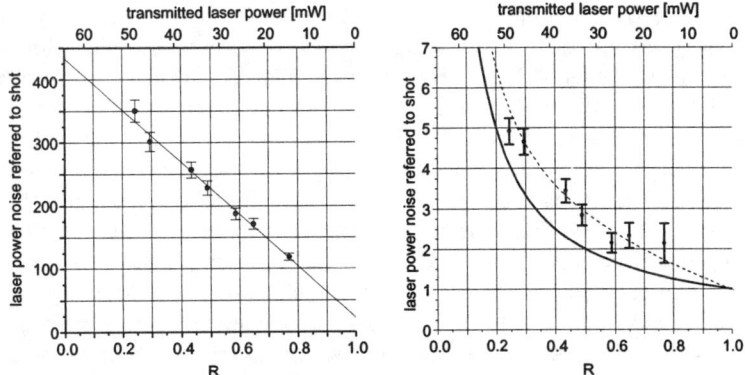

FIGURE 3. Laser intensity noise power spectrum integrated between 900 and $1100 Hz$ referred to the shot noise level with intensity noise reduction loop off (left) and on (right). Left: experimental points and fitting curve. Right: experimental points, finite loop gain fit curve (dashed line) and infinite loop gain limit (solid line).

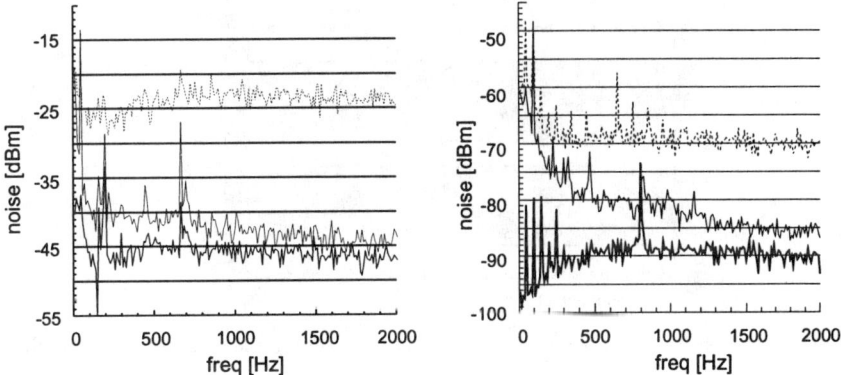

FIGURE 4. Performance of the laser intensity noise reduction; spectra refer to the out-of-loop beam (after an amplification stage). Left: free running laser (upper curve), corresponding shot-noise level (lower curve) and stabilized laser beam (middle curve). Right: free running laser transmitted by a singlemode optical fiber (upper curve), corresponding shot-noise level (lower curve) and fiber transmitted stabilized laser beam (middle curve).

REFERENCE CAVITY

As said above, the signal detection is performed at the sensing-cavity that acts as frequency reference. Therefore its own frequency stability should be better than that of the laser with respect to the transducer-cavity. With the values in Table 1 the shot noise limits the laser frequency locking to the transducer-cavity to $1 \cdot 10^{-4} Hz/\sqrt{Hz}$ (bilateral). A laser beam sent to a sensing-cavity with the characteristics described in Table 1 can be locked with a shot-noise limited frequency noise at the level of $3 \cdot 10^{-5} Hz/\sqrt{Hz}$. Obviously all the other noise sources must be negligible with respect to this level. The main sources come from thermal noise, lenght fluctuations induced by temperature ones, mechanical disturbancies and residual gas pressure variations. Presently we are planning to use a room temperature sensing cavity, which can be temperature tuned to resonance. Once the correct lenght is thus reached, the temperature is stabilized and the cavity is kept resonant by means of piezoelectric actuators (PZT) acting below the bar resonant frequency. The cavity is made out of a Zerodur spacer with optically contacted mirrors (measured finesse $4 \cdot 10^4$). Slow laser frequency variations are followed by changing the temperature. In order to prevent the electric and thermal noise of the actuators from compromising the frequency stability of the cavity, the actuators are mounted so that the effect of piezo motion is reduced down to the minimum necessary level. For a typical piezoelectric ceramic, a reduction by a factor 100 is accomplished by mounting the cavity as in Figure 5.

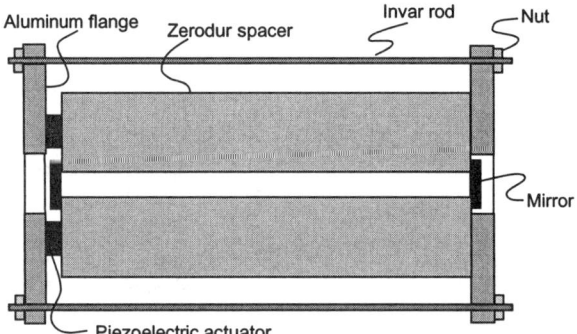

FIGURE 5. Schematic cross-section (not drawn to scale) of the sensing-cavity assembly.

The cavity is placed on the top of a double stage cantiveler suspension and is kept under vacuum by pumping with a ionic pump, which has the advantage of being vibration free. Following the calculations of ref. [8], the vacuum level for N_2 should be as good as $6 \cdot 10^{-7} mbar$. The cavity temperature can be changed from room temperature up to about $100°C$, enough for covering a free spectral range (the thermal expansion coefficient of Zerodur is $\sim 4 \cdot 10^{-8} K^{-1}$ [9]). The temperature is then actively stabilized and temperature variations are measured to be less than $1 mK$ for periods of several days: this is enough to stay within a cavity optical width just with the temperature control.

At present two reference cavities have beeen built in Florence in order to measure the frequency stability of one with respect to the other. The laser has been frequency locked to one of the two cavities with the necessary stability and work is in progress for completing the measurements.

THE TRANSDUCER

Experimental work on the mechanical resonator contituing the transducer has been described elsewhere [4,10]. A $50mW$ Nd:YAG laser beam was fed by means of a single-mode fiber to a ~ 1200 finesse Fabry-Perot cavity installed on a room temperature bar identical to the one used in AURIGA. The laser was frequency locked to this cavity for short time periods and the resonant modes of the bar and transducer assembly excited by thermal noise were observed in the correction signal. Present work is devoted to improving the mechanical Q of the assembly and the laser frequency locking to the cavity. Once this is reached and the signal compared to a reference cavity, a complete optical read-out for a bar detector, even if at room temperature, will be first realized and its sensitivity and reliability studied. A further effort is needed if the device is to be used in a cryogenic detector: optics and fiber behaviour at low temperature is to be investigated and a cryogenic system to guarantee the transducer-cavity alignment even during the cooling down has to be implemented.

ACKNOWLEDGMENTS

We gratefully thank prof. M. Inguscio for encouraging this research and assuring in Florence a suitable laboratory; L.C. also thanks the *European Laboratory for Nonlinear Spectroscopy (LENS)* for hospitality and for providing the technical support needed for the part of this work that was developed in Florence.

REFERENCES

1. Cerdonio, M., *et al.*, "Cryogenic resonant detectors of gravitational waves: current operation and prospects" in *Gravitation and Relativity: at the Turn of the Millenium, Proc. of the 15th Int. Conf. on General Relativity and Gravitation*, Pune, edited by N.Dadhich and J.Narlikar, IUCAA, 211-230 (1998).
2. Richard, J.P., *J. Appl. Phys.* **64**, 2202-2205 (1988).
3. Pang, Y., and Richard, J.P., *Appl. Opt.* **34**, 4982-4988 (1995).
4. Conti, L., *et al.*, "Optical Readout for the AURIGA detector" in *Proceedings of the XXXIVth Rencontres de Moriond on Gravitational Waves and Experimental Gravity*, Les Arcs, Savoie, France, (1999), in press.
5. Drever, R. W. P., *et al.*, *Appl. Phys. B* **31**, 97-105 (1983).
6. Wong, N.C., and Hall, J.L., *J. Opt. Soc. Am. B* **2**, 1527-1533 (1985).
7. Giacobino, E., Marin, F., Bramati, A., and Jost, V., *J. Nonlin. Opt. Phys. Mat.* **5**, 863-877 (1996).
8. Giazotto, A., *Phys. Rep.* **182**, 365-424 (1989).
9. Jacobs, S.F., *Opt. Acta* **33** 1377-1388 (1986).
10. Conti, L., *et al.*, *Rev. Sci. Instr.* **69**, 554-558 (1998).

MiniGRAIL, A 65 cm Spherical Antenna

Arlette de Waard & Giorgio Frossati

Kamerlingh Onnes Laboratory, Leiden University, Leiden
The Netherlands

Abstract. We intend to build a 65 cm diameter spherical, cryogenic gravitational wave antenna made of the CuAl(6%) alloy with a mass of 1168 Kg, a resonance frequency of 3.7kHz and a bandwidth around 300 Hz. The quantum-limited strain sensitivity $\delta L/L$ would be 2.2×10^{-21} at 10 mK. We believe that a sensitivity around 5×10^{-20} could be reached within the duration of the project, which is (at that frequency) comparable to that of the large interferometers LIGO and VIRGO presently being built. The sources we are aiming at are for instance non-axisymmetric dynamical instabilities of rotating neutron stars within our galaxy, where 10^8 neutron stars are expected to exist.

INTRODUCTION

The advantages of using a spherical detector are considerable and have already led to a number of projects, like GRAIL, the SFERA proposal (Rome), TIGA (LSU) and GRAVITON (Brazil). A sphere has 5 fundamental spheroidal quadrupole modes giving it omnidirectionality and equal sensitivity to both polarizations of the gravitational wave. It can detect the direction and the tensorial character of the wave and it has a very large cross section compared to bar detectors at the same frequency (1),(2),(3),(4).

THE MINIGRAIL

Facilities

The Minigrail facility will be located in the Kamerlingh Onnes Laboratory of the University of Leiden. In our experimental hall five concrete supports are built on vibration-free islands (figure1, left), which stand on pillars, planted 18 meters into the ground and are well isolated from the rest of the building. All the pumps are installed behind the wall on separate concrete blocks, all the pumping lines go through the concrete block on top of the fifth pillar. The Minigrail support (figure1, right) has a hydraulic system installed inside of the four concrete pillars, which carry the 3 tons concrete plate. The hydraulic system can lift the concrete plate with the sphere and suspension 1.5 meters to allow mounting the lower part of the dewar. This system is able to lift 20 tons. The dewar will be supported by a flange attached to the concrete plate.

FIGURE 1. Left: Picture of the experimental hall of the Kamerlingh Onnes Laboratry. All the concrete supports are build on 'vibration-free' islands. Right: Picture of the concrete support for the MiniGRAIL. Inside of the pillars a hydraulic system is installed to lift the concrete plate, sphere and suspension. The system is able to lift 20 tons.

Design of the MiniGRAIL Antenna

A schematic picture of the MiniGRAIL detector is shown in figure 2. The sphere will be made of CuAl(6%) because of its high quality factor of 15 million and good thermal conductivity at low temperatures since it doesn't become superconducting. This material can be cast in the Netherlands by the company LIPS in Drunen. The antenna will have a diameter of 65 cm and a weight of 1168 kg. It will be suspended from the center with a copper rod with a diameter of 10 mm. The vibration isolation stages consist of seven masses, the upper four made of CuAl(6%) and the last three made of copper. The masses will be connected to each other by springs or rods. Each stage will have a mass of about 120 kg and will give an attenuation of about 40 dB, so the total attenuation will be in the order of 300 dB at the resonant frequency of the sphere. The total system will be suspended by three titanium rods attached to the flange of the dewar. The dilution refrigerator (dr) will be off center, see figure 2, and the mixing chamber will be thermally anchored by copper 'spaghetti' to mass number 5, the first copper mass, of the vibration isolation system, so vibrations coming from the dr will be attenuated by 120 dB. The heat exchanger of the dr will be a large circular tube to avoid instabilities coming from convection of the ^{3}He inside of the dilute phase of the mixture. The dewar will be specially built for this purpose by Kadel (5). The He consumption will be less than 30 liter/day. The volume of the He reservoir is 300 liters, so He transfer will be needed every 10 days. The feed through of the 1K pot pumping line, the He and N_2 transfer lines are not shown in the picture.

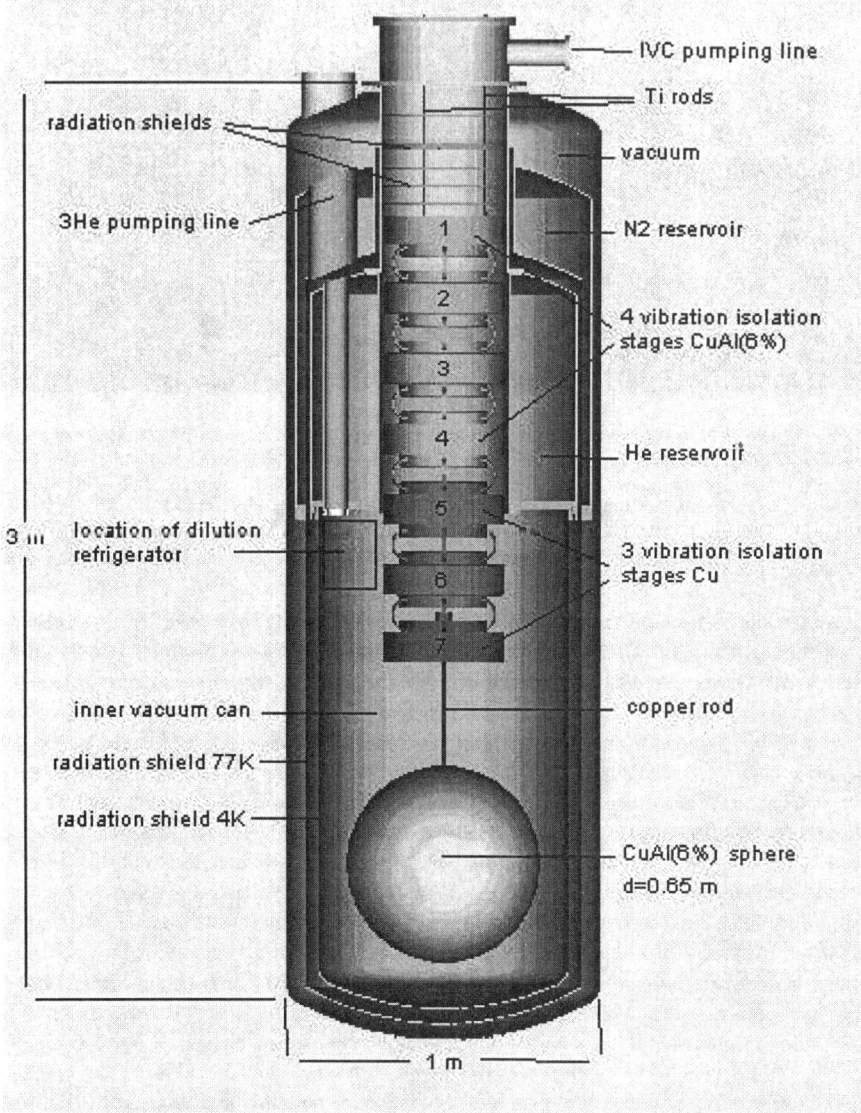

FIGURE 2. Schematic layout of the dewar with the antenna and the seven stages suspension. The upper four stages are made of CuAl(6%). The third copper mass will be thermally anchored to the mixing chamber of the dilution refrigerator. The dewar will be specially made by Kadel engineering.

MiniGRAIL sensitivity

The quadrupolar frequencies of a sphere are given by (4),(6)

$$\omega_n = \frac{c_n v_s}{R_s} \tag{1}$$

where n is the order of the quadrupole mode, the numerical coefficients $c_1=1.62$ and $c_2=3.12$, $v_s=4620$ m/s the sound velocity of CuAl(6%) and the radius of the sphere $R_s=0.325$ m, so the first quadrupolar frequency is $f_1 \cong 3665$ Hz.

The gravitational wave energy absorbed by a resonant antenna is expressed in terms of its cross section. For a spherical antenna, which has 5 fundamental spheroidal quadrupole modes the cross section is given by:

$$\sigma_n = F_n \frac{G}{c^3} M_s v_s^2 \tag{2}$$

$F_1=2.98$ and $F_2=1.14$ are dimensionless coefficients for the first and second quadrupole mode, $G=6.7 \times 10^{-11}$ m^3/kgs is the gravitational constant and $c=3.0 \times 10^8$ m/s is the speed of light, $M_s=1168$ kg is the mass of the sphere. The cross section for the first quadrupolar mode is $\sigma_1 \cong 1.04 \times 10^{-25}$ m^2/s. The initial goal would be to cool the sphere down to 20 mK and using a SQUID with an energy sensitivity of $E_N=100h\nu$ so that the noise temperature of the amplifier would be

$$T_N = \frac{E_N}{k_B} \cong 1.76 \times 10^{-5} K \tag{3}$$

The minimal energy a resonant antenna is able to detect, is given by in terms of the effective temperature, by (4)

$$T_{eff} \cong 2\sqrt{2} T_N \left(1 + \frac{2T}{\beta Q T_N}\right)^{\frac{1}{2}} \approx 4.99 \times 10^{-5} K \tag{4}$$

with $T \approx 20$ mK, the thermodynamic temperature, $\beta \approx 0.05$ is the ratio of the energy absorbed by the antenna and that transferred to the transducer and $Q \approx 1.5 \times 10^7$ (7) is the mechanical quality factor of CuAl(6%) at 20 mK.

We consider the sensitivity to a gravitational wave burst of a duration τ_g. The Fourier transform H(f) of the burst is assumed constant within a bandwidth Δf so

$$H(f) \cong H(f_0) = H_0 \cong \tfrac{1}{2} h_0 \tau_g$$

For a SNR=1 we have

$$H_0^{min} = \left[\frac{kT_{eff}}{\tfrac{1}{2} m l^2 (2\pi f_1)^4}\right]^{\frac{1}{2}} \cong 1.05 \times 10^{-23} \tag{5}$$

where $m=M_s=1168$ kg is the mass of the detector, $l=0.6R=0.195$m, f_1 is the resonant frequency of the first quadrupolar mode. The strain sensitivity to a gravitational wave burst of duration $\tau_g =1.5$ ms is

$$h_0^{min} \cong \frac{2H_0}{\tau_g} \approx 1.4 \times 10^{-20} \qquad (6)$$

The bandwidth is calculated with

$$\Delta f = 0.7 f_1 \beta \left(1+\frac{2T}{\beta Q T_N}\right)^{\frac{1}{2}} \cong 128 Hz \qquad (7)$$

and a spectral amplitude of

$$\tilde{h} = \sqrt{S_h(f_1)} = \sqrt{2\pi\Delta f_1}\tau_g h_0^{min} \approx 6.0 \times 10^{-22} \qquad (8)$$

We also calculated the values for the quantum limited SQUID sensitivity of $1\hbar\omega$ at a temperature of 10 mK. In the quantum limit the strain amplitude $h_0^{min} = 1.5 \times 10^{-21}$. When we look at the Heisemberg uncertainty principle

$$\Delta x \Delta p \geq \frac{\hbar}{2} \quad \Rightarrow \quad \overline{(\Delta x^2)} \geq \frac{\hbar}{2M_s\omega} \qquad (9)$$

$\Delta x_{QL}=1.4\times 10^{-21}$ m and the strain amplitude $h=(\Delta x/x) \cong 2.15 \times 10^{-21}$. From this we can see that the strain amplitude of a sphere with $\phi=65$ cm is limited by the Heisemberg uncertainty principle, although only by 20%. In table 1 there is an overview of all the parameters of the Minigrail for the initial goal, reaching a thermodynamic temperature of the antenna of about 20 mK and using a SQUID with an energy sensitivity of $100\hbar\omega$ and of the quantum limited case at a temperature of 10 mK.

TABLE 1. Minigrail parameters.

	Initial goal	Quantum limit
R	0.325 m	
Material	CuAl(6%)	
v_s	4620 m/s	
ρ	8121 kg/m³	
T	0.02 K	0.01 K
Q	1.5×10^7	
M	1168 kg	
M_1	230 kg	
M_2	2.3 kg	
M_3	0.023 kg	
F_0	3665 Hz	
β	0.05	0.3
T/βQ	2.6×10^{-8} K	2.2×10^{-9} K
E_N (hν)	100 hν	1 hν
$T_N = E_N/k$	1.76×10^{-5}	1.76×10^{-7}
T_{eff}	4.99×10^{-5} K	5.3×10^{-7} K
T/QT_N	0.76×10^{-4}	3.8×10^{-3}
Δf	128 Hz	779 Hz
h_0^{min} (τ=1.5ms)	1.4×10^{-20}	1.4×10^{-21}

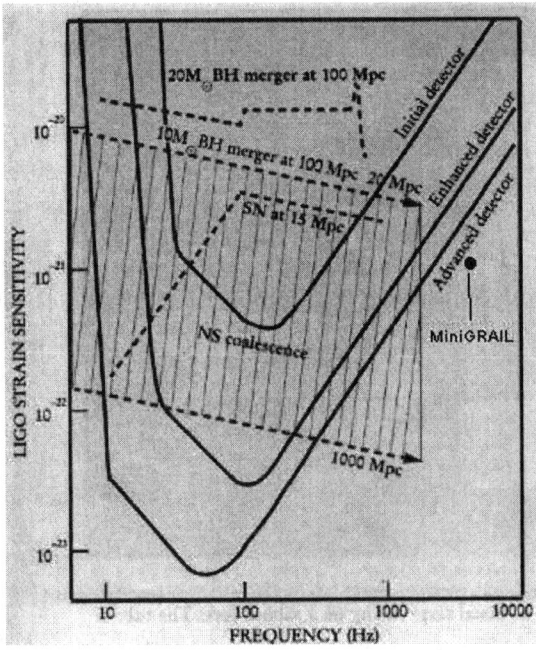

FIGURE 3. Minigrail sensitivity compared to the sensitivity of the advanced LIGO detector (8).

Cryogenics

Cooling from 300 K to 4.2 K

From the Nautilus and Auriga experiments we know that it takes about 20 days to cool down a 2.3 ton Al5056 antenna with a surface area of ~6m^2 from room temperature to 77K (9),(10). Because $c_{CuAl} \approx 2c_{Al}$, the total heat capacity of Minigrail is about the same as that of the bars, though the contact area is only 1.1m^2. Cooling the sphere with exchange gas would take more than 3.5 months (110 days). The total enthalpy that has to be removed is in the order of 10^8 J. Using a forced helium flow, which is able to remove 500W/m^2, the time needed to cool down the sphere to 77 K could be reduced to $t=(10^8/500)=2\times10^5$ sec$\approx$2.5days.

The enthalpy of copper at nitrogen temperature is $H_{77K}=6$J/g, so the total enthalpy of the sphere will be about 7×10^3 J, $t=7\times10^6/500=1.4\times10^4$ sec$\approx$4 hours. The total helium consumption from room temperature to 4.2 K will be around 600 liters.

Cooling from 4.2 K to 10 mK

Taking the specific heat of copper at 4K as $c_{Cu}=1.3\times10^{-5}$ J/cm^3

$$H = \int CdT = 10^{-4}\frac{J}{g} \Rightarrow H \sim 100 J \qquad (10)$$

The cooling power of a dilution refrigerator will be about 1 mW at 100 mK, so it will take about 30 hours to cool down to 100 mK. The enthalpy at 100 mK is about 0.1J and the cooling power of the dr 10μW at 10 mK and so t<3 hours.

Using a forced helium flow and a powerful dilution refrigerator, the Minigrail can be cooled from room temperature down to 10 mK in 4 days so tests and experiments can be done very quickly and at low costs.

ACKNOWLEDGMENTS

We thank the FOM for continuing to support this project. We also thank E. Coccia, M. Visco, V. Fafone, I. Modena, J.P. Zendri and M. Bassan for very stimulating discussions.

REFERENCES

1. R. Forward, *General Relativity Gravitation* 2, 1971, pp. 149.
2. R.V. Wagoner and H.J. Paik, in *Proceedings of the International Symposium on Experimental Gravitation*, Accademia Nazionale dei Lincei, Rome, 1977.
3. W.W. Johnson and S.M. Merkowitz, *Phys. Rev. Lett.* **70**, 1993, pp. 2367.
4. E. Coccia, in *14th International Conference on General Relativity and Gravitational Physics*, Florence, 1995.
5. Kadel Engineering
6. E. Coccia, J.A. Lobo and J.A. Ortega, *Phys. Rev. D* **52**, (1995), pp. 3735.
7. G. Frossati et al., *OMNI-1 Proceedings*, edited by W.F. Velloso Jr., O.D. Aguiar and N.S. Magelhaes (World Scientific Publishing Co., Singapore, 1997).
8. C.M. Will, Physics Today, (1995), pp. 38.
9. P. Astone et al., *Europhys. Lett.* **16**, 231 (1991).
10. P. Astone et al., *Astroparticle Phys.* (In press).
    P. Astone et al., *Technical Report No. LNF-97/005(P)*, INFN-Laboratori Nazionali di Frascati.
11. G. Frossati and E. Coccia, *ICEC 15 Proc.*, *Cryogenics* **34**, 9 (1994).

Detection of Cosmic Rays by NAUTILUS

P. Astone[1], M. Bassan[2], P. Bonifazi[3], P. Carelli[4], E. Coccia[2],
V. Fafone[6], S. D'Antonio[6], S. Frasca[5], A. Marini[6],
E.Mauceli[6], G.Mazzitelli[6], Y. Minenkov[2], I. Modena[2],
G. Modestino[6], A. Moleti[2], G. V. Pallottino[5], V. Pampaloni[1,6], M.A. Papa[6], G. Pizzella[7], F. Ronga[6], R. Terenzi[3], M. Visco[3], L. Votano[6]

[1] *Istituto Nazionale di Fisica Nucleare INFN, Rome*
[2] *University of Rome "Tor Vergata" and INFN, Rome*
[3] *IFSI-CNR and INFN, Rome*
[4] *University of L'Aquila and INFN, Rome*
[5] *University of Rome "La Sapienza" and INFN, Rome*
[6] *Istituto Nazionale di Fisica Nucleare INFN, Frascati*
[7] *University of Rome "Tor Vergata" and INFN, Frascati*

Abstract. The passage of cosmic rays has been observed to excite mechanical vibrations in the resonant gravitational wave detector NAUTILUS operating at temperature of 100 mK. A very significant correlation (more than ten standard deviations) is found.

Beron and Hofstander, already in 1969, carried out experiments aiming to detect oscillations of piezoelectric discs excited by a GeV electron beam. The results brought the authors to suggest that *a very large cosmic-ray event could excite mechanical vibrations in a metallic cylinder at its resonance frequency and they could provide an accidental background for experiments on gravitational waves* [1,2].

Later, a group at the University of Milan [3] estimated the possible effects of particles on a small aluminium cylinder and made an experiment which verified the calculations, although with rather large experimental errors.

The mechanical vibrations originate from the local thermal expansion caused by the warming up due to the energy lost by the particles crossing the material. The effect depends on the thermal expansion coefficient and the specific heat of the material. The ratio of these two quantities is the Grüneisen coefficient. It turns out that while both the expansion coefficient and the specific heat vary with temperature, the Grüneisen coefficient practically does not. In the case of aluminium, this is certainly true above 1 kelvin, but no data are available at lower temperatures when the aluminium becomes superconductor.

Subsequently, more refined calculations were made by several authors [4–8]. All these models agree in predicting, for the vibrational energy in the fundamental mode of an aluminium cylindrical bar, the following formula expressed in kelvin units:

$$\epsilon = 7.64 \cdot 10^{-9} \, W^2 \cdot f \qquad (1)$$

where W(in GeV) is the particle energy dissipated in the bar, and f is a geometrical factor of the order of unity.

The resonant-mass gravitational wave (g.w.) detector NAUTILUS [9], operating at the INFN Frascati Laboratory, consists of an aluminium 2300 kg bar cooled at 100 mK. The mechanical vibrations are converted by means of an electromechanical resonant transducer into an electrical signal which is amplified by a dcSQUID. The bar and the resonant transducer form a coupled oscillator system, which has two resonant modes, whose frequencies are f_1=906.40 Hz and f_2=921.95 Hz.

NAUTILUS is equipped with a cosmic ray (c.r.) detector system consisting of seven layers of streamer tubes for a total of 116 counters. Three superimposed layers, each one with area of 36 m^2, are located over the cryostat. Four superimposed layers are under the cryostat, each one with area of 16.5 m^2. The signal from each counter is digitized to measure the charge which is proportional to the number of particles. For extensive air showers (EAS) the efficiency is close to 100 %, but the systematic error on the absolute number of particles crossing the apparatus is of the order of 30 %. In addition, saturation begins to show for multiplicity greater then $1000 \frac{particles}{m^2}$. In the present data analysis we have put a lower threshold on the multiplicity M of the bottom layer detection, $M \geq 10^4$, because with this threshold a signal of the order of 1 mK is expected on the NAUTILUS detector [10].

The data regarding the vibrational energy of the bar have been correlated with the data obtained by the cosmic ray detector in the period October 1998 to January 1999.

The NAUTILUS data, recorded with a sampling time of 4.54 ms, are processed by a filter [11] optimized to detect impulsive signals applied to the bar.

For investigating the effect of c.r. we have selected the NAUTILUS data as follows:

a)for each c.r. event we have used 20,000 samples (for a total time of 90.8 s) centered at the time when the number of particles (due to the c.r. event) crossing the lower detector exceeded $M = 10^4$

b)the data stretches with noise temperature T_{eff} (obtained by averaging the filtered data over 6 minutes included the time of the cosmic ray event) larger than 5 mK were rejected, in order to select periods when the detector was properly working and the noise was of the order of the expected signals.

In this way we selected 93 stretches for $M \geq 10^4$ during a total time of 47.7 days.

One of the 93 stretches of data contained a large (about 0.5 K) mechanical excitation at a delay of + 6.2 s and was removed from the analysis, although there was no external veto. We average these data by superimposing the selected

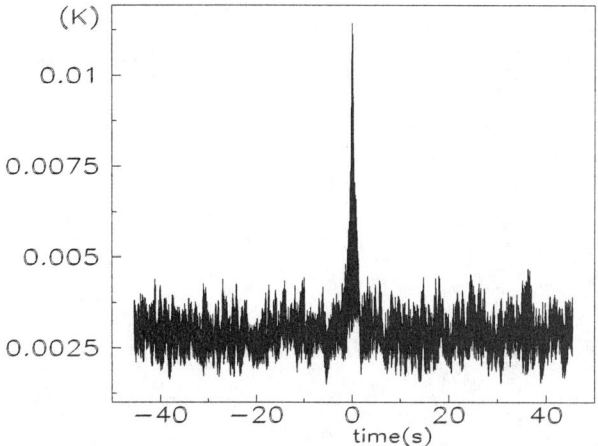

FIGURE 1. The average energy over 92 (for $M \geq 10^4$) stretches of NAUTILUS data versus time. A large signal appears at the cosmic ray arrival time.

TABLE 1. Results of the analysis for two thresholds of the multiplicity. The noise statistics were taken from fig.1 and fig.2 excluding data sampled between ± 2 s

$M \geq$	maximum [mK]	average [mK]	σ [mK]	excess [mK]	excess in σ units
10^4	11.5	2.89	0.43	8.6	20
$1.5 \cdot 10^4$	18.8	2.89	0.57	15.9	28

stretches of data taken at the same time relative to the cosmic ray trigger time. In this way we expect a noise variance that decreases with the number of stretches. The result of this analysis is shown in fig.1, where we plot the averages for each data sample (4.54 ms) versus the time relative to the cosmic ray trigger.

In order to increase our confidence that the observed effect is due to c.r. we have repeated the above procedure raising the threshold on the multiplicity to $15 \cdot 10^3$. In this way we have selected a subset of 46 stretches. The result of the analysis is shown in fig.2, where the signal at zero delay is higher, as expected. These results are also reported in table 1. For checking that the observed events are due to mechanical vibrations of the bar and not just to electrical noise we have performed the following three tests:

TEST I - We computed the average spectrum of the data both at the time of the c.r. trigger and at two other times off the c.r. trigger. More precisely, we computed spectrum 1 by averaging the 92 spectra obtained from 4096 samples (18.6s), centered at the trigger time, for each of the 92 stretches of data. We then repeated the same procedure for the 4096 samples from -45.4 s to -26.8 s, and for

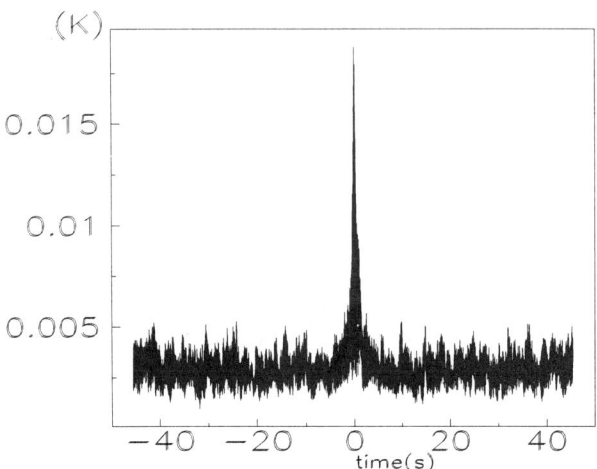

FIGURE 2. The average energy over 46 (for $M \geq 15\ 10^3$) stretches of NAUTILUS data versus time.

those from 26.8 to 45.4 s, obtaining the spectra 2 and 3, respectively. The plots of fig.3 show that only at the two resonances f_1 and f_2 the signal spectrum *1* differs from the background spectra *2* and *3*. This is a proof that at the times of the cosmic ray events a mechanical disturbance exited the detector resonances.

TEST II - We have checked that the observed signals are due to mechanical excitations of the g.w. detector also by examining in greater detail the time behavior of the signals near zero delay. The behavior of the signal energy with time for each sample averaged over the 46 events ($M \geq 15\ 10^3$) is shown, near the zero time, in fig.4. This behavior agrees very well with the behavior described in reference [11] for a simulated delta excitation and the envelope, as expected, follows the law

$$E(t) = E_o\ e^{-2\pi \Delta f |t - t_o|} \qquad (2)$$

where t_o is the time of the excitation and Δf is the bandwidth of the detector.

The periodicity we notice in fig.4 is in good agreement with the beat period due to the two resonance modes ($\frac{1}{f_2 - f_1} = 64\ ms$). Thus we are in presence of mechanical excitations and the data filter we are using behaves in the proper way.

We can estimate the bandwidth Δf from the data shown in fig.3. We find $\Delta f = 0.27 \pm 0.03\ Hz$ in very good agreement with the envelope obtained from fig.4 and described by eq.2.

TEST III - We have considered the possibility that the mechanical excitations be due to a backaction from the electronics. To eliminate this possibility we generated, very near the electronics of the transducer, strong electrical sparks, six orders of magnitude above any possible electrical disturbance which would excite the transducer and in turn the bar. No signal was observed.

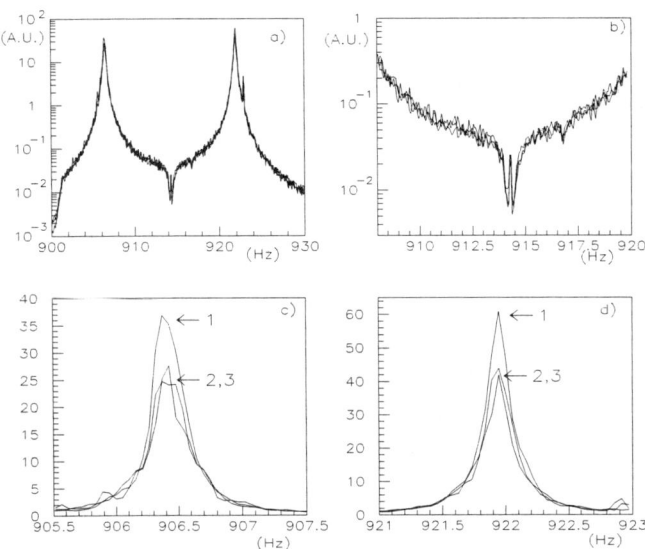

FIGURE 3. a) Power spectra 1,2,3, with arbitrary units on the ordinate scale. b) Zoom of fig.a off-resonance. c, d) Zoom of fig.a at the two resonances. Figure b shows that the three spectra off-resonance are equal (the dip is due to the calibration signal of the detector). Figures c and d show that the signal spectrum 1 is fairly larger than the background spectra 2 and 3 at the resonances.

Finally one could think about the possibility of a direct effect of c.r. on the transducer. This could occur in two ways. One way is to directly shake the transducer. It is easily recognized that this mechanical vibration due to few particles is extremely small, much smaller than that produced in the bar by thousand of particles, each particle releasing much more energy in a longer trajectory in the bar itself. The other possibility is that one particle strikes the transducer extracting some electrons which would be accelerated by the transducer electrical field, but with no electron multiplication because of the extreme high vacuum. This would give a few hundred eV for an impulsive excitation, orders of magnitude below the energy of several GeV released by an EAS in the bar. Thus we believe that the mechanical excitations are originated in the bar itself.

It is important to verify that the observed average effect is due to several events and not just to one. To this aim we have considered each cosmic ray event and taken the maximum energy value in the time range from -64 ms to 64 ms, obtaining 92 maximum values near zero time. We repeat this procedure for the time interval 10000 ± 64 ms obtaining a new set of maximum values. We determine the distributions of these two quantities for the 92 events and show them in fig.5. Using the Kolmogoroff test we find a probability of 0.016 that the two distributions are compatible. The upper one shows a spread of values. This verifies that the observed effect is due to several events.

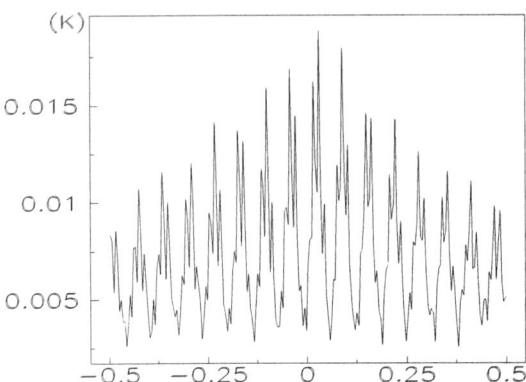

FIGURE 4. Zooming the average energy over 46 (for $M \geq 15 \; 10^3$) stretches of NAUTILUS data versus time.

The largest signals in NAUTILUS are associated to the largest c.r. events. The largest one is associated to a c.r. event with multiplicity $M = 32 \; 10^3$ (here the c.r. counters have a saturation effect). For the second largest we have $M = 22 \; 10^3$. Considering the six largest signals we have an average multiplicity of $M \sim 26 \; 10^3$, while for the remaining 86 signals the average multiplicity is $M \sim 17 \; 10^3$ and for the smallest fifty signals the corresponding average multiplicity is $M \sim 15 \; 10^3$. If we remove the two largest events the excess shown in table 1 reduces from 28 to 12 standard deviations, for $M \geq 15 \; 10^3$

The number of observed large signals agrees with the expectations. We expect in the 47.7 days (see eqs. 0.1 and 0.3) from one to four antenna signals due to c.r with energy grater than 50 mK. In view of the error on the absolute estimation of the released energy this result is acceptable.

The data of fig.5 allow us to estimate an upper limit for the energy delivered to the bar by the EAS. This is done by taking the average value of the maxima and subtracting the mean background energy 2.9 mK (see table 1).

Now we calculate the expected energy of the signals due to the EAS. At sea level, the rate of expected EAS (in agreement with previous experiments [10]) is given by [12]

$$H(\geq \Lambda) = k \; \Lambda^{-\lambda} \quad \frac{EAS}{day} \tag{3}$$

where
Λ is the particle density in the shower in units of number of charged particles per square meter, $\lambda = 1.32 + 0.038 \ln \Lambda$ and $k = 3.54 \cdot 10^4$.

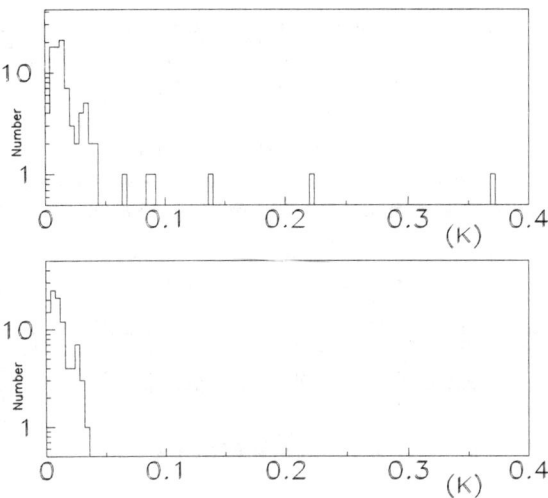

FIGURE 5. In the upper part we show the distribution of the maximum values in the time range from -64 ms to 64 ms (see text). In the lower part we show the distribution of the maximum values taken in the range $10000 \pm 64\ ms$

Using eq.1 and eq.3 we take into account distribution reported in reference [13] and the antenna geometry. With a threshold of 10^4 particles in the lower detector we obtain with this calculation 8 mK. Using the experimental multiplicity as measured by the lower detector we calculate 2.4 mK. We think the discrepancy is due to the saturation effects in the streamer tubes. All calculations are reported in the table 2 providing ranges of values that depend on the simplifications and on the systematic error in measuring the particle multiplicity. This experiment confirms the calculations on the cosmic ray effect made by various authors, both in terms of rate of occurrence and in terms of amplitude of the cosmic ray interaction with the resonant detectors.

Very important is the experimental demonstration of the overall well functioning of the apparatus with applied physical impulsive excitations and, in particular, the verification of the filtering algorithm performance, both in reducing noise and in extracting small signals. The experimental result indicates that, for the aluminium, the Grüneisen factor remains of the same order after the transition to the superconducting state. It also confirms previous conclusions [14,15] based on calculations, for a two detectors coincidence experiment. For the present detectors, there is no need to use an underground laboratory. For possible future more massive detectors operating near their quantum limit (noise temperature of the order of 10^{-7} K), it might be convenient to install just one of the them in an underground laboratory.

TABLE 2. Comparison of the theoretical predictions with the observations. ϵ_1 is the expected event energy averaged from the given multiplicity to ∞. The observed number of EAS per day (respectively 92 and 46 over a time of 47.7 days) is in agreement with the expectation. In the last column we report the lower and upper limits for the measured average excess E. The lower limit is obtained from table 1, the upper limit by summing the maxima of fig. 5 and subtracting the average background 2.9 mK.

$M \geq$	$\frac{particles}{m^2}$	calculated $\frac{number}{day}$	detected $\frac{number}{day}$	ϵ_1 [mK]	E [mK]
10^4	600 ± 200	$3.3 \div 0.95$	1.96	$2.4 \div 16$	$8.6 \div 21$
$1.5 \cdot 10^4$	900 ± 300	$1.6 \div 0.45$	0.98	$8 \div 26$	$16 \div 31$

REFERENCES

1. B.L.Beron and R. Hofstander: Phys.Rev.Lett., **23**, 184 (1969)
2. B.L.Beron, S.P.Boughn, W.O.Hamilton, R. Hofstander and T.W.Tartin: IEEE Trans. Nucl.Sci., **17**, 65 (1970).
3. A.M.Grassi Strini, G.Strini and G.Tagliaferri: J.Appl.Phys. **51**, 849 (1980)
4. A.M.Allega and N.Cabibbo, Lett.Nuovo Cimento **38**, 263 (1983)
5. C.Bernard, A.De Rujula and B.Lautrup, Nucl.Phys. B **242**, 93(1984)
6. A.De Rujula and S.L.Glashow, Nature **312**, 734 (1984)
7. E.Amaldi and G.Pizzella, Il Nuovo Cimento **9**, 612 (1986)
8. G.Liu and B.Barish, Phys.Rev.Lett. **61**, 271 (1988)
9. P. Astone et al (ROG Collaboration), Astroparticle Physics, **7**, 231 (1997)
10. E.Coccia, A.Marini, G.Mazzitelli, G.Modestino, F.Ricci, F.Ronga and L.Votano Nucl. Instr. and Methods A **335**, 624 (1995)
11. P.Astone, C.Buttiglione, S.Frasca, G.V.Pallottino, G.Pizzella, Il Nuovo Cimento **20**, 9 (1997)
12. G.Cocconi, Encyclopedia of Physics, ed. by S. *Flügge*, **46** 1, 228 (1961)
13. M. Aglietta et al., Phys Lett B , **333**, 555 (1994)
14. G. Pizzella, IR LNF-99 **001** (1999), http://wwwsis.lnf.infn.it/pub/LNF-99-001(IR).pdf
15. E. Coccia et al (ROG Collaboration), Nuclear Physics B (Proc. Suppl.) **70**, 461 (1999)

Niobe: Improved Noise Temperature and Back Ground Noise Suppression

Michael E. Tobar, Clayton R. Locke, Ik Siong Heng, Eugene N. Ivanov and David G. Blair

Department of Physics, the University of Western Australia, Nedlands, 6907, WA, Australia

Abstract. The calibration and sensitivity of the Niobe detector are presented. Typically the detector operates with a 1 mK noise temperature. A best noise temperature of 890 μK between 1300 to 2000 UTC for day 60 in 1997 is reported. The transducer has been upgraded with a new microwave amplifier, which has a measured electronic noise floor 40 dB lower than the previous amplifier, which is only 10 dB above the quantum limit. A detector noise temperature of 23 μk can be expected with this improvement. Also, we discuss a new filter to suppress accidental coincidences between two gravitational wave detectors. The filter is based on the amplitude ratio of events in pairs of detectors and improves the statistical significance of zero time delay coincidences.

INTRODUCTION

Currently there are five cryogenic resonant-mass gravitational wave antennas in long term operation around the world. They are Allegro[1], Auriga[2], Explorer[3], Nautilus[4] and Niobe[5]. Nautilus and Auriga are cooled to sub-Kelvin temperatures by a dilution refrigerator, while the other three are cooled by liquid helium. All the above with the exception of Niobe are made of 5056 aluminum alloy and use a transducer based upon a SQUID amplifier. Niobe is made from pure niobium, and consists of a 1.5 tone cylinder (fundamental acoustic frequency of 700 Hz) with a niobium bending flap tuned to the cylinder resonant frequency. The bending flap amplifies the displacement of the Niobium cylinder and is monitored by a superconducting re-entrant cavity transducer, which has been described in detail previously [6].

It is important to have an accurate method of calibration to determine the energy of possible gravitational wave events incident on the detector. Niobe is again unique due to the self-calibration properties of the transducer[7]. The other detectors require the measured response to a known force, which requires an auxiliary transducer to excite the antenna. To calibrate Niobe we only require the measurement of the transducer resonant frequency with displacement, df/dx, and the relationship between the measured displacement and the effective lumped oscillator displacement. For Niobe there is a displacement gain, G_{bf}, due to the cantilever action of the bending flap. Previously we estimated df/dx to be 345 MHz/μm and we assumed G_{bf}=1. Presently Niobe is undergoing a sensitivity upgrade and we have taken the opportunity to accurately determine these values by rigorously measuring the bending flap/re-entrant cavity electromechanical properties. We show that we had previously underestimated the sensitivity by the order of 30% and that we have a best noise performance of 890 µK, corresponding to a 1 ms burst strain sensitivity of 5 10^{-19}.

To upgrade the sensitivity of the detector we are replacing the GaAs FET amplifier with a new low noise HEMT amplifier. The amplifier has the potential to deliver a 40 dB improvement in the electronic noise added by the transducer. This will correspond to a detector noise temperature improvement to about 20 µK.

Despite low noise temperatures, resonant-mass detectors are susceptible to non-Gaussian disturbances that manifest as high-energy events at the output of the detector and are difficult to distinguish from gravitational events[8]. It has been shown that the distribution of the amplitude ratio of accidental coincidences between two detectors has a larger variance than the signal coincidences[9]. We discuss how an amplitude ratio window may be defined with negligible signal loss with a substantial reduction in accidental coincidences.

CALIBRATION AND SENSITIVITY

Determination of df/dx

The value of df/dx may be determined at room temperature. To first order the frequency does not change on cool down, the actual change varies depending on the accuracy of the bonding of the re-entrant cavity to the bending flap. In our last experimental run only a +0.5% second order change occurred.

A force was applied to the bending flap using a hook and pulley loaded with known masses. Figure 1 shows an exaggerated deflection plot of the finite element mesh of the bending flap and the position where the force was applied. The deflection was monitored by the re-entrant cavity, which had a Q-factor of order 200 at room temperature. The applied force (F) had a linear relation with the change in the re-entrant cavity frequency. Correspondingly we measured a df/dF of 75.6 MHz/N. It should be noted that the application of the force was at a different position to the re-entrant cavity. This was necessary as the bending flap was attached to the bar and the measurements were made *in situ*.

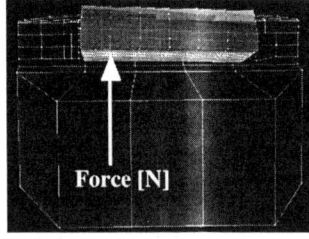

FIGURE 1. Schematic of a force applied to the bending flap acoustic resonator. A linear df/dF of 75.6 MHz/N was measured. The above diagram shows an exaggerated deflection of the corresponding finite element mesh of the bending flap.

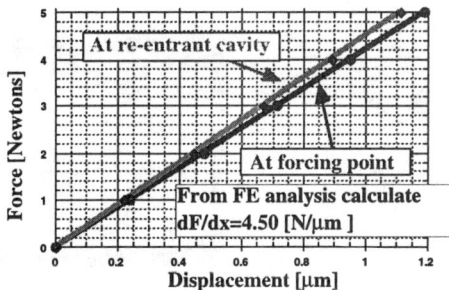

FIGURE 2. Calculation of the deflection at the forcing point and the re-entrant cavity transducer.

Finite element analysis was used to simulate the dynamic and static properties of the bending flap. A fixed boundary condition was assumed where the base of the bending flap was attached to the Niobium bar. The Young's modulus was determined by adjusting the value to calculate the correct bending flap frequency of 675 Hz at room temperature. It was determined to be 103.4 GPa for this particular piece of Niobium. Subsequently the static displacement of the bending flap as a function of applied force was calculated (see figure 2). Thus, a dF/dx of 4.50 N/μm was calculated at the re-entrant cavity. The value of df/dx was then calculated to be (df/dF)(dF/dx)=340 MHz/μm, which agrees very well with the previous calculation.

Once df/dx is known the calibration from voltage to displacement is achieved by sweeping the microwave frequency across the re-entrant cavity and measuring the output voltage response, dV/df. Thus, the relation to the voltage at the output, ΔV, to the displacement at the transducer, Δx_T, is calculated by ΔV=(dV/df)(df/dx)Δx_T. To further calibrate displacement to energy or noise temperature, T_n, the following relation must be used.

$$T_n = \left(\frac{\Delta x_T}{G_{bf}}\right)^2 \frac{m\omega^2}{k_B} \qquad (1)$$

Here, m is the normal mode effective mass, which is calculated experimentally *in situ* [7], ω is the normal mode resonant frequency, $\Delta x_T/G_{bf}$ is the effective lumped mass displacement and k_B is Boltzmann's constant.

Determination of G_{bf}

The displacement sensed by the transducer depends on its position along the profile of the bending flap. Figure 3 shows the profile of the forced sinusoidal response of the bending flap and the corresponding exaggerated finite element mesh. The re-entrant cavity senses the bending flap motion at position X=0.85 along the bending flap. At this point the normalized deflection is Y_{RC}=0.82. The effective lumped mass deflection may be determined by calculating the energy in the oscillating flap and equating it to the equation of the effective lumped mass oscillator. This was calculated to be Y_{LM}=0.64. The ratio gives the effective gain factor, which is calculated to be $G_{bf}=Y_{RC}/Y_{LM}$=1.3. Initially we had conservatively chosen to ignore this factor due to the uncertainty in our method of calculating df/dx.

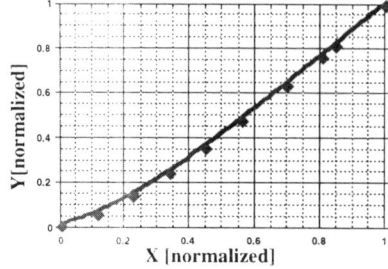 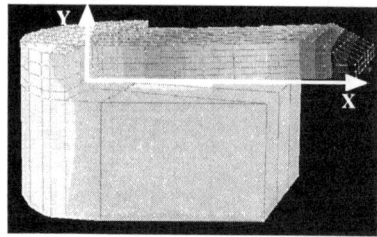

FIGURE 3. Left: Calculation of the normalized deflection profile due to a sinusoidal driving force at the resonant frequency and positioned at the re-entrant cavity transducer (X=0.85). Right: Finite element mesh of the forced resonant sinusoidal response of the bending flap.

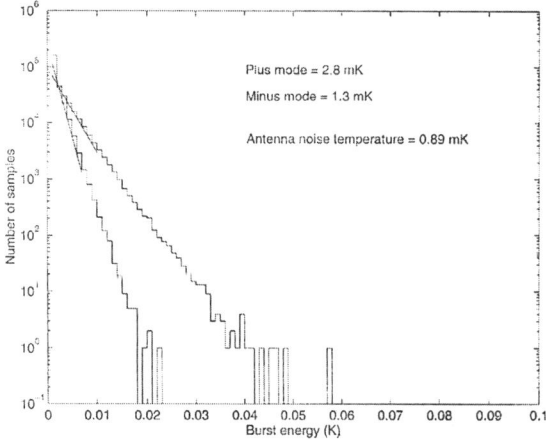

FIGURE 4. Histogram for day 60 in 1997 between 1300 to 2000 UTC.

Present Sensitivity

Typically the transducer operates at about 1 mK noise temperature. However our best period of operation is presented in figure 4. A noise temperature of 890 µK between 1300 to 2000 UTC for day 60 in 1997 was recorded.

SENSITIVITY UPGRADE

It has been shown that the noise temperature of the readout microwave amplifier is a major limitation to the sensitivity of Niobe[10]. A new HEMT amplifier has now been installed and figure 5 shows the measured improvement in amplifier noise temperature as a function of input power at 700 Hz Fourier frequency.

The interferometric readout of Niobe necessarily allows a high incident power on the transducer and a small input power to the amplifier by suppressing the carrier. Previously about a 40 dB suppression of the carrier was achieved corresponding to an

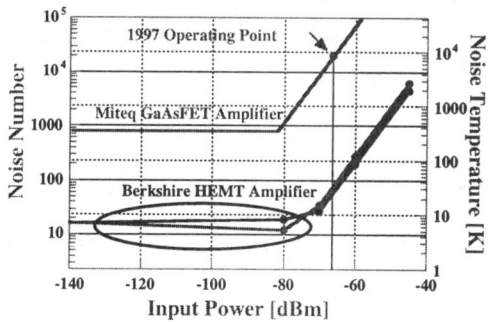

FIGURE 5. Measured noise temperature and noise number of the previous GaAs FET and new HEMT amplifier as a function of input power. The GaAs FET amplifier exhibited a large noise temperature of 7500 K with the input power operating condition of –70 to –60 dBm. In the new Niobe readout the HEMT amplifier will operate in the circled region of 6 to 8 K. This is possible due to the addition of an automatic carrier suppression system which will suppress the carrier frequency by up to 100 dB.

input power of –60 dBm to the amplifier. The old amplifier had a noise temperature of 7500 K, which corresponds to 2 10^4 (noise number) above the quantum limit of a microwave amplifier. The new operating point will give a amplifier noise temperature between 6 to 8 K and is only one order of magnitude above the quantum limit. With this improvement in the electronic readout noise the predicted sensitivity becomes 23 μK[10]. It is important to have a quantum limited transducer if we are to reach the quantum limit of detection which is 0.05 μK for a 700 Hz detector.

BACKGROUND NOISE SUPPRESSION

Despite being well-isolated from most sources of environmental noise, resonant mass antennas are still susceptible to noise sources such as seismic disturbances, electromagnetic pulses and cosmic rays[8,11]. To study the effect of non-Gaussian noise in the data, we simulated a series of 100 GW excitations with the same amplitude (SNR 10 in energy) in a 3.5 period of Niobe data. An adaptive Wiener-Kolmogorov (WK) filter then filtered these events[12] and the equivalent strain amplitude of each signal was calculated.

The amplitudes of the filtered signals were not identical despite having the same input amplitude because of the interaction between the signals and the noise[13]. Due to the uncertainty of the filtered signal energies, the amplitude ratios of the filtered signals produce a distribution with a non-zero variance. Accidental coincidences were obtained by randomly time shifting event lists exchanged between Niobe and Explorer in 1995. The data was filtered by the adaptive WK filter, with a total of 16971 Explorer events and 35672 Niobe events extracted above a threshold of 7 times the noise in each detector. The distribution of the amplitude ratios is shown in figure 6.

We define F(R) such that

$$F(R) = |log_{10}(R)| \qquad (2)$$

where R is the strain amplitude ratio between two coincidences. The amplitude ratio

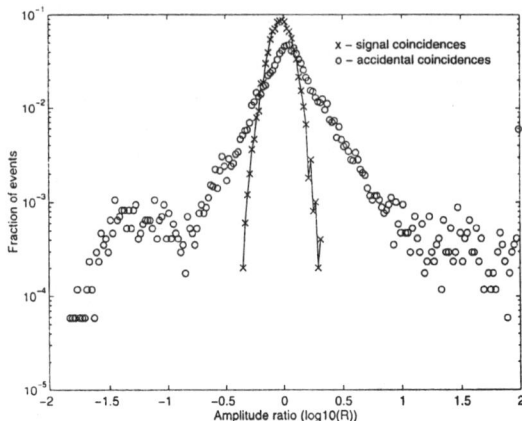

Figure 6. Distribution of the amplitude ratios of signal coincidences and accidental coincidences. The SNR of simulated signals is 10. The signals were simulated within Niobe data with a mean noise energy of 5.1 mK.

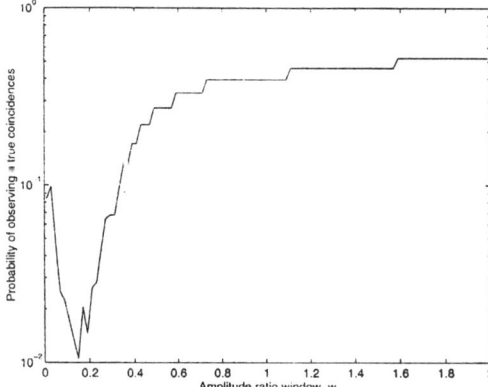

Figure 7. Probability of the number of signal coincidences, n, as a function of the amplitude ratio window. Both signal and accidental coincidences are removed when w < 0.4. The accidental coincidences are initially removed at a greater rate than the signal coincidences, so the probability improves until the amplitude ratio window gets so small that most of the signal coincidences are removed. This gives an amplitude ratio window at which the probability is optimized. In this case, for SNR = 10, the minimum occurs when 60% of the accidentals have been removed at an amplitude ratio window of 0.17.

distribution of the accidental coincidences, R_a, is clearly much broader than the amplitude ratio distribution of the signal coincidences, R_s. This arises because almost all the events, which are accidental coincidences are within the non-Gaussian tail arising from local impulsive events. By inspection of figure 4, it is evident that an amplitude ratio filter, which accepts coincidences with $F(R) < 0.5$ would ensure that all GW signal coincidences were retained while 20% of the accidental coincidences could be eliminated.

From the above example, it is clear that using a criterion whereby the amplitude ratio of coincident events must be within a certain window, w, such that $F(R) < w$, enables accidental coincidences to be removed from the data. To calculate the improvement in the statistical significance through the use of such a filter, 40 true coincidences and an average of 40 accidental coincidences were assumed. Using Poisson statistics, the probability was calculated to be 0.52. The probability of the number of true coincidences being accidental for decreasing amplitude ratio windows is shown in figure 7. The smallest probability of 1.1×10^{-2} occurs when $w = 0.17$. This is 50 times smaller than without the amplitude ratio window. However, at this window, there is a 12% chance that a GW signal coincidence will be excluded.

CONCLUSION

The accurate calibration of Niobe has been presented. We have shown a best noise temperature performance of 890 µK over a 7 hour period. The noise properties of a low noise HEMT microwave amplifier was measured. It was found to have a noise temperature only one order of magnitude above the quantum limit. With this upgrade in sensitivity a detector noise temperature of 23 µK is possible. Also, we discussed the application of the new amplitude ratio filter, which will improve the statistics of coincidence analysis of possible gravitational wave events between two detectors.

ACKNOWLEDGMENTS

This work was supported by the Australian Research Council.

REFERENCES

[1] E. Mauceli, Z. K. Geng, W. O. Hamilton, W. W. Johnson, S. Merkowitz, A. Morse, B. Price, and N. Solomonson, Phys. Rev. D **54**, 1264 (1996).

[2] M. Cerdonio et al, Proc. of the 1st Amaldi Conf. on Gravitational Wave Experiments (World Scientific Singapore), 176 (1995).

[3] P. Astone et al, Phys. Rev. D. **47**, 362-375 (1993).

[4] E. Coccia et al, Proc. of the 1st Amaldi Conf. on Gravitational Wave Experiments (World Scientific Singapore), 161 (1995).

[5] D. G. Blair, E. N. Ivanov, M. E. Tobar, P. J. Turner, F. V. Kann, and I. S. Heng, Physics Review Letters 74, 1908-1911 (1995).

[6] M. E. Tobar, D. G. Blair, E. N. Ivanov, F. v. Kann, N. P. Linthorne, P. J. Turner, and I. S. Heng, Aust. J. Phys., 1007-1025 (1995).

[7] M. E. Tobar, Journal of Physics D: Applied Physics **28**, 1729-1736 (1995).

[8] I. S. Heng, D. G. Blair, E. N. Ivanov, and M. E. Tobar, Phys. Let. A. **218,** 90 (1996).

[9] I. S. Heng, P. Bonifazi, D. G. Blair, M. E. Tobar, and E. N. Ivanov, submitted to GRG (1999).

[10] M. E. Tobar and D. G. Blair, Review of Scientific Instruments **66**, 108-110 (1995).

[11] E. Amaldi and G. Pizzella, Il Nuovo Cimento **9C**, (1986).

[12] P. Astone, P. Bonifazi, G.V. Pallottino and G. Pizzella, Il Nuovo Cimento C **17**, 713 (1994).

[13] A.D. Whalen, *Detection of Signals in Noise* (Academic, New York, 1971).

SUSPENSIONS AND THERMAL NOISE

Mechanical Loss Factors of Materials and Suspension Systems for Advanced Gravitational Wave Detectors

Sheila Rowan*, Alex Alexandrovski*, Gianpietro Cagnoli[†], Martin M. Fejer*, Eric K. Gustafson*, Jim Hough[†], Stephen McIntosh[†], Peter Sneddon[†], Roger Route*

*E.L. Ginzton Laboratory, Stanford University, Stanford, CA 94305-4085, USA
[†]Department of Physics and Astronomy, University of Glasgow, Glasgow G12 8QQ, UK

Abstract. The thermal noise of the fused silica test masses and their metal suspension fibers presents a significant limitation to the sensitivity of interferometric gravitational wave detectors. Advanced suspensions are likely to require the use of materials of substantially lower mechanical loss. Here we present measurements of the mechanical loss of fused quartz elements suitable for use as suspension fibers in an advanced suspension system and summarize measurements of the mechanical and optical loss of potential replacement test mass materials.

INTRODUCTION

The first generation of long baseline interferometric gravitational wave detectors will use fused silica as the substrate material for the mirrors, typically suspended on metal wires. The thermal noise from the materials of the test masses and their suspensions presents a significant limitation to detector sensitivity. Each suspension mode may be modeled as a lightly damped harmonic oscillator of mass m, at temperature T, and with resonant angular frequency ω_0 with a spectral density of thermally excited motion $x(\omega)$ given by: (1)

$$x^2(\omega) = \frac{4k_b T \omega_0^2 \phi(\omega)}{\omega m \left[\left(\omega_0^2 - \omega^2 \right)^2 + \omega_0^4 \phi^2(\omega) \right]} \quad (1)$$

where k_b is Boltzmann's constant and $\phi(\omega)$ is the mechanical loss of the mode.

It can be seen that the thermal motion is greatest at the frequency of a resonant mode; thus where possible, suspension systems are designed to have their resonances outwith the frequency band of interest for gravitational wave detection (a few 10's of Hz to a few kHz). It is thus the off-resonance thermal motion that must be minimized to improve detector sensitivity. One approach to this is to construct suspensions in which the resonant modes have as high a quality factor, Q, as possible, where Q is related to the mechanical loss at a resonant angular frequency ω_0 by $Q = 1/\phi(\omega_0)$.

Pendulum and Violin Mode Thermal Noise

Due to their low intrinsic loss (2-5) and high breaking stress, fused silica fibers are an attractive alternative to the carbon steel wires currently used in interferometric gravitational wave detectors. Pendulum mode Q factors of ~1-2 x 10^8 have been measured for quasi-monolithic fused silica pendulums (6-8) with synthetic fused silica or natural fused quartz fibers and a suspension design using fused silica suspension fibers is planned for the first upgrade of the LIGO interferometers (9). A combination of high Q factor and mode frequencies of ~ 1 Hz or less makes pendulum mode thermal noise significant only below a few 10's of Hz, however thermal noise from the violin modes of the fibers appears in the detection band of interferometric detectors. The first violin mode is typically at a few hundred Hz, and is accompanied by higher frequency harmonics. It is important that these modes are of high Q to ensure they occupy minimal frequency space. Like the pendulum modes, the violin modes have Q's related to but higher than the intrinsic Q of the fiber material (10). Previous measurements have concentrated on the violin mode Q of synthetic fused silica fibers (2). Experiments suggest that synthetic fused silica has a higher intrinsic Q than natural fused quartz in bulk form (11,12,5), but the measured values for pendulum mode Q's, after correction for known sources of excess loss, are comparable for pendulums suspended by fused quartz or fused silica fibers (6-8). The violin mode Q factors of fused quartz fibers are thus of interest and we report here the results of our measurements of these Q factors.

Internal Thermal Noise

The internal resonant modes of the test masses in interferometric gravitational wave detectors are typically at frequencies > 10 kHz. As described above it is desirable that the Q's of these modes are as high as possible. The level of Q depends on the material chosen for the test mass substrate; thus the intrinsic Q of fused silica, at best ~ few x 10^7 (11,12,5), sets a limit to achievable detector sensitivity in the frequency range from a few 10's of Hz to a few kHz. To improve on this it is likely that a different substrate material of higher intrinsic Q will be needed. In a later section we summarize our recent measurements of the Q of a range of potential replacement materials (13).

EXPERIMENTAL RESULTS

Measurements of Violin mode Q factor

Measurements were made of the Q factors of a number of violin modes of two fibers of natural fused quartz. A 5 mm diameter rod of natural fused quartz was heated in an oxy-hydrogen flame and pulled to form a fused quartz fiber of ~193 μm diameter. The thick ends of the fiber were cut off and the fiber welded to a 0.5 kg cylinder of fused quartz, ~ 7 cm long and ~ 6 cm in diameter, as shown in figure 1.

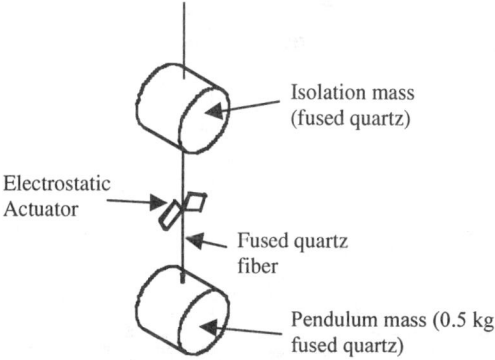

FIGURE 1. Experimental setup used in the measurement of the Q factors of the violin modes of fused quartz fibers.

A significant source of loss can be the coupling of energy from the violin modes of interest into resonances of the structure to which the fiber is attached. To minimize this a double pendulum arrangement was used in which the top of the fused quartz fiber was welded to a second fused quartz cylinder, itself suspended, via a fused quartz fiber, from an aluminum tripod. The intermediate mass isolates the vibrations of the lower fiber from the resonances of the surroundings, in a similar manner to that described in (2). The modes of the fiber were excited electrostatically using the fringe field of two electrodes placed on either side of the fiber. A shadow sensor, consisting of an LED and split photodiode, was used to sense the fiber motion. The Q factor of a given violin mode was calculated using measurements of the rate of decay of the amplitude of the fiber motion. It can be shown that the expected Q factor of the nth violin mode of a fiber of diameter d and length L, with an intrinsic mechanical loss $\phi_{mat}(\omega)$ is given by: (10)

$$Q_n^{-1} = \phi_{mat}(\omega) \frac{2}{k_e L}\left(1 + \frac{(n\pi)^2}{2 k_e L}\right) \quad (2)$$

where k_e is the wave number associated with the flexural stiffness of the fiber (10). Measurements were made of the unloaded mechanical loss of the fiber for resonances between 30 and 600 Hz. At frequencies above ~ 500 Hz, where thermo-elastic damping was insignificant (14) the loss factor of the material was found to be 6×10^{-7}. Inserting this and other appropriate values for the parameters shown into equation (2), produced the expected Q values shown as the upper points in figure 2.

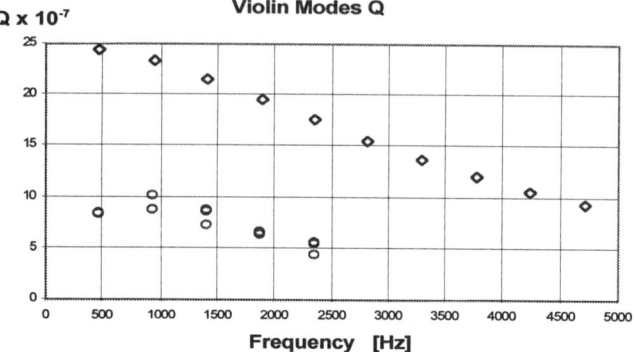

FIGURE 2. Expected and measured values for the Q of several violin modes of a fused quartz fiber of diameter 193μm and length 29.5 cm. The fiber diameter was calculated by substituting measured values for the frequencies of the violin modes into the appropriate equation found in (10).

The corresponding measured values for a range of violin modes are also shown in figure 2, and are clearly lower than predicted. The measurements were repeated using a second fiber of 229 μm and the results are shown in figure 3. In both cases there is some excess loss, which, above 1 - 2 kHz increases in size. It is interesting to note that the maximum violin Q's measured for these fused quartz fibers were in each case ~ 1 x 10^8. This is approximately the same as the best reported values for the violin mode Q factors of fused silica fibers (2). The source of the excess loss observed here is under further investigation, but even in the presence of this loss the Q factors measured are high enough that the thermal noise from these modes should not significantly degrade the sensitivity desired for advanced interferometric detectors.

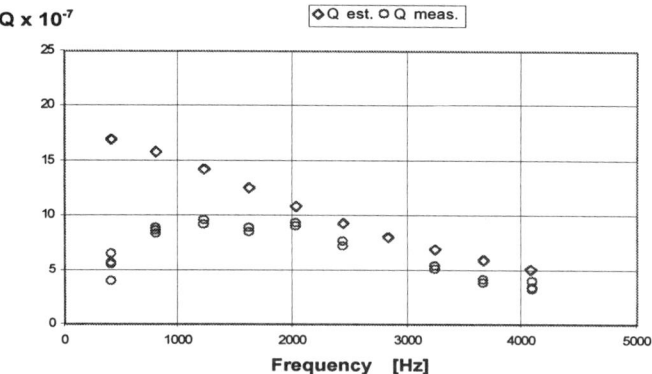

FIGURE 3. Expected and measured values for the Q of several violin modes of a fused quartz fiber of diameter 229 μm and length 29 cm. The fiber diameter was calculated by substituting measured values for the frequencies of the violin modes into the appropriate equation found in (10).

Mechanical Loss of Crystalline Test Mass Materials

As described above, it is of interest to have comparative measurements of the Q factors of materials with the potential to replace fused silica as a substrate material. Measurements have shown that sapphire, (Al_2O_3), grown by the method of horizontal oriented crystallization can have Q factors of $\sim 3 \times 10^8$ (15); however this growth technique appears unsuitable for producing crystals of the size and quality needed for mirror substrates in advanced detectors. The HEM (Heat Exchanger Method) used by Crystal Systems (16) has the potential to be suitable and thus the Q of a sample of HEM sapphire was measured. In addition, some optically isotropic materials, YAG ($Y_3Al_5O_{12}$) and spinel ($MgAl_2O_4$), are known to have acoustic losses comparable to or lower than sapphire at very high (GHz) frequencies (17). The Q of samples of these materials was also measured. Cylindrical samples of each material were suspended using either a single loop of fine silk (~ 120 μm diameter) or polished tungsten wire of diameter 50 μm in a vacuum chamber evacuated to 10^{-3} torr or better. The fundamental longitudinal mode of each mass was excited electrostatically and the resulting motion of the sample sensed interferometrically. A fine coating of grease was applied to each mass where it touched the suspension fiber. This suspension technique is very similar to that described in (18). Using silk suspension fibers the Q's of the spinel, YAG, and HEMEX sapphire samples were found to be $\sim 8 \times 10^6$, 2.9×10^7 and 1.3×10^8 respectively. On replacing the silk fibers by tungsten wire the sapphire Q increased to $(25.9 +/- 0.5) \times 10^7$ (13). It is possible that losses associated with the suspensions of the samples were contributing to the measured values. The surface roughness of the samples was not identical and investigation of the effects of this on the measured Q's are continuing. The measured Q factor of $\sim 2.6 \times 10^8$ is to our knowledge the highest Q factor obtained for sapphire of this type.

Measurements of Optical Absorption

High light powers are required to achieve desired levels of interferometer sensitivity, with ~ 10 kW of 1064 nm light needed at the interferometer beamsplitter to reach the performance required for LIGO II (9). Absorption in the coatings of reflective optics or in the substrate material of transmissive optics can result in thermally induced wavefront distortions which can reduce the overall interferometer sensitivity (19). The optical absorption of the material used for any transmissive optics should be as low as possible at the wavelength of light used to illuminate the interferometer. Measurements were thus made of the optical absorption at 1064 nm of the samples of YAG, spinel and HEMEX sapphire described above.

The method used was photothermal common path interferometry, where the absorption of a sample is determined through monitoring how the absorption of a pump beam transforms to the distortion of a probe beam. A tightly focussed pump beam results in local heating of the sample through absorption of the pump light. The resulting localized temperature change and subsequent change in refractive index of the sample causes a weak distortion of the wavefront of a probe beam. The emerging probe beam can be considered as consisting of an undistorted wave and a weak wave representing the distortion; the interference of these waves corresponds to an intensity fluctuation in the detected probe beam, the magnitude of which can be used to calculate the absorption of the sample. This is described in more detail by A. Alexandrovski in (20). The measured absorption of samples of YAG, spinel and HEMEX sapphire is shown in table 1.

TABLE 1. Optical absorption at 1064nm of candidate test mass materials

Material	absorption @ 1064 nm
YAG	100 - 130 ppm/cm
Spinel	50 - 100 ppm/cm
HEMEX sapphire	40 – 100 ppm/cm

The absorption levels of these samples are higher than that achievable for the best fused silica, (< 1 ppm/cm (21)) however no special efforts had been taken in the growth of these samples to produce material with particularly low loss at 1064 nm; thus it may be possible to improve on these numbers. It should also be noted that the thermal conductivity of these materials is higher than fused silica (22), thus higher absorption levels can be tolerated before thermal distortions become significant (19).

The spinel sample showed localised areas of high absorption, apparently due to inclusions or defects in the crystal; however of particular interest was the variation in the absorption characteristics of the sapphire, with highest absorption measured near the surface of the piece, which had been annealed in oxygen. The reasons for this are under further investigation.

CONCLUSIONS

Violin mode Q factors of $\sim 1 \times 10^8$ were measured for fused quartz fibers similar to those proposed for use as suspension fibers in advanced interferometric gravitational wave detectors. Our measurements also show that commercially available sapphire has a Q factor high enough to make it an acceptable substrate material in terms of thermal noise performance in advanced detectors. Samples of YAG and spinel had lower measured Q's, but the surface quality of the samples may have contributed to this, and is an issue under further investigation. The optical absorption at 1064 nm in sapphire, YAG and spinel was found in each case to be higher than that of the best fused silica, however no special efforts had been taken in the growth of these samples to produce material with particularly low absorption.

ACKNOWLEDGMENTS

We would like to thank colleagues in the University of Glasgow, the Ginzton Laboratory, Stanford University, and the Department of Physics, Moscow State University, for their interest in this work and are grateful for financial support provided by PPARC, University of Glasgow and the National Science Foundation.

REFERENCES

1. Saulson P.R. *Phys. Rev. D* **42**, 2437, (1990)
2. Braginskii V.B., Mitrofanov V.P., Tokmakov K.V., *Physics - Doklady* **40**, 11, (1995)
3. Kovalik J., Saulson P.R., *Rev. Sci. Instr.* **64**, 10, (1993)
4. Rowan S., Hutchins R., McLaren A., Robertson N.A., Twyford S.M., Hough J., *Phys. Lett. A*, **227**, 153-158 (1997)
5. Gretarsson A.M. and Harry G.M., *Rev. Sci. Instr* **70**, 4081, (1999)
6. Braginsky V.B., Mitrofanov V.P., Tokmakov K.V., P*hys. Lett. A*, **218**, 164-166, (1996)
7. Rowan S.,Twyford S.M., Hutchins R., Logan J.E., McLaren A.C., Robertson N.A., Hough J., *Phys. Lett. A* , **233**, 303-308, (1997)
8. Tokmakov K.V., Mitrofanov V.P.,Braginsky V.B., Rowan S., Hough J., "Bi-filar pendulum mode Q factor for silicate bonded pendulum" *Ibid*
9. Gustafson E., Shoemaker D., Strain K., Weiss R., LSC White Paper on Detector Research and Development, available at http://www.ligo.caltech.edu/~ligo2/ (1999)
10. Gonzalez G.I., Saulson P.R., *J. Acoust. Soc. Am.* **96**, 1, (1994)
11. Danchevskaya M.N., Lunin B.S., Batov I.V., 13 VII$_{th}$ All-Union Scientific-Technical conference on Fused Silica, Reports, St. Petersburg, 19-20, (1991)
12. Gillespie A., "Thermal Noise in the Initial LIGO Interferometers", PhD Thesis, California Institute of Technology (1995)
13. Rowan S., Cagnoli G., Sneddon P., Hough J., Route R., Gustafson E.K., Fejer M.M., and Mitrofanov V., *Phys. Lett. A in press* (1999)
14. Nowick A.S., Berry B.S. *Anelastic Relaxation in Crystalline Solids*, Academic Press (1972)
15. Mitrofanov V.P. and Shiyan V.S., *Sov. Phys. Crystallogr.* **24** , 174, (1979)
16. Crystal Systems Inc. http://www.cysys.com/hem.html
17. Auld B., *Acoustic Fields and Waves in Solids, Vol II.* J. Wiley and Sons (1993)
18. Braginsky V.B., Mitrofanov V.P., and Panov V.I., *Systems with Small Dissipation*, University of Chicago Press, Chicago and London, (1985)
19. Winkler, W., Danzmann, K., Rudiger, A., Schilling, R., *Phys. Rev. A* , **44** , 11, 7022 (1991)
20. Alexandrovski. A, Fejer M.M. *in preparation*
21. Boccara C., Loriette V., *VIRGO note* VIR-NOT-PCI-1380-65 (1993)
22. Musikant S., *Optical Materials, An Introduction to Selection and Application*, Marcel Dekker Inc, New York and Basel, (1990)

Active Seismic Isolation for Enhanced LIGO Detectors

Joseph Giaime*, Brian Lantz[†], Daniel DeBra[†], Jonathan How[†], Corwin Hardham[†], Sam Richman[‡], and Robin Stebbins[§]

*Department of Physics and Astronomy, Louisiana State University, Baton Rouge, LA 70808
[†] Ginzton Laboratory, Stanford University, Stanford, CA 94305
[‡] Center for Space Research, Massachusetts Institute of Technology, Cambridge, MA 02139
[§] JILA, University of Colorado, Boulder, CO 80309

Abstract. The levels of seismic isolation needed for LIGO II will require a dramatic technological shift from the systems used in the initial LIGO detector. To take advantage of the improved thermal noise of a 30 kg test mass made of high-Q material and suspended with fused silica fibers, one must attenuate the ground motion by more than 10 orders of magnitude at 10 Hz. Aggressive active isolation of ground motion to reduce the root-mean-squared ground displacement and the displacement noise in the gravitational wave band, coupled with multiple pendulum suspensions, can make this possible. We will describe the mechanical design for such a system, and discuss the issues of active control that confront this endeavor.

INTRODUCTION

LIGO II [1], a set of Laser Interferometer Gravitational-wave Observatory (LIGO) detector upgrades currently being contemplated, would require a test mass root-mean-squared (RMS) displacement of 10^{-13} m, and a displacement spectral density as low as 10^{-19} m/$\sqrt{\text{Hz}}$ at 10 Hz. To reach this instrumental noise level, the test mass suspension, to be designed by the GEO 600 group at the University of Glasgow, is likely to be a triple or quadruple pendulum. The test mass, which is planned to be made of sapphire, and the next-up mass are to be connected with fused silica ribbons, allowing extremely low thermal noise from both the pendulum and internal test mass modes. [2,3]

To take advantage of this, the seismic noise transmitted to the suspension point must be reduced significantly. The multiple pendulum provides seismic isolation itself; its displacement transmission is approximately 10^{-7} horizontally and 10^{-5} vertically at 10 Hz. [4] Since we allow for about 10^{-3} vertical to horizontal coupling in the suspension and from the Earth's curvature, the seismic isolation noise requirements on the platform that supports this suspension are 10^{-12} m/$\sqrt{\text{Hz}}$ at

10 Hz in the horizontal direction, and slightly more in the vertical direction.

We propose a two-component seismic isolation system (SEI) to meet LIGO-II's goals: coarse, low-bandwidth external actuators and a two-stage stiff-sprung in-vacuum active platform. The system described here is sized for the "Basic Symmetric Chamber" (BSC) that currently houses LIGO's test masses, but can also fit in vacuum chamber type that houses LIGO auxiliary optics, the HAM, underneath the present optics table surface.

I COARSE ACTUATION

The entire system is supported (external to the vacuum system) on the existing (LIGO-I) SEI piers through set of hydraulic actuators, which will replace the LIGO-I coarse and fine actuation components. The new actuators support the existing support tubes which in turn carry a new space frame above the active isolation platform. The outermost stage of this design is a laminar flow "quiet hydraulic"

FIGURE 1. (left to right) Drawings of the hydraulic actuator system in its paired configuration at one corner, a single actuator, and a schematic of the hydraulic circuit.

system. A 2-D actuator will be located at the top of each of the four piers, allowing control and alignment of all 6 degrees of freedom of the internal support structure.

The designed actuation range is ±1.5 mm, which is sufficient to remove tidal and microseismic motion, and should be enough to compensate for differential changes caused by seasonal variations

The system runs with a viscous working fluid such as 150 weight oil or glycerin at 150 psi. The fluid flow remains laminar, and the hydraulic resonances are kept above 40 Hz. Differential bellows (Fig. 1) are used as actuators to eliminate the stiction associated with standard pistons. The right panel in Figure 1 depicts the continuous laminar flow, differential actuation scheme. This system follows extensive work by D. DeBra and others, for actuators that are quiet and precise to be used for precision machine tools.

The 2-D actuator incorporates two of the single actuators, also shown in the left panel of Figure 1. The arrangement of the flexure rods allows the system to be very stiff along the 2 actuation directions, but soft in the other 4 degrees of freedom.

At present, this stage is intended for use primarily to take up the large dynamic range motion, allowing the inner stages to stay near the centers of their range.

II ACTIVE SEISMIC ISOLATION PLATFORM

The in-vacuum system is a two-stage active isolation platform on which, for each suspended optic, we mount a quadruple pendulum. The inner platform stage of the stiff active component is built around a 1.5 m diameter optics table. For each suspended optic, the suspension components are hung from this flat table, and may

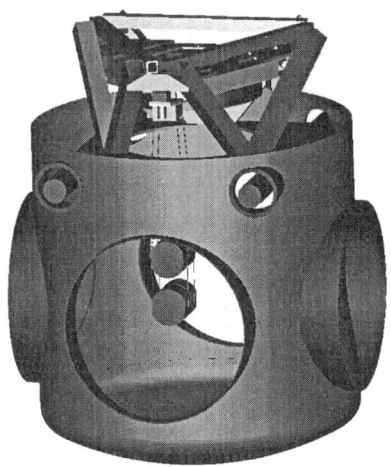

FIGURE 2. Sketch of the active platform and quadruple pendulum as they could be mounted in the LIGO BSC chamber. The outer space frame attaches to the support tubes and supports the first active stage through small blade springs and vertical wires. The inner active stage terminates in a round optics table and is suspended from the outer stage. Pods are provided where the stages are adjacent for the actuators and sensors.

be positioned and oriented as desired.

Global control signals for both optic position and angle can be distributed to each of seismic isolation and suspension stage components, large slow motions to the hydraulic outer system and progressively faster and smaller ones to the inner layers. Figure 3 shows the main noise and control pathways.

Local damping and noise-reduction control signals are generated for each DOF in the in-vacuum active platform by a combination of short- and long-period seis-

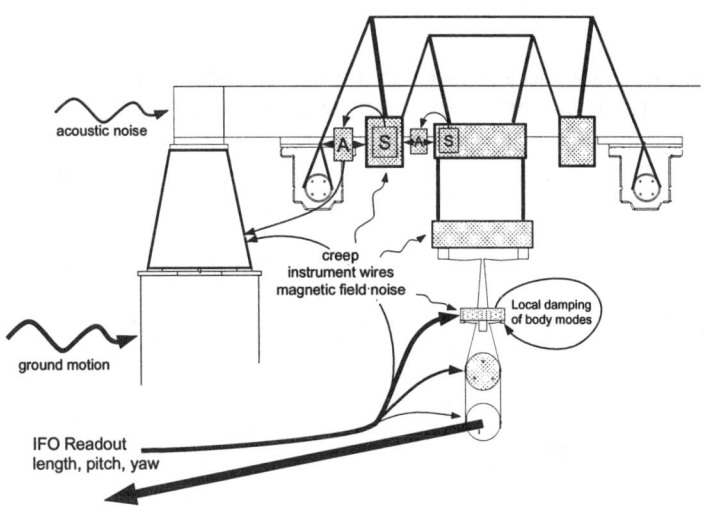

FIGURE 3. Sketch of control pathways available in the system.

mometers and relative position sensors. The long-period device currently being considered is the Streckeisen STS-2 triaxial seismometer; three are needed per platform, and relevant frequencies are between approximately 20 mHz and several hertz. Geophones are to be used for the short-period signals between several hertz and about 100 Hz. At the very low frequency end, there is a crossover to position sensing at 20 mHz. (The inner active stage does not use the long-period seismometer.) The actuators are electromagnetic non-contacting forcers, a coil in a closed magnetic field, available from several venders, although modified for our UHV needs. The actuators apply forces between adjacent stages. In a similar manner, position sensor signals in the suspension system are used to damp resonances in the pendulum suspension system, leading to its good behavior as an actuator. The suspension and isolation systems both benefit from all degrees of freedom (DOFs) being under active control, allowing controlled recovery from accidental excitations.

The SEI component of the BSC scheme (Figs. 3 and 2) resembles the pre-prototype built and tested at JILA over the past decade [5], an active isolation platform with two active stages. The JILA development effort was directed towards a much more ambitious goal, to detect GWs at 1 Hz, so our scheme can benefit from the JILA experience and 3-D modeling effort without the need for the ultra-high-performance seismometers that were part of JILA's development work. Instead, we can use commercial, off-the-shelf seismometers, greatly simplifying the engineering and reducing risk to the project. In addition, the design candidate presented here has a stiffer suspension (uncoupled natural frequencies in the 2 to 5 Hz range) to minimize the effects of thermal drift and signal-wire stiffness.

III PERFORMANCE ESTIMATES

Using a 3-D state-space model based on that used for the JILA platform, we can estimate the noise level at the suspension mounting point on the inner active stage. A number of assumptions are made. The loop gain for each DOF is approximately 50, with upper unity gain frequencies of about 60 Hz. The ground noise is approximately that at Livingston, LA, somewhat noisier than Hanford, WA, at low frequencies. We also use the advertised noise floors of the STS-2 and GS-13 seismometers, added at the appropriate points in the model. These levels are shown in Fig. 4. The noise model reported here **does not** include any noise reduction that might be achieved using the outer (hydraulic) stage, or feed-forward from, for example, floor-mounted seismometers.

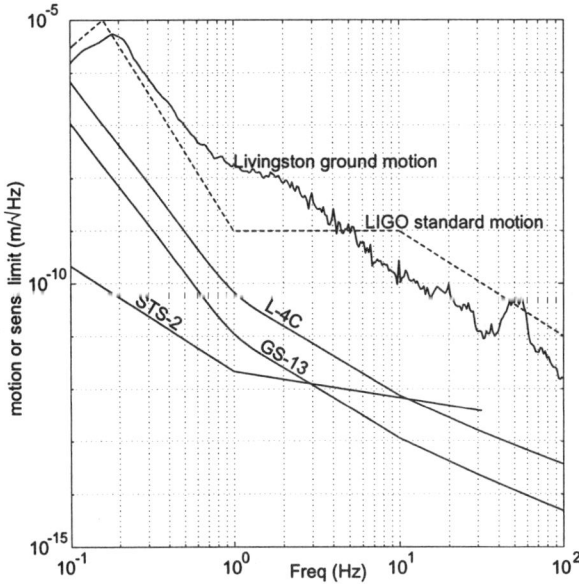

FIGURE 4. Equivalent displacement noise levels in various seismometers considered for use in the active isolation scheme. The combination to be used in the inner active stage (STS-2 and GS-13) is several orders of magnitude quieter than the typical ground motion.

Figure 5 shows the modeled noise levels at the suspension point from the expected ground noise and seismometer seismometer noises. It can be seen that the 10^{-12} m/$\sqrt{\text{Hz}}$ goal is likely to be met.

The control of RMS displacement to the required level relies heavily on the global control servo loops that command test mass position with respect to the interferometer signals. Simple estimates of the remaining RMS displacement at

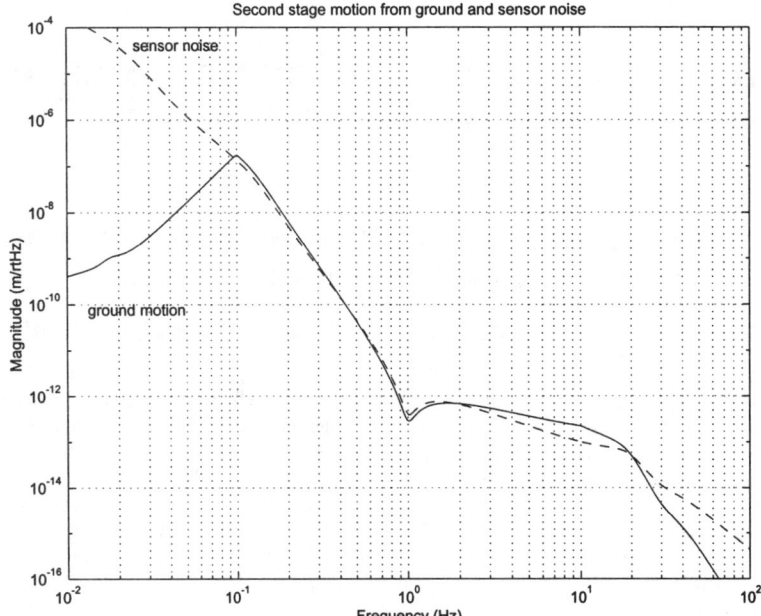

FIGURE 5. Calculated noise performance of two-stage active system, using the JILA 3-D model. The two curves show the horizontal noise at the suspension mounting table due to seismometer noise and due to transmitted ground noise.

the test mass indicate that the 10^{-13} m goal can obtained with a global servo upper unity gain frequency of about 30 Hz.

REFERENCES

1. D. Shoemaker & D. Coyne, "LIGO-II Conceptual Project Book," 9/10/99 (LIGO M990288-A-M); "LIGO-II Seismic Isolation Design Requirements Document," 8/3/99 (LIGO E990303-02-D). Also, see http://www.ligo.caltech.edu/~ligo2/.
2. Rowan, S., "Mechanical Loss Factors of Material and Suspension Systems for Advanced Gravitational Wave Detectors," this volume.
3. Robertson, N., "Suspension Design for GEO 600," this volume.
4. Strain, K., private communication
5. Richman, S.J.; Giaime, J.A.; Newell, D.B.; Stebbins, R.T.; Bender, P.L.; Faller, J.E.; "Multi-stage active vibration isolation system," Review of Scientific Instruments, **69:6**, 2531. (June '98)

Effect of Optical Coating and Surface Treatments on Mechanical Loss in Fused Silica

Andri M. Gretarsson, Gregory M. Harry, Steven D. Penn, Peter R. Saulson, John J. Schiller, and William J. Startin

Department of Physics, Syracuse University, Syracuse, NY 13244-1130, U.S.A.

Abstract. We report on the mechanical loss in fused silica samples with various surface treatments and compare them with samples having an optical coating. Mild surface treatments such as washing in detergent or acetone were not found to affect the mechanical loss of flame-drawn fused silica fibers stored in air. However, mechanical contact (with steel calipers) significantly increased the loss. The application of a high-reflective optical coating of the type used for the LIGO test masses was found to greatly increase the mechanical loss of commercially polished fused silica microscope slides. We discuss the implications for the noise budget of interferometers.

I INTRODUCTION

In samples made of high Q materials, such as fused silica or sapphire, a damaged or optically coated surface can be the dominant source of mechanical loss and could limit our ability to reduce thermal noise in interferometers. We apply a general method for quantifying surface loss to measurements of samples with optical coatings and differing surface treatments. This enables us to estimate the effect of coatings on the internal mode thermal noise of interferometer test masses as well as the effect of suspension filament surface damage on the pendulum mode thermal noise.

II QUANTIFYING SURFACE LOSS

Surface loss may be quantified by the dissipation depth d_s, defined by[1]

$$\phi = \phi_{bulk}(1 + \mu \frac{d_s}{V/S}), \tag{1}$$

where $\phi = 1/Q$ is the measured loss angle of the sample when all sources of extrinsic loss (such as recoil damping or clamping friction) have been eliminated, ϕ_{bulk} is the loss angle of the bulk material, V is the volume of the sample, and S is the surface area. The unitless μ is a geometrical factor that takes into account the relative amount of elastic deformation occurring at the surface and hence the emphasis placed on the condition of the surface due to the sample geometry and mode of oscillation. The geometrical factor μ is of order unity for simple geometries so that, as a rule of thumb, surface loss tends to dominate when d_s is greater than the volume to surface area ratio. For fibers in transverse oscillation $\mu = 2$, while for ribbon or microscope slide geometries in transverse oscillation $\mu = 3$. Although ϕ_{bulk} and d_s may in general be functions of frequency, no frequency dependence was seen in our measurements.

III SURFACE TREATMENT OF UNCOATED SAMPLES

For uncoated samples the dissipation depth provides a quantitative measure of the physical condition of the surface. By measuring the quality factor Q of samples before and after different types of surface treatment, we calculated the dissipation depth associated with each treatment. We measured the quality factors of untreated and treated fused silica (Suprasil 2) fibers drawn in a natural gas and oxygen flame. We also measured the quality factor of a fused silica (Suprasil 2) microscope slide, both as supplied (mechanically polished) and as subsequently etched. Using an apparatus specifically designed for the purpose of reducing extrinsic sources of loss (Fig. 1a) we were generally able to reduce extrinsic losses sufficiently so that the dominant sources of loss remaining were thermoelastic loss, bulk loss, and surface loss.[1] In each case we measured the quality factors at frequencies where thermoelastic loss was negligible. In this regime the quality factors were frequency independent, although in a minority of cases random mode-to-mode differences in Q were apparent. This was most likely due to residual sources of excess loss. To reduce the systematic error due to such residual sources of excess loss we took the highest Q mode to be indicative of the quality factor resulting from bulk loss and surface loss alone.

To investigate the effects of washing surfaces in solvents we wiped a fiber with paper wipes (Kimwipes^{Tm}) saturated with acetone. We also agitated a fiber in an ultrasonic bath of detergent and warm tap-water for a half hour, followed by a half hour ultrasonic bath of warm tap-water, followed by a second rinse with a stream of distilled water. After measuring the Q we then waited 14 days with the fiber under vacuum ($\approx 10^{-6}$ Torr) and re-measured the Q. In an attempt to simulate the effects of hydroxy-catalysis bonding[2] (silicate bonding) of fused silica surfaces we washed a fiber with ethyl alcohol and then submerged it in a 0.5 Molar solution of KOH and distilled water for 24.6 hrs, then rinsed in distilled water. Also, to investigate the effect of mechanical damage we lightly pinched two fibers at 1 cm intervals with stainless steel measurement calipers.

To remove the outer surface entirely (and with it any mechanical surface damage) we etched three fused silica fibers in solutions of hydrofluoric acid (HF) and distilled water. After etching, the fibers were rinsed with distilled water. The first etch was performed on a fiber of diameter 120 ± 20 μm and the etch removed 1.5 ± 0.5 μm from the surface. The second etch was performed on one of the fibers previously

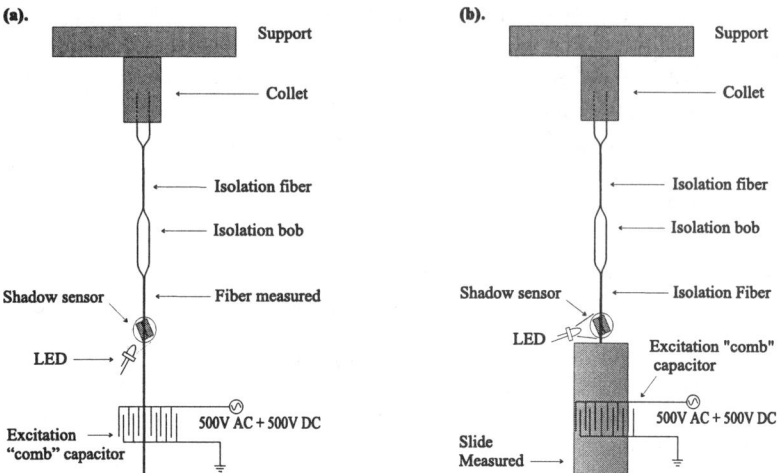

FIGURE 1. Schematic diagram of the experimental setups. (a). Setup for measuring fiber Qs. (b). Setup for measuring slide Qs. The isolation bobs and fibers prevent the measured sample Q from being degraded by rubbing in the clamp and from recoil in the lower Q support structure.

pinched with calipers. It had a pre-etch diameter of 840 ± 50 μm and the etch depth was 45 ± 3 μm. The third etch was performed on a fiber of pre-etch diameter 350 ± 60 μm. The etch depth was 90 ± 28 μm. Finally, we etched the microscope slide. As supplied, the microscope slide surface had received a commercial 80-50 (scratch-dig) polish. The etch removed 100 μm from this surface.

Table 1 summarizes our results. For three of the fibers, with surfaces as drawn, the dissipation depth is around 200 μm. Fiber I has a surface that is initially worse (higher d_s) while Fiber M has a surface that is initially better. Although an effort was made not to touch the surfaces of the fibers with fingers or other objects during handling, conditions were not stringently uniform. The fibers were also stored for varying durations in clean glass tubes and could come into light contact with the inner surface of the tubes. Depending on the storage time or amount of contact, some deviation in d_s can be expected.

Figure 2 shows how strongly different treatments affected the surface of fibers. Most of the treatments either produced no change in the condition of the surface or they made it only slightly worse. However, pinching the surface at regular (1 cm) intervals with calipers significantly increased the dissipation depth, possibly due to small cracks formed by the mechanical contact. Similarly, the mechanically polished microscope slide had the highest measured d_s. It is interesting to note that the surface of Fiber I, after being significantly damaged by calipers, was restored to a condition better than as drawn (or perhaps more appropriately better than "as stored and handled") by the 45 μm HF etch. The resulting dissipation depth agrees with the best as drawn case, having a value of about 100 μm.

The question arises why most samples undergoing HF etches did not show sig-

TABLE 1. Dissipation depth for different surface treatments.

sample	treatment	d_s [μm]	Δd_s [μm] [a]
Fiber B	as drawn	180 ± 50	
Fiber B	acetone	200 ± 50	17 ± 10
Fiber C	as drawn	190 ± 30	
Fiber C	calipers	310 ± 40	124 ± 20
Fiber F	as drawn	190 ± 40	
Fiber F	detergent solution, rinse	190 ± 40	−1 ± 9
Fiber F	after 14 days in vacuum	160 ± 30	−25 ± 10
Fiber F	1.5 μm HF etch	220 ± 50	31 ± 12
Fiber I	as drawn	340 ± 50	
Fiber I	calipers	620 ± 100	281 ± 74
Fiber I	45μm HF etch	100 ± 20	−244 ± 30
Fiber L	as drawn	210 ± 50	
Fiber L	KOH, 0.5 M solution, rinse	310 ± 80	95 ± 29
Fiber M	as drawn	100 ± 20	
Fiber M	90 μm HF etch	180 ± 40	86 ± 25
Slide C	as supplied (polished)	860 ± 140	
Slice C	100 μm HF etch	850 ± 160	10 ± 140

[a] Change in d_s from the as drawn or as supplied state. The uncertainty in Δd_s is not the root of the quadratic sum of uncertainties in d_s since not all the variables involved in calculating d_s are independent between treatments.

nificant surface improvement. In the case of the severely damaged slide, we believe the etch was too shallow. After etching, hairline scratches on the slide were visible to the naked eye. Etching opens up microscopic cracks imparted by the polishing process and their presence, post-etch, is evidence that the surface was still damaged. As for the fiber etches, only one (the etch of Fiber I) resulted in an improved dissipation depth. This may be due to the fact that Fiber I had an as drawn dissipation depth somewhat higher than any other fiber and may have been inadvertently damaged between drawing and installation in our apparatus. Mechanical damage can be repaired by HF etching[3,4] (though the etches must be sufficiently deep). The HF etch may thus have removed the damaged surface of Fiber I, reducing the dissipation depth from the initially measured value.

We should not neglect the possibility that chemical contamination of the surface, in particular contamination with atmospheric water,[4] may also lead to increased loss. The ubiquity of d_s values in the range 100-200 μm could be due to the difficulty of isolating samples from atmospheric water. This would also explain the failure of the etches to reduce the dissipation depth below this range.

IV SURFACE LOSS DUE TO OPTICAL COATING

The surface loss due to optical coatings was investigated by measuring the quality factors of the modes of fused silica slides. We measured the Qs for three slides

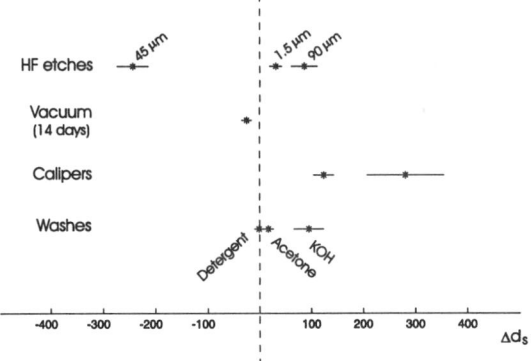

FIGURE 2. Change in the condition of fiber surfaces as compared to the initial surface as drawn, Δd_s vs treatment. Horizontal lines mark the uncertainty in Δd_s.

of dimensions 76 mm × 25 mm × 1 mm. The slides were suspended below a monolithic, fused-silica isolation system, as shown in Fig. 1b. The slides' vibration was monitored by positioning the LED and shadow sensor around the suspending fiber directly above the slide.

Two of the three slides, A and B, were optically coated while the third slide, C, was retained uncoated as a control. (Slide C was later etched as reported in the preceding section.) As supplied, the slides had received a commercial 80-50 polish. The optical coating applied to Slide A and Slide B was a high-reflective (HR) coating 2.4 µm thick consisting of 14 layers of alternating SiO_2 and Ta_2O_5. The slides were coated by ion beam sputtering in the same coating run as optics for LIGO by Research Electro-Optics Corporation in Boulder Colorado. After the coating run they were baked at 450°C to relieve stress.

The quality factors for each measured resonant mode of slides A and B and equivalent dissipation depths are shown in Table 2. The quality factors and equivalent dissipation depths for the measured modes of the uncoated Slide C (as supplied) are given for comparison.

The coated Slide B was suspended from the center of one of its short edges, as shown in Fig. 1b. When the supporting fiber was connected to the slide using a hydrogen-oxygen torch the coating became visibly damaged. Where the flame from the torch contacted the coating, the coating took on a milky appearance. This occurred in a crescent shape approximately 3 mm across at the top of the slide. The high value of d_s for the second mode is believed to be due to this damage. To test this, the top 5 mm of this slide were immersed in a 50% solution (by weight) of HF and water for about 6 hours. Rinses with distilled water were applied periodically to remove flakes of the coating. The etch removed most of the damaged part of the coating and the Q was re-measured. The Q and dissipation depth of the second

TABLE 2. Resonant Qs and equivalent dissipation depth in coated slides.

Slide	Surface treatment	Mode	Frequency	Q	d_s [μm]
A	HR-coating with no visible damage	2	1022 Hz	$1.1 \pm 0.5 \times 10^5$	$46 \pm 21 \times 10^3$
		3	1944 Hz	$1.6 \pm 0.1 \times 10^5$	$32 \pm 3 \times 10^3$
		4	2815 Hz	$1.6 \pm 0.1 \times 10^5$	$32 \pm 3 \times 10^3$
B	HR-coating damaged at top by flame.	2	952 Hz	$3.1 \pm 0.2 \times 10^4$	$160 \pm 15 \times 10^3$
		3	1851 Hz	$1.6 \pm 0.1 \times 10^5$	$32 \pm 3 \times 10^3$
B	Damaged region removed	2	962 Hz	$1.3 \pm 0.1 \times 10^5$	$39 \pm 4 \times 10^3$
C	Uncoated, as supplied ("80-50" polish)	2	1188 Hz	$4.0 \pm 0.2 \times 10^6$	$1.1 \pm 0.2 \times 10^3$
		3	2271 Hz	$4.9 \pm 0.3 \times 10^6$	$0.86 \pm 0.14 \times 10^3$

mode was now of the same magnitude as that measured for the third mode and for all modes of Slide B.

The coated Slide A was hung from a corner rather than from the center of the top edge. This was because, in the corner, the fused silica substrate was masked (by the supports) during the coating process. This left a region with no optical coating about 1 mm in radius and centered on the corner. The fiber was very carefully welded to the slide at this point. While some heat from the torch certainly reached the coated region, no damage to the coating could be seen afterwards. Both modes of Slide A showed similar Qs and similar dissipation depths as the modes of Slide B after the damaged region was removed. Since the uncoated Slide C has significantly less dissipation than the coated slides, and since the coated slides all show approximately the same level of dissipation, we conclude that the high dissipation depth associated with the coated slides, $d_s \approx 3$ cm, is a result of the HR optical coating.

If our measurements are characteristic of the coatings for LIGO, this would lead to noticeably increased thermal noise for the LIGO test masses. However, the surfaces of the slides did not receive the same treatment prior to the coating as the LIGO test masses. They were not superpolished and no particular efforts were made to ensure the absolute cleanliness of the surfaces. It is possible that the interface between the coating and the silica is more lossy than a polished surface interface would be. Superpolished samples of fused silica have been obtained and research is continuing to determine the loss in superpolished and coated samples.

V IMPLICATIONS FOR THERMAL NOISE

Surface loss in the filaments suspending LIGO test masses could have implications for the interferometer noise budget.[5] Surface loss associated with fibers implies a lower limit on the level of pendulum mode thermal noise achievable using thin ribbon suspensions. While dissipation dilution implies reduced pendulum mode thermal noise as the ribbon thickness is reduced, the effects of surface loss are increased. The result is a diameter-independent lower limit for the pendulum mode thermal displacement noise spectral density

$$x_{\min}^2(\omega) = \frac{24k_B T g}{ML^2\omega^5} \sqrt{\frac{Y}{12\sigma}} d_s \phi_{\text{bulk}}, \qquad (2)$$

where ω is the angular frequency, k_B is Boltzmann's constant, T is the temperature, g is the acceleration due to gravity, M is the suspended mass, L is the length of the suspension, Y is Young's modulus, and σ is the stress in the suspending ribbons. For typical values of the parameters and $d_s = 200$ μm, we have

$$x_{\min}(\omega = 2\pi \times 10 \text{ Hz}) \approx 6 \times 10^{-20} \text{ m}/\sqrt{\text{Hz}}.$$

While this is sufficient for the goals of LIGO II, it is clear from the dependence on d_s that mechanical surface damage such as is induced by calipers must be prevented.

Surface loss due to optical coatings may significantly increase the thermal noise due to internal modes of the test masses. To relate the dissipation depth measured for an optical coating to the internal mode thermal noise we follow the work of Levin[6] and Bondu et al.[7] This enables an approximate calculation of the relevant μ. Using Eq. 1 we obtain after some analysis a preliminary estimate for the test-mass loss angle,

$$\phi \approx \phi_{bulk}(1 + 0.4\frac{d_s}{r_0}), \qquad (3)$$

where r_0 is the Gaussian radius of the laser beam. Since $r_0 \approx 2$ cm for LIGO II, it is clear that if $d_s \approx 3$ cm, as measured for the coated slides, then the HR coating will be a significant contributor to test mass thermal noise.

REFERENCES

1. A. M. Gretarsson, G. M. Harry, Rev. Sci. Instr., **70** 4081 (1999).
2. S. Rowan, S.M. Twyford, J. Hough, D.-H. Gwo and R. Route Phys. Lett. A **246** 471 (1998).
3. Uhlman and Kreidl ed., *Elasticity and Strength in Glass*, Academic Press, New York, 1980.
4. R. H. Doremus, *Glass Science*, Second Edition, John Wiley, New York, 1994.
5. A manuscript describing these effects is in preparation by the authors and by S. Rowan, G. Cagnoli, and J. Hough of University of Glasgow.
6. Y. Levin, Phys. Rev. D **57** 659 (1998).
7. F. Bondu, P. Hello, J. Y. Vinet, Phys. Lett. A **246** 227 (1998).

Suspension Design for GEO 600 - An Update

N.A. Robertson, G. Cagnoli, J. Hough, M.E. Husman, S. McIntosh,
D. Palmer, M.V. Plissi, D.I. Robertson, S. Rowan*, P.Sneddon,
K.A. Strain, C.I. Torrie, H. Ward

Department of Physics and Astronomy, University of Glasgow, Glasgow G12 8QQ, Scotland, UK
* also at Ginzton Laboratory, Stanford University, Stanford, CA 94305-4085, USA

Abstract. The GEO 600 gravitational wave detector (1) is currently under construction at Ruthe, near Hannover in Germany. The design of the suspension system for the main mirrors in the detector has been chosen such that thermal noise due to the internal modes of the mirrors is expected to set the sensitivity limit from 50 Hz to ~200 Hz. Thus the design must be such that the effects of seismic noise and thermal noise from the suspensions are lower than the "internal" thermal noise at and above 50 Hz. To achieve this, a triple pendulum suspension incorporating fused silica fibres in the lowest stage forms the major part of the overall suspension and isolation system. In this paper, recent work on developing several aspects of the triple pendulum design is discussed.

GEO 600 DESIGN SPECIFICATION AND REQUIREMENTS

GEO 600 has been designed to operate with good sensitivity in a frequency band from 50 Hz to a few kHz. At the lower end the sensitivity is expected to be limited by the effects of thermal noise from the internal modes of the test masses. The test masses will be made of Suprasil 1, a type of fused silica which has been measured to have a quality factor of ~5 x 10^6 in experiments in Glasgow and Hannover (2). Using this value, one can calculate the expected thermal noise level at each test mass (~6 kg in mass), due to the sum of the internal modes, to be ~7 x 10^{-20} m/√Hz at 50 Hz. Thus the aim is to design a suitable suspension and isolation system so that the level of the other main sources of noise at low frequency, namely seismic noise and thermal noise associated with the pendulum suspensions of the mirrors, are lower than the internal thermal noise level at each mirror. For the pendulum thermal noise, it can be shown that this aim requires a quality factor greater than 7 x 10^6 for the longitudinal pendulum modes, and greater than 3.5 x 10^6 for the violin modes. Given the level of seismic noise at the GEO 600 site, which closely approximates ($10^{-7}/f^2$) m/√Hz at frequencies, f, from ~10 to 200 Hz, the isolation required is ~6 x 10^9 in the horizontal, and ~6 x 10^6 in the vertical direction, where a cross-coupling factor of 0.1% from vertical to horizontal has been assumed. These isolation factors would result in a residual seismic noise level a factor of 10 below the internal thermal noise figure.

THERMAL NOISE MEASUREMENTS AND RELATED INVESTIGATIONS

For several years we have been investigating the use of fused silica fibres for suspending the fused silica mirrors in GEO 600 to achieve the required quality factors for the suspension modes (pendulum and violin). These investigations have covered several aspects including: pulling fibres of suitable cross section, testing breaking strengths of fibres, measuring intrinsic losses of the fibre material, investigating bonding techniques of fibres to silica masses and measuring resulting quality factors of pendulums suspended with such fibres. Our investigations to date have yielded results which clearly indicate that the use of such monolithic fused silica suspensions and masses can achieve the require suspension quality factors as quoted above.

Pendulum and Violin Mode Quality Factors

We have achieved a Q of $\sim 9 \times 10^7$ for a 0.1 kg fused silica mass welded to a fused silica fibre (3). This demonstrated clearly that such high Qs can be achieved. However good welded joints were very difficult to achieve with larger masses which, after welding, were found to exhibit cracks which grew over time, spreading out from the welding point. Thus we started to investigate an alternative method of bonding to the mass, using the technique of hydroxy-catalyis bonding (so-called silicate bonding), first developed at Stanford for the GPB satellite project. We have investigated extending this technique for use in gravitational wave detectors (4). Our method involves silicate bonding fused silica ears onto the flat sides of a mass, where to achieve an optimum bond, the flats should be figured to lamba/10. An example of ears bonded to a GEO sized silica mass is shown in figure 1. The suspension fibres are pulled from rods typically 5 mm in diameter and short stubs are left at each end. These stubs are then welded to the ears attached to the mass. A 3 kg mass suspended on two silica fibres, assembled in this way in Glasgow, was taken to the VIRGO lab in Perugia and has been hanging in a vacuum system there for well over a year. The suspension Q has been measured to be $\sim 2.4 \times 10^7$. (5)

Recently in work done in collaboration with Moscow State University, a 0.5 kg mass on a fused silica bifilar suspension, assembled in Glasgow and suspended in Moscow, has yielded a pendulum Q for the torsional mode of $\sim 2.3 \times 10^8$. (6)

In work carried out in the Glasgow laboratory, violin mode Qs of $\sim 10^8$ have been measured for several modes, where the fibres have been pulled from standard quartz rods.

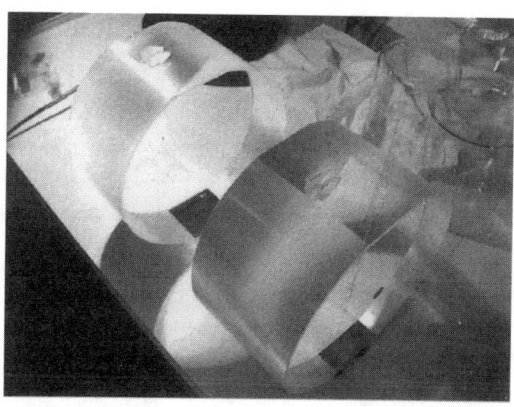

FIGURE 1. GEO sized silica masses with bonded ears for attachment of fibres.

All of these measurements have yielded quality factors in excess of the GEO 600 requirements. Further work is now in progress to extend the measurements to GEO sized masses.

Fibre Pulling and Fibre Strength Tests

We have investigated several techniques for producing fibres of the desired cross-section and length, including hand pulling fibres in an oxy-hydrogen flame, and using an automated pulling system within an RF oven. Recently we have developed a third technique where the rod to be pulled is supported vertically, such that it is symmetrically surrounded by 4 oxy-hydrogen flames emitted from needle apertures. The breaking stresses of many fibres produced by this technique have been measured as shown in figure 2. We found that our first measurements using an old batch of fused quartz were typically much lower than those from a new batch, suggesting our first batch was of poorer quality material. This was backed up by our measurements on material supplied to us from the Perugia lab which again yielded much better results. Treating the fibres with a solution of Caro's acid (so-called Piranha solution) tended to enhance the results further. It is now clear that breaking stresses of up to 3.5 GPa can be achieved, and this is as good as carbon steel. A high breaking stress implies that thinner fibres can be used for the suspensions, which in turn raises the violin mode frequencies. Since the first few modes will lie within the frequency band of operation of the detector, it is advantageous to have these as high as possible to minimise the number of such modes.

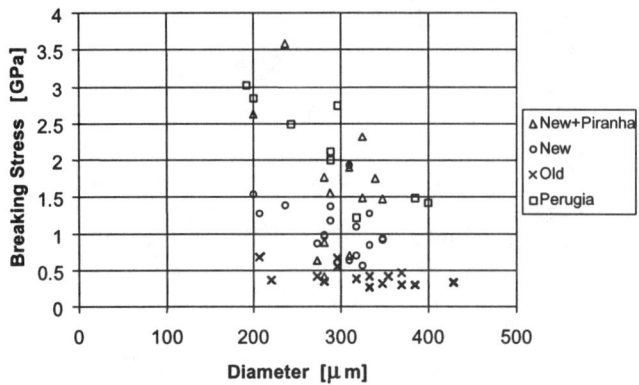

FIGURE 2. Breaking stresses of various batches of fused quartz fibres.

OVERVIEW OF THE ELEMENTS OF THE SUSPENSION AND ISOLATION SYSTEM

A schematic of the overall design for the GEO 600 suspension system is shown in figure 3. The fused silica mirror is the bottom element of a triple pendulum in which 4 vertical fused silica fibres are attached (as described above) to the mirror and the mass above, which is also made of fused silica. The triple pendulum contains two stages of cantilever blades of a design similar to those used in VIRGO to improve the overall vertical isolation. The triple pendulum is hung from a yoke, or rotational stage, which rests on the "stack stabiliser". Both of these pieces are made from box section stainless steel lined with damping material, and between them are 3 ruby and ceramic bearings to allow crude rotational alignment of the mirror. The stack stabiliser links the three stack legs which are mounted at their base to the inside of the vacuum tank. Each stack contains one active stage and one passive stage of isolation, the elements of which are encased in steel bellows to avoid contamination of the mirrors by hydrocarbon materials. The active stage, which uses geophones for sensing and piezoelectric actuators as the feedback elements, is primarily for use at low frequencies around the micro-seismic peak, and does not contribute to the overall isolation at the operating frequency band of the detector. The passive stack consists of a steel block resting on graphite loaded RTV rubber, and enhances both horizontal and vertical isolation, so that the combination of the stack plus triple pendulum gives the desired overall isolation. Where feedback forces have to be applied for global control of the interferometer, a reaction triple pendulum similar in design to the main pendulum is included so that forces can be applied from a seismically isolated platform.

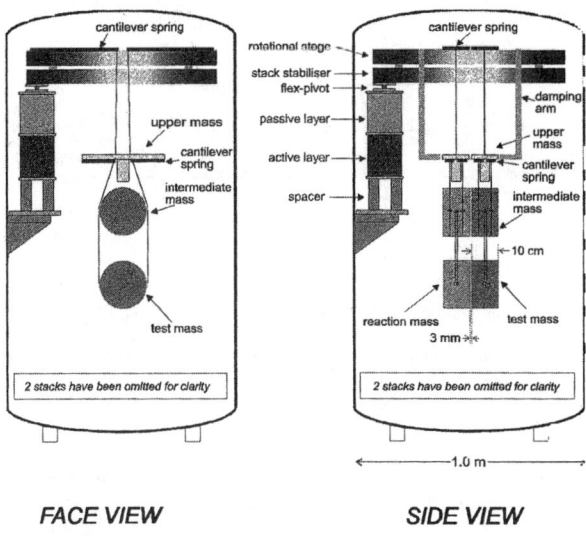

FIGURE 3. Schematic diagram of GEO 600 suspension system.

DEVELOPMENT OF TRIPLE PENDULUM MODEL

In order to investigate the likely performance and characteristics of the triple pendulum design, two computer models were written. One was written using MATLAB, starting from the equations of motion, and keeping first order terms. This model proved very useful for testing the variation of parameters to optimise the design and for modelling servo systems to damp all the low frequency (< 5Hz) pendulum modes. A second model was also written, using Maple, and starting from Lagrange's equations for the system. This second model, which is much more comprehensive, since it does not, a priori, drop all terms above first order, or assume symmetries in the likely design, proved very useful for analysing the magnitude of cross-coupling due to misalignments in construction, and also could be used for investigating thermal noise issues in the triple pendulum. These types of analysis are done at the expense of a large amount of computational time, and thus there were advantages to having both models available for different types of investigation. Having both also served as a good cross check that each was indeed correctly formulated.

An example of the type of cross-coupling investigation done with the Maple model is shown in figure 4. Here it can be seen how the magnitude of cross-coupling in the final stage of the triple pendulum (the fused silica suspension stage) varies as the spring constant and the offset in attachment of one of the fibres is varied. From these and other such investigations, we concluded that the assumed figure of 0.1% cross-coupling between vertical and horizontal motion was an appropriate value for GEO 600.

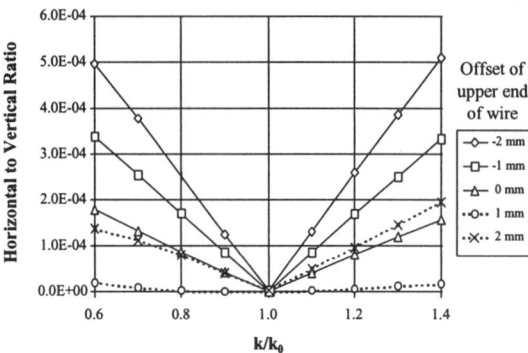

FIGURE 4. Cross-coupling of vertical motion into horizontal motion in final stage of the GEO triple pendulum, for a point 1 mm below the centre of mass. The variation in coupling is shown for changes in the spring constant, k, of one of the 4 fibres, and the offset in position of its top point of attachment.

INVESTIGATIONS WITH PROTOTYPE PENDULUM

To test the predictions of our model in terms of the values of mode frequencies and the performance of the electronic damping of modes, we constructed a prototype triple pendulum in Glasgow in which the fused silica intermediate mass and test mass (mirror) were replaced by annular aluminium masses of same mass and outer dimensions.
Comparison of the theoretical and experimental mode frequencies for the prototype showed very good agreement for those modes which could easily be observed. It should also be noted that all the mode frequencies, except one roll mode and one vertical mode, lay within the desired range 0.5 Hz to 4 Hz, thus ensuring that these modes could easily be damped by the local control loops, as discussed below. The high frequency modes are associated with the vertical and roll motion of the mirror on its fused silica fibres, and these cannot be reduced in frequency using cantilever blades for example without compromising the thermal noise performance.
The design philosophy for the local control is to use 6 co-located shadow sensors and coil/magnet actuators on the top mass of the triple pendulum. By careful design of the triple pendulum to ensure good coupling to the top mass of all the low frequency modes, they can all be damped to quality factors of a few with 6 loops of essentially the same design except for minor gain variations. By applying the feedback at the top mass of the triple pendulum, we also ensure that the electronic noise associated with the sensing and actuation is sufficiently attenuated by the lower stages so that this noise does not compromise the overall sensitivity.
Fuller details of the modelling and experimental investigation of the isolation and control aspects of the triple pendulum can be found in references (7) and (8).

PRESENT STATUS

The GEO suspension design is now essentially complete, and from measurements of performance of the various elements we expect to achieve an overall seismic noise level a factor of ~3 below the target thermal noise level at 50 Hz, and more than adequate isolation above this frequency. While this has not quite met our factor of 10 target, it is quite acceptable. In addition the results of internal, pendulum and violin mode quality factors and fibre breaking stresses have been shown to be good enough to meet our requirements. At the time of writing the suspension systems are beginning to be constructed at the GEO 600 site near Hannover.

ACKNOWLEDGEMENTS

We would like to thank our colleagues in GEO 600 for their help and interest in this work. This research was supported by PPARC and the University of Glasgow.

REFERENCES

1. Lück, H. et al., ibid
2. Traeger, S., Willke, B. and Danzmann, K., Phys. Lett. A **225**, 39-44 (1997)
3. Rowan, S. et al., Phys. Lett. A **233**, 303-308 (1997)
4. Rowan, S. et al., Phys. Lett. A **246**, 471-478 (1998)
5. Cagnoli, G. et al., submitted to Phys. Rev. Lett.
6. Tokmakov, K. et al., ibid
7. Husman, M. et al., accepted for publication in Rev. Sci. Instrum.
8. Plissi, M. V. et al, submitted to Rev. Sci. Instrum.

New Seismic Attenuation System (SAS) for the Advanced LIGO Configurations (LIGO2)

Alessandro Bertolini[1], Giancarlo Cella[1], Erika D'Ambrosio[2], Riccardo DeSalvo[2], Virginio Sannibale[2], Akiteru Takamori[3], Hiroaki Yamamoto[2]

I. Universita' di Pisa,	piazza Torricelli 2,	56100- Pisa,	Italy
II. California Institute of Technology,	1200 E. California Bl.	Pasadena, CA 91125,	USA
III. Tokyo University,	7-3-1 Hongo Bunkyo-ku	Tokyo 113-0033,	Japan

Abstract. A new passive seismic attenuation system is being developed to replace the current passive attenuation stacks in LIGO 2, it is expected to drive the seismic contribution to the interferometer noise below any other noise source. The SAS will be effective completely starting at about 5 Hz, well inside the (uncompensated) gravity gradient noise wall.

INTRODUCTION

The SAS is a passive chain of low frequency attenuation filters preceded by an Ultra Low Frequency (ULF), passive and active pre-filtering stage and followed by a multiple pendulum mirror suspension system. The multiple pendulum controls are performed, like in Virgo[1] and GEO[2] from recoiling masses. These masses are suspended, together with the mirror and its upper bobs, from the SAS last filter. The two top multiple pendulum masses are expected to have a GEO-like configuration and would be acted on, like in the Virgo "marionetta", from a rigid structure mounted on the last SAS filter that supports suitable electromagnetic control actuators. The multiple pendulum configuration, on top of allowing better control of suspension thermal noise, is expected to allow better control authority distribution to lighten, or perhaps virtually eliminating, the control load of the mirror actuators. To this end it is also necessary to deliver a very low level of residual r.m.s. motion which is obtained by means of the ULF pre-filtering stage and its active inertial damping of resonances[3,4,5].

The ULF stage, tuned well below 100 mHz, has the triple function to provide:
- Passive attenuation in frequency regions of the micro-seismic peak, where the passive filter chain has no attenuation capabilities, and of the passive filters internal resonances.
- DC and ULF positioning and control.
- A soft support for the active inertial damping of the SAS passive filters resonances and residual r.m.s. motion reduction.

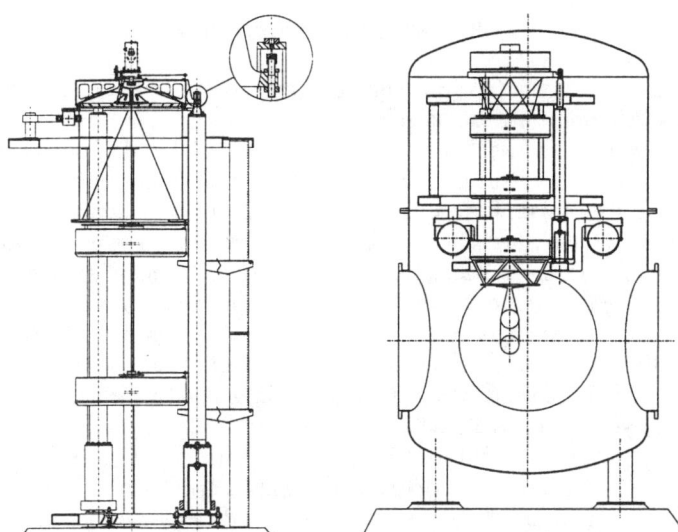

FIGURE 1. Present prototype of SAS (left) and tentative design to fit within the LIGO BSC chambers (right). The Filter Zero (at the top of the figure) is supported by three IP legs bolted on a rigid platform fixed on ground. The soft hinges (insert) allow full x, y and ϕ movement of the IP. The IP is controlled by actuators and position sensors (one shown on the left top corner) mounted on a rigid safety/reference structure. It will also carry horizontal and vertical accelerometers for inertial active damping of resonances. The passive chain suspension wire hangs down from the center of F0. F0 also carries a disk, suspended by means of an hexa-wire, loaded with magnets which, generating eddy current on the first passive filter surface, provide effective velocity damping of all resonances. Two or three passive filters (see figure 2) carry the payload. The triple pendulum, visible in the BSC assembly on the right, is suspended and acted on from the bottom filter body.

GEOMETRICAL ANTI SPRING FILTERS (GASF)

The core of SAS is the Geometric Anti Spring Filter (GASF)[6]. In a GASF low frequency vertical oscillations are obtained by directing part of the cantilever suspension springs return forces in an antagonist way that reduces the effective vertical spring constant without significantly reducing the necessary suspension force. The effective spring constant, and hence the vertical resonant frequency can be tuned at will, conceptually down to zero, by adjusting the level of spring antagonism. As a result a low and stable vertical resonant frequency can be factory tuned. The chosen frequency varies little with temperature and vertical working point choice; as a result an extremely large linearity is obtained with negligible up-conversions even with large oscillation excitations. Also the GASF require no precise tuning of their vertical working point.

Passive attenuation in the other five degrees of freedom is obtained by standard low frequency pendulum action[7].

Creep and Creak Immunity

Great care is taken in the GASF design to eliminate from the stress path any joint subject to shear to eliminate creak noise[8].

In all the three joints, between the suspension wires and the filter bodies, and at the two ends of the cantilever suspension blades, the stress is perpendicular (or quasi perpendicular) to the contact surfaces and a low melting alloy coating will braze these joints together during the filter's bake-out. To avoid internal noise production we also eliminated all helicoidal springs that, on top of their low frequency resonances, present problems of stress concentration or stick and slip at their extremities.

Frozen dislocation materials (precipitation hardened alloys like Maraging Steel[9]) used at carefully controlled stress levels are used to eliminate the creep noise from highly stresses components (cantilever blade springs and suspension wires) while bonded steel alloys are used in all low stress parts.

Additionally a bake-out at full load is foreseen to burn out all residual dislocation and potential creep noise before operations. This same bake-out procedure will performed the brazing of all the joints under stress and clean-out the filters for Ultra High Vacuum (UHV) operation. The simplicity of the filter's structure and it's all metal nature makes it easy to clean them to UHV requirements.

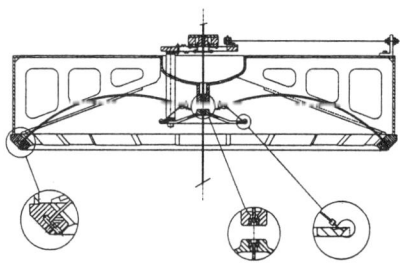

FIGURE 2. Standard Filter design. The filter body is suspended from a suspension wire, it carries an even number of prestressed cantilever spring blades (the ached components) which work in partially antagonist way. This configuration allows an easy tuning of the vertical resonant frequency. All contacts between the filter components (see inserts) are made with zero or negligible efforts in the contact planes to avoid creak noise.

Some SAS Technical Advantages

SAS, being a passive system is stable even in case of power failure. The great softness and extremely large geometrical dynamic range of its ULF stage[10] allows essentially unlimited operation using very low in vacuum power levels to compensate for tidal and other daily movements of the interferometer baseline. Additionally these control forces are applied very far from the payload, thus minimizing possible thermal perturbations to the mirror suspensions.

SAS is designed as a self supporting, transportable unit that can rapidly installed in the existing LIGO vacuum vessels[11,12]. Additionally SAS has large payload capabilities (more than 0.5 tons without modifications) and two independent SAS chains can be mounted inside the existing LIGO BSC inner diameter.

SAS ADVANCEMENT STATUS

Individual GASF and the behavior of complete chains have been simulated and complete engineering design have been produced.

UHV grade GASF and Inverted Pendulum prototypes have already been built.

The attenuation performances of both isolated GASFs and doublets have been partially tested. The measured behavior perfectly matched the earlier simulated one thus validating the simulations of longer chains. These simulations show that SAS can generate seismic attenuation crossing the mirror thermal noise floor below 6 Hz, exceeding the 10 Hz LIGO 2 specifications by such a margin to give good confidence that the seismic noise will not be a limiting factor.

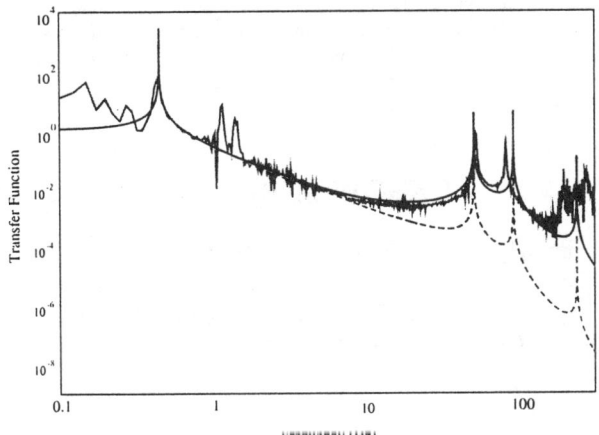

FIGURE 3. Attenuation performance of a single standard filter. Please note the agreement between simulation (red) and later measured data (black). The two peaks above 1 Hz are payload tilt resonances. In this measurement the mode dampers designed to eliminate the peaks at 55 and 80 Hz have not been implemented yet. The measurement is instrument noise dominated above 150 Hz.

The fact that SAS is effective at such low frequencies that gravity gradient noise already dominates may seem a waste; however this frequency domain overkill comes for free and shields the test mass from possible up-conversions and nonlinear effects from the control system and transient perturbations. Additionally it may come useful if gravity gradient noise cancellation techniques were to prove viable.

The best performance of SAS are obtained with tall towers, still fitting below the LIGO hall ceiling. The vacuum vessel for these towers would be obtained introducing a barrel extension between the BSC body and its hat. Even a scaled down SAS version, fitting inside the present short vacuum envelopes, largely outperforms any other proposed seismic attenuation scheme and, thanks to the SAS modularity, can later on be extended to the high performance tall tower configuration with simple addition of units and replacement of elastic joints and Inverted Pendulum legs.

High sensitivity accelerometers are being developed for the SAS inertial damping[13].

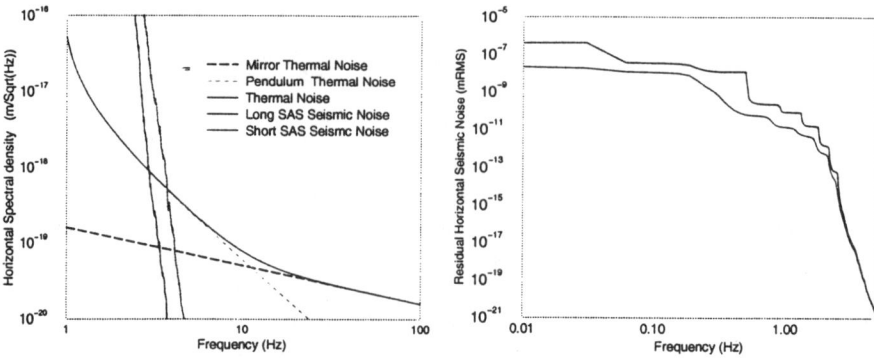

FIGURE 4. Simulated performance of the SAS chains: residual mirror seismic noise on the left and integrated residual r.m.s. motion on the right. In both cases the ground seismic activity of the Livingstone site have been used as an input to the simulation. The short SAS simulation corresponds to the system of figure 2. The long SAS simulation is for a longer chain mounted in a BSC extension barrel designed for rapid installation procedures. The integrated r.m.s. motion on the right is calculated for the short chain.

ACKNOWLEDGMENTS

We like to acknowledge the enthusiastic contribution of Henry Lubatti, David Ahkavan and Nicolas Viboud, students that recently joined the SAS group. We also acknowledge the professionality of Ricardo Paniagua and his crew of the Caltech machine shop, Gianni Gennaro, master designer and Carlo Galli, passionate builder of the SAS system.

REFERENCES

1. Virgo Collaboration VIRGO final design report. VIR-TRE-1000-13 (May 1997)
2. M.V. Plissi, et al., Rev. Sci. Instrum., 69, 3055-3061, 1998
3. G. Losurdo: Thesis: "Ultra-low Frequency Inverted Pendulum for the Virgo Test Mass Suspension", Scuola Normale Superiore di Pisa, Classe di Scienze, 1998.
4. G Losurdo, et al. "Inertial control of the VIRGO Superattenuator", this conference
5. G. Losurdo, et al: Review of Scientific Instruments, 70, 2507, 1999.
6. A. Bertolini, at al., Nucl. Instr. and Meth. in Phys. Res. A 435 3 (1999) pp. 475-483.
7. M. Beccaria, et al: Nucl. Instr. and Meth. in Phys. Res. A, 397, 1999.
8. R DeSalvo, et al., conference proceedings, Moriond rencontres ... January 1999.
9. S. Braccini, et al., Nucl. Instr. and Meth. in Phys. Res., A : 404, 455-479 - (1998).
10. R. DeSalvo, et al: Nucl. Instr. and Meth. in Phys. Res. A 420, 316, 1999.
11. M. Barton, et al.: "Proposal of a Seismic Attenuation System for ...", Note, LIGO-T990075-00-D
12. M. Barton, et al.: "Answers to 'Isolation criteria for the LIGO II...' ", Note, LIGO-T990076-00-D
13. A.Bertolini, et al., "High sensitivity accelerometers for high performant seismic attenuators", this conference.

Reducing Low-Frequency Residual Motion in Vibration Isolation to the Nanometre Level.

J. Winterflood, Z.B. Zhou†, and D.G. Blair

Department of Physics, University of Western Australia, Perth 6907, Australia.
†*Department of Physics, Huazhong University of Science and Technology, Wuhan, 430074, P.R. China.*

Abstract. If the low frequency residual motion of a vibration isolated test mass can be reduced to the nanometre level, then the forces required to control its position may be applied directly to the test mass without injecting noise and the control and acquisition of lock in an interferometer is rendered trivial. We show that two stages of already demonstrated pre-isolation elements, combined with tilt control and a novel self-damped isolator structure, are capable of achieving this performance. The designs and performance requirements of the individual components making up this system are discussed and various experimental measurements presented.

1 INTRODUCTION

Gravitational wave (GW) detection requires the reduction of all sources of vibration to the lowest level possible. Isolation from seismic motion is critical - not only in the GW detection band, but also at frequencies below this because even low frequency disturbances make locking of high finesse cavities very difficult and call for large servo control forces which easily inject noise (from limited dynamic range of electronics and through seismic coupling). The main effect necessitating large control forces to be applied near the test mass, is the residual horizontal low frequency motion which feeds through an isolation stack below its intended cutoff frequency (a few Hz).

Figure 1 illustrates how this residual motion may be successively reduced by the various techniques that we present. The graph on the left shows the spectral densities in m/√Hz, while the one on the right has been integrated to give the RMS motion in metres for all frequencies above that shown on the axis. The dotted lines show a typical seismic level of $10^{-6}/f^2$ m/√Hz above 1Hz which levels out below the microseismic peak at about 0.2Hz.

Curve (a) shows the response of a typical 5 stage, 2.3m high, all-metal isolation stack with high Q-factor resonances (Q's ~ 1000). The residual motion spectrum is dominated by these internal resonant modes and the RMS residual motion is almost two orders of magnitude above the seismic level. Adequate damping of these resonances as in curve (b) can immediately provide an order of magnitude reduction in this residual motion. There are various approaches to obtain this damping, and in section 2 we present a new and entirely passive scheme which we have termed "self-damping".

An additional approach to reduce the residual motion is to precede the stack with an ultra-low frequency (ULF) pre-isolator stage. The typical result of a 25sec pre-

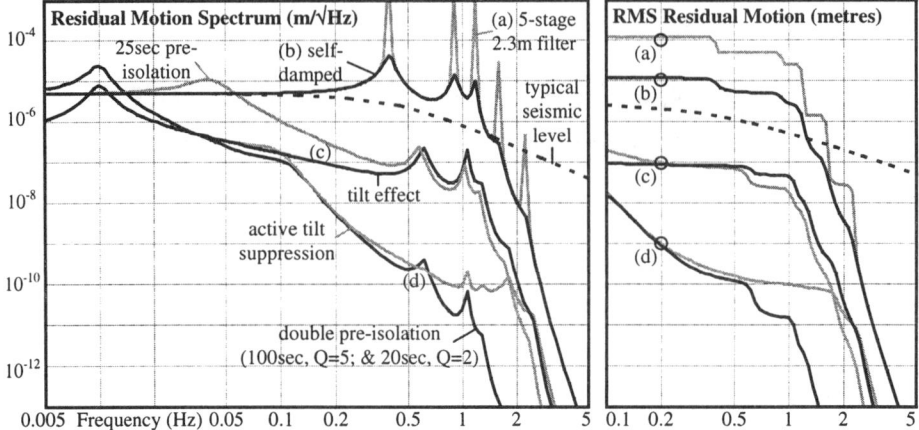

FIGURE 1. Successive reduction of residual motion - (a) high Q-factor internal modes resonantly amplify the seismic motion at the normal mode frequencies, (b) these modes can be passively damped for a factor of 10 reduction in residual motion, (c) ULF pre-isolation can provide an additional factor of 100 reduction, (d) dual ULF pre-isolation and active tilt suppression can gain another factor of 100.

isolation stage for translational seismic motion (ie no tilting) is shown by the grey curve (c). This is no longer a new idea and has been reported on and implemented by several groups[1,2,3,4]. Such a pre-isolator may have little or no effect in the GW detection band, but can provide two orders of magnitude reduction in the seismic drive to the stack resonances and thus in the residual motion at these frequencies. It also provides a structure before the isolation stack to which very low frequency drift correction forces may be applied - avoiding the problems associated with applying these forces near the test mass. For this reason frequencies below approx 0.2Hz are not considered important in this discussion.

Once a single horizontal ULF stage is in place, seismic tilting motion becomes strongly coupled to horizontal motion at the top of the stack and rapidly becomes the dominant source of residual motion. The result of this coupling together with the expected relationship between translational and tilt seismic motion (due to wavelength of seismic motion travelling with velocity ~500m/s) is shown as the black curve (c). The total residual motion in this case is the RMS sum of the two (c) curves - at the 100nm level. It can be seen from the level of this tilt effect (black curve (c) compared to grey curve) that further improvement in pre-isolator performance beyond a single stage of ~25sec period, will not provide any reduction in the residual motion. This is about the best that can be achieved with passive isolation techniques alone. In order to benefit vibrationally from a higher performance pre-isolator of say 100 seconds period, or even two stages of ULF pre-isolation as we propose in section 4, it is necessary to sense and actively counteract the tilt. In section 3 we present a design for a 2-D tilt sensor together with some preliminary measurements towards providing this sensing.

With two stages of ULF pre-isolation the residual motion from *translational* seism may be reduced to the black curve (d). If tilt control is implemented as described in section 3, it should also be possible to reduce the tilt effect to the grey curve (d).

2 SELF-DAMPED STACK

With ULF pre-isolation in place, disturbance at the top of the stack is already far below seismic and so damping cannot be achieved with frame mounted motion sensing. This self-damping approach seems to be the best of the frame independent methods. The basic approach is to viscously cross-couple modes on the same isolator stage which have different degrees of freedom. Thus the rocking (or tilting) modes which are inevitably coupled to a small extent to the horizontal translational modes, are deliberately engineered to be strongly and viscously coupled and to have mode frequencies that differ sufficiently to optimise damping. This approach is illustrated with progressive schematics in figure 2. The normal pendulum link in 2(a) is made rigid in 2(b) and the large mass that is normally suspended at the end of the pendulum link is made into a dumbbell shape to accentuate its large rotational inertia, and is flexibly attached by a soft spring pivot. This separates the majority of the mass from the now rigid pendulum rod so that it is not forced to tilt as the pendulum swings, but can rock or tilt relatively independently at another frequency. Then an angular dashpot is added between these two motions to optimally damp them against each other. This structure can use x-y pivots and a doughnut shaped mass to operate in both horizontal degrees of freedom, damping x pendulum motion against θ_y rocking and y motion against θ_x rocking.

The vertical suspension included in 2(b) as a telescopically extensible pendulum rod is far from practical and any realisation of it is an engineering challenge if the centre of percussion (COP) is to remain at a pivot point regardless of extension. The solution shown in 2(c) is to split the pendulum into two sections - one being an almost

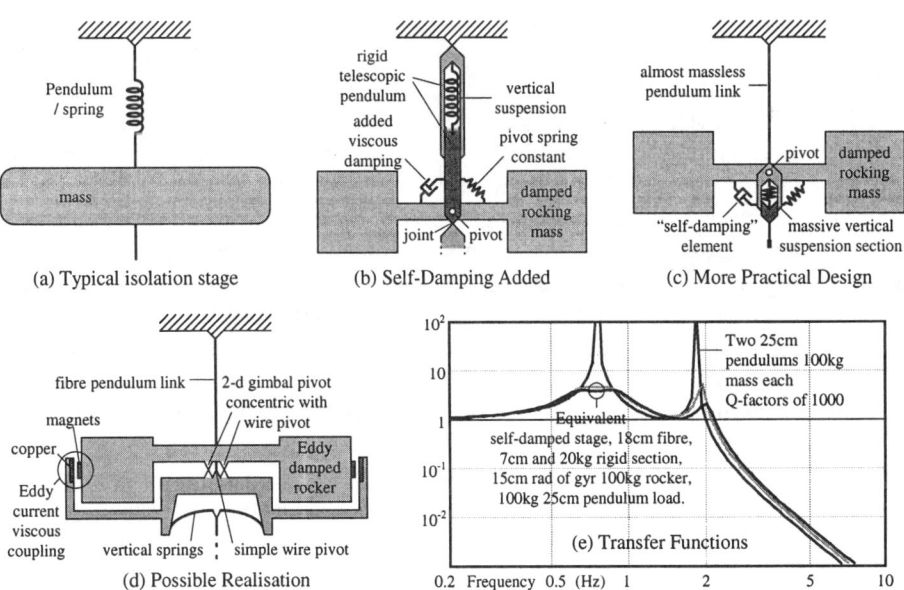

FIGURE 2. (a)-(d) Development of self-damped stage, (e) Comparative transfer functions.

massless suspension fibre (and hence no problems with COP tuning), and the second being a shorter and massive vertical extension section in which the horizontal COP effect may be neglected (thanks to the fibre section) and vertical effects given maximum consideration. In 2(d) the rocking mass is pivoted on this vertical extension section and the angle between them viscously damped by magnetic eddy current damping at their extremities. Most methods of vertical suspension can be incorporated since size and mass is not an issue. Cantilever springs have been shown to give good performance and initially straight ones which are bending under stress are shown in 2(d). Loop spring versions are shown in figure 4(g).

There are a considerable number of free parameters in such a design which may be adjusted for best damping performance. The ratios of the fibre length to the rigid section length and the ratio of tilting mass to rigid section mass are somewhat restricted, but the pivot spring rate and viscous coupling coefficient are almost unconstrained. A couple of solutions found by manually manipulating these parameters are shown in 2(d) and are compared with a simple two stage pendulum.

Both of the self-damped structures consist of a single 25cm self-damped stage loaded with a 25cm simple pendulum of infinite Q-factor. The normal mode Q-factors are reduced from ~1000 down to 3 for the 0.7Hz peaks and 10 (or 20 for middle curve) for the 2Hz peaks. The difference between the two damped solutions is due different heights of the centre of mass of the 20kg rigid pendulum section. The 2-d damping of these solutions is obtainable from 750g of NIB magnets at 50cm lever arm.

3 TWO DIMENSIONAL TILT SENSOR

In order to obtain full benefit from high-performance or dual ULF pre-isolation, it is necessary to sense and counteract tilt in the region of 0.1Hz to 5Hz. A possible tailored loop gain that was used to obtain the tilt suppression of figure 1(d) is shown in figure 3(e). A couple of angular accelerometers aimed at this task have appeared in the literature recently[5,6] but the design shown in figure 3(a) has novel features including measuring both tilts with a single device allowing simple central mounting.

Precision electric discharge machining allows very thin and wide (<50um thick × cms wide) flexures to be made from a monolithic block of low loss metallic material. With an arrangement of cuts similar to figure 3(b) it is possible to make x-y flexures with intersecting axes. Suspending a disk shaped test mass very near to its centre of mass with such a flexure allows sensing both θ_x and θ_y tilts. Utilising two of these x-y flexures with a pendulum link between provides continued sensitivity down to DC as well as improved immunity to translational motion. Sensitivity to DC is not required in this case but may reduce the requirement for very low resonant frequencies.

The readout of angle may be achieved by various means but we propose to bounce a laser from a mirrored surface onto a quadrant photo-diode thereby obtaining both angles with a single readout device. The laser beam displacement with test-mass angle can be greatly magnified by mounting another mirror on the frame almost parallel to the test mass mirror surface, and bouncing the laser beam back and forth between them multiple times before landing on the detector. We have termed this angle sensing

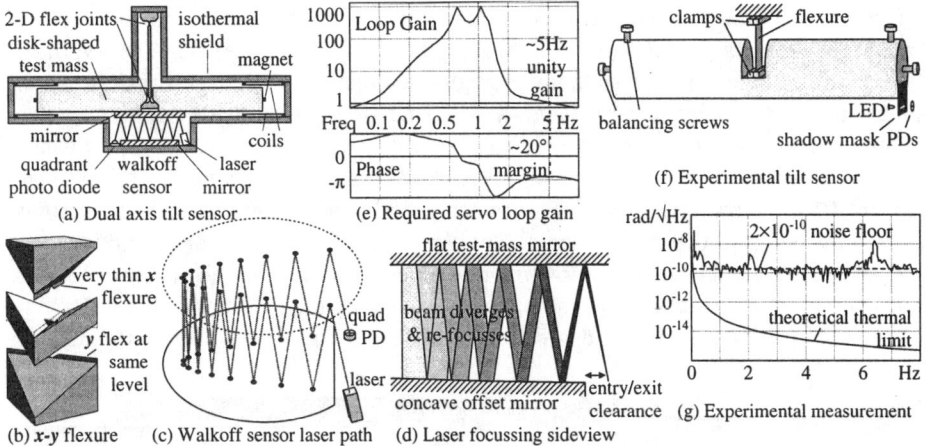

FIGURE 3. (a) dual axis tilt sensor, (b) exploded view of monolithic dual-axis super-flexible joint; (c) maximising sensitivity with direction reversal and (d) beam focussing, (e) servo loop gain requirement, (e) experimental single axis tilt meter using shadow sensor, (f) measured and predicted noise levels.

technique a "walk-off" sensor. The displacement with angle increases as the number of bounces squared, but due to finite beam divergence the sensitivity gain obtained typically only goes as the number of bounces. The number of bounces obtainable from a given mirror size can be increased fourfold over a simple parallel arrangement by presetting the fixed mirror to a small offset angle so that the laser traverses its width twice as illustrated in figure 3(c). If in addition the fixed mirror is made slightly concave so as to re-focus the beam back at the detector as in 3(d), then sensitivity is greatly increased and noise from laser beam pointing fluctuations is minimised.

Figure 3(f) shows an experimental prototype that we built to obtain preliminary tilt measurements. With a simple 3mm illuminated aperture and split photo-diode we obtained the sensor noise floor shown in 3(g) of 2×10^{-10} rad/$\sqrt{Hz}$. The sensitivity we require for the tilt suppression of figure 1(d) is 10^{-11} rad/$\sqrt{Hz}$. We expect this to be readily obtainable with a walk-off sensor operated in a vacuum.

4 TILT-RIGID 3-D ULF PRE-ISOLATOR STAGE

The need to measure the transfer function performance and null out percussion effects in a high-performance pre-isolator structure in-situ has shown the importance of providing a suitable tilt-rigid but 3-D translatable shake-test frame prior to the entire suspension structure. If this frame is designed to allow reasonably small forces to produce significant motion, then we have nothing less than a *second* pre-isolation stage which ideally should be left permanently in place for additional isolation. We have recently built and tested such a 3-D ULF shake-test frame which is designed to function cascaded with the high-performance 3-D pre-isolation stages[3,7] in 4(e) and 4(f). It is constructed as cascaded horizontal and vertical sections.

The simplest method of tilt-rigid horizontal suspension seems to be inverse pendulums supporting a platform as a "wobbly table" in figure 4(a). Tilt-rigid vertical suspension can be obtained from a linear guidance structure which rigidly constrain all degrees of freedom except vertical translation, and sprung to provide support against gravity with a low spring rate. We chose the Wilmore suspension 4(c) over guidance wires for its greater rigidity and linear spring rate, and added springs in the LaCoste manner[8] 4(b) for low spring rate. This is achieved by attaching diagonal coil springs between two sides of the shearing parallelograms formed by the pivot arms and frames to give the structure in 4(d). The low internal mode frequency of coil springs (~20Hz) is not a problem in this case as isolation operation is only required up to about 3Hz.

We used springs with a high degree of pre-tension to give a good approximation to

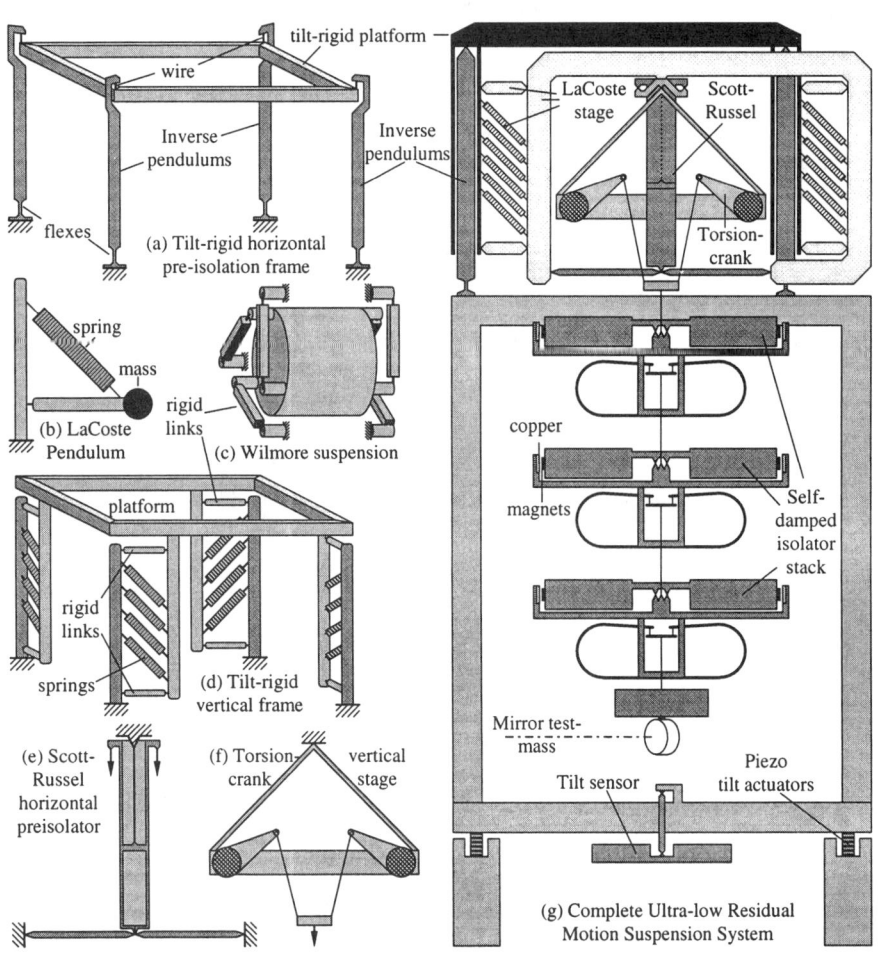

FIGURE 4. An ultra-low residual motion suspension system: (a) Tilt-rigid horizontal and (d) vertical frames, (e) second horizontal and (f) vertical pre-isolation stages, (g) complete system including self-damping and tilt control. The vertical frame is derived from (b) LaCoste and (c) Wilmore suspensions.

the "zero-length" requirement of the LaCoste geometry[8], each spring only providing about 10Kg of supporting force - making it very easy to add or remove springs to support different loads. Adjustments smaller than a whole spring are made with mass ballast. The separation between the pivot joints on the horizontal arms is made the same as the separation between the flexing points on the rigidly clamped springs so that adding and subtracting springs does not alter the vertical spring rate. To negate the significant spring rate of the flexure joints in the horizontal arms, several additional springs on each side (not shown) are stretched horizontally to a much greater width than the separation between the flexure joints. This inverse pendulum effect adds a negative spring rate to the arm flexure joints and making this width adjustable allows the vertical spring rate to be adjusted to obtain very low resonant frequencies.

The structure we have built fits into a vacuum tank of 1.5m diameter and is about 1m high. It only uses the perimeter volume of the tank leaving a clear central volume with a diameter of 0.85m for suspending other equipment. It is designed to support a load of up to 1 tonne and provide vibration isolation from 0.2 to 3Hz. We obtained vertical resonant periods in excess of 25secs with Q-factor>1, and horizontal resonant periods of ~40secs. The isolation will be bypassed vertically by spring resonances at 17Hz and horizontally by leg bending resonances at 80Hz.

5 CONCLUSION

With the complete system shown in figure 4(g) incorporating active tilt control with sensitivity and loop gain as discussed it should be possible to reduce the RMS residual motion above 0.2Hz down to the nanometre level, requiring only micronewtons of locking force at the test mass and a time to drift through a micron wavelength fringe in the order of 100 seconds. This allows simple frame mounted direct actuation near the test mass and no reaction mass or wires down the isolation stack.

ACKNOWLEDGMENTS

This work was supported by the Australian Research Council.

REFERENCES

1. Barton M.A., Kanda N., and Kuroda K., Rev. Sci. Instrum. **67** 3994- (1996).
2. Winterflood J., Blair D.G., Phys. Lett. A, **222** 141-7 (1996) 141.
3. Winterflood J., Blair D.G, Phys. Lett. A, **243** 1-6 (1998).
4. Losurdo G, et al, Rev. Sci. Inst. **70** 2507-15 (1999).
5. Speake C.C. and Newell D.B, Rev. Sci. Instrum. **61** 1500- (1990).
6. Luiten A.N, et al, Rev. Sci. Instrum. **68** 1889-93 (1997).
7. Winterflood J, Losurdo G, Blair D.G, "Initial results from a long-period conical pendulum vibration isolator with application for gravitational wave detection", Phys. Lett. A, in press (1999).
8. LaCoste L.J.B, Physics **5** 178-80 (1934).

Inertial control of the VIRGO Superattenuator

Giovanni Losurdo* for Pisa and Florence VIRGO Groups

Istituto Nazionale di Fisica Nucleare - Sezione di Pisa
Via Livornese 1291 - 56010 - S.Piero a Grado (Pisa) - Italy
e-mail: losurdo@galileo.pi.infn.it

Abstract. The VIRGO superattenuator (SA) is effective in depressing the seismic noise below the thermal noise level above 4 Hz. On the other hand, the residual mirror motion associated to the SA normal modes can saturate the dynamics of the interferometer locking system. This motion is reduced implementing a wideband (DC-5 Hz) multidimensional control (the so called *inertial damping*) which makes use of both accelerometers and position sensors and of a DSP system. Feedback forces are exerted by coil-magnet actuators on the top of the inverted pendulum. The inertial damping is successful in reducing the mirror motion within the requirements. The results are presented.

I INTRODUCTION

The test mass suspension of the VIRGO detector, the superattenuator (SA) [1], has been designed in order to suppress the seismic noise below the thermal noise level above 4 Hz. The expected residual motion of the mirror is $\sim 10^{-18}$ m/$\sqrt{Hz}$@4 Hz. At lower frequencies, the residual motion of the mirror is much larger (~ 0.1 mm RMS), due to the normal modes of the SA (the resonant frequencies of the system are in the range 0.04-2 Hz).

To lock the VIRGO interferometer the RMS motion of the suspended mirrors must not exceed 10^{-12} m (to avoid the saturation of the read-out electronics). VIRGO locking strategy is based on a hierarchical control: feedback forces can be exerted on 3 points of the SA (inverted pendulum (IP) [2], *marionetta* and mirror). The control on the 3 points is operated in different ranges of frequency and amplitude. The maximum mirror displacement that can be controlled from the marionetta without injecting noise in the detection band is ~ 10 μm. Therefore, a damping of the SA normal modes is required for a correct operation of the locking system. An active control of the SA normal modes, using sensors and actuators on top of the IP, capable of reducing the mirror residual motion within a few microns, has been successfully implemented.

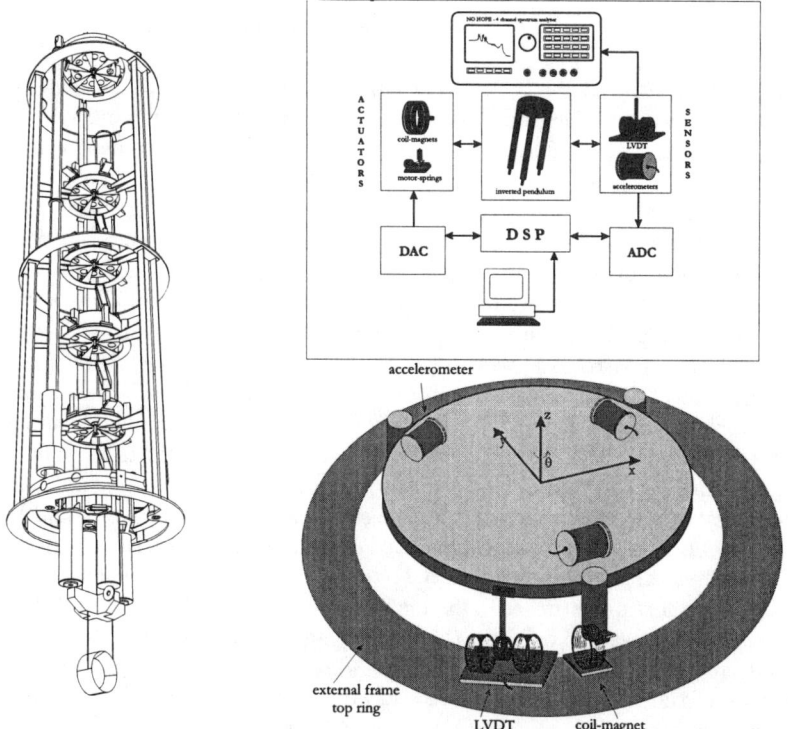

FIGURE 1. LEFT: the superattenuator; RIGHT TOP: logical scheme of the setup for the local active control; RIGHT BOTTOM: simplified view of the IP top table, provided with the 3 accelerometers. One LVDT position sensors and one coil-magnet actuator are also shown.

II EXPERIMENTAL SETUP

The setup (fig. 1) of the experiment is composed by a full scale superattenuator, provided with 3 accelerometers (placed on the top of the IP), 3 LVDT position sensors (measuring the relative motion of the IP with respect to an external frame), 3 coil-magnet actuators. The accelerometers work in the range DC-400 Hz and have acceleration spectral sensitivity $\sim 10^{-9}$ m s^{-2} Hz$^{-1/2}$ below 3 Hz [3]. The sensors and actuators are all placed in *pin-wheel* configuration. The sensors and actuators signals are elaborated by a computer controlled ADC (16 bit)-DSP-DAC (20 bit) system. The DSP allows to handle the signals of all the sensors and actuators, to recombine them by means of matrices, to create complex feedback filters (like the one of fig. 5) with high precision poles/zeroes placement and to perform a large amount of calculations at high sampling rate (10 kHz). The suspended mirror is provided with an LVDT to measure its displacement with respect to ground.

III THE CONTROL STRATEGY

The active control of the SA normal modes is defined *inertial damping*, because it makes use of inertial sensors (accelerometers) to sense the SA motion. The advantage of using accelerometers is that they perform the measurement with respect to the "fixed stars", while position sensors do it with respect to a reference frame which is not seismic noise free. Therefore, inertial sensors are to be used so that no seismic noise is reinjected by the feedback. Actually, in the real SA control both sensors are used: position sensors provide a low frequency (DC - 10 mHz) control of the SA position (in order to avoid drifts), while accelerometers allow a wideband reduction of the noise in the region of the SA resonances (10 mHz - 2 Hz).

The object to control is a MIMO (multiple in-multiple out) system: each sensor (accelerometer/LVDT) is sensitive to the 3 modes (X,Y,Θ) of the IP and each actuator excites all the modes. To simplify the control strategy the sensors outputs and the actuators currents are digitally recombined to obtain independent SISO (single in-single out) systems (fig. 2): the system is described in the normal modes coordinates (for a description of the diagonalization procedure see [5,6]). Each normal mode is associated to a so called *virtual sensor* (sensitive to that mode and "blind" to the others) and to a *virtual actuator* (acting on one mode only, leaving the others undisturbed). In this way one is able to implement independent feedback loops on each d.o.f., greatly simplifying the control strategy. Fig. 3 shows the output of the virtual accelerometers X and Θ. In the X plot, the 40 mHz resonance of the IP translation mode and all the modes of the SA chain are visible (as pole/zero structures). In the Θ plot, only the rotation mode of the IP is visible. The two plots show that different feedback strategies have to be implemented on

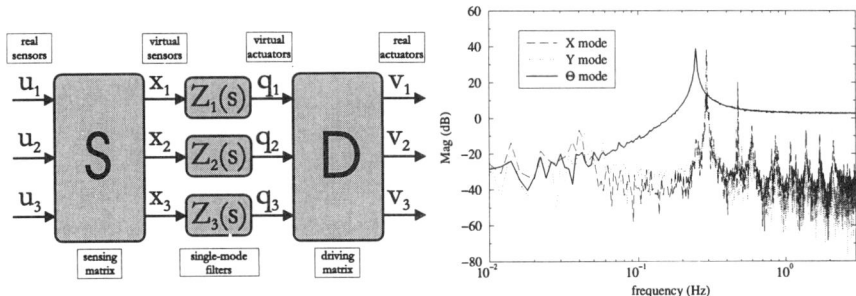

FIGURE 2. LEFT: The logic of the diagonalisation: the output u_i of the sensors are linearly recombined by a matrix **S** in order to produce 3 *virtual* sensors outputs (x_i), sensitive to pure modes. Three independent feedback filters $Z_i(s)$ are designed for the pure modes and 3 generalized forces q_i are produced. The q_i are turned into real currents (v_i) to feed the actuators via the matrix **D**. RIGHT: Effect of the digital diagonalization: the 3 modes are uncoupled, the Θ mode is excited.

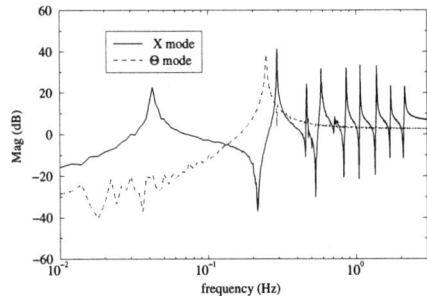
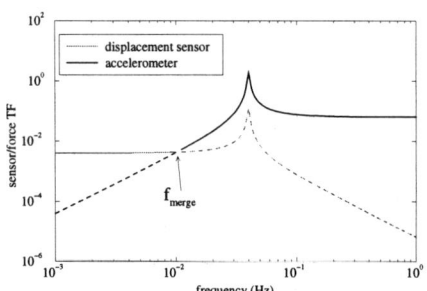

FIGURE 3. The output of the virtual accelerometers X and Θ are compared. Different feedback strategies are required for the two modes

FIGURE 4. *Merging* of displacement and acceleration sensors (simulation for a simple pendulum).

the different d.o.f.. The basic idea of inertial damping is to use the accelerometer signal to build up the feedback force. Actually, if the control band is to be extended down to DC, a position signal is necessary. Our solution was a *merging* of the two sensors: the virtual LVDT and accelerometer signals are combined in such a way that the LVDT signal $(l(s))$ dominates below a chosen cross frequency f_{merge} while the accelerometer signal $(a(s))$ dominates above it (see fig. 4 and ref. [4]). The feedback force has the form[1]:

$$f_{\text{fb}} = G(s)\left[a(s) + \epsilon l(s)\right] \qquad (1)$$

where $G(s)$ is the digital filter transfer function (see fig. 5) and ϵ is the parameter whose value determines f_{merge}. We have chosen $f_{\text{merge}} \sim 10$ mHz (corresponding to $\epsilon \sim 5 \cdot 10^{-3}$). This approach allows to stabilize the system with respect to low frequency drifts at the cost of reinjecting a fraction ϵ of the seismic noise via the feedback.

IV   INERTIAL CONTROL PERFORMANCE

The result of the inertial control (on 3 d.o.f.) is shown in figure 6. The measurement has been performed in air. The noise on the top of the IP is reduced over a wide band (10 mHz - 4 Hz). A gain > 1000 is obtained at the main SA resonance (0.3 Hz). The RMS motion of the IP (calculated as $x_{\text{RMS}}(f) = \sqrt{\int_f^\infty \tilde{x}^2(\nu)d\nu}$) in 10 sec. is reduced from 30 to 0.3 μm. The closed loop floor noise corresponds to the fraction of seismic noise reinjected by using the position sensors for the DC control

---

[1] Actually, the LVDT signal $l(s)$ is properly filtered in order to preserve the feedback stability at the cross frequency and in order to reduce the amount of reinjected noise at $f > f_{\text{merge}}$.

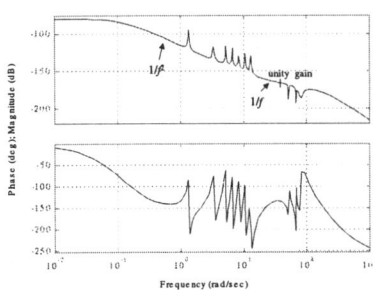 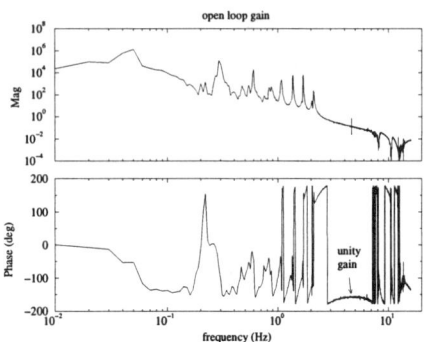

FIGURE 5. LEFT: Digital filter used for the inertial damping of a translation mode (X). The filter slope is f^{-2} in the range 10 mHz$< f <$3 Hz, f^{-1} for $f >$3 Hz. The unity gain is at 4 Hz. The peaks in the digital filter are necessary to compensate the dips in the mechanical transfer function (see the transfer function of the X mode in fig. 3). RIGHT: open loop gain function (measured). The phase margin at the unity gain frequency is about 25°.

and can, in principle, be reduced by a steeper low pass filtering of the LVDT signal at $f > f_{\text{merge}}$ and by lowering f_{merge}: both this solution have drawbacks and need a careful implementation.

Preliminary measurements of the displacement of the mirror with respect to ground have been performed in air, using an LVDT position sensor. The residual RMS mirror motion in 10 sec. is[2]:

$$x_{\text{RMS}}(0.1\,\text{Hz}) \leq 3\,\mu\text{m}. \tag{2}$$

When the damping is on such a measurement can provide only an upper bound because the LVDT output is dominated by the seismic motion of the ground.

V FURTHER DEVELOPMENTS

Several ways of improving the inertial damping performance have been identified:

- a steeper low pass filtering of the LVDT output above f_{merge} may reduce the amount of reinjected seismic noise. In doing this one has to be careful to preserve proper phase difference between the LVDT and accelerometer signals;

- the lower f_{merge} the smaller the amount of reinjected noise. Lowering f_{merge} is difficult due to the mechanical tolerance on the parallelism of the IP legs: if

---

[2] This number has been obtained with a feedback design less *aggressive* then the one of fig. 5: the gain raised as $1/f$, the cross frequency was 30 mHz and no compensation of the dips was needed.

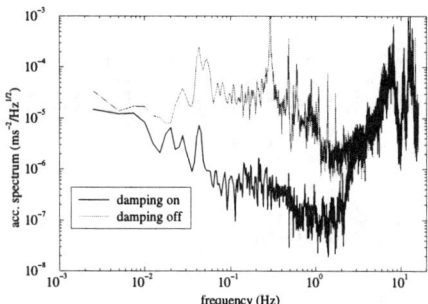 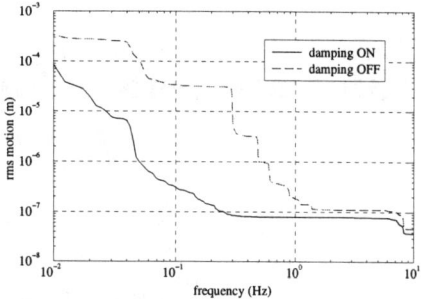

FIGURE 6. Performance of the inertial control (X, Y, Θ loops closed) of the superattuenuator, measured on the top of the IP: the left plot shows the acceleration spectral density as measured by the *virtual* accelerometer X (translation). The right plot shows the effect of the feedback on the RMS residual motion of the IP as a function of the frequency.

the legs are not perfectly parallel, the top table tilts slightly as it translates. Therefore, the accelerometer signal is dominated by the tilt below 15-20 mHz, making thus impossible to use the accelerometers at very low frequencies. A technique for subtracting the effect of the tilt (using the information provided by the displacement sensors) has been defined and used to obtain the results here described [7]. Cancelling the tilt effect down to $\sim$ 5 mHz makes us able to use the accelerometers down to 10 mHz [7]. Stricter requirements on the IP legs machining and improvements in the tilt subtraction technique may allow a lower cross frequency.

REFERENCES

1. F.Frasconi, et al., Performances of the R&D superattenuator chain of the VIRGO experiment, to be published in the proceedings of the *XXXIVth Rencontres de Moriond* (1999).
2. G.Losurdo, et al., *Rev. Sci. Instrum.*, **70** (5), 2507-2515 (1999).
3. S.Braccini, et al., *Rev. Sci. Instrum.*, **66** (3), 2672-2676 (1995).
4. S.J.Richman, J.A.Giaime, D.B.Newell, R.T.Stebbins, P.L.Bender, J.E.Faller, *Rev. Sci. Instrum.*, **69** (6), 2531 (1998).
5. A.Gennai, et al., *A control model for the inverted pendulum*, VIRGO Note VIR-TRE-PIS-4900-102 (1997).
6. G.Losurdo, *Ultra-low frequency inverted pendulum pre-isolator for the VIRGO test mass suspension*, PhD Thesis, Scuola Normale Superiore, Pisa (1998).
7. A.Gennai, G.Losurdo, A.Marin, D.Passuello, *Inertial damping of the superattenuator*, VIRGO Note VIR-TRE-PIS-4900-104 (1999).

Measurements of Mechanical Q in Levitated Paramagnetic Crystals

S. J. Augst and R. W. P. Drever

*Department of Physics, Mathematics, and Astronomy
California Institute of Technology, Pasadena, CA 91125*

Abstract. Thermal noise from test masses, arising both from internal noise in the test mass material and from losses in the suspension wires and their attachments, is a significant factor limiting sensitivity of interferometric gravity-wave detectors. To investigate ways of reducing these noise sources we are using magnetic levitation in place of suspension wires. A search for high-Q crystals with magnetic properties allowing tests in moderate field strengths has led us to paramagnetic crystals, and we report preliminary results with small levitated samples of Gadolinium Gallium Garnet (GGG) and Terbium Gallium Garnet (TGG). The technique seems the first to allow Q measurements with no mechanical contact, and may facilitate work aimed at reducing thermal noise.

INTRODUCTION

Interferometric gravitational-wave detectors may experience noise from a number of sources, several of which are due to noise from wires or fibers used to suspend and isolate the test masses, or are transmitted by these from the surroundings to the test masses. In the program of which the present work is a part, magnetic levitation of test masses is being investigated as a possible alternative to wire or fiber suspensions. Potential benefits that might be obtained include: 1) the possibility of achieving lower suspension resonance frequencies, 2) elimination of "violin mode" resonances of wires or fibers, 3) improved seismic isolation, 4) possible reduction of suspension thermal noise, and 5) avoidance of damping of internal modes of the test mass by the wires or fibers.

There are some potential disadvantages also, including possibilities of other noise sources, and these are being investigated. In earlier experiments[1] we have levitated test masses by attaching to them non-conducting ceramic permanent magnets, and applying lifting forces by interaction with fixed ceramic permanent magnets located above them. Relatively long pendulum-mode relaxation times (~15 hours) were achieved, and long suspension periods (~8 seconds or more). However the ceramic magnets on the test masses have higher mechanical damping than typical test mass materials, and this could lead to significant internal thermal noise unless there is effective isolation between the magnets and the test masses themselves. It would be preferable to find materials which themselves combine high mechanical quality factors, Q, with sufficiently strong magnetic properties to make levitation practicable without extreme magnetic fields, and use these to obtain low thermal noise for both pendulum and internal modes. We have been looking for suitable materials for several years, and recently have been considering paramagnetic crystals, which might also possess good optical qualities. Two materials which we found to give encouraging results in preliminary Q tests using piezoelectric transducers were Terbium Gallium Garnet (TGG) and Gadolinium Gallium Garnet (GGG).[2] These materials are used commonly in optical Faraday rotators.

EXPERIMENTAL PROCEDURE AND RESULTS

Here we report on mechanical Q measurements made with cylindrical samples of TGG and GGG, 15 mm in diameter and 8 mm thick. It was found that with these small samples there could be serious damping from suspension wires and piezoelectric drivers, so we have developed and used a levitation technique with non-contacting excitation and sensing of internal resonances. A simplified diagram of the arrangement is shown in Figure 1.

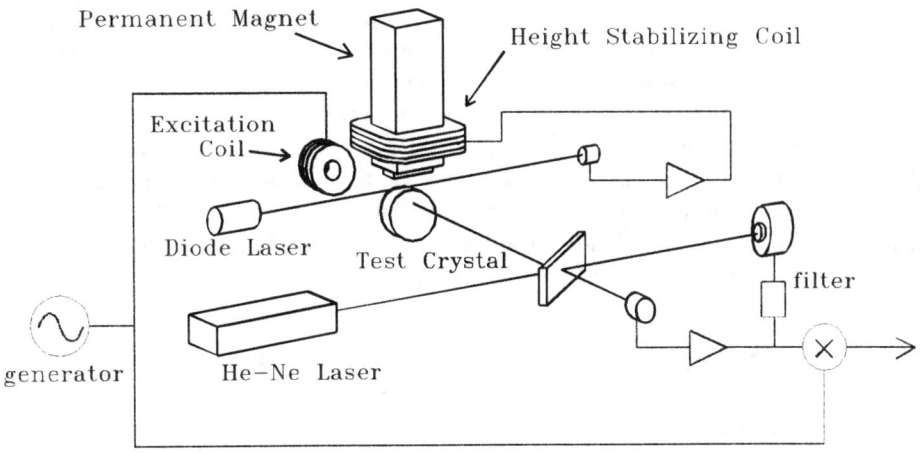

FIGURE 1. Experimental setup for Q measurement. The cylindrical paramagnetic sample is suspended in the field of a permanent magnet, with its height stabilized using a shadow sensor whose output is fed back to a coil which adjusts the field. A mechanical resonance in the sample is excited by another coil and the resulting crystal vibration is sensed with the coherently demodulated signal from a Michelson interferometer. Light in one arm of the interferometer is reflected by an uncoated surface of the sample, while a mirror in the other arm is servo-controlled to lock the interferometer to a single fringe. Most of the system is in vacuum.

The paramagnetic sample is placed in the field of a permanent magnet which creates an induced magnetism in the crystal which interacts with the field gradient to provide lift. The induced magnetism makes it convenient to excite resonances in the sample by a high-frequency current in an adjacent coil, whose frequency is scanned through the resonances, while the vibrations are sensed with a Michelson interferometer. A short focus "cats-eye" lens (not shown) and a slow servo-controlled mirror makes the signal insensitive to misalignment and slow motions of the levitated crystal. For precise Q determination the excitation coil can be suddenly disconnected and the decay of the undriven signal measured. This technique provides a means to measure the Q of a sample in a totally non-contacting way, with minimal external damping.

To maintain stable levitation the output of a shadow sensor is fed back to a height stabilizing coil which adjusts the field of the permanent magnet. The upward force on a paramagnetic sample is proportional to $B_z \times \chi_m \times dB_z/dz$ where B_z is the vertical component of the magnetic field, and χ_m is the paramagnetic susceptibility. Typical products of field times field gradient in these samples were approximately 7 kGauss^2/cm for TGG and 10 kGauss^2/cm for GGG. Rare-earth magnets were used to obtain the fields in these measurements. These magnets are electrically conductive, and thus could be a source of damping from eddy currents. Checks of the pendulum mode relaxation time with these samples indicate, however, that in our present situation and for the levels of Q found here eddy current damping is not a significant factor.

The crystals used in the present experiment are cylindrical in shape and initially had an inspection polish on the two flat surfaces with a fine grind finish on the circumference. Under these conditions the value for Q found for the GGG sample was 6.7×10^5 at a resonance frequency of 221.5 kHz. It was expected that the ground finish was limiting Q, and the same sample was subsequently repolished after the Meeting, and was then found to give significantly higher Q.[3]

The TGG sample available initially had a small visible crack in its interior and a chip on the exterior. With this sample we obtained a Q value of 3.7×10^4 at 219.6 kHz. After the Meeting, we obtained an undamaged and finer-polished sample of TGG with the same dimensions, and this also gave a significantly higher value for Q.[3] It is clear that further work is required, and these preliminary results should only be regarded as encouraging lower limits to the material Q values.

POTENTIAL APPLICATIONS AND FURTHER WORK

The long term aim in this research has been to find ways of suspending test masses with reduced thermal noise. Present results are encouraging, at least for small test masses, but further work is needed for larger masses since high fields and gradients are required. In the long term this technique may prove to be particularly convenient at cryogenic temperatures where paramagnetic susceptibilities are significantly higher (e.g. χ_m is about 30 times larger at 10 K than at room temperature).

A potentially more immediate possibility is the use of paramagnetic crystals as replacements for the orientation and length control magnets that are currently attached to test masses in interferometers. The initial LIGO test mass Q's are damped by the low-Q nature of these magnets and they must not be present in the LIGO II upgrade if it is to achieve its projected noise levels. Potential control technologies that have been suggested are electrostatic transducers or photon pressure from a laser. A possible alternative solution is to bond high-Q paramagnetic crystals to the test masses and use conventional coil drivers to provide driving forces. It is unlikely that this arrangement would damp the internal test mass Q. Additionally the control coils could be operated with a high-frequency driving field (rather than a DC field as is required with permanent magnets) thus making the test masses less prone to noise from external magnetic field fluctuations (e.g. from the earth's field or from electrical equipment).

As far as we know these experiments are the first measurements of high Q single crystals made without any mechanical contact. This may provide a means of achieving higher values of Q than possible with any other arrangement. Comparison of results using this technique with results from conventional wire suspensions may be useful in studies to identify practical sources of loss.

The application of this measurement technique to larger samples will be used to determine values of Q for lower frequencies. In addition samples that are super-polished on all surfaces will provide further insight into the importance of surface losses versus bulk material losses. Larger crystals require a larger volume of high magnetic field so an electromagnet is now being used to levitate these. Further tests should show whether TGG or GGG crystals could be competitive alternative test mass materials to fused silica or perhaps even to sapphire. Unfortunately, levitation of these crystals requires large magnetic fields and scaling the crystal samples up to LIGO-sized test masses might require superconducting magnets for levitation, and this is an area that requires further study.

The techniques of magnetic levitation for these Q measurements is important in itself since it provides the possibility of studying suspension losses. A comparison of Q measurements between a fiber suspended mass and the same mass suspended magnetically might provide important insights into the loss mechanisms present in conventional LIGO suspension designs. Additionally, this technique could prove to be useful for small, very-low noise systems such as for QND experiments, and these ideas are being considered further.

ACKNOWLEDGMENTS

We would like to acknowledge valuable discussions with J. L. Hall and R.E. Cowen, and H. J. Kimble for some early material samples. This research is based upon work supported by the National Science Foundation under Grant number PHY-9722112, and the California Institute of Technology.

REFERENCES

1. Drever, R. W. P. "Some New Concepts for Laser Interferometer Gravitational Wave Detectors," *Proceedings of the Moriond Workshop on Dark Matter & Cosmology, Quantum Measurements and Experimental Gravitation*, Les Arcs, January 1996, ed. J. Tran Thanh Van (Editions Frontieres, 1996).
2. These particular materials were suggested to us by J. L. Hall (JILA).
3. New results have been obtained since the conference took place. A new polished TGG crystal was obtained and the existing GGG crystal was polished. Both faces, as well as the circumference of each crystal were polished to 60/40 scratch/dig and $\lambda/1$. Measured values of Q for the polished crystals were 6.0×10^6 and 9.7×10^6 respectively. It is anticipated that a "super polish" would raise the values of Q even further.

SIGNAL PROCESSING AND DATA ANALYSIS

Validation Of Data In Operating Resonant Detectors

G.A.Prodi[1], L.Baggio[2], M.Cerdonio[2], V.Crivelli Visconti[2], V.Martinucci[1], A.Ortolan[3], L.Taffarello[4], G.Vedovato[3], S.Vitale[1], J.P.Zendri[4]

[1] *Dipartimento di Fisica, Università di Trento and INFN, I-38050 Povo, Trento, Italy*
[2] *Dipartimento di Fisica, Università di Padova and INFN, I-35131 Padova, Italy*
[3] *INFN, Laboratori Nazionali di Legnaro, I-35020 Legnaro, Padova, Italy*
[4] *INFN, Sezione di Padova, I-35131 Padova, Italy*

Abstract. Assessing the confidence of detection for candidate signals of gravitational waves is a particularly subtle matter. A fundamental step toward this achievement is the validation of the output data of the detectors involved. Here we present how this is accomplished in the operating resonant detector AURIGA by discriminating between satisfactory and unsatisfactory periods of operations on the basis of data self consistency. In particular, the statistics of the operating noise is checked against its simple model and the compliance to the expected shape of the candidates for burst gravitational wave events is assessed by means of a χ^2 test. This approach helps in reducing the false alarm rate of each operating detector and moreover ensures the correctness of the estimated parameters of the candidate events.

INTRODUCTION

The operating gravitational wave observatory made by the five resonant bar detectors ALLEGRO[1], AURIGA[2], EXPLORER[3], NAUTILUS[4] and NIOBE[5] is currently searching for burst signals by means of a time coincidence analysis between the candidate events provided by each detector under the International Gravitational Event Collaboration[6]. The confidence of detection relies mainly on the reduction of the false alarm rate to low values. This is helped by a careful determination of the time periods of satisfactory operation of each detector and of the compliance of the candidate signals to the expected shape. A future step to get the signature of single candidates of gravitational wave detection will require the measurement of peculiar properties of the gravitational wave, such as its propagation speed, the source location and the transversality and tracelessness of the Riemann tensor. New capabilities of the bar detectors –such as the submillisecond timing resolution[7] and the χ^2 test on the shape of candidate events[8]– can play a crucial role in ensuring the confidence of detection.

In this paper we will present how the AURIGA detector is able to discriminate between satisfactory and unsatisfactory time periods of operation by testing the statistics of the measured noise and the self-consistency of the data analysis. We will briefly review the relevance of a statistical test of the compliance of each candidate signal to the expected shape from the viewpoint of validating the estimated parameters

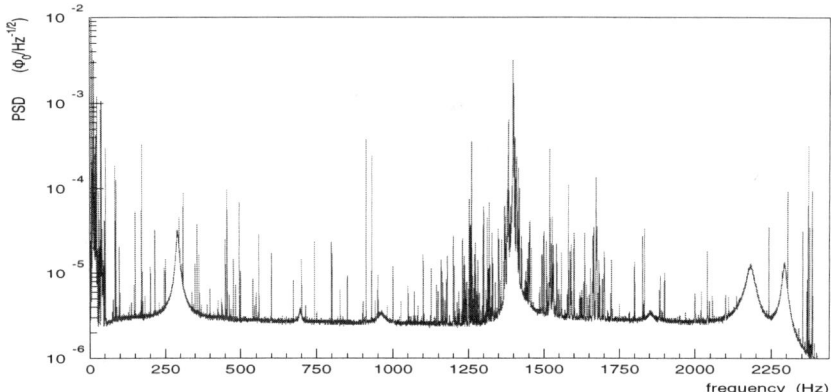

FIGURE 1. Noise power spectral density of AURIGA raw data averaged over 1 hour. The two peaks at 911 Hz and 930 Hz which dominates the spectrum around 1 kHz are the bar-transducer modes. The white noise level comes from the d.c. SQUID noise amplifier, corresponding to an energy resolution of about 4000 $\hbar$.

of the signal. Finally, we point out the near future opportunities given by the measurement of the arrival time of the g.w. bursts in the array of detectors and by testing the consistency of candidate g.w. coincidences with respect to the detected amplitude parameters at different detectors.

DATA ANALYSIS AND NON STATIONARY BEHAVIOUR

The system of data acquisition and analysis of the AURIGA detector has been recently presented[9]. Let us recall here that the data acquisition is based on a signal sampling at 4.9 kHz and is synchronized to the Universal Time Coordinate within 1 µs by means of Global Positioning System clock. Fig.1 shows a sample of the raw data noise power spectrum for the year 1999; the bar-transducer mechanical resonances sensitive to the g.w. signals, show up as the highest peaks present in the 1 kHz region, respectively at 911 Hz and 930 Hz. The noise performance for impulsive signals recently achieved by the AURIGA detector is presented elsewhere in these proceedings[10] and corresponds to a minimum detectable Fourier transform of the g.w. amplitude of about 2×10^{-22}/Hz and a minimum detectable energy of 1 mK. The full raw data are archieved to allow for data reprocessing.

The AURIGA data analysis is fully numerical and is based on a Wiener-Kolmogoroff (WK) filter matched to δ-like signals. This filter assumes a very simple model for the noise of the detector in the bandwidth close to the bar modes, i.e. that the system is described by two coupled harmonic oscillators plus a white noise contribution from the redout SQUID amplifier. The noise spectral density can be written down as

$$S(\omega) = S_0 \prod_{k=1}^{2} \frac{(i\omega - q_k)(i\omega + q_k)(i\omega - q_k^*)(i\omega + q_k^*)}{(i\omega - p_k)(i\omega + p_k)(i\omega - p_k^*)(i\omega + p_k^*)}. \quad (1)$$

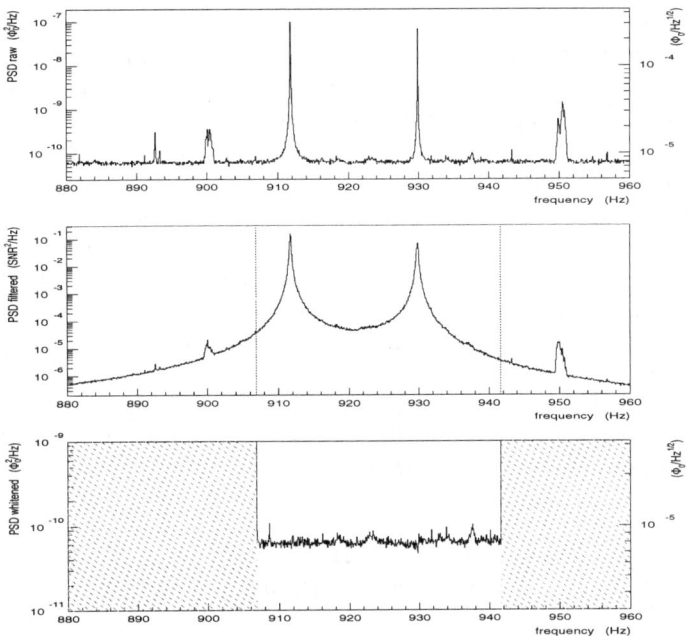

FIGURE 2. Noise power spectral density around the bar-transducer modes of AURIGA for: 1) raw data (upper); 2) W.K. filtered data (middle); 3) whitened decimated data (lower) for the same time period.

where the poles p's and zeros q's are 4 complex parameters, while S_0 is a (positive) real number. These parameters have to be estimated to build the WK filter, but due to the unavoidable non-stationarieties of the system some of their values changes significantly in time. If the non-stationarity is slow, i.e. occurs on a time scale longer than the relaxation time of the modes, the data analysis has to track them by a slowly adaptive algorithm.

We use different techniques to estimate each parameter:
- $\omega_k \equiv \text{Im}(p_k)$ the mode's frequencies are monitored by two fully digital lock-ins.
- $\Delta\omega_k \equiv \text{Re}(p_k)$ the pre-detection bandwidth is known not to be a critical parameter, and is left to the value measured at the beginning of a data taking period
- $\omega_k^{opt} \equiv \text{Im}(q_k)$ the post-filtering frequency is practically identical to the pre-filtering one
- $\Delta\omega_k^{opt} \equiv \text{Re}(q_k)$ the post-filtering bandwidth is adapted so to keep flat the whitened data spectrum (see fig. 2d in ref. 11, these proceedings)
- S_0 the level of the amplifier's white noise is monitored by a lock-in displaced from the detector modes.

In practice, this model works well within a reduced bandwidth of about 35 Hz around the bar-transducer modes. The frequency domain expression for the whitening filter and for the W.K. filter matched to δ-like signals are ratios of polinomials as well, as shown elsewhere in these proceedings[11]. The actual implementation of the data filtering in AURIGA is however made in the time domain[9] by means of an Auto-

Regressive Moving Average. The whitening filter is applied only after the bandwidth of the data stream is narrowed around the bar-transducer modes where the noise model proves to work and, in the presence of noise alone, the whitened data in the reduced bandwidth do show a white noise power spectrum (see Fig. 2).

To track the slow non-stationarieties we use moving averages to smooth the parameters estimates over time scales of the order of the system's proper relaxation time $(\Delta\omega_k)^{-1} \sim 1000$s, much longer than the Wiener filter time $(\Delta\omega_k^{opt})^{-1} \sim 1$s. A non stationary behaviour faster than $(\Delta\omega_k)^{-1}$ does not allow to estimate correctly the noise parameters, and therefore the analysis should point out this difficulty to the experimentalists, as discussed in the next section. A possible signal would instead show up as a very fast variation of the detector output noise, limited in time and not too long with respect to the Wiener time.

When a signal affects the noise the whitened data show a residual color even if the parameters estimates are correct, so the estimates of the noise parameters cannot be trusted, as is shown elsewhere in these proceedings[11].

SATISFACTORY AND UNSATISFACTORY DETECTOR OPERATION

A first level of vetoes on time periods of detector operation is set by the experimentalists, who judge what operations are known to affect the detector output, namely some of the manteinance operations, calibration procedures, failures and so on. These vetoed periods can never be considered as useful observation time of the detector. The AURIGA data analysis provides also for automatic vetoing of time periods in which the statistics of the observed noise is not as expected. In particular, the analysis tests if the filtered and whitened data follow a gaussian statistic and checks if the whitened data are uncorrelated. When the noise is not found to be as expected, then the analysis marks that time period as one affected by "excess noise" or overimposed "signals". In case this condition occurs too frequently, we conclude that the noise model to which the filter is matched is not the right one and consequently that the filtered data of the detector are unsatisfactory. These time periods are vetoed from the observation time of the detector unless some new filtering attempt is succesful in better matching the measured noise.

In more detail, the analysis groups the data streams of both filtered and whitened data in 2 minutes long buffers. The analysis tests if the statistics of each buffer is gaussian buffer by buffer, both by looking at higher moments of the distribution, namely the Kurtosis, and by performing a Chauvenet selection. The latter procedure consists in eliminating those data samples which fall outside 3 times the calculated root mean square value of the buffer, then iterating the same procedure until no other data sample is eliminated. A data buffer is considered to be gaussian if its Kurtosis is compatible with that obtained in Monte Carlo simulatons within 99.7% Confidence Level and if the Chauvenet algorithm converges within a few steps eliminating at most a few percent of the data samples. Moreover, the whitened data buffers whose correlation exceeds a 99.7% CL threshold with respect to Monte Carlo simulations are not used for the estimation of the noise parameters.

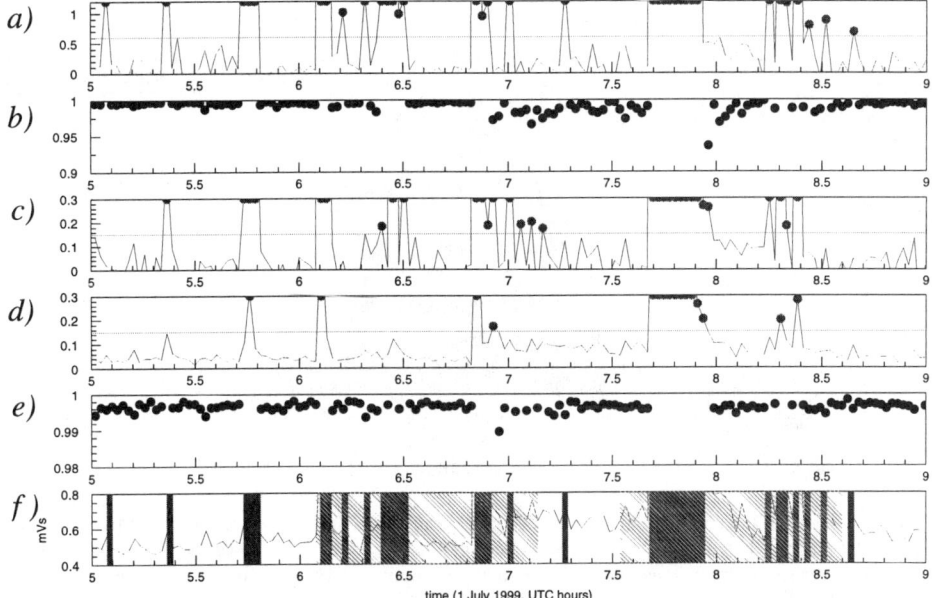

FIGURE 3. The statistical tests performed on filtered and whitened data in four hours of AURIGA operation. For each buffer are shown the Kurtosis and the fraction of data kept after the Chauvenet selection. Filered data: a) Kurtosis, b) Chauvenet selection. Whitened data: c) Kurtosis, d) Chauvenet selection, e) correlation. Thin horizontal lines stand for the 3 sigma threshold on estimates as calculated from the modeled statistical fluctuations. The bottom graph f) shows: 1) the standard deviation of the filtered data buffers (thin line); 2) the time periods corresponding to data buffers with bad statistical properties (gray shadows); 3) the vetoed periods of operation due to too frequent bad data buffers (dashed areas).

The discrimination between good and bad data buffers by requiring strict compliance of the noise with a parametric model is a cornerstone feature of the AURIGA data analysis. Sample pictures of good and bad buffers are reported elsewhere in these proceedings[11]. Fig.3 shows the result of this procedure for four hours of AURIGA operation. When either a filtered data or a whitened data buffer is bad, the corresponding time period is not used to estimate the noise parameters of the detector and the W.K. filter is "freezed" to the previous condition. The bottom graph of Fig. 3 shows bad buffers times as gray areas.

Two main situations may then arise: either there is a dominant contribution of the modeled quasi-stationary noise with short time periods showing unmodeled excess noise and/or signals, or the data are dominated by unmodeled excess noise. In the first case, the bad buffers are rare enough so that the analysis is able to reliably estimate the noise parameters and to adapt the W.K. filter to any slow non-stationary noise behaviour by using good buffers only. The estimate of the noise parameters is therefore performed in a reliable way by using only the periods when the modeled noise is dominating and disregarding the bad buffers, which instead contain some "signals".

DAY 182, THURSDAY Jul 1, 1999

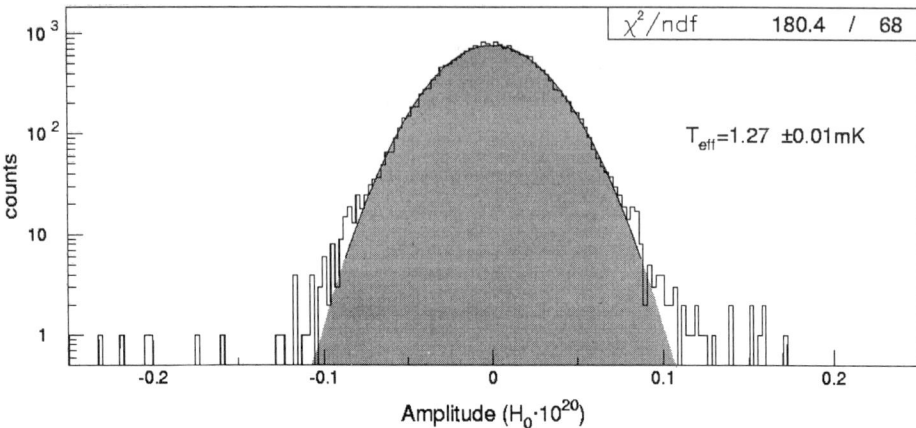

FIGURE 4. Amplitude histogram of filtered data during one day of satisfactory operation of the AURIGA detector. The data has been subsampled one per second to get almost independent samples. The small deviations from gaussian statistics at high amplitudes are due to the presence of signals ovrimposed on the noise.

This is therefore a satisfactory condition for detector operation, corresponding to the time periods in Fig. 3 where the bottom graph is not dashed. Fig.4 shows a sample amplitude histogram of the filtered data during the satisfactory operation of the detector during the same day of Fig. 3; the statistics is gaussian with small excess counts in the tails due to the "signals" present in the rare bad buffers.

The other main operating condition, i.e. that the data are dominated by unmodeled excess noise, occurs when at least half of the buffers are found to be bad within a fixed time window, ten buffers long in our case. In this condition, the W.K. filter is badly matched to the noise –in fact it could be that WK linear filter theory is not applicable. The analyzed data therefore lack of self-consistency and the output data is vetoed, as shown in the bottom graph of Fig.3 with dashed areas. Under this condition the detector is not necessarily blind and with different noise models and/or parameters one could recover some of the vetoed observating time.

GOODNESS OF THE FIT TESTS (AND COMPLIANCE OF CANDIDATE SIGNALS WITH THE EXPECTED SHAPE)

Once that the W.K. filtered output is validated as described above, the compliance of the shape of each candidate signal to the one that the W.K. filter was looking for is to be discussed. By definition, during satisfactory AURIGA detector operation the noise statistic is purely gaussian, so the W.K. filter is a maximum likelihood fit[13] and the goodness of the fit must be checked by means of a χ^2 test. In all the operating resonant bar detectors, the filter is matched to δ-like g.w. and the fitting parameters are the wave amplitude and arrival time. In particular, the best estimate of event amplitude is given by the local maximum value of the interpolated W.K. filter output and its corresponding time coordinate is the best estimate of the arrival time.

In the AURIGA filtered data, a δ-like g.w. would show up as a beating signal with a carrier frequency centered between the bar-transducer modes sampled at 4.9 kHz and with a rise and fall time given by the Wiener time, as is shown elsewhere in these proceedings[11]. The sampled signal is reconstructed in the continuum to search for the local maximum within 3 Wiener time by a max-hold algorithm. Then the mean square differencies are computed between the sampled data and the template function evaluated at each sample, and this number is used as a test statistic. In fact it is just the standard χ^2 test. The details of its implementation in the AURIGA data analysis is described in ref. 8. Such a test provides a mean to discriminate between candidate g.w. bursts exciting the bar and other kinds of spurious excitations entering the detector at different ports and/or not δ-like shaped. Here let us state its relevance from the data validation point of view.

As the WK filter theory depends on the compliace of both the noise and the signal with the models, it is clear why it is so important to have checked the first to make accurate statements on the second. For a candidate event in a satisfactory AURIGA operation, passing the χ^2 test is a necessary and sufficient condition for assessing the reliability of the estimates of the event parameters. If the test fails we reject these estimates as biased and state there was no event with the expected shape. Of course, an signal whatever was there indeed, and we can infer from it a template to build a new filter, in order to find out if similar events has been or will be present in the detector output. An efficient way to perform this task is to project the signal on a set of noise autocorrelation matrix eigenfunctions and store just the coefficients of the chief terms of the expansion[12].

An overall consistency check of the detector operation can be realized by applying the χ^2 test to mechanical calibration signals applied to the bar, which are anyway a natural step to complete the calibration procedure. As for the AURIGA detector the work is in progress and up to now we tested the system in this definitive way only from the acquired data on, that is by providing software calibration signals overimposed to the real raw data stream after the ADC.

Fig. 5 shows the histogram of the calculated χ_e^2 for the events exchanged by AURIGA under the IGEC Collaboration from Sept. 1997 to March 1998. AURIGA exchanged all found events with Signal to Noise Ratio (SNR) in amplitude larger than 5 and with a 141 degrees of freedom calculated χ_e^2 smaller than 1.5, corresponding to a confidence level of about $1 - 1 \times 10^{-4}$. The effective distribution of the calculated χ_e^2 of the events was found to be in agreement with the χ_r^2 distribution at least for the low SNR events which are the great majority of the exchanged ones. Also in Fig.5 most of the events fall inside the expected χ_r^2 distribution.

We can point out another useful application of goodness-of-the-fit tests[8] for the currently operating array of detectors. Since all the detectors are almost parallel and, apart from NIOBE, their operating frequencies fall within a 40 Hz bandwidth around 900 Hz, the amplitudes of g.w. signals detected in a N-fold coincidence must be consistent. If each detector implements an analysis capable of discriminating the

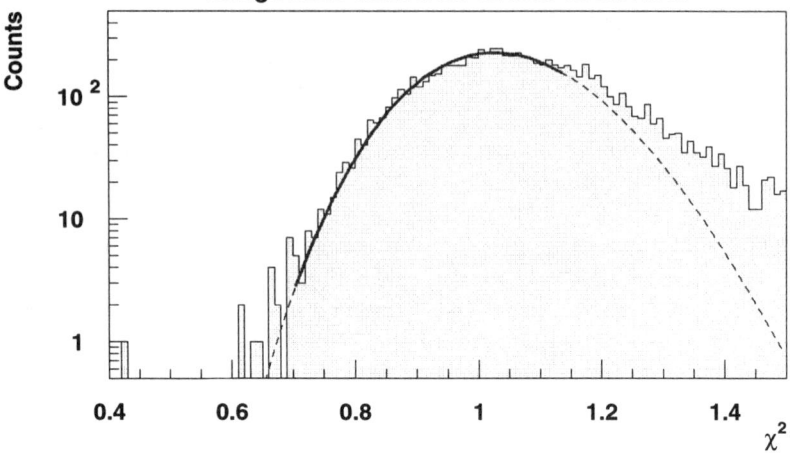

FIGURE 5. Histogram of the calculated χ_e^2 of the candidate events of the AURIGA detector under the IGEC Collaboration from Sept. 1997 to March 1998. These 15854 events have amplitude SNR>5 and are below a χ_e^2 threshold which corresponds to a false dismissal of about 1×10^{-4}. The χ_r^2 distribution is also shown as a continuous line.

satisfactory periods of operation as AURIGA does, then the amplitude estimate A_i of the i-th detector follows a gaussian distribution of variance σ_i^2 and

$$A = \frac{\sum_{i=1}^{N} \frac{A_i}{\sigma_i^2}}{\sum_{i=1}^{N} \frac{1}{\sigma_i^2}} \qquad \chi_A^2 = \sum_{i=1}^{N} \frac{(A_i - A)^2}{\sigma_i^2} \qquad (2)$$

are respectively the optimal amplitude estimate for the g.w. and the corresponding value of the experimental χ^2 with N-1 degrees of freedom.

As an exercise to probe the efficiency of this χ_A^2 test we randomly grouped the AURIGA events of Fig.5 in random triplets and quintuplets. The result is shown in Tab. 1. The relatively low fractions of rejected random coincidences are likely a lower limit of the method, because AURIGA showed a stable noise performance in that time period and therefore most of its exchanged events are close to SNR=5 and do have the same amplitude within the proper σ. The efficiency of rejection should increase significantly for higher SNR events and/or in the case that the detectors are not setting the same amplitude thresholds. Especially in dealing with high SNR events, care must be taken to account for inaccuracies of the detector calibrations.

TABLE 1. Implementation of χ_A^2 on randomly chosen triplets and quintuplets from AURIGA candidates events.

Confidence Level	Random triplets Fraction rejected	Random quintuplets Fraction rejected
0.9	0.23	0.30
0.99	0.11	0.17

This goodness-of-the-fit method can be generalized to the case of detectors with different antenna patterns at the cost of solving also for the two parameters describing the source location, unless this is known by other means. It has recently been proposed also a different approach aimed at testing amplitude consistency in 2-fold coincidences between parallel detectors[14]. The approach consists in checking how the value of the logarithm of the ratio of the amplitudes of the events in coincidence compares with the distribution of the values calculated for spurious coincidences, as those generated by time shifting the response of one detector with respect to the other.

FINAL REMARKS

We have shown how a satisfactory level of data validation can be accomplished for an optimal WK filter operating on a real detector. The key point is the ability to check the compliance of the filter with the noise of the detector and use this test to discriminate between satisfactory and unsatisfactory time periods of detector operation. In the satisfactory periods then the filter performance is validated and goodness-of-the-fit tests can be performed to check also the compliance of the observed signals with the signal shape to which the filter is matched, therefore assessing the reliability of the estimates of the signal paramenters.

The confidence of signal detection would be further improved if some peculiar properties of the incoming gravitational waves can also be detected. The capability of measuring arrival times with submillisecond resolution has been demonstrated on a resonant bar prototype[7] and would provide, once implemented on improved versions of the operating detectors with a larger post-filtering bandwidth, to measure the wave propagation speed and the source location[15].

ACKNOWLEDGMENTS

We wish to acknoledge the precious work of the other members of the AURIGA collaboration who helped us in setting up the detectors. This work has been supported in part by a grant from M.U.R.S.T.-COFIN'97.

REFERENCES

1. E. Mauceli et al., Phys. Rev. D 54, 1264 (1996)
2. G.A. Prodi et al., "Initial operation of the gravitational wave detector AURIGA" in Gravitational Waves, edited by E. Coccia et al., Proceedings of the second Edoardo Amaldi Conference, CERN, Geneva 1997, Singapore: World Scientific, 1998, pp. 148-158.
3. P. Astone et al., Phys. Rev. D 47, 362-375 (1993).
4. P. Astone et al., Astroparticle Phys. 7, 231-243 (1997).
5. D.G. Blair et al., Phys. Rev. Lett. 74, 1908-1911 (1995).
6. IGEC agreement of July 4th, 1997: http://igec.lnl.infn.it/igec
7. V. Crivelli Visconti et al., Phys. Rev. D 57, 1 (1998).
8. L.Baggio et al., submitted to Phys. Rev. D.

9. A. Ortolan et al., "Data analysis for the resonant Gravitational wave detector AURIGA" in Gravitational Waves, edited by E. Coccia et al., Proceedings of the second Edoardo Amaldi Conference CERN Geneva 1997, Singapore: World Scientific, 1998, pp.204-215.
10. J.P. Zendri et al., "Status Report of the Gravitational Wave Detector AURIGA" elsewhere in these Proceedings
11. L.Baggio et al., "Noise and Signal Reconstruction and Characterization in the AURIGA Detector" elsewhere in these Proceedings
12. S. Vitale et al., Phys. Rev. D 50, 4737 (1994).
13. See for instance: C. W. Helstrom, "Statistical Theory of Signal Detection", Oxford: Pergamon, 1968.
14. I.S. Heng et al., submitted to Gen. Rel. Grav.
15. M.Cerdonio et al., "Cryogenic detectors of gravitational waves: current operations and prospects" in Gravitiation and Relativity, edited by N. Dadhic et al., Proceedings of the GR-15 Conference, Pune, India 1997, Pune: IUCAA, 1998, pp.211-230

SNEWS: The SuperNova Early Warning System

Kate Scholberg*

*Boston University Department of Physics, 590 Commonwealth Ave., Boston, MA 02215

Abstract. World-wide, several detectors currently running or nearing completion are sensitive to a prompt core collapse supernova neutrino signal in the Galaxy. The SNEWS system will be able to provide a robust early warning of a supernova's occurrence to the astronomical community using a coincidence of neutrino signals around the world. This talk describes the nature of the neutrino signal, detection techniques and the motivation for a coincidence alert. It describes the implementation of SNEWS, its current status, and its future, which can include gravitational wave detectors.

THE EXPECTED NEUTRINO SIGNAL

When the core of a massive star at the end of its life collapses, nearly all of the total gravitational binding energy of a neutron star is emitted in the form of neutrinos, some $E_b \sim 3 \times 10^{53}$ ergs. Less than 1% of this energy is expected to be released in the form of kinetic energy and optically visible radiation. The remainder is radiated in neutrinos, of which approximately 1% will be electron neutrinos from an initial "neutronization" burst and the remaining 99% will be neutrinos from the later cooling reactions, equally distributed among flavors. Average neutrino energies are expected to be about 12 MeV for electron neutrinos, 15 MeV for electron antineutrinos, and 18 MeV for all other flavors. The neutrinos are emitted over a total timescale of tens of seconds, with about half emitted during the first 1-2 seconds, and with the spectrum eventually softening as the proto-neutron star cools. Reference [1] summarizes the expected neutrino signal. The basic features of neutrino emission models were well confirmed in 1987A with the observation of neutrinos from SN1987A. We await the next Galactic supernova to learn more.

NEUTRINO DETECTORS

There are several classes of detectors capable of detecting a burst of neutrinos from a gravitational collapse in our Galaxy. Table 1 gives a brief overview; more details can be found via reference [2]. Table 2 lists some specific supernova neutrino detectors and their capabilities.

TABLE 1. Supernova neutrino detector types.

Detector type	Material	Energy	Time	Point	Flavor
scintillator	C,H	y	y	n	$\bar{\nu}_e$
water Cherenkov	H_2O	y	y	y	$\bar{\nu}_e$
heavy water	D_2O	NC: n	y	n	all
		CC: y	y	y	$\nu_e, \bar{\nu}_e$
long string water Cherenkov	H_2O	n	y	n	$\bar{\nu}_e$
liquid argon	Ar	y	y	y	ν_e
high Z/neutron	NaCl, Pb, Fe	n	y	n	all
radio-chemical	^{37}Cl, ^{127}I, ^{71}Ga	n	n	n	ν_e

EARLY SUPERNOVA OBSERVATION

The neutrino burst produced by the core collapse emerges promptly from the stellar envelope. However, the the shock wave produced by the collapse takes some time to travel outwards from the core to the photosphere of the star. The time of first shock breakout of a supernova is highly dependent on the nature of the stellar envelope, and can range from minutes for bare-core stars to hours for red giants. For SN1987A, first light was observed about 2.5 hours after the neutrino burst; the first observable photons probably occurred about one hour earlier than that.

The observation of very early light from a supernova just after shock breakout is astrophysically very interesting [4], and rare for extragalactic supernovae. The environment immediately around the progenitor star is probed by the initial stages of the supernova. For example, any effects of a close binary companion upon the blast would occur very soon. In addition, shock breakout may also be accompanied by a UV and soft x-ray flash. The tail of such a flash was observed by the EUVE satellite for SN1897A. And of course, an observation of very early supernova light

TABLE 2. Specific supernova neutrino detectors.

Detector	Type	Mass (kton)	Location	# of events @8.5 kpc	Status
Super-K	H_2O Ch.	32	Japan	5000	online
MACRO	scint.	0.6	Italy	150	online
SNO	H_2O,	1.4	Canada	300	running
	D_2O	1		450	
LVD	scint.	0.7	Italy	170	online
AMANDA	long string	Meff $\sim$0.1/pmt	Antarctica		running
Baksan	scint.	0.33	Russia	50	running
Borexino	scint.	0.3	Italy	100	2001
KamLAND	scint.	1	Japan	300	2001
OMNIS (Pb/Fe)	high Z	5	USA	2000	2000+
LAND (Pb)	high Z		Canada		2000+
Icanoe	liquid argon	9	Italy		2000+

could also yield entirely unexpected effects.

It is possible that a core collapse event will not yield an optically bright supernova, either because the explosion "fizzles", or because the supernova is in an optically obscured region of the sky. In the latter case there may still be an observable event in some wavelengths, or in gravitational radiation.

COINCIDENCE OF NEUTRINO SIGNALS

There are several benefits from a system which coordinates neutrino signals from two or more different detectors. All detectors are subject to false alarms due to bursts of events due to detector pathologies or other non-Poissonian phenomena (for example, flashing phototubes or other sources of spurious light, electronic noise, correlated radioactivity events due to muon spallation of nuclei, etc.). Therefore, if an individual experiment is to issue an alarm, a human operator must first check the event burst to confirm its supernova-like nature, which can take significant time even when a fast-response human alert system is set up. Requiring a coincidence between independent detectors will add great confidence to the detection of a supernova neutrino burst, to the extent that a completely automated alert may be possible. The automation could save enough time that important early observations would not be lost.

THE SNEWS SYSTEM: IMPLEMENTATION OF A COINCIDENCE MONITOR

Software for a prototype international supernova watch coincidence system has been designed by Alec Habig and Kate Scholberg. It is written in standard C and uses a standard UDP protocol client/server setup to make direct network connections using sockets. Dedicated phone lines could be used to increase reliability if it proves necessary. Figure 1 shows the setup.

A central machine runs a "server" program, which sits and waits for input from the outside. The individual experiments participating in the project run "client" programs. Whenever an experiment detects a candidate burst, the client program makes a connection to the server machine and sends it an alarm datagram via direct socket connection. The alarm message contains information about which experiment observed the burst, along with the time stamp information. The datagram will be expanded in the future to include information about the significance and size of the burst.

When the server receives an alarm message from any experiment, it places the alarm in a queue sorted by UT time, and searches through all alarm messages in the queue for a coincidence within a given time window (currently 10 seconds). If there are two or more different experiments in coincidence, it sends out an alarm. A test coincidence server has been set up at the Super-K site in Mozumi, Japan.

Currently, MACRO, Super-K and LVD are connected. Privacy is maintained, and security precautions are taken. Additional servers can be set up at other sites.

WHAT DO ASTRONOMERS WANT?

What astronomers want from an early supernova alert can be summarized by the "three P's": "prompt", "pointing" and "positive". This section describes how SNEWS can address these "P's".

A "Prompt"

The alert must be as prompt as possible to catch the early stages of shock breakout. All detectors currently in the coincidence can provide an alert datagram within 30 minutes (worst case, and to be improved) of the time of the first event in the detector, and in most cases within only a few minutes. Delays are usually due to buffering at the detectors. The coincidence itself and resultant alarm message take only the time needed for a network connection. It will be entirely feasible to have a coincident alarm message produced within about 15 minutes of the neutrino signal.

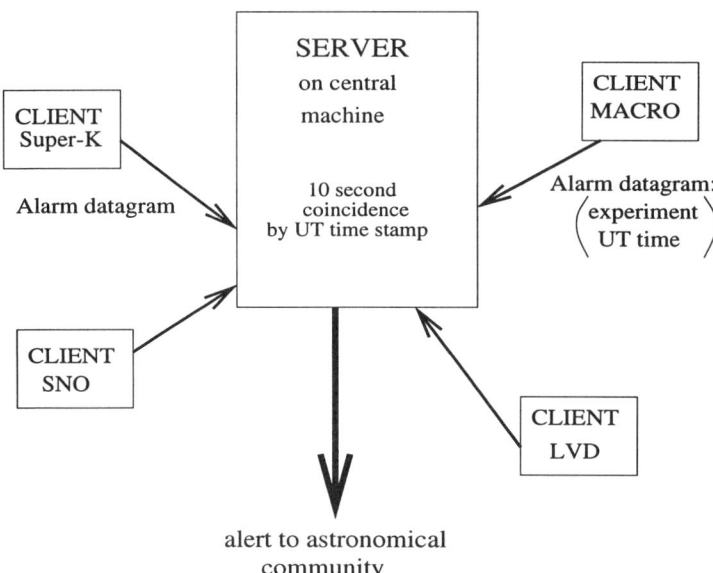

FIGURE 1. SNEWS setup.

B "Pointing"

Clearly, the more accurately we can point to a core collapse event using neutrino information, the more likely it will be that early light turn on will be observed by astronomers. Even for the case when no directional information is available (e.g. for a single scintillator detector online) it is still useful for astronomers to know that a gravitational collapse event has occurred. However any pointing information at all is extremely valuable. The question of pointing to the supernova using the neutrino data has been examined in detail in reference [3].

- **Asymmetric reactions:** Water Cherenkov detectors can exploit the neutrino-electron elastic scattering reaction, to point back to the supernova source. For a collapse at the center of the Galaxy, a few hundreds of elastic scattering events are expected in Super-Kamiokande and tens are expected in SNO. The recoil electrons follow the neutrino direction with an opening angle of about 25°. One can make a rough, optimistic estimate of the pointing resolution from $\delta\theta \sim \frac{25°}{\sqrt{n}}$, where n is the number of observed events; for 200 elastic scattering events, $\delta\theta \sim 2°$. However, the problem is really that of finding the center of a peak on top of background, and more realistic estimates of the resolution yield somewhat worse results. Reference [3] estimates a correction factor of 2-3, giving 5 degree pointing for Super-K and 20 degree pointing for SNO.

- **Triangulation:** In principle source direction information can be deduced from the timing of neutrino events at the different detectors. Since flight time across the Earth is of order tens of ms, for successful triangulation the time of the neutrino pulses at the individual detectors must be tagged to milliseconds or less. Since the neutrinos in the pulse are emitted over tens of seconds, the individual detectors must perform a pulse registration with limited sampling statistics. Reference [3] has studied the statistical problem in detail, concluding that with the current generation of detectors (Super-K and SNO), concluding that triangulation is not promising even in the best case. There are additional practical difficulties: a prompt triangulation requires immediate and complete exchange of event-by-event information, which is difficult in practice. However, for a very close supernova, or if the neutrino pulse comprises unexpectedly sharp features, triangulation may still be feasible. Any information is better than none and the triangulation may at least provide a cross-check of the elastic scattering pointing.

C "Positive"

SNEWS must not disseminate any false supernova alarms to the astronomical community. One cannot realistically decrease the false alarm rate to zero, since individual experiments will usually have a residual rate of false alarms from Poissonian

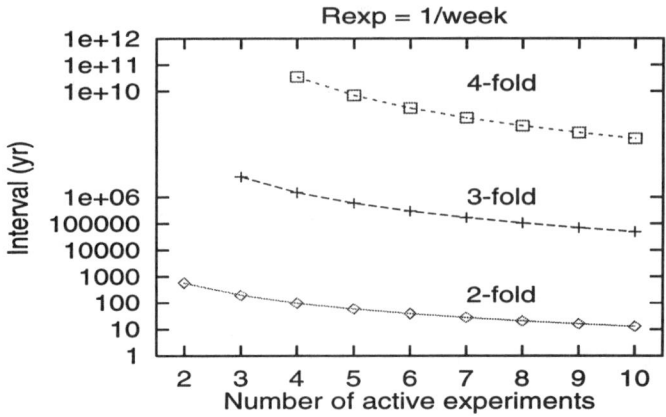

FIGURE 2. Average time between accidental coincidence alarms between experiments, as a function of number of experiments online, for a 10 second coincidence time window, for 2-fold, 3-fold and 4-fold coincidences. The assumed individual experiment alarm rate is $R = 1$/week.

and non-Poissonian sources; there will then be accidental coincidences between signals from the individual detectors. One must weigh the increased sensitivity from lowering of the alarm thresholds (both for the individual experiments and for the coincidence) against potential waste of resources (and loss of credibility) from issuance of false alerts.

We have chosen the nominal acceptable average false alarm rate to be one per century. Assuming equal, constant, uncorrelated alarm rates for each experiment, and a 10 second coincidence window, one can calculate the average interval between accidental alarms for an n-fold coincidence of N experiments. Figure 2 shows the result for an individual experiment background alarm rate of 1 per week: this rate is acceptable only if fewer than 4 experiments are online, or if a 3-fold coincidence if required; otherwise a lower individual experiment rate is required. We are also investigating any possibility of non-Poissonian alarms correlated between experiments.

THE ASTRONOMICAL ALERT

The astronomical alert from a SNEWS coincidence will be sent out to a mailing list of interested parties. In an ideal case, the coincidence network provides the astronomical community with an event time and an error box on the sky at which interested observers could point their instruments. In a realistic case, the error box, which is dependent on the location of the supernova and experiments which are online, may be very large. However, members of the mailing list with wide-angle viewing capability (satellites, small telescopes and amateurs) should be able to pinpoint an optical event quickly.

AN ADDITIONAL ROLE FOR SNEWS?

So far, SNEWS has been intended to provide an early warning for astronomers. However, it has another potential role. If a core collapse supernova happens in the Galaxy, it will be an unprecedented opportunity for science, and all possible data – neutrinos, electromagnetic, gravitational waves, perhaps other kinds – would be extremely valuable. But many detectors which are capable of providing useful information are not necessarily capable of triggering themselves on a supernova burst and may not be continuously archiving information. They may be noisy and/or may not know what kind of signal to look for from a supernova. Some examples of detectors in this category would be: some of the long string detectors (ANTARES, Baikal), gravitational wave detectors (if not all data is archived), and surface neutrino-sensitive detectors with a high rate of cosmic ray background. The SNEWS neutrino coincidence will be a high confidence indication that a supernova has occurred. Noisy SN detectors could therefore arrange to use the SNEWS coincidence as an input – they could set up a buffering system to record data (for hours or days, depending on resources available) that would routinely be overwritten, but which could be saved to permanent storage in the case of a SNEWS coincidence. This approach would greatly enrich the world's supernova data sample.

CURRENT STATUS AND FUTURE

Currently, a test coincidence server is running at the Super-K site in Mozumi. Three experiments are online (Super-K, MACRO and LVD), sending alarm datagrams in test mode. SNO and AMANDA are expected to join within about 6 months. There is no automated alert to astronomers yet; we expect the automated alert to be activated after a test period.

ACKNOWLEDGMENTS

The author wishes to thank all the members of the SNEWS inter-experiment working group (in particular Alec Habig) and John Beacom.

REFERENCES

1. Burrows A. *et al.*, *Phys. Rev.* **D45**, 3362 (1992).
2. The SNEWS web page, http:/hep.bu.edu/~snnet/.
3. Beacom J. F. and P. Vogel, *Phys. Rev.* **D60** 033007 (1999). astro-ph/9811350
4. J. Bahcall (P.I.), "Observing the next nearby supernova", HST proposal 8404.

Removing Instrumental Artifacts: Suspension Violin Modes

Soma Mukherjee[1],[2] and Lee Samuel Finn[1],[2],[3]

(1) *Center for Gravitational Physics and Geometry, The Pennsylvania State University;* (2) *Department of Physics, The Pennsylvania State University;* (3) *Department of Astronomy and Astrophysics, The Pennsylvania State University.*

Abstract. We describe the design of a Kalman filter that identifies suspension violin modes in an interferometric gravitational wave detectors data channel. We demonstrate the filter's effectiveness by applying it to data taken on the LIGO 40M prototype.

I MOTIVATION

The wire suspensions of the interferometer test masses are a conduit for environmental noise to enter the gravitational wave data channel; additionally, they are a source of noise themselves. Both the pendulum mode (whose frequency is out of the interferometers band) and the suspension wire violin modes (whose fundamental mode frequencies are in-band) are energized by their contact with the thermal bath. Owing to their weak damping this energy is strongly concentrated about the mode resonant frequencies. At the fundamental violin mode frequencies the thermal noise dominates the other noise sources by approximately 50 dB. In addition to this thermal noise component, non-thermal excitations of the suspension wires (e.g., sudden creep events) can lead to excitations in the interferometer output.

These narrow band features are instrumental artifacts, not gravitational waves: along with other artifacts they should be removed from the data before it is studied for the presence of other signals. As instrumental artifacts, however, they carry important information about the instrument's state. Thermal and technical noise that disturbs the suspension excites these modes and move the mirrors, leading to an artifact in the gravitational wave channel. Gravitational waves give rise to a signal by changing the distance between the mirrors, but do not move or otherwise excite the suspension modes. Correspondingly, if we can determine the mode state as a function of time we have a way of eliminating a broad class of technical noise sources that might otherwise masquerade as gravitational wave bursts.

How can we identify the state of the suspension violin modes, given only the gravitational wave channel? This is a classic problem in data analysis, generally addressed by a Kalman filter. We have developed such a filter for use in LIGO and shown, using data taken in November 1994 at the LIGO 40M prototype, that it enables us to follow the state of the violin modes independently of the other noise sources — both technical and fundamental — that contribute to the gravitational wave channel.

II THE KALMAN FILTER

The Kalman Filter [1] is a mechanism for predicting the multi-dimensional *state* of a dynamical system from a multi-dimensional *observable*. The system is assumed to evolve linearly and the observable is assumed to be linearly related to the state. denoting the system state $\mathbf{x}$ we have (for discrete time series):

$$\mathbf{x}[k] = \mathbf{A} \cdot \mathbf{x}[k-1] + \mathbf{w}[k-1]. \tag{1}$$

We assume that the system is driven by a stochastic force, referred to as the *process noise* and denoted $\mathbf{w}$. The state dynamics determine the linear operator $\mathbf{A}$.

The state contributes to the observation $\mathbf{y}$, which also includes a stochastic, additive *measurement noise* $\mathbf{v}$:

$$\mathbf{y}[k] = \mathbf{C} \cdot \mathbf{x}[k] + \mathbf{v}[k]. \tag{2}$$

In the classic Kalman filter the process and measurement noises are assumed to be Normal processes with known co-variances $\mathbf{W}$ and $\mathbf{V}$.[1]

Now suppose that we have an estimate $\hat{\mathbf{x}}[k-1]$ of the state, and also an estimate of the error co-variance $\mathbf{P}[k-1]$ in the estimate, at sample $k-1$. The Kalman filter uses these estimates, the observation $\mathbf{y}[k]$ at sample k, and $\mathbf{A}, \mathbf{C}, \mathbf{W}$ and $\mathbf{V}$ to form an estimate of the state and its error co-variance at sample k:

$$\hat{\mathbf{x}}[k] := \mathbf{K}[k] \cdot (\mathbf{y}[k] - \hat{\mathbf{y}}[k]) \tag{3}$$

$$\hat{\mathbf{P}}[k] := (\mathbf{I} - \mathbf{K}[k] \cdot \mathbf{C}) \cdot \tilde{\mathbf{P}}[k] \cdot (\mathbf{I} - \mathbf{K}[k] \cdot \mathbf{C})^T \tag{4}$$

where

$$\hat{\mathbf{y}}[k] := \mathbf{C} \cdot \mathbf{A} \cdot \hat{\mathbf{x}}[k-1] \tag{5}$$

$$\mathbf{K}[k] := \tilde{\mathbf{P}}[k] \cdot \mathbf{C}^T / \left(\mathbf{V} + \mathbf{C} \cdot \tilde{\mathbf{P}}[k] \mathbf{C}^T \right) \tag{6}$$

$$\tilde{\mathbf{P}}[k] := \mathbf{A} \cdot \hat{\mathbf{P}}[k-1] \cdot \mathbf{A}^T + \mathbf{W}. \tag{7}$$

The estimated system state $\mathbf{x}[k]$ (*e.g.*, the generalized coordinate and conjugate momentum of the violin mode normal mode of the wire) is thus completely determined by the observation $\mathbf{y}[k]$, the estimated state at sample $k-1$, the wire dynamics,

---

[1] Even when $\mathbf{w}$ and $\mathbf{v}$ are not Normal the Kalman filter estimates of the state $\mathbf{x}[k]$ can be shown to have the smallest mean-square error of all linear state estimators that depend only on the co-variances.

and the statistical properties of the process and measurement noise. The error in the estimate $\mathbf{x}[k]$ falls with k, converging upon a limiting error covariance that is fully determined by $\mathbf{A}$, $\mathbf{C}$, $\mathbf{W}$ and $\mathbf{V}$; correspondingly, we can choose any initial estimate of $\mathbf{x}$ and $\mathbf{P}$ and the filter will, after several iterations, adjust the state estimate and error accordingly.

From the state estimate at each sample we can, through the measurement equation, estimate the contribution of the system to the actual observation. This estimated contribution can be subtractively removed from the actual observation, leaving a residual that is as free from the contaminating influence of the process as we can make it.

III MODELING THE VIOLIN MODES

To describe a Kalman filter for our system we need a model for the wire state dynamics, the relationship between the wire state and the appearance of the mode in the detector data channel, and estimates of the process and measurement noise.

The wire motion contributes significantly to the data channel only in a narrow band about the violin mode resonant frequency; correspondingly, we can focus attention this narrow band and model the wire dynamics as a viscously damped harmonic oscillator driven by white noise[2]:

$$\ddot{\psi} = -\omega_0^2 \psi - \gamma \dot{\psi} + N(t). \tag{8}$$

We assume that the measurement is of the state variable ψ plus white measurement noise.

There are many violin modes, corresponding to the many wires that are used to suspend the interferometer mirrors: for each wire there are separate state variables and a separate equation describing the dynamics of that mode.

Using standard lock-in techniques we mix the ψ and the data channel with a local oscillator whose frequency is near that of the violin modes and band-limit the output of the lock-in to the narrow band over which our model is accurate. The in-phase and quadrature-phase components of mixed-down ψ become the state $\mathbf{x}$ used in the Kalman filter, and the in-phase and quadrature-phase components of the data channel become the observation $\mathbf{y}$. The matrices $\mathbf{A}$ and $\mathbf{C}$, describing the state dynamics and the relationship between the state $\mathbf{x}$ and observation $\mathbf{y}$ are derived from the our model in light of the lock-in, band-limiting, and discrete time sampling operations.

---

[2] Over the relevant, narrow band near the resonant peak there is no distinction between viscous and structural damping.

IV RESULTS AND DISCUSSION

A Removing the artifact

To explore the effectiveness of the Kalman filter in identifying the contribution of the violin modes to the detector output we have applied it to data taken in November 1994 at the LIGO 40M prototype detector. In this instrument the fundamental violin mode resonances are all in the (571.6, 605.425) Hz band.

The upper and lower panels of figure 1 show the power spectra of the interferometer output in a 45 Hz band between 565.0 and 610.0 Hz before and after the subtractive removal of the Kalman filter estimate of the violin mode contribution to the detector output. The filter identifies the contribution of the mode to the detector output, allowing us to suppress this artifact by 40 dB. The residual bumps positioned in the wings of the removed lines, are non-linear artifacts: the violin mode amplitudes are so large that they modulate the detector transfer function, up-converting other detector noise frequency-modulating the violin mode signal itself.

B Statistics of the artifact and residual

Some simple exploratory statistics show the value of identifying and removing the known instrumental artifacts from the data stream.

Figure 2 shows a histogram of the sample amplitude relative to the RMS sample amplitude for the data channel before (top) and after (middle) removal of the Kalman filter estimate of the violin mode contribution, and (bottom) for the estimated violin mode contribution itself. In an ideal world the measurement noise and the process noise are Normal; correspondingly, each of these distributions should be Rayleigh and appear, in these figures, as straight lines. Departures from a straight line thus imply non-Gaussian noise statistics.

Comparing the three panels in figure 2 shows that the violin mode artifact contributes significantly to the non-Gaussian component of the noise in the detector data channel. Since gravitational waves do not excite the violin modes, this excess noise component is strictly technical. Generalizing to the full-scale detector, removing artifacts like these, together with their associated excess noise, from the data stream before analysis thus strengthens our ability to make significant statements regarding the detection of gravitational waves in the residual.

C Violin modes vs. gravitational waves

The Kalman filter estimates the violin mode contribution to the gravitational wave channel. If that estimate is influenced by the presence of a gravitational wave signal then removing the estimated contribution may distort evidence of the wave in the output.

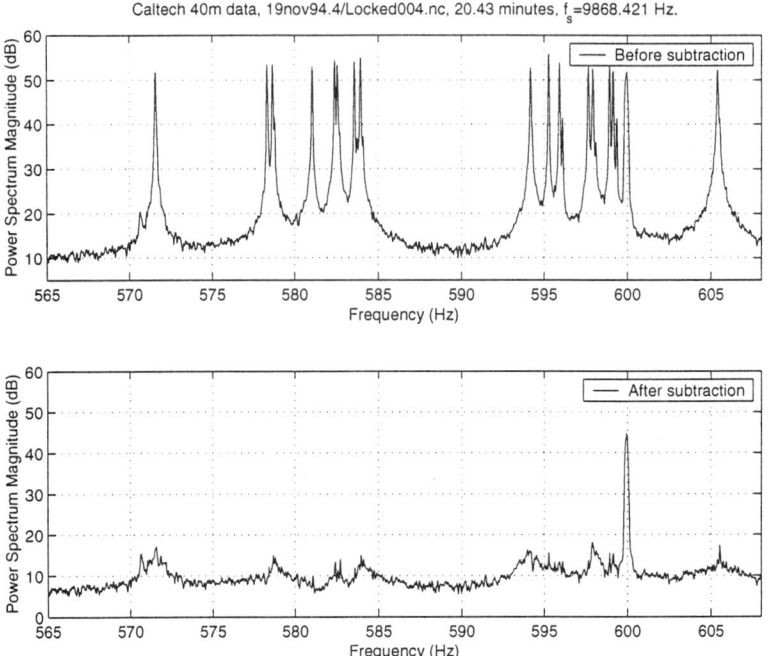

FIGURE 1. Power spectrum of the 40 meter data showing the violin modes between 571 and 605 Hz before and after subtraction of the Kalman estimates for all the modes. The remaining line feature is the 9th harmonic of the 60 Hz power main. Features like these are dealt with in other ways [2–5].

The Kalman filter identifies the violin mode contribution through its dynamics. Since the evolution with time of expected gravitational wave signals is different than the dynamics of the mode contribution to the detector output, we expect that the Kalman estimates of the violin mode contribution will not be influenced by the presence of a gravitational wave signal.

It is useful to consider two different kinds of sources: burst sources, such as inspiraling neutron star binary systems, and periodic sources, such as a pulsar at a frequency near to but not identical with the violin mode frequency.

Nearly all the signal to noise ratio signal-to-noise from an inspiraling binary is contributed when the signal is in the band from about 70 Hz to 250 Hz [6,7]: what happens in the band near the violin mode is is inconsequential. Similarly, the filter output is entirely unaffected by what happens in the band where the S/N is deposited. We have verified this by forming estimates of the violin mode state from the LIGO 40M prototype data, and from the same data set but with an added, simulated gravitational wave signal corresponding to a coalescing neutron star binary. Even for a very strong signal there is no difference in the predicted

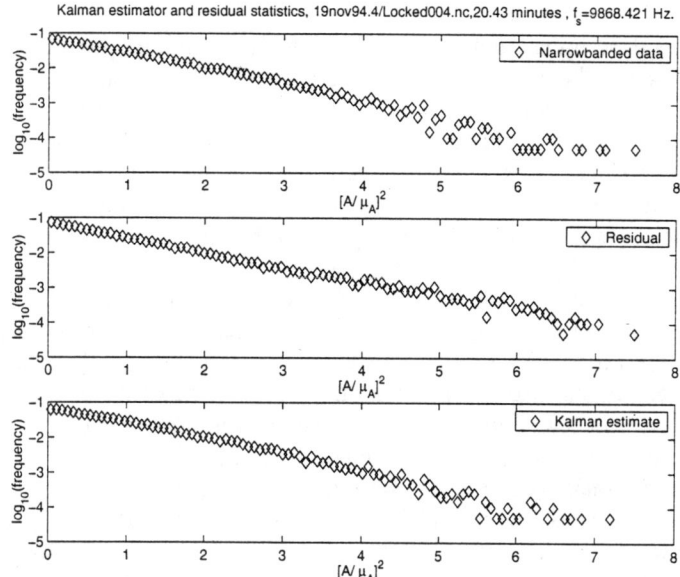

FIGURE 2. Histograms of the square of Kalman estimated amplitude divided by mean amplitude for (top) data in the narrow-band (570-599.5 Hz) having the fundamental violin modes, (middle) residual after subtraction of the Kalman estimates of the modes from the data and (bottom) the Kalman estimated violin modes. Deviations from Gaussianity occur with much lower frequency in the residual than in the mode contribution to the observation.

mode amplitudes when the signal was in the relevant band.

In separate experiments we have looked at how well the Kalman filter rejects nearby monochromatic signals, such as might arise from a pulsar. As long as the signal is greater than a linewidth away from the violin mode, the estimated state is unaffected by the periodic signal; correspondingly, the periodic signal is unaffected the subtractive removal of the estimated contribution of the mode to the output.

D Monitoring the wire state

The Kalman filter estimates separately the state of each violin mode; correspondingly, we can monitor the state of each of these wires separately. It is convenient to represent the state in terms of its amplitude and phase (as opposed to generalized coordinate and momentum); additionally, it is convenient to express the amplitude as an instantaneous measure of the energy in the mode, expressed in terms of temperature. Excess noise will raise the effective temperature of the mode, while the projection of the coresponding mirror motion normal to the optical axis will lower the effective temperature.

V SUMMARY

Thermal and technical noise that disturbs the violin modes of the substrate suspension moves the mirrors, leading to a strong, narrow band artifact in the gravitational wave channel. Gravitational waves, on the other hand, also change the distance between the mirrors, but without moving or otherwise exciting these modes. The Kalman filter described here distinguishes between excitations due to gravitational waves and those due to thermal or other excitations of the violin modes, allowing us to eliminate a broad class of technical noise sources that might otherwise masquerade as gravitational wave bursts.

A Kalman filter uses the known dynamics of the modes to distinguish between the mode "signal" and other contributions to the measured detector output: *i.e.*, it *detects* the violin modes. This distinguishes it from other methods (*e.g.*, multitaper methods, linear notch filters [8]) which purport to characterize or remove artifacts but that in fact simply suppress contributions to the noise within a narrow band, not distinguishing violin mode from other contributions.

The computational cost of identifying and removing the violin modes using the Kalman filter described here is negligible: an interpreted Matlab [9] implementation on a low-end workstation runs at greater than 20× the detector's real-time sample rate. A compiled implementation, with attention paid to optimization, an additional speed-up of 10 or more can be expected.

We thank Albert Lazzarini for drawing our attention to Kalman Filtering and the LIGO Laboratory for its hospitality during the 1997/8 academic year and for the use of the 40 meter prototype data. SM thanks S. Mohanty for many valuable insights and LSF thanks P. Fritschel for valuable discussions. This work was supported by NSF awards PHY 98-00111 and 99-96213.

REFERENCES

1. R. Kalman, Trans. ASME, J. Basic Eng. **82D**, (1960).
2. A. Sintes and B. Schutz, Phys. Rev. D **58**, 122003 (1998).
3. L. S. Finn *et al.*, Toward gravitational wave detection, 1999, gr-qc/9911001; to appear in the proceedings of the Third Edoardo Amaldi Conference on Gravitational Waves.
4. L. S. Finn and M. F. Huq, Regressing power line artifacts in gravitational wave detector data, 1999, in preparation.
5. B. Allen, W. Hua, and A. Ottewill, Automatic cross-talk removal from multi-channel data, gr-qc/9909083.
6. L. S. Finn, Phys. Rev. D **53**, 2878 (1996).
7. L. S. Finn, Technical Report No. T970167, California Institute of Technology, LIGO Laboratory, California Institute of Technology, Pasadena CA 91125 (unpublished).
8. B. Allen, GRASP: a data analysis package for gravitational wave detection, 1998, available from <http://www.lsc-group.phys.uwm.edu/~ballen/grasp-distribution>.
9. Matlab, a technical computing environment for high-performance numeric computations in linear algebra, is a product of The MathWorks, Inc.

Search for gravitational radiation with the Allegro and Explorer detectors

P. Astone[1], M. Bassan[2], P. Bonifazi[3,1], P. Carelli[4], E. Coccia[2],
C. Cosmelli[5], V. Fafone[6], S. Frasca[5], K. Geng[7], W.O. Hamilton[7],
W.W. Johnson[7], E. Mauceli[6], M.P. McHugh[7], S. Merkowitz[6],
Y. Minenkov[2], I. Modena[2], G. Modestino[6], A. Moleti[2], A. Morse[7],
G. V. Pallottino[5], M.A. Papa[6], <u>G. Pizzella</u>[2,6], N. Solomonson[7],
R. Terenzi[2,3], M. Visco[2,3] and N. Zhu[7]

[1] *Istituto Nazionale di Fisica Nucleare (INFN), Rome, Italy*
[2] *University of Rome "Tor Vergata" and INFN, Rome, Italy*
[3] *IFSI-CNR and INFN, Frascati, Italy*
[4] *University of L'Aquila and INFN, Rome, Italy*
[5] *University of Rome "La Sapienza" and INFN, Rome, Italy*
[6] *Laboratorio Nazionale INFN, Frascati, Italy*
[7] *Louisiana State University, Baton Rouge, Louisiana, U.S.A.*

Abstract. The results of a search for short bursts of gravitational radiation coincident between the Allegro and Explorer cryogenic resonant mass detectors with strain amplitudes greater than 3×10^{-18} are reported for data taken from June until December of 1991. While no significant excess of coincident events was found, an improved upper limit to the rate of gravitational wave bursts incident on the Earth has been set.

I INTRODUCTION

We report on the search for coincident events between the cryogenic resonant gravitational radiation detectors Allegro at Louisiana State University (LSU) and Explorer at CERN (operated by the University of Rome) such as might be produced by the collapse of a massive star. The search involved data taken in 1991. The detectors and methods of data analysis used by each group to search for burst signals are described in detail elsewhere [1-3]. The search involved an exchange of lists of candidate events and an independent coincidence search by each group.

	Position (latitude, longitude, orientation)	Mass (kg)	Frequencies (Hz)	Temperature (K)	Sampling Time (s)
Allegro	30.2 N, 91.2 W, 40.4 W	2300	896.7 920.2	4.2	0.080
Explorer	46.2 N, 6.1 E, 39.3 E	2300	904.7 921.3	2.6	0.2908

TABLE 1. The detectors involved in the search. Orientation is given in degrees from North in the direction indicated.

II DATA EXCHANGED

The data analyzed in this paper began at the start of June 19, 1991 (UTC day 170) and ended on December 16, 1991 (UTC day 350). There were a total of 2035 hours of coincident operation over this time span.

The detectors (see Table 1) are essentially identical aluminum alloy cylinders with their primary quadrupole resonance near 910 Hz. Both detectors are cooled to cryogenic temperatures to reduce the thermal noise. The antennas are oriented so that their bar axes are close to parallel, and both bar axes are perpendicular to local vertical. This results in nearly identical signal reception patterns, so gravitational waves from any direction are expected to produce similar sized signals in each detector. Both detectors use resonant transducers to convert up the vibrational amplitude of the bar. The Allegro detector uses a single coil inductive transducer [4], while the Explorer detector uses a capacitive transducer [5]. Each coupled bar-transducer system has two normal modes of vibration with the resonant frequencies given in Table 1.

Each group has its own methods of data analysis to search for burst gravitational waves. Such a signal is expected from, for example, the collapse of a massive star in a supernova. The result in each case is a list of candidate events characterized by an arrival time and a signal amplitude. The arrival time is reported in UTC. The signal is modeled as having a constant Fourier spectrum over a frequency range $\delta\nu = 1/\tau_b$, where τ_b is roughly the duration of the burst. As a convention we adopt $\tau_b = 10^{-3}$ s, so the Fourier spectrum of the gravitational wave is also constant across the detection bandwidth. The gravitational wave is assumed to be incident with the optimal polarization and direction. The reported signal amplitude is then given by

$$h_c = \frac{|H(\omega_R)|}{\tau_b} \qquad (1)$$

where $H(\omega_R)$ is the Fourier component of the burst at the detector resonant frequencies. The data exchanged consisted of lists of event arrival times and event amplitudes. The number of candidate events in each data set exchanged are listed in Table 2.

Each group used a different method of optimal filtering to extract small signals from the detector noise. The Explorer data was generated by applying a Weiner-

	no. of events	time span (UTC days)
Allegro	18412	170-349
Explorer	25086	170-349

TABLE 2. The number of events above threshold from each detector for 1991.

Kolmogorov filter to data sampled at 0.2908 s. The Wiener-Kolmogorov filter was designed to minimize the mean-square error between the signal and its estimation. This filter is adaptive, updating parameters based on the calculated Explorer noise spectrum every two hours.

The Explorer data were thresholded so that only those events with amplitude greater than $h_c = 2.3 \times 10^{-18}$ were included. Since the target signal was a short duration burst, it was expected that the same signal amplitude would be registered in each of the resonant modes. Therefore, those events where the ratio of the measured signal strength in each mode was greater than 1.5 were vetoed from the Explorer data. Periods when the event rate exceeded 60 events/hour were also eliminated from the data set, as such a high event rate was considered a sign of poor detector operation. This reduced the total operational time for Explorer from 180 days to roughly 123 days. Finally, those events which could be correlated to a seismic disturbance or other housekeeping measure were eliminated.

The Allegro data were filtered in the time domain by a non-adaptive filter designed to maximize the signal to noise ratio for a burst signal. The filter operated on data sampled at 80 ms. A moving threshold of 11.5 times the stationary noise level was calculated every 6 minutes and applied to the Allegro data. This level was chosen so that there would be roughly 100 events per day above threshold. Except for times of detector maintenance, such as liquid helium refilling, no other vetos were applied.

III ANALYSIS

Definition of what constitutes a coincident excitation between the two detectors involves a number of considerations, but the result is usually at least an order of magnitude larger than the light travel time between the detectors. This is due in part to uncertainty in the clocks used for timing in 1991, and to the effect of the filtering performed to extract small signals from the noise. At the time of the data exchange, it was decided that two events should be considered coincident if an event from one detector occurred within ±1.00 s (referred to as the coincidence "window") of an event recorded by the other detector. This result was arrived at by applying different filtering methods to the same data set and examining the resulting coincidences from the two analyses of the same data [6]. A more recent analysis [7] has led to the conclusion that the coincidence window should be taken as small as possible. The limiting factor is the largest sampling time of the participating detectors. For this experiment that was 0.2908 s for the Explorer data, resulting in a coincidence window of ±0.29 s. For the sake of completeness and for consistency with previously reported results [8], we present the results for both choices of the coincidence window.

The difficulty in searching for coincident events is that for purely random data there are going to be coincidences which are not produced by gravitational waves. If the event rates in each detector are stationary, then the average number of these "accidental" coincidences are accurately estimated by

$$n_{acc} = n_1 n_2 \frac{\Delta t}{T_{obs}} \qquad (2)$$

where n_1 and n_2 are the number of events from each detector, T_{obs} is the total observing time, and Δt is the coincidence window (equal to twice the stated time since each window is defined as ± some time interval). Gravitational wave signals present in the data would cause the observed number of coincidences to exceed the number of accidentals.

However the event rates were not stationary in the two detectors over the 6 months of data taking. One then expects Eq. 2 to be only a rough approximation to the expected number of coincidences. The following section describes the analysis techniques used here (see also [9] and [10]) to address this issue.

The standard analysis technique to search for coincident events has been referred to in the literature as the "experimental probability" [6]. Instead of using Eq. 2 to estimate the number of accidental coincidences, one measures it experimentally by shifting the event times in one of the two data sets by an amount δt (called the time delay) and determining the number of coincidences $n(\delta t)$. A plot of $n(\delta t)$ vs. δt is referred to as a "time delay histogram". Repeating for N different values of the time delay, the expected number of coincidences is simply the sample mean of the $N - 1$ values of $n(\delta t)$ at delays other than $\delta t = 0$,

$$\bar{n} = \frac{1}{N-1} \Sigma_{\delta t} n(\delta t). \qquad (3)$$

Since there is no signal at delays other than $\delta t = 0$, the number of coincidences at these delays can be considered an experimental estimation of the parent distribution from which the accidental coincidences are drawn. For detectors with stationary event rates the parent distribution is Poisson with a mean given by Eq. 2. The number of coincidences at zero delay can then be compared to the experimental distribution. If enough gravitational waves are present in the data, $n(0)$ will lie outside of the distribution.

By counting the delays at which the number of accidental coincidences equals or exceeds the number at zero delay ($n_>$), one can measure experimentally the probability that the coincidences at zero delay occurred by chance:

$$p_{exp} = n_> / (N - 1) \qquad (4)$$

and hence the term "experimental probability" for this type of analysis.

data set	$n(0)$	$\bar{n}$	n_{acc}	p_{exp}
Allegro-Explorer (±0.29 s)	19	17.1	17.2	0.36
Allegro-Explorer (±1.00 s)	70	59.3	59.4	0.11

TABLE 3. The number of coincidences at zero delay $n(0)$, the measured ($\bar{n}$) estimated (n_{acc}) values for the accidental coincidences, and the experimental probability that the coincidences at zero delay were drawn from the accidentals distribution (p_{exp}).

IV RESULTS

The 1001 time delays were chosen to run from -1000 s to 1000 s in increments of 2 s. The sample mean ($\bar{n}$), and the number of coincidences at zero delay ($n(0)$) for each choice of the coincidence window are given in Table 3.

Also listed are the number of accidentals (n_{acc}) calculated from Eq. 2 and the values listed in Table 2, and the experimental probability calculated by Eq. 4.

The measured parent distribution of the accidentals for each choice of coincidence window is shown in Fig. 1, along with a Poisson distribution generated using the measured sample mean. As can be seen, even though the event rates in each detector are not constant over time, the measured accidentals distribution is well

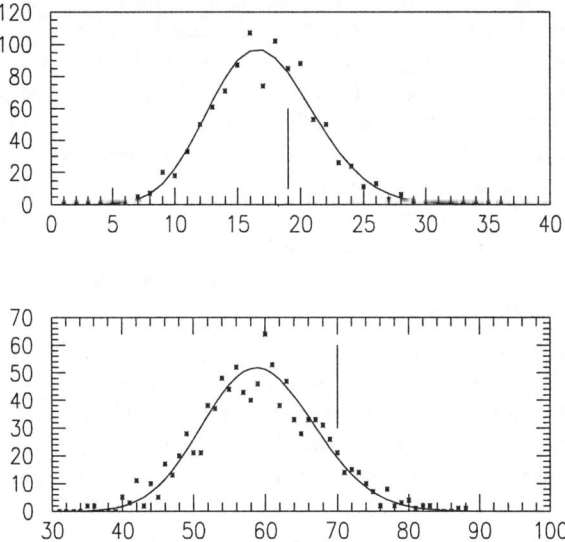

FIGURE 1. (a) The upper figure shows the accidentals distribution derived from the window ±0.29 s. (b) The lower figure shows the accidentals distribution for the window ±1.0 s. The two lines show the number of real coincidences for the two windows.

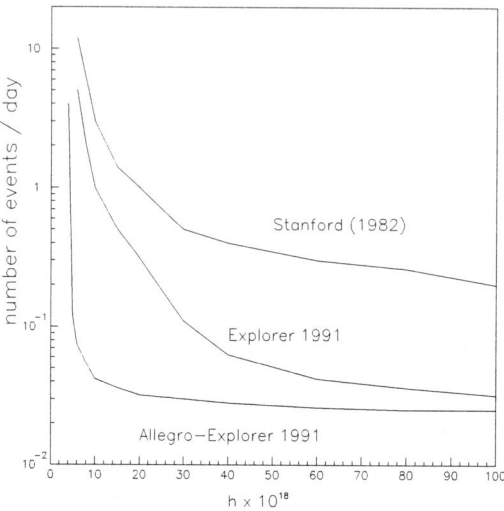

FIGURE 2. Upper limit to the rate of gravitational wave bursts incident on the Earth.

matched by the Poisson distribution. This is also supported by the good agreement between the measured number of accidentals ($\bar{n}$) and the number expected from purely Poisson statistics (n_{acc}). For a window of ±0.29s, the number of coincidences at zero delay are near the mean of the distribution shown in Fig. 1(a) with an experimental probability of 36%. We therefore infer that for this choice of coincidence window, the coincidences at zero delay are samples drawn from the accidentals distribution shown in Fig. 1(a). For a window of ±1.00 s, the number of coincidences at zero delay is slightly greater than the mean of the distribution. The corresponding experimental probability is 11%, still too large to claim a discovery. Again we infer that the slight excess of coincidences at zero delay are drawn from the accidentals distribution shown in Fig. 1(b).

Interpreting these conclusions as a null result, we used this information to calculate the upper limit to the possible flux of gravitational wave bursts incident on the Earth. The procedure is described in detail in [9]. The number of coincidences at zero delay are assumed to be drawn from one of two distributions. If these coincidences are due to noise alone, $n(0)$ is a sample from a Poisson process with mean $\bar{n}$ determined by Eq. 3. If a signal is present, $n(0)$ is a sample drawn from two independent Poisson processes, events due to detector noise and events due to coincident signals. The mean of this distribution is given by the sum of $\bar{n}$ and the mean number of coincident events due to gravitational waves (explicitly the rate of coincidences due to gravitational waves multiplied by the observation time). By setting false alarm and false dismissal levels (both were set to 0.05), the mean rate of coincidences due to gravitational waves was determined. This is not yet the desired quantity, as not all incident gravitational waves will cause coincident detection. A Montecarlo simulation was used to determine the probability that a gravitational wave of a particular amplitude would cause coincident events in the detectors. The distribution of sources was assumed to be isotropic with a random distribution of polarization. Combining these two pieces of information resulted in the upper limit, at a 95% confidence level, that is shown as a function of signal amplitude in Fig. 2 [8].

For comparison we also show the upper limit obtained with the Stanford antenna alone in 1982 [11] and with the Explorer detector alone in 1991 [1].

V CONCLUSION

No statistically significant coincident excitations were observed between the Explorer and Allegro gravitational wave detectors from June until December of 1991. From this result we have set an upper limit on the rate of gravitational wave bursts incident on the Earth that is significantly lower than has been previously observed.

REFERENCES

1. P. Astone et al. Physical Review D, 47, 362 1993.
2. P. Astone et al. Nuovo Cimento C, 17, 713 1994.
3. E. Mauceli et al. Physical Review D, 54, 1996.
4. N. Solomonson, W. O. Hamilton, and W. Johnson. Rev. Sci. Instrum., 65(1):174, 1994.
5. P. Rapagnani. Nuovo Cimento C, 5, 385, 1982.
6. P. Astone, G. Pizzella, and ROG collaboration. Coincidences in gravitational wave experiments. *Laboratori Nazionali di Frascati*, Jan 1995.
7. P. Astone et al. Astroparticle Physics, 10, 83, 1999.
8. P. Astone et al, in *Proceedings of the Tenth Italian Conference on General Relativity and Gravitational Physics*, (World Scientific Publishing Co., Singapore, 1993).
9. E. Amaldi et al. Astronomy and Astrophysics, 216, 1989.
10. J. Weber. Physical Review Letters, 25, 1970.
11. S.P. Boughn et al. *The Astrophysical Journal*, 261, L19, 1982.

Gravitational Wave Detection - the Way Forward

James Hough

Department of Physics and Astronomy, University of Glasgow, Glasgow G12 8QQ, UK

Abstract. Source predictions and detector developments described at this meeting strongly suggest that gravitational waves will be observed with an upgraded LIGO 2 detector system. This paper is a summary of some of the exciting work presented.

INTRODUCTION

For the first time the Amaldi meeting has been held in its new role as the focus for the gravitational waves community worldwide. With its plenary format it has allowed the most important developments across the field to be presented to all sections of the community.
The meeting was targeted towards two main areas of research:
- Gravitational Wave Sources and Signals
- Development of Gravitational Wave Detectors

and significant encouragement was to be derived from recent developments in both areas. On the one hand it does seem that that if detectors with an observing range of greater than 200MPc can be built, then a healthy, detectable rate of compact binary coalescences is to be expected. On the other hand it now looks possible to design and build such detectors.

SOURCES AND SIGNALS

The predictions from Sterl Phinney and Vicky Kalogera for event rates from coalescing compact binaries, both NS/NS and NS/BH, are highly encouraging with ~ 3 NS/NS events/year out to 200Mpc being expected. NS/BH and indeed BH/BH rates are much less certain, but with the increased distance from which such coalescences may be seen, there is much promise for the future of gravitational wave detection with ground based detectors.

Exciting predictions for massive BH mergers were also presented by Sterl Phinney, with AlbertoVecchio outlining the possibilities for using LISA, the proposed space-borne detector, to provide demographic information about such objects. Clearly, however, the questions as to possible low event rate raised by Omer Blaes cannot be ignored.

The r modes of rapidly rotating neutron stars are unstable to gravitational radiation reaction under astrophysically realistic conditions and look a real possibility for detection (Ben Owen) provided saturation effects can be understood. Further, the Rossi X-ray Timing Explorer observations of rapidly accreting, weakly magnetic neutron stars in low -mass X-ray binaries in our galaxy suggest that such stars are rotating with a very narrow range of frequencies around 300 Hz. Such buffering of the frequency suggests that the accretion spin up may be balanced by gravitational radiation being emitted at a very significant level (Gregory Ushomirsky). While such sources present a significant challenge on the data analysis front due to uncertainties in frequency and phase of the signals, developments are underway to allow us to seriously look forward to their detection.

PROGRESS IN DETECTOR DEVELOPMENT

Bars and Interferometers

Over the last few years there appears to have been a significant improvement in the performance and prospects for both bar and interferometric gravitational wave detectors.

Currently there are five low temperature bar detectors in operation or under development - Explorer, Nautilus, Auriga, Allegro, Niobe. In almost all cases there are plans for upgrading the sensitivity and some excellent new results from Auriga have been reported at this meeting by Jean-Paul Zendri, a detector noise level equivalent to a pulse sensitivity of about 2×10^{-19} having been demonstrated.

On the interferometer front, LIGO (USA), VIRGO (France/Italy), GEO (Germany/UK), and TAMA (Japan) are now under construction and are reasonably up to schedule. Indeed first displacement noise levels from TAMA - currently operating without power recycling - are available (Kimio Tsubono and Masaki Ando)! While considerable improvement is to be expected over the next few months, the operation of such a detector with 300m arm lengths and Fabry-Perot cavities in the arms is a real step forward for the gravitational wave field.

Developments for the Next Interferometers

In both the short and the long term most 'experimental work relevant to interferometers is aimed at achieving better sensitivity. As outlined by K. Strain, in the short term this is centred round power recycling where the power at the beam-splitter of an interferometer is enhanced by the placing of a mirror between the laser and the beam-splitter so that the whole system behaves as a resonant cavity. In the longer term - for planned upgraded LIGO 2 detectors for example - there are a number of other areas of development.

Advanced Optical Configurations

These are based on Signal Recycling (Gerhard Heinzel) or Resonant Sideband Extraction (Mal Gray, Jim Mason) where a mirror is placed at the output port of the interferometer to reflect the side-bands induced on the light by a gravitational wave signal back into the interferometer. These side-bands are effectively resonated in a cavity formed between the output mirror and elements of the interferometer, and thus signal enhancement takes place around a frequency which is set by the positioning of the mirror. These techniques have significant possibility for improving photon noise limited sensitivity at frequencies above a few hundred Hz.

Higher Power Lasers

Increasing single frequency laser power from the present level of 10W by more than an order of magnitude seems relatively straightforward either by utilising further amplifier stages fed from the current lasers (Bill Tulloch) or by using injection locked systems with the final slave laser based on unstable resonator topology (Peter Veitch). Consequent reductions in photon noise limited sensitivity are to be expected.

Seismic Isolation

Developments in Caltech (Riccardo DeSalvo) and in Perth, Western Australia, (David Blair) follow VIRGO's lead for implementing multi-pendulum isolation systems which will allow useful sensitivity at frequencies down to 10Hz or lower. Also, interest in active isolation systems to allow similar performance in a more compact way has been increasing again particularly at JILA, MIT (Joe Giaime) and Stanford (Brian Lantz). However cancellation of gravity gradient noise seems essential to get full benefit from these schemes.

Thermal Noise and Material Issues

These issues were clearly reported as being among the most significant factors limiting the performance of gravitational wave detectors at frequencies below a few hundred Hz (Luca Gammaitoni) Of particular note were;
- **Q factor measurements on all-silica pendulum systems** being carried out in Moscow (Kirill Tokmakov), in the GEO group in Glasgow (Sheila Rowan) and in the VIRGO group in Perugia (Joe Kovalik and Michele Punturo). Results from these experiments which are to a large extent being performed in a collaborative way between the groups are very encouraging indeed, effective pendulum Quality factors of above 10^8 being measured.

- **Q factor measurements on crystalline materials.** Recent measurements (Sheila Rowan) by the Stanford and GEO groups working with V. Mitrofanov from MSU on HEMEX sapphire, YAG and spinel are providing really exciting results. For example a sample of HEMEX has been shown to have a Q greater than 2.5×10^8 at room temperature - a value close to the best ever obtained for crystals grown (and measured) in Russia by the method of horizontal crystallisation. This is very encouraging as it seems that the HEM (heat exchange) method of Crystal Systems is the only way to make sapphire crystals large enough for gravitational wave detectors. However worries about higher than desirable optical loss in sapphire need addressed (Roger Route, Li Ju).

CONCLUSION

From the above discussions it can be deduced that given the technical advances already well under development it should be possible for the next versions of gravitational wave detectors to reach out to distances greater than 200Mpc for neutron star/neutron star binaries, thus engendering a high level of confidence that important observations will be made with these instruments. But what is appearing on the horizon to follow such advances?

- Advent of cooled test masses in ground based gravitational wave detectors. Such techniques are already being proposed for the Japanese LCGT detector (Kazuaki Kuroda); and it may be that thermal noise will apparently be reduced to a level below the quantum limit. Cooling will become particularly important when the techniques for getting past the quantum limit, as discussed in the talk by Farid Khahili, become practicable. Thus we might expect to see a further extension of LIGO and a new European ground based detector utilizing such ideas.
- LISA, the space antenna which with arm lengths of 5×10^6 km is aimed at low frequency sources of gravitational radiation such as black hole interactions of various types (Karsten Danzmann). LISA will significantly increase the scope of gravitational wave detection as it will open up new horizons for black hole astronomy, and there is a real possibility that LISA could be launched within the next ten years.

PART II
CONTRIBUTED PAPERS

ADVANCED CONFIGURATIONS

All-Reflective Interferometry For Gravitational-Wave Detection

S. Traeger[+], P. Beyersdorf, E. Gustafson, R. Beausoleil, R.K. Route, R.L. Byer, and M.M. Fejer

Ginzton Laboratory, Stanford University, Stanford, CA 94305-4085, U.S.A.
[+]*Email: Traeger@loki.stanford.edu*

Abstract. The circulating power within an interferometric gravitational-wave detector is limited by thermal lensing in transmissive optical elements [1]. All-reflective interferometers have been proposed to overcome this limitation [2,3]. In this paper we discuss the use of a low-efficiency diffraction grating as an input coupler for a high-finesse cavity. Residual thermal effects due to absorption in the high-reflection optical coatings are discussed.

The rapid development of high-power coherent light sources and optical materials with high mechanical quality factor has enabled the design of next generation gravitational-wave detectors operating at the standard quantum limit [4,5] within the measurement band. To reach the proposed sensitivity limit, signal-recycling techniques have to be applied, narrowing the bandwidth of the detector. In an all-reflective interferometer thermal lensing is eliminated making the detector able to handle much higher circulating power and therefore to reach the same sensitivity within a broader measurement bandwidth.

Sun and Byer [6] used the specular order of a reflection grating in Littrow configuration to couple light into a cavity, and the diffracted order to close the cavity. The best high-efficiency multilayer dielectric diffraction gratings today have a first order diffraction efficiency of around 0.96 [7]. This limits the cavity power gain to approximately 25. We suggest here the use of the diffracted order of a weak grating to couple light into a cavity, and the reflected order to close the cavity. This allows a much higher finesse, and therefore a much higher power gain.

Assume a reflection grating coupling 1 percent of the incoming power into the first order at a diffraction angle Φ. The diffracted field is coupled into a ring cavity, which is closed on the reflected order of the grating. The ring cavity is designed so that the return beam is incident on the grating at the same angle Φ, but on the opposite side of the grating normal. Using a symmetric grating 99 percent of the light is used to close the cavity loop.

A weak grating can be written into the top layer of a low-loss mirror. As the coupling efficiency of the grating is weak, the scatter due to the grating is small. This means that the losses of the cavity are dominated by the output coupling efficiency of the grating, and a high finesse can easily be realized. The power gain in the described system is approximately 100.

Diffracting a light field influences the shape of the beam. A circular Gaussian laser beam becomes elliptical by an amount depending on the difference of Φ and Θ (angle of incidence). However, by using the grating so that $\Phi \approx \Theta$, the effect is small and can be treated as a loss due to non-perfect input coupling.

Although there is no thermal lens effect in this system, residual absorption in the high-reflection coatings results in thermo-elastic deformation of the optical surfaces. Winkler et al. [8] give an approximation for the normal surface distortion. The change in the shape of the optics changes the cavity eigenmodes, resulting in less input coupling efficiency, as well as possibly increasing the losses in the cavity. Both effects would reduce the power gain. The amount of surface deformation for a given absorbed optical power depends on material parameters of the substrate. A good indication is the ratio of thermal expansion and thermal conductivity, which should be as small as possible. By far the best of the materials currently under discussion for the test masses of advanced gravitational-wave detectors is Silicon ($\alpha/\kappa = 0.1 \times 10^{-7}$ m/W), followed by Sapphire ($\alpha/\kappa = 2 \times 10^{-7}$ m/W), and Fused Silica ($\alpha/\kappa = 3.4 \times 10^{-7}$ m/W). Assuming coating absorption of 1 ppm, a Silicon cavity as described above can handle circulating light powers of 10 MW without reducing the effective power gain. It should be noted that other material parameters, e.g., the mechanical Q, also influence the choice of the test mass material.

S. Traeger thanks the German Alexander von Humboldt-Stiftung for his Feodor-Lynen fellowship. R.G. Beausoleil is also at Hewlett-Packard Laboratories, 13837 175th Pl. NE, Redmond, WA 98052-2180. This research is supported by the NSF (PHY-9630172).

REFERENCES

1. Strain, K. et al., *Physics Letters A* **194** 124 (1994).
2. Byer, R.L. et al., *Stanford Advanced Gravitational-Wave Laser Interferometer Program - GALILEO*, Ginzton Laboratory, Stanford University (1995), pp. 14 - 16.
3. Drever, R., in: *Proceedings of the 7th Marcel Grossmann Meeting on General Relativity*, eds. M. Keiser, R.T. Jantzon, World Scientific, 1995, pp. 1401 - 1406.
4. Braginsky, V.B. and Khalilli, F.Y. *Quantum Measurements*, Cambridge University Press, 1992.
5. Jennrich, O. et all., in: *Gravitational Waves:Proceedings of the 2nd Edoardo Amaldi Conference*, eds. E.Coccia, G. Pizzella, G. Veneziano, World Scientific, 1998, pp. 556 - 559.
6. Sun, K. and Byer, R. *Optics Letters* **23** (8) 567 (1998).
7. Perry, M.D. et al., *Optics Letters* **20** (8) 940 (1995).
8. Winkler, W. et al., *Phys. Rev A* **44** (11) 7022 (1991).

LASERS AND OPTICS

GEO 600 Slave Laser Prototype II

I. Zawischa[*], O.S. Brozek[†], V. Quetschke[‡], C. Fallnich[*], B. Willke[‡], K. Danzmann[†‡] and H. Welling[*]

[*] *Laser Zentrum Hannover, Hollerithallee 8, D-30419 Hannover*
[†] *Max-Planck-Institut für Quantenoptik, Außenstelle Hannover, Callinstraße 38, D-30169 Hannover*
[‡] *Universität Hannover, Institut für Atom- und Molekülphysik, Callinstraße 38, D-30169 Hannover*

Abstract. Key-points of the design of the quasi-monolithic, longitudinally diode-pumped 13W Nd:YAG ring-laser that serves as slave in the GEO 600 injection locked laser source are reported. Performance data of the 2nd stage prototype are given as well as a short outlook to the next steps.

The light source for the GEO 600 gravitational wave detector will be an injection locked laser system. A diode pumped, monolithic non-planar ring oscillator (NPRO) is taken as master oscillator. It is pre-stabilized to a vacuum suspended ULE reference resonator. Intensity fluctuations are reduced by a feedback to the pump diode current. In order to increase the output power from about 800 mW to the GEO requirement of about 10 W a power oscillator is injection locked to the NPRO emission. The slave laser cavity is kept within the locking range to the NPRO emission by the Pound-Drever-Hall technique. Further information about the whole laser system can be found in reference 1.

The GEO slave laser is designed as a longitudinally, mode selectively pumped two rod bow tie ring resonator (fig. 1). This design was chosen because its high efficiency minimizes energy supply requirements as well as waste heat removal requirements. The latter affords high passive stability of the resonator where as the first promises good passive stability and active control properties in the power supply.

In the 2nd stage prototype two fused silica Brewster plates compensate for the astigmatism of the resonator and the hot laser crystals and enforce p-polarized emission. Together with the depolarizing effects in outer radii of the rods the plates also efficiently suppress higher transverse modes.

As a special feature in active resonators the laser employs a quasi-monolithic resonator design, i.e. three of four mirrors are pressed and glued directly to the rigid aluminum spacer. The fourth, piezo mounted mirror is glued into a special jig and once again fixed in proper position to the spacer without any later alignment and self-misalignment possibilities. Thus a very rugged, disturbance insensitive, and potentially long-term stable laser could be realized.

At 34 W pump power the laser system emits radiation at 1064 nm in excess of 13W in a 99% p-polarized TEM_{00} mode. With spherical lenses 94.5 % of the externally attenuated beam could be coupled through a ring mode cleaner cavity of finesse 200 (fig. 2 left). As the attenuation polarizers are believed to cause some mode distortion, and residual ellipticity is not corrected even better values are expected in the final setup.

In order to estimate the short term stability of the slave resonator the piezo actuator signal within the locking servo bandwidth was recorded and converted to the corresponding free-running frequency correction (fig.2 right). During this measurement the master laser was kept locked to its reference cavity. The frequency correction signal is almost identical to the frequency fluctuation of a free running NPRO. Though being a very good result, the coincidence requires further investigation and renders the data preliminary.

As future steps the resonator spacer will be cut in Ni36Fe64 alloy reducing the sensitivity to thermal fluctuations of the environment. Furtheron focus will be put on an intensity noise reduction by both passive pump diode current filtering and active feedback from an optical power monitor. Applicability of pump current modulation as part of the frequency servo (as it works well with the NPRO) will be considered.

The authors wish to thank I. Freitag, S. Knoke, for their support during the development of the laser.

FIGURE 1. Photograph of the GEO 600 Slave Laser Prototype II.

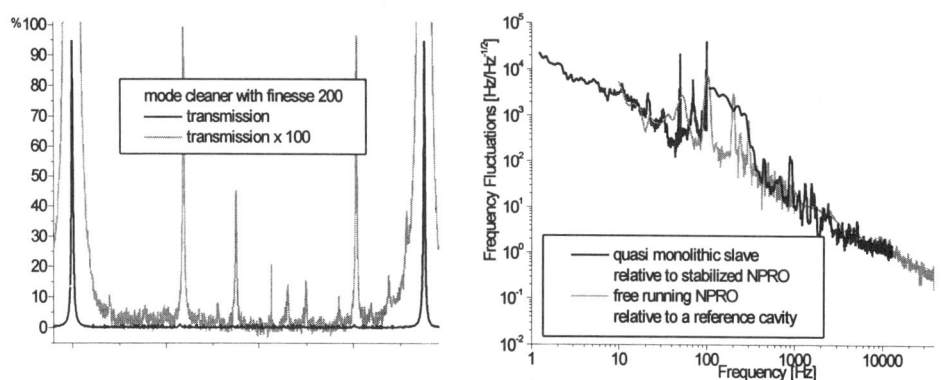

FIGURE 2. Performance of the GEO 600 Slave Laser Prototype II.

1. B. Willke et. al., The GEO600 Stabilized Laser System and the Current-Lock Technique, this proceedings

An Injection Locked Nd:YAG Laser For The Glasgow 10m Interferometric Gravitational Wave Detector

D. A. Clubley, K. D. Skeldon and G. P. Newton

Department of Physics and Astronomy
University of Glasgow, Glasgow G12 8QQ, Scotland

Abstract. We report on recent work to develop a new injection-locked Nd:YAG laser for the Glasgow 10m prototype gravitational wave interferometer. We have stabilised the master laser frequency to a rigid reference cavity and measured the residual frequency noise using one of the arm cavities of the 10-m interferometer as an independent discriminator. We have demonstrated shot-noise limited performance of the stabilisation scheme over our frequency range of interest and its transferal to the frequency noise behaviour of the injection-locked slave.

INTRODUCTION

The laser source for the 10m-interferometer is required to have an ultimate frequency noise of $<10^{-5}$ Hz/$\sqrt{Hz}$ to allow a sensitivity goal of $<10^{-18}$ m/$\sqrt{Hz}$ to become a realistic aim. This is required primarily over the frequency band within which gravitational wave detection will occur; from tens to thousands of Hz. The laser we have developed is an injection-locked system using a Laser Zentrum Hannover NPRO as the master oscillator and a slave laser based on a Nd:YAG crystal obtained from the Ginzton lab at Stanford University (1). The frequency stabilisation is carried out on the master oscillator only (2) and uses a high gain servo (~10^2 at 1kHz) to reduce the approximate $10^4/f$ frequency noise profile of the free-running NPRO. A number of measurements of the frequency noise of the master, and injection locked slave, were made as the system evolved, all of which were based on locking one of the arm cavities of the 10m-interferometer to the stabilised light. The interferometer arm cavity is a very low noise frequency discriminator for various reasons: Its mirrors have optical losses around 40ppm, giving a high finesse of around 15000. The mirrors are suspended as double pendulums giving isolation from seismic noise, while the entire system is maintained under high vacuum, shielding it from effects of air pressure fluctuations and acoustic noise.

EXPERIMENTAL SETUP

Some light from the NPRO is split off and directed into a vacuum chamber containing a triangular reference cavity of 42cm round-trip length and linewidth around 1MHz. The process of rf-reflection locking (3) is used with a modulation

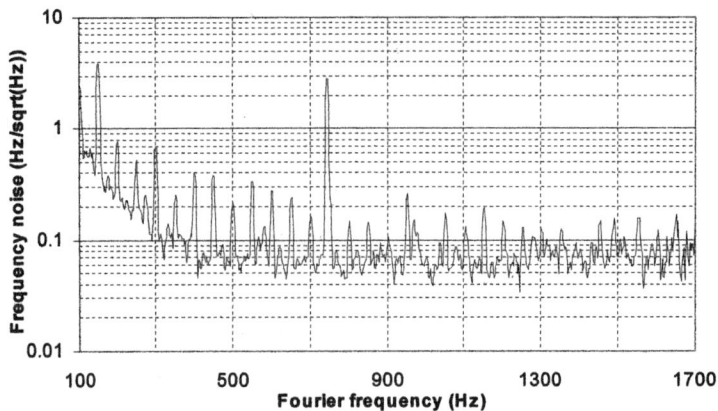

Figure 1. Frequency noise of slave laser measured with the 10m arm cavity.

frequency of 23.7MHz to lock the laser frequency to the reference cavity. The loop has a unity gain frequency of ~60kHz and uses a PZT bonded to the NPRO crystal for frequency tuning. The slave laser crystal is side-pumped using a 25W diode array and produces a maximum of 4W in a TEM_{00} mode, but in these experiments the nominal output power was closer to 2W. The slave laser is held within the injection locking range using a second rf reflection lock, but with the same 23.7MHz modulation, and using an intra-cavity PZT to act on the slave cavity frequency. The injection-locked laser light is passed through a single-mode optical fibre to suppress beam geometry fluctuations before being mode-matched into the 10m interferometer cavity. Another rf-reflection lock scheme is employed to keep the 10m cavity resonant with the laser light. The feedback signals are applied to coil-magnet actuators behind one of the suspended mirrors of this cavity, and form the basic signal which is then measured and calibrated to give the frequency noise. An example trace is shown in Figure.1 which shows a typical level of $5 \times 10^{-2} Hz/\sqrt{Hz}$ around 1kHz, matching the expected shot-noise limited performance of the stabilisation loop.

ACKNOWLEDGEMENTS

We would like to thank the rest of the Gravitational Waves Group in our department for their support and interest in this work. We acknowledge the PPARC, SHEFC and the University of Glasgow for the provision of financial support. One of us, K.D.S is supported by a Royal Society of Edinburgh BP/RSE Fellowship.

REFERENCES

1. Farinas, A.D., Gustafson, E.K. and Byer, R.L., Opt. Lett 19, 114-116 (1994)
2. Farinas, A.D., Gustafson, E.K. and Byer, R.L., J. Opt. Soc. America B 12, 328-334 (1995)
3. Drever, R.W.P., Hall, J.L., Kawalski, F.V., Hough, J., Ford, M., Munley, A.J. and Ward, H., Appl. Phys B 31, 97-105 (1983)

The Laser System For The LISA Technology Demonstrator

O.S. Brozek[1], M. Peterseim[1,2], K. Danzmann[1],
I. Freitag[3], C. Fallnich[2], H. Welling[2]

1) Max-Planck-Institut für Quantenoptik, Callinstrasse 38, D-30167 Hannover
2) Laser Zentrum Hannover, Hollerithallee 8, D-30419 Hannover
3) Innolight GmbH, Vahrenwalder Strasse 7, D-30165 Hannover

Abstract. Future space science missions rely on qualified interferometric position and length measurement technology. Laser metrology is one of the key issues. We describe the laser system for the LISA technology demonstrator. This laser system consists of a laser-diode pumped monolithic non-planar Nd:YAG ring laser with a design optimised to withstand the severe space environment.

1. INTRODUCTION

Key technologies for the laser interferometer space antenna - LISA [1] - are drag free control, micro-Newton thrusters and laser metrology with highly stabilised lasers in the detection frequency-band between 1 mHz and 1 Hz [2,3]. It is of great importance to validate these technologies in space. The aim of the LISA technology demonstrator, currently studied by both ESA and NASA, is to prove these techniques in space in a GTO, geosynchronous or other suitable orbit. The laser metrology shall be demonstrated with a measurement accuracy of better than 10 pm/$\sqrt{Hz}$.

2. THE LASER SYSTEM

Due to their compactness and high efficiency laser-diode pumped monolithic non-planar Nd:YAG ring lasers (NPRO) [4,5] show excellent frequency and power noise performances and are therefore best suited for this mission. The laser (Fig.1) is fabricated in a way that the laser resonator and the active medium are made of one monolithic piece of Nd:YAG. The resonator geometry is enforced by total internal reflections and the dual band coated front surface. The laser is pumped by two laser diodes at 808 nm, which are operated at 50 % of their nominal output

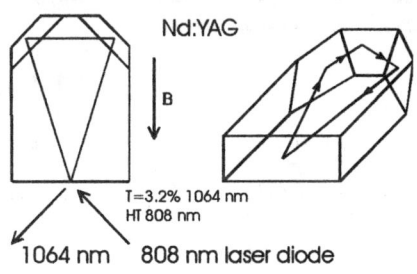

FIGURE 1: *Ray traces in the monolithic Nd:YAG ring laser.*

power to achieve a redundant pump scheme in case of the failure of one of the diodes. An optical output power of 1 W is required for LISA, but a few mW are sufficient for the technology demonstrator. Active power stabilisation and frequency stabilisation to a Fabry-Perot ring cavity will be used to achieve $5 \times 10^{-4}/\sqrt{Hz}$ of relative power noise and 1.5×10^{3} $Hz/\sqrt{Hz}$ of frequency noise in the frequency-band between 1 mHz and 1 Hz [2]. A compact laser head (Fig.2) has been designed that fulfils the power and mass constraints and withstands thermal and mechanical stresses as encountered, e.g., during launch. All components are included in a quasi-monolithic design with no movable parts and are contacted to a massive mono-block made of fused silica. The light from the NPRO is matched into a polarisation preserving single-mode fibre. The fibre directs the laser light to the optical bench of the main experiment, where it is used to verify the performance of the drag free control system [1,2].

The selection, space qualification and mechanical design studies of the individual components of the laser head are under progress.

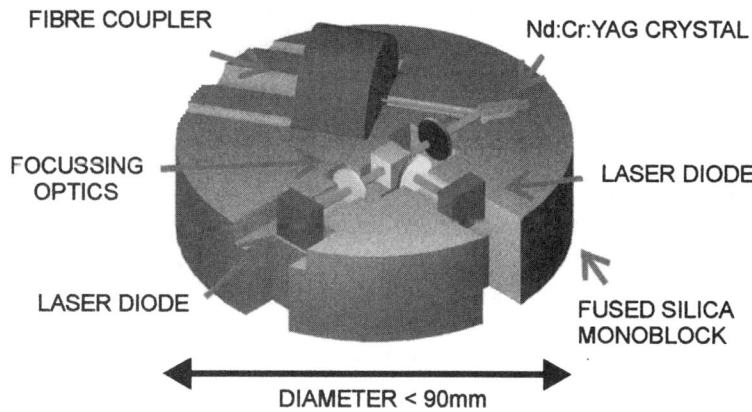

FIGURE 2: *Schematic layout of the quasi-monolithic laser head.*

ACKNOWLEDGMENTS

This work was supported by a grant of the Wernher von Braun Stiftung zur Förderung der Weltraumwissenschaften.

REFERENCES

1. K. Danzmann et al., *LISA: Pre-Phase A Report*, 2nd Edition, MPQ **233**, 1998.
2. M. Peterseim et al., *Advances in LISA technology*, these proceedings.
3. M. Peterseim, O.S. Brozek, A. Tünnermann and K. Danzmann, *Earthbound and deep space testing of laser metrology*, AIP Conference Proceedings **456**, 391-395 (1998).
4. T. Kane and R.L. Byer, *Monolithic, unidirectional single-mode Nd:YAG ring laser*, Opt. Lett. **10**, 65-67 (1985).
5. I. Freitag, A. Tünnermann and H. Welling, *Power scaling of diode-pumped monolithic Nd:YAG lasers to output powers of several watts*, Opt. Comm. **115**, 511-515 (1995).

Effect of annealing on the light absorption in sapphire

Alexei L. Alexandrovski, Martin M. Fejer and Roger K. Route

Ginzton Laboratory, Stanford University, Stanford, CA, 94305-4085

Abstract. Photothermal common-path interferometry has been used to measure optical absorption at 1064 nm and at 514 nm in high purity sapphire crystals. In as-grown and hydrogen-annealed material, uniform optical absorptions of 40-50 ppm/cm were found throughout. In a sample that had been oxygen-annealed after polishing, spatial variations in absorption and scattering/fluorescence revealed a ~4 mm thick 'skin' of lossy material surrounding a core having lower losses typical of as-grown material. A 'valley' with optical absorption about 2-4 times lower than as-grown material (at both 1064 nm and 514 nm) was found at the interface between the oxidized surface layer and the core. The mechanism by which the 'valley' formed is not known, but it does suggest that a better understanding of the oxidation process may lead to a reduction in the optical absorption of sapphire crystals.

A recent study has reported high-purity sapphire crystals (1) with optical absorption in the near IR (1064 nm) as low as a few ppm/cm. We have carried out absorption measurements on sapphire crystals of the same grade (2) and have not found any with bulk absorption at 1064 nm lower than ~40 ppm/cm. We have encountered localized regions between supposedly oxidized surface layers and as-grown core material with absorption values in the 10-20 ppm/cm range, which has raised questions about the origin of this anomaly, and suggested possible ways to fabricate sapphire crystals having lower values of optical absorption than currently obtainable. This study focused therefore on the effects of redox treatments (annealing in reducing or oxidizing atmosphere) on the optical absorption in high-purity sapphire crystals, specifically, Hemex Ultra grade sapphire produced by Crystal Systems Inc.

Two cylindrical samples, 10 mm Ø x 20 mm in length with their axes aligned along the optical axis (crystallographic c-axis) were prepared from the same boule. One sample was annealed in hydrogen, the other in oxygen. Optical absorption was measured at 1064 nm and at 514 nm using photothermal common-path interferometry (3), a highly sensitive modification of the thermal lensing technique. A crossed-beam pump/probe measurement configuration permitted three-dimensional scanning with spatial resolution of <100 µm. The absolute magnitude of the optical absorption coefficient was evaluated relative to a Ti-doped sapphire reference crystal. Scattering of the 514 nm (green) pump and red fluorescence due to a background Ti^{3+} impurity in the ppm range were readily detectable by eye.

No distinguishing differences between as-grown and hydrogen-annealed sapphire were detected. Uniform optical absorptions (50 ppm/cm at 1064 nm and 600 ppm/cm at 514 nm) were seen along with a weak, uniform red Ti^{3+} fluorescence. No localized scattering of the green pump was observed. The fluorescence intensity corresponded to few ppm of Ti^{3+}, as determined by comparison with a Ti-doped standard.

The oxygen-annealed sapphire specimen exhibited both scattering and fluorescence patterns that correlated with the measured absorption: there was no scattering in the core of the sample while strong scattering was observed from the surface regions. Fluorescence in the core region was of the same intensity as seen in the hydrogen-annealed sample: fluorescence was not observed from the surface regions. Results of the absorption measurements on the oxygen-annealed sample are given in Fig.1. While the absorption in the core region was the same as for hydrogen-annealed sapphire, absorption levels near the surfaces were high both at 514 nm and at 1064 nm.

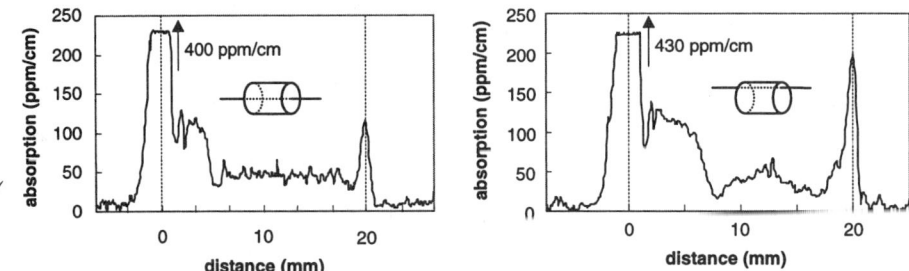

FIGURE 1. Oxygen-annealed sapphire: absorption at 1064 nm, scan from surface to surface (locations shown dashed) through the central part of the sample (left) and closer to the cylindrical surface (right).

The most interesting finding from the oxygen-annealed sample was the existence of a 'valley' at the interface between the core and surface regions with both green and IR absorption levels lower than that in the as-grown core. One possible explanation is that moderate oxidation reduces green and IR absorption by converting Ti^{3+} to Ti^{4+} (4). The reason for scatter in the surface region is unclear, but is likely due to precipitation of an impurity phase, a likely candidate being TiO_2. We have no explanation for the increased optical absorption found in the surface region. Additional experiments are needed to understand how scattering and fluorescence behave in the transition layer.

REFERENCES

1. Blair, J. D., Cleva, F., and Man, C. N., *Optical Materials* **8**, 223-236 (1997).
2. Alexandrovski, A. L., Fejer, M. M., and Route, R. K., "Absorption measurements in fused silica and sapphire," presented at the Annual LIGO meeting, Hanford, WA, 1998. (unbpublished)
3. Alexandrovski, A. L. and Fejer, M. M., "Photothermal common-path interferometer for the measurement of weak absorption in solids," in preparation.
4. Aggarwal, R. L., Sanchez, A., et al, *IEEE JQE* **24**, 1003-1008 (1988).

LISA

Double Test-Mass Control for L.I.S.A.

Peter C. E. Roberts

College of Aeronautics, Cranfield University,
Cranfield, Bedfordshire MK43 0AL, United Kingdom

Abstract. Drag-free control of the two test-masses within the each L.I.S.A (Laser Interferometer Space Antenna) spacecraft is considered. The baseline concept of actuating the masses to counter the gravity gradient produced by the spacecraft is reviewed. It is shown that this method causes instability or excessive transference of acceleration noise to the masses. An alternative approach is suggested by the author that fulfills the necessary performance requirements.

INTRODUCTION

Drag-free control for the 1995 L.I.S.A. baseline (1), where each spacecraft contains a single test-mass, has been considered by different authors. Touboul *et al* (2) describe the basic control of a simple one translational dimension model of the system. Klotz *et al* (3) consider a complete translation and attitude model. This author has also considered this system (4). All the above have shown that control of a single test-mass within each L.I.S.A. spacecraft at the performance levels required appears possible.

The effect of the local gravity gradient produced by the spacecraft is not considered in these studies because it does not effect gravitational wave measurement. The effect it does have is to cause the test-mass to accelerate uniformly, the spacecraft follows the test-mass due to the drag-free control system, and so the two accelerate together. This acceleration, to first order, will have zero frequency and, as it does not act in the measurement bandwidth, can be ignored.

When one considers two test-masses in the same spacecraft for the 1998 L.I.S.A. baseline (5), the gravity gradient becomes a problem as the test-masses are not collocated. The different accelerations they feel due to the gradient means one or both of the masses must be actuated to remove the difference. This actuation is problematic.

BASELINE SOLUTION

The baseline control philosophy (5) has a master-test-mass, the center of which is set at the "drag-free null", and a slave-test-mass that is actuated to follow the master. This actuation will have some component in the sensitive axis of the slave payload. The total acceleration required will be around 10^{-10} m s^{-2} (6). Let us assume a spring like actuation force, the exact size of the acceleration required being unknown. To avoid excessive movement of the test-mass relative to the spacecraft a spring constant

$$k = \frac{mass \times (gravitational\ acceleration)}{maximum\ allowable\ displacement} \quad [N\ m^{-1}] \qquad (1)$$

is required. This will be several orders of magnitude above the residual stiffness of the electrostatic position sensing system. Thus, it will transfer several orders of magnitude more acceleration noise to the test-mass, an untenable position. Using a spring-like force that only acts below the measurement band introduces instability as it cannot follow the high frequency dynamics of the system. A different approach is needed.

AXIALLY CONSTRAINED DRAG-FREE CONTROL

The alternative approach suggested by the author avoids actuation of either test-mass in the sensitive axis of their respective payloads. Let us consider only the plane containing the sensitive axes of the two payloads. The difference in gravitational acceleration that acts upon the slave-test-mass can be split into two non-normal components perpendicular to the sensitive axes of the slave and master payloads. The first of these components can be actuated directly without effecting the slave sensitive axis. If the spacecraft acts in a drag-free manner following the slave-test-mass along the second component, again the sensitive axis of the slave remains unaffected. The master-test-mass, for this later actuation, will have to be actuated perpendicular to its sensitive axis to follow the spacecraft, but this is not a problem.

This method avoids actuation of both test-masses in their sensitive axes whilst overcoming the effect of the spacecraft gravity gradient. A model of this system has been produced using MATLAB and Simulink. The approach described above can be applied very simply by increasing the off-sensitive-axis electrostatic spring strengths to make the test-masses follow the spacecraft. The sensitive axes remain unaffected. Further information on the approach and models described can be found in (7).

REFERENCES

1. LISA Study Team. *LISA. Pre-Phase A Report.* Max-Plank-Institut, Germany, 1995.
2. Touboul, P., Rodrigues, M. and Le Clerc, M. *Classical & Quantum Gravity*, **13**, A259-A270 (1996).
3. Klotz, H., Strauch, H., Wolfsberger, W., Sumner, T., Speake, C., Andrews, P., Marcuccio, S., Genovese, A. and Paolucci, F. *Drag-Free Satellite Control.* ESA CRP-4229. 1998.
4. Roberts, P. *Space Technology – Industrial and Commercial Applications.* Pre-publication, 1999.
5. LISA Study Team. *LISA. Pre-Phase A Report – 2nd Edition.* Max-Plank-Institut, Germany, 1998.
6. ELITE Proposal Team. *ELITE - European LISA Technology Demonstration Satellite for the LISA Mission in ESA's Space Science Programme.* ESA Proposal. 1998.
7. Roberts, P. *Drag-Free Control and Technological Risk Assessment for the L.I.S.A. Gravitational Wave Space Antenna.* EngD Thesis - Cranfield University. 1999.

Computing LISA far field phase patterns

Eugene Waluschka
NASA/Goddard Space Flight Center/551.0, Greenbelt, Maryland 20771

INTRODUCTION

Long baseline interferometers can have internal optical path lengths which range from centimeters to millions of kilometers. The requirements on the knowledge of the path length differences, between two arms, of such a interferometer can be very stringent, of the order of a few picometers. Designing and analyzing such long optical systems is not a straightforward task if standard ray trace computer codes are used. This is because all the optical design codes use double precision arithmetic. With double precision the machine epsilon, the smallest number which when added to one will produce a result different from one, is about 10^{-16}.

Double precision machine epsilon is sufficiently small for most applications. However, it was not small enough for the (straightforward) analysis of the (proposed) Laser Interferometer Space Antenna(LISA) for the detection of gravitational waves. It is expected that gravitational waves will produce strains ($\Delta L/L$) of the order of 10^{-21}. This means that in an interferometer with an arm length of five million kilometers, as in the LISA mission, the expected gravity wave displacements are only about five picometers. Adding 5 picometers to 5 million meters will result in identically 5 million kilometers and as such is a meaningless addition when performed using standard double precision arithmetic. This disparity in magnitudes does not preclude the analysis of such large optical systems. It only makes it less straightforward to analyze them with double precision. Simplicity of implementation is the main reason why all optical codes are double precision now and not single precision.

QUADRUPLE PRECISION

On Digital Alpha computers converting a FORTRAN program (QRAYPKS) to REAL*16 is a fairly simple matter because the compiler supports this option. Run times are also not increased because the chip set performs this extended arithmetic in hardware. The machine epsilon in this case is about 10^{-34}. This epsilon is sufficiently small for the straightforward modeling of a gravitational wave induced strains and very long baseline interferometers.

A simple example of the utility of REAL*16 arithmetic is provided by the numerical approximation of the diffraction integral

$$A(X,Y,Z) \cdot e^{i\frac{2\pi}{\lambda}R} \cdot e^{i\frac{2\pi}{\lambda}\phi(X,Y,Z)} = \int\int E(x,y,z) \frac{e^{i\frac{2\pi}{\lambda}(Z_n + S)}}{S} dxdy$$

CP523, *Gravitational Waves: Third Edoardo Amaldi Conference,* edited by S. Meshkov
2000 American Institute of Physics 1-56396-944-0

Where we integrate over an exit pupil (254 mm in diameter) containing a Gaussian amplitude, but a flat phase and a central obscuration with supporting struts to determine

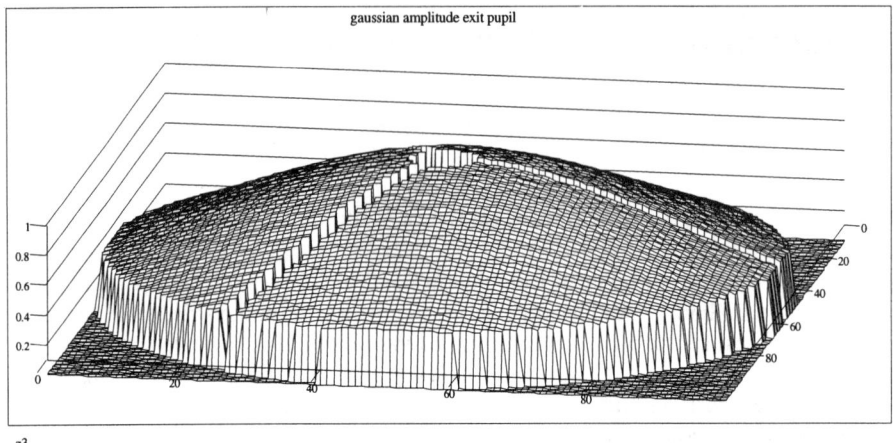

the change (from the case of no obscurations) in the phase distribution on a five million kilometer sphere (centered on the exit pupil) as shown.

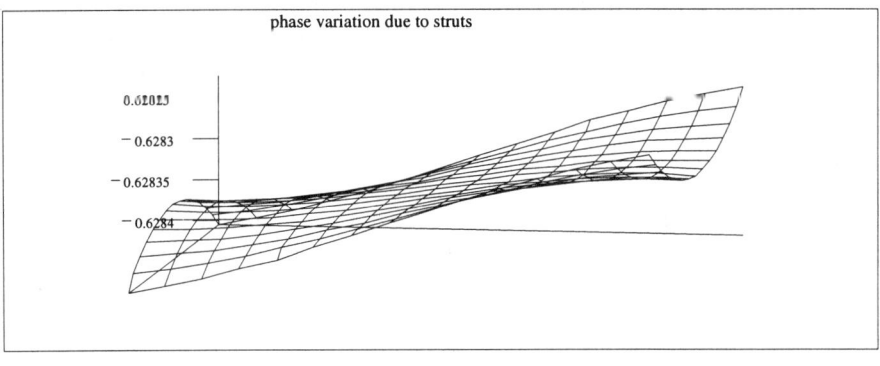

Where the far-field spatial extent is -2.64 to +2.64 kilometers and the phase (vertical axis) is in radians

ACKNOWLEDGMENTS

This work was performed at JILA, a joint institute of the National Institute of Standards and Technology and the University of Colorado at Boulder and supported by a Goddard Space Flight Center's Fellowship.

LISA observations of massive black hole binaries using post-Newtonian waveforms

Alicia M. Sintes and Alberto Vecchio

Max-Planck-Institut für Gravitationsphysik, Am Mühlenberg 1, D-14476 Golm, Germany

Abstract. We consider LISA observations of in-spiral signals emitted by massive black hole binary systems in circular orbit and with negligible spins. We study the accuracy with which the source parameters can be extracted from the data stream. We show that the use of waveforms retaining post-Newtonian corrections not only to the phase but also the amplitude can drastically improve the estimation of some parameters.

The strongest sources of gravitational waves for LISA [1,2] are likely to be binary systems of massive black holes (10^4-10^7 $M_\odot$). So far, parameter estimation of in-spiral signals for LISA has been investigated within the so-called restricted post-Newtonian approximation [3,4]: PN corrections are taken into account in the phase of the waveform, whereas the amplitude is retained at the lowest Newtonian order. Here, we investigate the implications for parameter estimation of the introduction of PN corrections also to the amplitude. Going to higher PN order in the amplitude implies the use of several multipole components and not just the quadrupole one. The gravitational radiation is then described by eleven independent parameters associated with distance, masses, spins, position and orientation of the source in the sky, and instant and phase of the final collapse. We compute the results regarding the expected errors associated with measurement of the parameters that characterize the source using two different waveform approximations [5]: (a) The standard restricted 2 PN approximation: $h(t) = \mathrm{Re}\left[h_2^0\, e^{i2\Phi}\right]$; (b) A signal where we consider radiation emitted not only at twice the orbital frequency f_{orb} but also at f_{orb} and $3f_{orb}$; in this case we retain only the 0.5 PN correction to the amplitude, the signal reads: $h(t) = \mathrm{Re}\left[h_1^{0.5}\, e^{i\Phi} + h_2^0\, e^{i2\Phi} + h_3^{0.5}\, e^{i3\Phi}\right]$. The results depend strongly on the actual values of the source parameters, in particular location and orientation of the source. Therefore, we perform a Monte-Carlo simulation, keeping the source distance and masses fixed and varying randomly $\hat{N}$ and $\hat{L}$, and we construct histograms to show the distributions of the different parameter measurement errors. We checked that the angular resolution Ω_N is basically unaffected by adding amplitude corrections. Even though in the waveform (b) there is more information about position and orientation of the source, this turned out to be too weak to

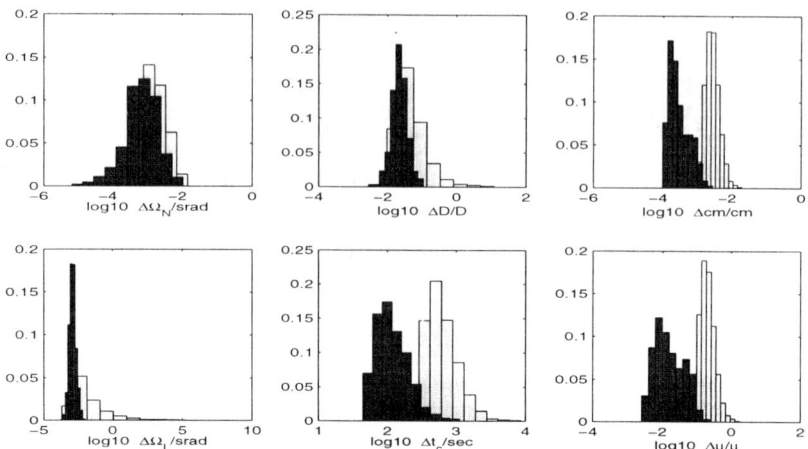

FIGURE 1. Probability distribution of the angular resolution, orientation, distance measurement errors, timing accuracy and mass measurements errors for LISA observations of the final year of massive black hole in-spiral binaries ($m_1 = 10^7 \, M_\odot$, $m_2 = 10^6 \, M_\odot$, $z = 1$). The light and dark colors refer to waveform (a) and (b) respectively (see text). The histograms show the error distribution for 1000 random values of $\hat{N}$ and $\hat{L}$.

provide an improvement in the angular resolution. For the distance $\Delta D/D$, we have the same values on average but smaller dispersion of the errors around the mean. On the contrary, the determination of physical parameters such as time of coalescence t_c, chirp mass $\mathcal{M}$ (referred as cm in the figure), and reduced mass μ is strongly improved by a factor of 10 ore more: Within model (a) information about t_c, $\mathcal{M}$ and μ are conveyed only by the GW phase 2Φ; in model (b) the evolution of all three phases and amplitudes is controlled in a different way by those physical parameters providing therefore more information.

We conclude that according to our results, more accurate (i.e., beyond the standard restricted PN approximation) in-spiral waveforms, as those including the radiation emitted at different multiples of the orbital frequency, do play an important role for LISA as GW observatory; in fact they provide extra information that improve the measurements of the physical source parameters.

REFERENCES

1. Bender P., et al., *LISA Pre-Phase A Report*, Garching: **MPQ 233** (1998).
2. Vecchio A., "Deep surveys of massive black holes with LISA", in this volume.
3. Cutler C., *Phys. Rev. D* **57**, 7089 (1998).
4. Vecchio A., Cutler C., in *Laser Interferometer Space Antenna*, proceedings of 2nd International LISA Symposium, (Ed.) W. Folkner, AIP, 101-109 (1998).
5. Blanchet L., Iyer B., Will C.M., Wiseman A.G., *Class. Quant. Grav.* **13**, 575 (1996).

Globular Cluster Distance Measurements with LISA

Matthew Benacquista

Montana State University - Billings
Billings, Montana, 59101

Abstract. The dynamics of globular cluster evolution is thought to favor the production of hard binary systems in the cluster core. Binaries containing two degenerate objects will be detectable by LISA with a signal to noise ratio above 10. Very short period binaries with orbital periods below about 500 s will have a measurable frequency change (or chirp) due to inspiral during a one year integration. Using the measured chirp, it is possible to determine the distance to the binary with better than 10% precision if the angular position is known. Consequently, it is shown that globular cluster distances can be measured accurately using gravitational radiation.

In the cores of globular clusters collisions and multiple-body encounters serve to enhance the number of binary systems with very short orbital period. In this way, it is expected that a population of neutron star-white dwarf binaries should be created with many systems having short orbital period. The known double-degenerate binaries in globular clusters are interacting binaries and are detected through emissions arising from accretion processes. It is reasonable to expect that there are many non-interacting systems which will be quite bright in gravitational radiation. One model for the population of such systems predicts about 10 binaries with orbital period below $2000s$ per globular cluster [1]. These sources will be easily detectable by LISA and it should be possible to determine their angular position to within the size of their host globular cluster. If the orbital period decreases measurably during a one-year observation, it will be possible to determine the distance to that cluster to less than 10% precision.

The estimation of both position and distance of a source of gravitational radiation depends upon the precision with which a parameterization of the expected signal can be matched with the actual signal in the presence of noise. This review of signal analysis is from Cutler [2] and is intended to present the notation. The signal is described by $h_\alpha(\lambda_i)$ where $\alpha = I, II$ indicates the interferometer the signal is taken from and λ_i are the parameters describing the signal. The Fourier transform is defined as $\tilde{h}_\alpha(f) = \int_{-\infty}^{\infty} e^{2\pi i f t} h_\alpha(t)\, dt$. The signal to noise ratio is found from $\rho^2 = 4 \int_0^\infty \sum_\alpha (\tilde{h}_\alpha^* \tilde{h}_\alpha / S_n) df$, where S_n is the spectral density of the instrument noise

TABLE 1. Properties of the model binary sources used.

Binary Type	M_1 ($M_\odot$)	M_2 ($M_\odot$)	f_c (mHz)
NS-NS	1.4	1.4	3.7
NS-CO	1.4	0.6	4.6
NS-He	1.4	0.3	5.4
CO-CO	0.6	0.6	5.5
CO-He	0.6	0.3	6.4
He-He	0.3	0.3	7.5

and the Fisher information matrix is defined as $\Gamma_{ij} = 2\int_0^\infty \sum_\alpha (\partial_i \tilde{h}_\alpha^* \partial_j \tilde{h}_\alpha / S_n) df$ with $\partial_i = \partial/\partial \lambda_i$. If the best-fit parameters differ from the true parameters by an amount $\delta \lambda_i$, then the variance-covariance matrix is approximated by $\langle \Delta \lambda_i \Delta \lambda_j \rangle \approx \Gamma_{ij}^{-1}$, where $\langle\ \rangle$ indicates an expectation value.

Using the six model binary types given in Table 1, and assuming that the binaries are located in either NGC 6752 or 47 Tucanae (at $3.9 kpc$ and $4.3 kpc$, respectively), the signal-to-noise ratios were calculated and found to be above 50 for all but the least massive He-He systems. In most cases $\rho > 100$. Using the results of Cutler [2] and Peterseim et al. [3] it can be shown that the angular position of the binaries of Table 1 can be determined with an angular resolution comparable to the angular size of the host cluster. The angular position of these sources can then be identified with the known position of the host cluster. Binaries with orbital frequency above the chirp frequency (f_c in Table 1) which have been identified with the host cluster can then be used to determine the distance to the cluster. An upper bound to the relative uncertainty in the measurement of the distance is found using:

$$\frac{\Delta d}{d} \approx \frac{\Delta \dot{f}}{\dot{f}} + \frac{11}{3} + \frac{\Delta A}{A}. \tag{1}$$

The relative uncertainty in the distance to chirping NS-NS binaries is less than 10% for all orbital frequencies above $3.7 mHz$. At orbital frequencies above $8 mHz$ all binaries except He-He will give distance measurements with better than 5% precision. These measurements would be a significant improvement over the current measurements made using electromagnetic radiation.

REFERENCES

1. Benacquista, M., *Ap J* **520**, 233 (1999).
2. Cutler, C., *Phys. Rev. D* **57**, 7089 (1998).
3. Peterseim, M., Jennrich, O., Danzmann, K., and Schutz, B.F., *Class. Quant. Grav.* **14**, 1507 (1997).

BAR ANTENNAE

High Sensitivity Accelerometers for High Performance Seismic Attenuators

A. Bertolini[1], R. De Salvo[2], F. Fidecaro[1], M. Francesconi[1]
V. Sannibale[2], A. Takamori[3]

[1] *Dipartimento di Fisica dell'Università di Pisa and INFM*
Via Buonarroti, 2 I-56127 Pisa, Italy
[2] *LIGO Project, California Institute of Technology*
1200 E. California Blvd., 91125 Pasadena, CA
[3] *TAMA Project, University of Tokyo*
7-3-1 Hongo, Bukyo-ku, Tokyo, 113-8654, Japan

Abstract. We present concepts and features of a new horizontal accelerometer whose mechanical design and machining process aim to improve the sensitivity in the frequency region between 10 mHz and 1 Hz. The expected sensitivity, less than 10^{-11} m/s$^2/\sqrt{\text{Hz}}$ around 100 mHz, is a couple of orders of magnitude below the state of art limits. This accelerometer could be integrated in the active control of the LIGO II mirror seismic isolators.

We present a new highly sensitive, low frequency, horizontal accelerometer to be used in a second level of active control in an actively pre–damped (1) seismic isolator for interferometric gravity wave detectors. Our effort has been focussed on the mechanics of the device in order to reduce dissipative and thermal effects which limit the performances of the accelerometers (2) at frequencies between 0.01 and 1 Hz. A two active stage seismic attenuator, equipped with these new accelerometers, could freeze the rms residual motion of the interferometer mirror within few nanometers, about a factor 10 better than the state of art (1).

The mechanics

The accelerometer mechanics is shaped as a small folded pendulum (3). A folded pendulum is essentially a mass suspended on one side by a simple pendulum and on the other by an inverted pendulum working antagonistically (see Figure 1a). In this way the straight pendulum positive gravitational spring constant is balanced by the inverted pendulum's negative one; by carefully shaping the mass it is also possible to lower arbitrarily the resonant frequency. In a folded pendulum the only dissipation is in the mechanics elastic flexures and in the readout/actuation system. If the elastic flexure spring constant is minimized, the mechanical losses are minimal. Q–factors larger than the intrinsic one of the material are achievable (4). The thermal noise acceleration spectral density is

$$a_{\text{tn}}(\omega) = \frac{1}{Ml}\sqrt{\frac{4K_B T \kappa_{\text{flex}}}{Q\omega}}$$

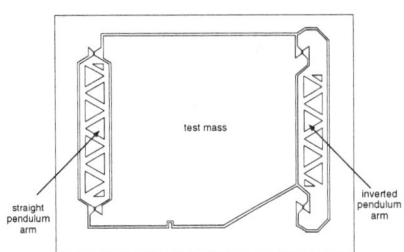 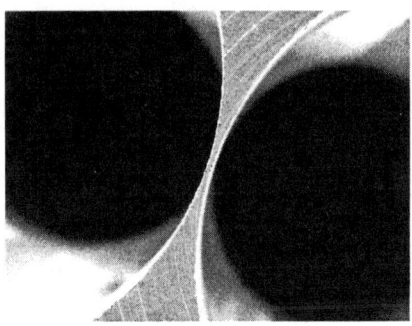

Fig. 1a. Sketch of the mechanical design of the accelerometer.

Fig. 1b. SEM image of a Cu-Be link 50 μm thick machined by EDM.

where M, l, Q and κ_{flex} are respectively the mass, the pendulum arm length, the material Q-factor and the flexure angular stiffness. By using a high strength, low-losses material like Cu-Be one can design very low κ_{flex} flex joints (10-20 μm thick) to support masses of few kilos. In this system one can estimate a thermal noise level around 10^{-12} m/s^2/$\sqrt{Hz}$ at 0.01 Hz, with a resonant frequency of 0.2 Hz and a Q factor around 1000. To assure high Q factors it is necessary to eliminate all the clamps from the mechanics eliminating the structural damping due to the so-called "stick-and-slip" mechanism. This effect is generated by boundary shear stress between contact surfaces of distinguishable mechanical parts. For this reason the accelerometer is completely machined from a single piece of Cu-Be alloy by electric-discharge-machining (EDM). EDM allows to pre-shape the flex joints thickness to 50 μm (see Figure 1b); the final thickness of 10 μm will be reached by electropolishing.

Control and Readout

The instrument will be equipped with a low noise position sensor; the signal from the sensor is filtered by a PID controller and fed back to the mass through a capacitive force actuator for feedback closed-loop operation. The sensing element is a capacitive sensor working at 10 MHz, simple, rugged and with a dynamic range between 10^6 and 10^7; the sensor noise matches the expected thermal noise performances, 10^{-12} m/$\sqrt{Hz}$, with a measuring range of few microns. This results have been obtained by using the resonant phase shift readout technique (5).

REFERENCES

1. Losurdo, G., et al., *Rev. Sci. Instrum.* **70**, 2507-2520 (1999).
2. Braccini, S., et al., *Rev. Sci. Instrum.* **66**, 2672-2677 (1995).
3. Blair, D. G., et al., *Phys. Lett. A* **193**, 223-226 (1994).
4. Bertolini, A., et al., LIGO Note T990777-00-D (1999).
5. Miller, G. L., et al., *Rev. Sci. Instrum.* **61**, 1267-1272 (1990).

Errors on the inverse problem solution in a noisy spherical GW antenna

J. Alberto Lobo* and Stephen M. Merkowitz[†]

*Departament de Física Fonamental, Universitat de Barcelona
Diagonal 647, 08028 Barcelona, Spain
[†]Nuclear Physics Laboratory, University of Washington
Seattle WA 98195-4290, USA

Abstract. The inverse problem can be solved using only linear algebra in the case of a spherical GW detector, even in the presence of noise. The simplicity of such solution enables one to explore the error on the solution using standard techniques. We derive the error on the direction and polarization measurements of a gravitational wave, and show that the solid angle error and the uncertainty on the wave amplitude are direction independent. We show that the polarization amplitudes can too be determined with isotropic sensitivity for any given gravitational wave source.

The so called *inverse problem* for a spherical GW detector consists in the determination of the two amplitudes h_+ and $h_\times$, as well as the incidence direction of the wave, in terms of the detector's output data. Thanks to the multimode nature of the spherical antenna [1], this is possible with a *single* instrument. Merkowitz [2] has recently proposed an *analytic* procedure to estimate the GW parameters in a realistic, *noisy* system; it is applicable when the detector's readout data can be combined in the form of so called *mode channels*. This requires suitable locations of the motion sensors on the surface of the sphere, and there are two layouts known to enable such possibility: *TIGA* [3] and *PHC* [4].

Merkowitz's proposal is based on the algebraic analysis of a "detector response" matrix $\boldsymbol{A}$ which, in the absence of noise, is equal to the GW strain tensor, expressed in lab frame coordinates. Noise however gets added to the mode channels, thereby destroying the equivalence between $\boldsymbol{A}$ and the GW tensor. A *best fit* procedure shows that the matrix $\boldsymbol{A}$ must be diagonalised, whence the *best estimate* for the GW amplitude $h \equiv (h_+^2 + h_\times^2)^{1/2}$ is the semi-difference $(\lambda_1 - \lambda_2)/2$, while the incidence direction is best approximated by the *third* eigenvector $\boldsymbol{v}_3$.

This communication is concerned with the *errors* associated to those estimates, as well as with the actual GW amplitudes h_+ and $h_\times$, as a function of the signal-to-noise ratio [5]. Other authors have attempted to calculate such errors before [6,7], but they have failed to either go beyond first order terms in 1/SNR, or to properly assess un-natural claims of spherical detector anisotropies.

The following expressions are found for errors in the GW amplitude h and incidence direction (in solid angle), respectively,

$$\sigma_h^2 = \left(1 - \frac{3}{2} SNR^{-1}\right) \sigma^2 , \quad \Delta\Omega = \frac{2\pi}{SNR} \left(1 + SNR^{-1}\right)^{-1}$$

and are valid to next-to-linear order in SNR^{-1} down to signal to noise ratios of about 2 or 3, as can be seen by comparison of the above equations with numerical simulations —see [5] for full details.

The last question is the assessment of errors in the GW amplitudes h_+ and $h_\times$. We find at this point that a rotation of angle α, say, around the incidence direction v_3 must be applied to the diagonal form of A in order to identify those amplitudes. In the past, strongly *observer dependent* criteria have been set up to calculate α, which resulted in very suspect anisotropies in the sensitivity for a *spherical* antenna.

The characterisation of h_+ and $h_\times$ is most naturally associated to the GW *source*, which unambiguously defines the polarisation axes of the wave. The determination of h_+ and $h_\times$ cannot therefore be accomplished without proper reference to the source. Overlook of this essential fact is the origin of past mistakes. The trouble, though, is that the nature and position of the source may not be known ahead of time. To overcome this difficulty a *hypothesis* must be made about it, then maximum likelihood methods applied to assess its goodness, as well as the actual estimation of the signal parameters.

It is very important to stress that this is *very different* from what has been done with h and v_3, which can be determined *independently* of the source properties. The process at this stage starts with the construction of a maximum likelihood funcion of the form

$$\Lambda = \Lambda(h; \alpha; \mathcal{K})$$

where $\mathcal{K}$ stands for the set of (unknown) source parameters. Standard manipulations of Λ yield both best estimates of the signal parameters $\mathcal{K}$ *and* of the polarization angle α, and errors and cross correlations between any pair of these. Since Λ is obviously *direction independent*, we recover the expected result that a spherical GW detector produces isotropic estimates of *all* the GW parameters [5].

REFERENCES

1. Lobo J. A., *Phys. Rev. D* **52**, 591 (1995).
2. Merkowitz S. M., *Phys Rev D* **58**, 062002 (1998).
3. Johnson W. W., and Merkowitz S. M., *Phys Rev Lett* **70**, 2367 (1993).
4. Lobo J. A., and Serrano M. A., *Europhys Lett* **35**, 253 (1996).
5. Merkowitz S. M., Lobo J. A., and Serrano M. A., *Class Quant Grav* **16**, 3035 (1999).
6. Zhou C., and Michelson P. F., *Phys Rev D* **51**, 2517 (1995).
7. Stevenson T. R., *Phys Rev D* **56**, 564 (1997).

The First Phase of the Brazilian Graviton Project: The Mário Schenberg Detector

Odylio D. Aguiar, Nadja S. Magalhães, José Carlos N. de Araújo,
Oswaldo D. Miranda, José Luiz Melo, Kilder L. Ribeiro,
Luiz Alberto de Andrade, Karla Beatriz M. Salles, Sérgio R. Furtado,
Nei F. Oliveira Jr.*, Walter F. Velloso Jr.**, Carlos Frajuca***,
Rubens M. Marinho Jr.°, Giorgio Frossati°°

*Divisão de Astrofísica, Instituto Nacional de Pesquisas Espaciais,
Av. Astronautas 1758, São José dos Campos, SP 12227-010, Brazil*
**Instituto de Física, Universidade de São Paulo,
C.P. 66318, São Paulo, SP 05315-970, Brazil*
***Instituto Astronômico e Geofísico, Universidade de São Paulo,
C.P. 3386, São Paulo, SP 01060-970, Brazil*
****Centro Federal de Educação Tecnológica de São Paulo,
Rua Pedro Vicente 625, São Paulo, SP 01109-010, Brazil*
*°Instituto Tecnológico de Aeronáutica, CTA, Departamento de Física
São José dos Campos, SP 12228-970, Brazil*
*°°Kamerlingh Onnes Laboratorium,
Postbus 9506, 2300 RA LEIDEN, The Netherlands*

Abstract. The construction of a CuAl (94%-6%) 0.6-meter diameter truncated icosahedral gravitational wave antenna is currently being proposed in Brazil. The system is planned to feature a sensitivity better than $h = 10^{-21} Hz^{-1/2}$ at both (4.10.4) kHz and (7.90.8) kHz bandwiths.

I- THE GRAVITON PROJECT

Gravitational wave astronomy is expect to begin in Brazil with the GRAVITON project. The major goal of this project is to construct an array of supercryogenic resonant-mass gravitational wave antennas with the shape of solid buckyballs (truncated icosahedrons). A sufficient number of transducers [1], will be used to monitor both the first and second quadrupole modes, and the first monopole mode of the antenna [2,3]. Such a system is omnidirectional, which means that it has the same sensitivity in all directions in space, and it will be able to determine the direction and the polarization of the gravitational waves [4].

II – THE PRESENT PROPOSAL : GRAVITON – 1ST PHASE

This first phase of the GRAVITON Project will be divided in two sub-phases: GRANTED (Gravitational Radiation ANtenna TEchonology Demonstration) and actual Detector (Mário Schenberg Detector).

In the GRANTED sub-phase, we will test the feasibility to construct a heavy ($\sim$ 800kg) Cu-Al (94%-6%) buckyball mass with a high ($Q_m > 10^7$) mechanical Figure of Merit below 50mK, the feasibility to quickly cool it down to $T \sim$ 50mK, the feasibility to protect such a buckyball mass from laboratory vibrations, and to extract and analyze data using a single transducer.

After completing the technology demonstration phase (GRANTED), we intend to construct and place a set of high sensitivity transducers attached to the surface of the buckyball, transforming the above prototype into an effective gravitational wave detector, the Mário Schenberg detector.

The antenna would be covering a range (3.9-4.3 kHz and 7.5-8.3 kHz) of astrophysical significance and would be operating in coincidence with some large laser interferometer detectors, such as LIGO and VIRGO, which can also cover these high frequencies with similar sensitivities [5].

The astrophysical events that we anticipate as probable candidates for detection are: neutron stars going to hydrodynamical instability [6]; excitation of the f, p and w modes of a neutron star [7], excitation of the quasi-normal modes of three solar masses black holes [8]. We also can speculate some "exotic" sources such as sub-milisecond rotation of bosonic or strange matter stars, and inspiralling of mini-black holes.

ACKNOWLEDGMENTS

We acknowledge FAPESP (São Paulo, Brazil) for financial support.

REFERENCES

1. N.S. Magalhães, O.D. Aguiar, W.W. Johnson, and C. Frajuca, "Possible Resonator Configurations for the Spherical Gravitational Wave Antenna", *General Relativity and Gravitation* **29** (12), 1511-25 (1997).
2. D.M. Eardley et al., "Gravitational-Wave Observations as a Tool for Testing Relativistic Gravity", *Phys. Rev. Lett.* **30** (18), 884-6 (1973).
3. S.M. Merkowitz, private communication (1997).
4. N.S. Magalhães, W.W. Johnson, C. Frajuca, and O.D. Aguiar, "A Geometric Method for Location of Gravitational Wave Sources", *Astrophysical Journal* **475**, 462-8 (1997).
5. G.M. Harry, T.R. Stevenson and H.-J. Paik, "Detectability of gravitational wave events by spherical resonant-mass antennas", *Physical Review* **D54** (4), 2409 (1996).
6. J.L. Houser, J.M. Centrella, and S.C. Smith, "Gravitational Radiation from Nonaxisymmetric Instability in a Rotating Star", *Phys. Rev. Lett.* **72** (9), 1314-7 (1994).
7. N. Anderson and K.D. Kokkotas, *Phys. Rev. Lett.* **77** (20), 4134 (1996).
8. N. Anderson, M.E. Arajo and B.F. Schutz, "Generalized Bohr-Sommerfeld Formula for Schwarzschild Black Hole Normal Modes", *Class. Quantum Grav.* **10** (1993), 757-765.

An Advanced Inductive Transducer for Resonant Mass Gravitational Wave Detectors

Gregory M. Harry[1], Insik Jin, Ho Jung Paik,
Thomas R. Stevenson[2], and Frederick C. Wellstood

*Center for Superconductivity Research, Department of Physics, University of Maryland,
College Park, Maryland 20742*

Abstract. We report on the sensitivity of a transducer which uses a low noise SQUID amplifier. The transducer is made of niobium and was electropolished and heat treated to improve the Qs. The mechanical Q was found to be 3.15×10^6 at 4 K and the electrical Q was 2.52×10^5. With a low noise SQUID designed for this transducer, the noise temperature of the transducer was found to be 1.1 mK, of which 0.7 mK was due to the SQUID. This is about 200× larger than the expected noise from the SQUID. Noise from flux trapped on the chip is the most likely cause.

We have built and tested an inductive transducer with a resonant frequency near 900 Hz. The transducer uses a niobium proof mass between two coils of superconducting coils. The gap between the coils and the mass was measured to be 10 μm at 4.2 K and 25 μm at room temperature. The proof mass was electropolished and then heat treated to a maximum temperature of 1500°C. The Q measured in the transducer at 4.2 K was

$$Q_{\text{tot}} = 2.60 \times 10^6, \tag{1}$$

which can be expressed as a mechancial Q, 3.15×10^6, an an electrical Q, 2.52×10^5.

The additive noise comes mostly from the first stage amplifier. We used a two SQUID chip, the first acting as a preamp for the second. The first SQUID was kept in a flux-locked loop by modulating the second SQUID with a 500 kHz square-wave signal. A second feedback loop was employed to keep the flux gain between the SQUIDs at a maximum [1]. Noise measurements were made, with both loops operating, in a band of 62.5 Hz around the transducer resonance of 892 Hz (which corresponds to 6.5 A stored in the coils). The resulting noise spectrum, expressed

---

[1] Present address: Department of Physics, Syracuse University, Syracuse, New York 13244-1130
[2] Present address: Dept. of Applied Physics, Yale University, PO Box 208284, New Haven, Connecticut 06520-8284

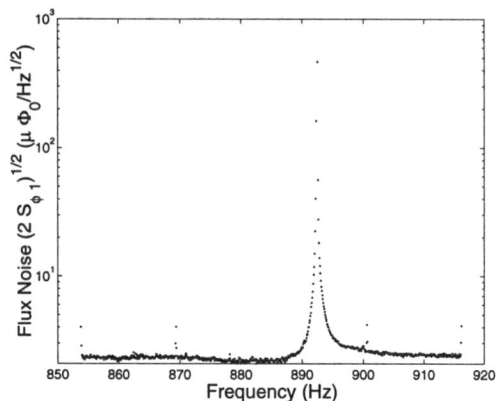

FIGURE 1. Noise spectral density near the transducer resonance expressed as flux in the first SQUID.

as flux in the first SQUID, is shown in Figure 1. The asymmetry in the flux-noise peak near the resonance indicates an excess of back-action noise from the SQUID.

The total noise was fit to the expected signal-to-noise ratio density [2] and the transducer noise temperature was found to be

$$T_n = 1.08 \times 10^{-3} \text{ K}. \qquad (2)$$

The thermal force noise was then subtracted off and the noise due solely to the SQUID amplifier was found:

$$T_s = 6.99 \times 10^{-4} \text{ K}. \qquad (3)$$

The expected noise for the first SQUID from CTG theory [3] would be

$$T_{\text{CTG}} = 3.5 \times 10^{-6} \text{ K}. \qquad (4)$$

Most of this excess noise can be understood as excess voltage, rather than current, at the input of the first SQUID. Calculations are consistent with the conclusion that this noise is due to excess flux trapped in the impedance matching transformer at the input of the SQUID chip [4].

REFERENCES

1. T. R. Stevenson, oral presentation at Fourteenth International Conference on General Relativity and Gravitation (Florence Italy, 1995).
2. J. C. Price, Phys. Rev. D **56**, 564 (1987).
3. J. Clarke, C. D. Tesche, and R. P. Giffard, J. Low Temp. Phys. **37**, 405 (1979).
4. G. Harry, Ph.D. thesis, University of Maryland, College Park, 1999.

Perspectives on Transducers for Spherical Gravitational Wave Detectors

Carlos Frajuca

*Centro Federal de Educação Tecnológica de São Paulo,
Rua Pedro Vicente 625, São Paulo, SP 01109-010, Brazil*

Odylio D. Aguiar, Nadja S. Magalhães,
Kilder L. Ribeiro and Luiz Alberto Andrade

*Divisão de Astrofísica, Instituto Nacional de Pesquisas Espaciais,
Av. Astronautas 1758, São José dos Campos, SP 12227-010, Brazil*

Abstract. At present, transducers of three different types are being developed for gravitational wave detectors: the so-called passive transducer (inductive or capacitive), the parametric microwave transducer and the optical transducer. In this work we simulate the performance of a spherical detector equipped with transducers of each type separately. We calculated the noise temperature and the burst strain sensitivity for each case.

INTRODUCTION

This work is part of the design of the "Mario Schenberg Detector", a Spheroidal Gravitational Wave Detector of 0.6 meter diameter, which is the goal of the first phase of the Graviton Project.

We simulate using a noise model [1] and reasonable values for the system parameters, the performance of the detector equipped with six two-mechanical mode transducers on the antenna surface in a dodecahedron distribution [2].

GENERAL CARACTERISTIC OF THE SPHERICAL ANTENNA

Size: 60 cm diameter Material: CuAl – 94%-6%
Sphere: mass = 800 kg, mechanical quality factor (Q_m) = 10 M
First Mode Transducer: effective mass = 0.41 kg , Q_m = 10 M
Second Mode Transducer: effective mass = 0.001 kg, Q_m = 50 M
Resonant Frequency: 4200 Hz, bandwidth ~ 300 Hz
Thermodynamic Temperature: 0.05 K

Parametric Microwave Transducer [3]

Pump Frequency: 10^{10} Hz
Electrical Coupling: 0.3
df/dx: 3×10^{14} Hz/m
Amplifier Temperature: 6 K
Microwave Power: 10^{-7} W
Pump Phase Noise: -145 dBc/Hz
Pump Amplitud Noise: -165 dBc/Hz
Electrical Quality Factor: 1 Million
Thermodynamic temperature = 0.05 K
→ Noise Temperature: Tn = 4.53×10^{-6} K
Burst strain sensitivity: 8.32×10^{-20}

Inductive Transducer [4]

SQUID Noise Number: 35
Reduced Inductance: 200 pH
Coupling (sensor SQUID): 0.65
Shunt Resistance per Junction: 6 ohms
Thermodynamic temperature = 0.05 K
→ Noise Temperature: Tn = 5.33×10^{-6} K
Burst strain sensitivity: 9.03×10^{-20}

Optical Transducer [5]

Transducer-Cavity Length: 0.01m
Transducer-Cavity finesse: 3×10^5.
Sensing-Cavity Length: 0.2 m
Sensing-Cavity finesse: 3×10^5
Photodiode Sensitivity: 0.6 A/W
Laser Frequency: 2.82×10^{14} Hz
Laser Power: 0.01 W
Laser Power Spectral density
   at 1 kHz: 10^{-18} W^2/Hz
Piezoelectric const.: 8×10^{-11} m/V
Piezo driver voltage noise:
   4×10^{-17} V^2/Hz
Piezo impedance (real part)
   at 1 kHz: 20 Ω
→ Noise Temperature:
Tn = 1.42×10^{-6} K
Burst strain sensitivity:
4.65×10^{-20}

ACKNOWLEDGMENTS

C.F. and O.D.A. acknowledge FAPESP (Brazil) for financial support, and K.L.R. and L.A.A. acknowledge CAPES and CNPq for financial support.

REFERENCES

1. Norbert Solomonson, W.W. Johnson, and W.O. Hamilton, "Comparative Performance of Two-, Three-, and Four-mode Gravitational Radiation Detectors", *Phys. Rev.* **D46** (6), 2299 (1992); Carlos Frajuca, private communication (1999).
2. Warren W. Johnson and Stephen M. Merkowitz, "Truncated Icosahedral Gravitational Wave Antenna", *Physical Review Letters*, **70**, 16 (1993).
3. Michael Edmund Tobar and David Gerald Blair, "Sensitivity Analysis of a Resonant Mass Gravitational Wave Antenna with a Parametric Transducer", *Rev. Sci. Instrum.* **66(4)**, 2751-2759 (1995).
4. I. Jin, A. Amar, T. R. Stevenson and F. C. Wellstood, "35 h-barr Two- Stage SQUID System for Gravity Wave Detection", *IEEE Transactions on Applied Superconductivity*, **7**, 2 (1997).
5. L. Conti et al., "Optical Transduction Chain for gravitational Wave Bar Detectors", *Review of Scientific Instruments*, **69**,2 (1998).

The Auriga Ultracryogenic Test Facility: A New Capacitive Resonant Transducer.

V. Crivelli-Visconti[†], J.-P. Zendri[¥], L. Taffarello[¥], G.A. Prodi[•], S. Vitale[•],
M. Cerdonio[†], M. Bonaldi[*•], P. Falferi[*•], R. Mezzena[•], A. Mattioli[•]

[†]*Phys. Dep.Univ. Padova and INFN, sec. Padova –Italy. http://axln01.lnl.infn.it/*
[¥]*INFN sec. Padova - Italy*
[•]*INFN Gr. Coll. Trento – Italy*
[*]*CEFSA, ITC-CNR, Trento - Italy*

Abstract. Sensitivity of presently working gravitational wave resonant detectors is limited by the transduction chain – resonant transducer, impedance matching circuits and SQUIDs. For this reason the Auriga[1] collaboration has started in 1998 the construction of an ultracryogenic test facility where the entire transduction chain can be tested and improved. Mechanical isolation will easely allow for intrinsic thermal noise measurement. In this poster we present the design and first measurements of a new resonant capacitive transducer, intended to widen the detector bandwidth up to 50 Hz with the available new generation of SQUIDs.

The capacitive resonant transducer that we present here was the result of a long work of theoretical optimization of the transduction chain used in resonant bar gravitational wave detectors. In particular we built an equivalent electrical model of the system[2] that includes the bar, the transducer, a high Q RLC resonator for impedance matching and a DC-Squid. We could move parameters of the transducer like its effective mass[3] and resonant frequency, and of the electric resonator, like its central frequency and Q. We explored the space of these parameters using two different kind of Squid's, both available and beeing tested by the Auriga Group in Trento.

Calculations are algebraic. Bandwidth has been calculated using the following equations.

$$\Delta v = \frac{I_1^2}{2\pi I_2}$$

$$I_1 = \frac{1}{2\pi} \int_{-\infty}^{+\infty} SNR(\omega) d\omega \qquad (1)$$

$$I_2 = \frac{1}{2\pi} \int_{-\infty}^{+\infty} SNR^2(\omega) d\omega$$

Bandwidth is the parameter we used most to rank possible configurations. This is because we found out that sensitivity is a much more tolerant parameter and it stays high in a much wider parameter area. Results of simulations are summarized here below, for what concerns the best transducer configuration and relative achievable bandwidth and sensitivity.

TABLE 1. Transducer Optimal Configuration.

Parameter	Optimal Value
Transducer Mass	3.5 kg
Transducer Frequency	887 Hz
Electrical Mode ($Q=10^5$)	952 Hz
Bandwidth	25 Hz

Based on results from simulations we designed and built a new capacitive transducer. We modified the usual 'mushroom' shape of the resonator in order to increase mass without moving frequencies too much. The model was simulated in detail using Pro/Mechanica[4] to make sure no dangerous resonance were introduced, that the central mode was were needed and that stresses were not too concentrated to avoid thermoelastic dissipations. Below is a picture showing the final transducer design.

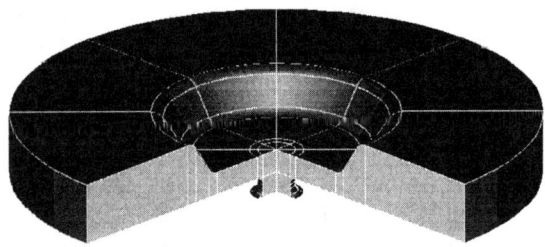

FIGURE 1. The final transducer design.

In a room temperature test we could measure the transducer central frequency at 885 Hz, which is more than satisfactory given our goal and all the numerical and mechanical error involved. Mechanical Q is less than 3000 and needs to be increased. More measure are scheduled as soon as the ultracryogenic facility will be operative.

REFERENCES

[1] See J.P. Zendri et al., elsewhere in these Proceedings
[2] M. Cerdonio et al., Physica B, 194-196 (1994) 3-4
[3] P. Rapagnani, Y. Ogawa, Nuovo Cimento 7C, N.1 (1984), pg. 21
[4] See http://www.ptc.com/

Status Report of the Gravitational Wave Detector AURIGA

J.P. Zendri#, L. Baggio†, M. Bonaldi*, M. Cerdonio†, L.Conti °, V. Crivelli Visconti†, P. Falferi*, P.L. Fortini×, V. Martinucci°, R. Mezzena°, A. Ortolan∇, G.A. Prodi°, G. Soranzo#, L. Taffarello#, G. Vedovato∇, A. Vinante°, S. Vitale°

#INFN sec. Padova - Italy
†Phys. Dep. Univ. Padova and INFN, sec. Padova -Italy
*Cefsa and INFN Gr. Coll. Trento - Italy
° Phys. Dep. Univ. Trento and INFN Gr. Coll. Trento - Italy
× Phys. Dep. Univ. Ferrara and INFN sec. Ferrara-Italy
∇INFN Lab. Naz. Legnaro - Italy

Abstract. We present the status of the ultracryogenic gravitational wave detector AURIGA, which is taking data since may 1997 with an energy sensitivity in the mK range and bandwidth greater than 1 Hz. The typical detector output is summarized in daily reports which are important tools for detector diagnostic and for checking the vetoes of periods of unsatisfactory operation of the detector.

THE AURIGA DETECTOR

The ultracryogenic gravitational wave detector AURIGA [1] is data taking since May 1997 [2]. The recorded data cover a fraction of about 80% of the elapsed time. The rest of the time has been employed for detector ordinary (He4 main bath refill and 1K-pot filling) and extraordinary (cryogenic or electronic failures) maintenance or for calibration operations. An automatic vetoing procedure [3] is applied to the data in order to select the time periods when the detector noise is almost Gaussian and stationary and the assumption of a two mode system dynamics is accurate. The resulting overall duty cycle is of the order of 25%. However it could be consistently increased once filtering procedures are able to handle the presence of spurious peaks appearing around the two relevant modes.

The acquired data are summarized in daily reports which cover the detector noise, the analysis output, the efficiency of the filter adaptive procedure, the vetoes and the monitor of ambient disturbances (seismometers). These reports are consulted for the certification of the AURIGA events list before sharing these data with the

IGEC [4] collaboration.

Recent Improvements and Current Performances

During the data taking period many attempts have been performed to increase the duty cycle and improve the sensitivity. In cryogenic circle a continuous liquid helium transfer from an external dewar with a capacity of 3500 liters to the main AURIGA cryostat has been implemented. This operation does not affect the detector sensitivity and increases helium consumption by less than 10%. The liquid helium refill procedures are thus reduced to one external dewar substitution each $20 \div 25$ days. Many attempts to continuously fill the 1K-pot has been also performed: in this case the filling procedures turn out to be very noisy unless the liquid level in the pot ranges in a very narrow interval. In order to keep the level almost constant in time we are developing a computer controlled refilling procedure.

A sensitivity enhancement has been obtained substituting the commercial room temperature electronic controls of the SQUID with an up-graded version. This model allow the SQUID noise parameters optimization without any stability problem. The resulting white noise power spectrum is $4\mu\Phi_0/\sqrt{Hz}$ at $1.8\ K$ and $3.5\mu\Phi_0/\sqrt{Hz}$ at $0.5\ K$, about a factor two better than in the previous configuration. These values are approximately equal to the noise amplitude we measured for the SQUID alone in a separated bench test before the AURIGA final assembling. As a consequence of the amplifier noise improvements, the detector bandwidth increased up to about 2 Hz and the equivalent noise strain power spectrum is less than $6 \times 10^{-21}\ /\sqrt{Hz}$ in a frequency span of 40 Hz around the two modes, while the minima, at the mode frequencies ($911\ Hz$ and $929\ Hz$), were as low as $4 \times 10^{-22}\ /\sqrt{Hz}$.

In terms of energy the present SQUID amplifier sensitivity is about $4000\hbar$ that corresponds to a noise temperature of $T_n = 0.2\ mK$. Thus according to the Giffard limit the expected detector best effective temperature should be $T_{eff} = 2T_n = 0.4\ mK$ while the measured AURIGA best energy sensitivity is only slightly less than $1\ mK$. The reason of this excess can't be attributed to the oscillators thermal noise since, by a simple estimation, the thermal contribution accounts for less than one half of the excess. The origin of this unknown noise source is still under investigation.

REFERENCES

1. M. Cerdonio et al. in *First E. Amaldi Conf. on Gravitational Waves*, edited by E. Coccia et al. et al. (World Scientific, Singapore, 1996), p. 176
2. G. A. Prodi et al. in *Second E. Amaldi Conf. on Gravitational Waves*, edited by E. Coccia et al. (World Scientific, Singapore, 1998), p. 148.
3. G. A. Prodi et al. and L. Baggio et al. in these proceedings.
4. see the Web site "http://igec.lnl.infn.it".

Noise and Signal Reconstruction and Characterization in the AURIGA Detector

L. Baggio [a], M. Cerdonio [a], V. Martinucci [c], A. Ortolan [b], G.A. Prodi [c], L. Taffarello [d], G. Vedovato [b], S.Vitale [c], J.P. Zendri [d]

[a]*Univ. of Padova and INFN (Sezione di Padova)* [b]*INFN (Laboratori Nazionali di Legnaro)*
[c]*Univ. of Trento and INFN (Gruppo Collegato Sezione di Padova)* [d]*INFN (Sezione di Padova)*

Abstract. Both the noise power spectrum and signal transfer function must be well known to reliably extract candidates of gravitational wave (GW) signals using linear Wiener filters. We review AURIGA data analysis techniques relative to post-filtering statistical tests and validation.

The experimentally measured power spectrum density (PSD) of the noise $S(\omega)$ in AURIGA detector is closely fitted by assuming a model of two coupled harmonic oscillators (the bar and the transducer) plus the amplifier wide band electronic noise. The complex poles and zeros $\{p_k, q_k\}$ are just what we need to build the *whitening filter* $L(\omega)$ for this noise (defined by $S(\omega) \equiv S_0 L(\omega) L^*(\omega)$), and also the complete Wiener-Kolgomorov (WK) filter $F^{(\delta)}(\omega)$ matched to a δ-like gravitational event:

$$L(\omega) = \prod_{k=1}^{2} \frac{(i\omega - p_k)(i\omega - p_k^*)}{(i\omega - q_k)(i\omega - q_k^*)}, \quad F^{(\delta)}(\omega) = N(q_k) L(\omega) \prod_{k=1}^{2} \frac{i\omega}{(i\omega + q_k)(i\omega + q_k^*)}, \quad (1)$$

where $N(q_k)$ is a normalization factor.

Granted that a two-poles-and-zeroes model is appliable in the first place, we need to check that we guessed the true parameters. The mode frequencies $\omega_k = \operatorname{Im} p_k \approx \operatorname{Im} q_k$ are followed with digital lock-ins in the raw data, while the quality factors of the modes ($Q_k \approx \omega_k / 2 \operatorname{Re} p_k$) are just measured once per acquisition run, as they depend on major setup parameters of the detector. The post-filtering bandwidths ($\approx 2 \operatorname{Re} q_k$) are corrected on an hourly basis by a feedback on the residual 'color' around the modes in the whitened data PSD (see FIGURE 1d). A big short-lived excitation that enters the system –either GW signal or spurious– spoils the whitened noise PSD estimate, but also the histogram of WK filtered data (FIGURE 1e,f), so we take care of this by freezing the parameters estimation when non-gaussian behaviour is detected.

Event search is model dependent as well. A candidate δ-like event is a pattern in the WK filter output with a specific mix of an exponential decay ($\tau \equiv -\max_k\{\operatorname{Re}(q_k)\}$) a beat modulation ($\omega_* \equiv \operatorname{Im}(q_2) - \operatorname{Im}(q_1)$) and a carrier wave ($\omega_0 \equiv \frac{1}{2}[\operatorname{Im}(q_2) + \operatorname{Im}(q_1)]$):

$$f_{WK}(t) \approx A e^{-t/\tau} \cdot \cos(\omega_* t) \cos(\omega_0 t) \qquad (2)$$

We locate precisely its maximum by interpolation (FIGURE 2), and then wait at least 3 decay times before accepting a new event.

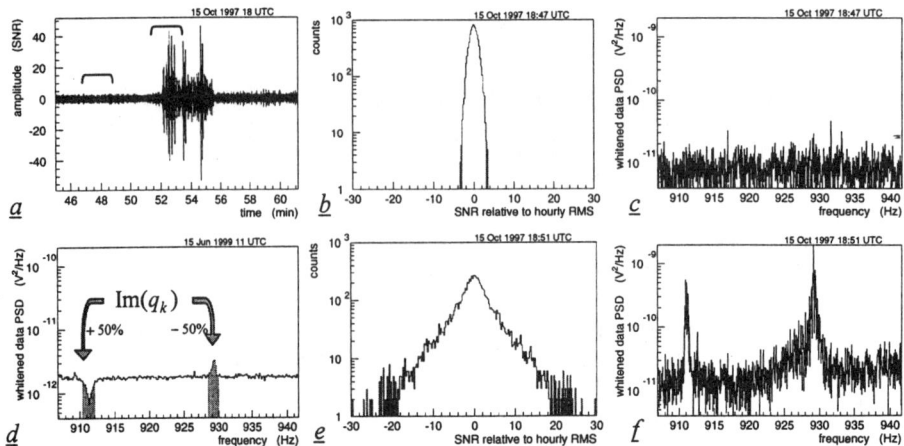

FIGURE 1. The filtered data ($\underline{a}$) are divided into buffers of 2 minutes. The two marked with brackets have quite different statistical distributions (resp. $\underline{b}$ and $\underline{e}$), particularly on tails beyond 3 times the Root Mean Square (*in gray*). The non-gaussian buffer is not let enter the effective noise temperature estimate (which is a RMS moving average). Notice that in the same 'bad' buffer it seems that the whitening filter is no more working properly (see $\underline{c}$ and $\underline{f}$), in particular it mimics a displacement of the zeros q_1 and q_2. Compare it with the effect on whitened PSD of a ±50% error on the post-filtering bandwidth ($\underline{d}$).

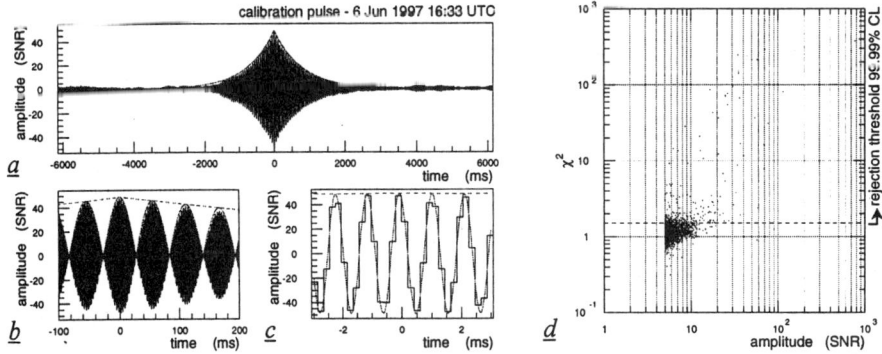

FIGURE 2. – $\underline{a},\underline{b},\underline{c}$: A high SNR event extracted in AURIGA normal operation data. A good match with the model is found at different scales (*gray line*). – $\underline{d}$: Thresholds on SNR and χ^2 are necessary to select from candidate events those with reliable time of arrival and amplitude estimates.

WK filtering is a maximum-likelihood fit based on models for both the noise and the signal, so if we can trust the first one, then passing a χ^2 test is a necessary and sufficient condition for time-of-arrival and amplitude estimation to make sense. The χ^2 test is used to discriminate between fast mechanical (or gravitational) bursts on the bar and other spurious events –e.g. electromagnetic pulses on the amplifier– with an efficiency which has a quadratic dependence on Signal-to-Noise-Ratio (FIGURE 2$\underline{d}$).

REFERENCES

1. G.A. Prodi *et. al.* elsewhere in these proceedings.
2. A. Ortolan *et. al.* in 2nd *E. Amaldi Conf. On Grav. Waves* (World Scientific, Singapore, 1998), p.204.

Recent results of NAUTILUS

P. Astone[1], M. Bassan[2,3], P. Bonifazi[4,1], P. Carelli[5,3],
E. Coccia[2,3], S. D'Antonio[6], V. Fafone[6], A. Marini[6], E. Mauceli[6],
Y. Minenkov[2,3], I. Modena[2,3], G. Modestino[6], A. Moleti[2,3],
G.V. Pallottino[7,1], M.A. Papa[6], G. Pizzella[2,6], F. Ronga[6],
R. Terenzi[4,3], M. Visco[4,3], L. Votano[6]

[1] *INFN, Sezione di Roma 1, Piazzale A. Moro 2, 00185 Rome, Italy*
[2] *Dip. Fisica, Università di Roma "Tor Vergata", V. Ricerca Scientifica 1, 00133 Rome, Italy*
[3] *INFN, Sezione di Roma 2, V. Ricerca Scientifica 1, 00133 Rome, Italy*
[4] *CNR, Istituto Fisica Spazio Interplanetario, V. Fosso del Cavaliere, 00133 Rome, Italy*
[5] *Dip. Ingegneria Elettrica, Università de L'Aquila, 67040 Monteluco di Roio (AQ), Italy*
[6] *INFN, Laboratori Nazionali di Frascati, V. E. Fermi 40, 00044 Frascati (Rome), Italy*
[7] *Dip. Fisica, Università di Roma "La Sapienza", P. A. Moro 2, 00185 Rome, Italy*

Abstract. NAUTILUS started its second run of continuous data taking in june 1998. The measured strain sensitivity at the two resonances is $4\,10^{-22}/\sqrt{Hz}$ over a bandwidth of 1 Hz and better than $3\,10^{-20}/\sqrt{Hz}$ over a band of about 25 Hz, with a duty cycle mainly limited by cryogenic operations. A summary of the most recent results obtained with the NAUTILUS data is given.

NAUTILUS, the ultracryogenic resonant detector operated by the ROG collaboration at the INFN Frascati National Laboratories [1], has undergone, in the first half of 1998, a partial overhaul of its mechanical suspensions and thermal contacts. The results obtained since June 1998 show a considerable improvement in the rejection of non stationary noise and in the sensitivity of the apparatus. We show in Figure 1 the strain sensitivity of the detector, expressed in units of $Hz^{-1/2}$, and in Figure 2 the detector noise temperature over three days in september 1999. This sensitivity allows the detection at $SNR = 1$ of an impulsive gravitational wave (GW) signal of duration 1 ms at amplitudes $h \simeq 4\,10^{-19}$.

Analyses using the NAUTILUS data, just published or in progress, are the following:

− A search for coincidences among three resonant mass detectors: EXPLORER operated by the Rome group at CERN, NIOBE in Perth of the UWA group and NAUTILUS [2]. The analysis is based on the comparison of candidate event lists recorded by the detectors in the period December 1994 through October 1996. Due to the different periods of data taking it was not possible to search for triple coincidences. The results have been: a weak coincidence excess with respect to the accidental ones between EXPLORER and NAUTILUS and no coincidence excess between EXPLORER and NIOBE.

– The first crosscorrelation analysis between two cryogenic detectors for the measurement of the GW stochastic background [3]. The results obtained with the data of EXPLORER and NAUTILUS, tuned in February 97 at 907 Hz gives a value of $\Omega(f) \leq 60$.

– The passage of cosmic rays has been observed to excite the NAUTILUS resonant mode [4]. A very significant correlation (more than 10 standard deviations) is found (see P. Astone et al, this conference).

– A search for monochromatic signal is in progress. The present sensitivity of NAUTILUS allows the detection at $SNR = 1$ of a continuous GW signal around 1 kHz of amplitude $h \simeq 5\,10^{-26}$ with an abservation time of 100 days.

– Analysis of the 1997-1998 data released under the IGEC (International Gravitational Event Collaboration) protocol is also in progress. The results of the search for concidence due to short GW bursts among all the five IGEC resonant detectors will be reported in the forthcoming GWDAW conference (Rome, December 1999).

REFERENCES

1. Astone P. et al. (ROG collaboration), *Astroparticle Phys.* **7**, 231-243 (1997).
2. Astone P. et al. (ROG collaboration), *Astroparticle Phys.* **10**, 83-92 (1999).
3. Astone P. et al. (ROG collaboration), *Astron. Astrophys.* **351**, 811-814 (1999).
4. Astone P. et al. (ROG collaboration), *Phys. Rev. Lett.* (1999).

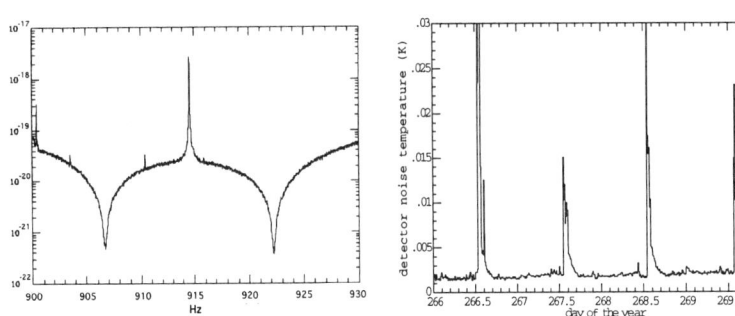

FIGURE 1. Experimental strain sensitivity of NAUTILUS (input noise spectral amplitude in units of $Hz^{-1/2}$). The sensitivity at the two resonances is about $4\,10^{-22}\,Hz^{-1/2}$. The spectral amplitude is better than $3\,10^{-20}\,Hz^{-1/2}$ over a band of about 25 Hz. The peak at 914.6 Hz is a calibration reference signal fed into the dcSQUID amplifier to monitor the gain of the electronics.

FIGURE 2. Detector noise temperature versus time, averaged every ten minutes, over a period of three days. A noise temperature of 2 mK corresponds to a minimum detectable amplitude h of a GW burst of duration 1 ms $h = 4\,10^{-19}$. The peaks are due to a daily cryogenic operation.

Improving the Sensitivity of Resonant Detectors with Advanced Linear Capacitive Transducers

M.Bassan and Y.F.Minenkov
for the ROG Collaboration

Dipartimento di Fisica, Universita' Tor Vergata and I.N.F.N.- Sezione Roma2
00133 Roma - Italy

Abstract. A new resonant transducer has been assembled on the Explorer detector. Some of its characteristiscs are recalled, and the perspectives for sensitivity gain are discussed

Introduction

Resonant transducers are essential component of high sensitivity resonant g.w. antennas. They have been developed over the years, starting from the pioneering work of H.J.Paik, in several varieties:
-diaphragm, mushroom, or beam mechanical resonators
-inductive, capacitive, parametric or optical electromechanical coupling
In most cases, sensitivity (in terms of both detection noise temperature and bandwidth) depends crucially on the size of the gap **d** between the resonator and the opposite surface (electrode, coil or cavity), where the e.m. field is stored.
In a linear transducer, as the gap is reduced (and the coupling α increased) a larger resonating mass m_t is required to optimally match the antenna to the amplifier, according to the equation [1]:

$$m_t = 2\pi \frac{\alpha}{V_n} \sqrt{\frac{2k_B T M_a}{\omega_a^3 Q}} \qquad (1)$$

with the usual meaning of symbols. Eq.(1) shows that a heavier transducer is also desirable when better amplifiers (lower V_n) are used.

The New, Small Gap Transducer of The ROG Group

We have developed [1,2,3] an improved linear capacitive resonant transducer with a gap **d** $\approx$ **10μm**, based on the "rosette" design, to be used on the cryogenic antennas EXPLORER and NAUTILUS of the ROG group. We recall some of the relevant features of this device:

THE RESONATOR has a resonating mass that can be adjusted between 0.3 and 2 kg without major design changes. Its frequency, mass and surface finish can be independently adjusted with precision machining on different surfaces. The vibrating surface moves of parallel motion, so that the effective dynamical mass coincides with the physical mass. The resonating mass and support rim are e.d.m. machined in a single plane layout, to minimize machining tolerances on flatness and parallelism.

THE FIXED ELECTRODE has no insulating spacers in the (axial) direction of elastic wave; it is assembled with thermal shrink fit of the insulated electrode, to minimize the number of bolts used. Its stray capacitance is small (about 2% of the active capacitance) and measured with precision.

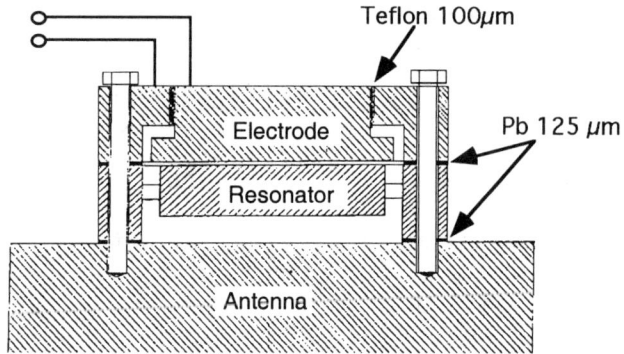

FIGURE 1. Schematic of the rosette transducer as assembled and mounted on the antenna. In actual operation the assembly is rotated by 90°, so that the antenna axis is horizontal.

THE ASSEMBLED TRANSDUCER has a small gap ($10\mu m$), obtained via precision milling and careful hand finish of the two electrode surfaces. Its Q is rather good ($>1.5 \; 10^6$), obtained via attentive cleaning and careful contact of mating metal surfaces via soft Pb interface shims. It shows excellent breakdown properties, as it well stands a field larger than $2 \cdot 10^7 MV/m$.

The New Transducer on the EXPLORER Detector

In the first months of 1999 Explorer underwent an upgrade of its suspensions and the change of its readout system with the new, high coupling transducer, nicknamed RE2, and a new, high sensitivity SQUID [4]. In a trial run, recently completed, the transducer has shown excellent stability and tuning features, with Qs $\sim 10^6$ even with large bias field. However, we found that the advanced double SQUID configuration needs further developing. Explorer will be cooled again in early 2000, but we can try and predict (see figure 2) the spectral sensitivity it might reach, based on the measured parameters of the transducer, as a function of the attainable wide-band noise level of the SQUID amplifier.

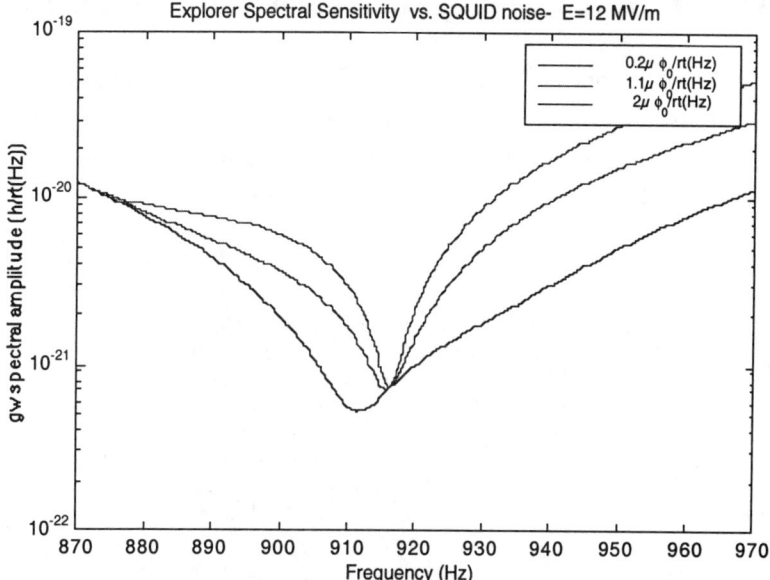

FIGURE 2. Spectral sensitivity of Explorer, depending on the final SQUID flux noise we shall achieve.

The lowest curve corresponds to a remarkable burst sensitivity $T_{eff} = 0.15$ mK, and a bandwidth of 9 Hz.

Note that, due to high coupling and low SQUID noise, the usual normal mode picture, with two separate sensitivity peaks, breaks down and only one SNR peak is present. Note also that, on the resonance ($f_a = 915$ Hz) the value of S_h does not depend on amplifier noise.

REFERENCES

[1] M.Bassan, Y.Minenkov, G.Zaccarian: *"Resonators of novel geometry for large mass resonant transducers"* in Proc. of the first Edoardo Amaldi Conference - E.Coccia, G.Pizzella, F.Ronga editors, World Scientific 1995
[2] M.Bassan, Y.Minenkov, R.Simonetti: *"Advances in Linear Transducers for Resonant Gravitational Wave Antennas"* in Proceedings of the Virgo Conference on Gravitational Waves: Sources and Detectors, Cascina, Mar.1996- World Scientific 1997
[3] M.Bassan, Y.Minenkov, V.Fafone P.Bonifazi, M.Visco P.Carelli, M.G.Castellano- For the ROG Collaboration: *"Advances and Perspectives in Capacitive Transducers and Associated Readout "* - in Proceedings of the MG8 Conference - Jerusalem 1997
[4] P. Carelli M.G. Castellano, R. Leoni and G. Torrioli Appl. Phys. Lett.72,115-117 (1998).

Tests on a prototype spherical detector

P.Astone[2], M.Bassan[5,3], P.Bonifazi[6,2], P.Carelli[7,3], E.Coccia[5,3],
S.D'Antonio[4], V.Fafone[1], G.Frossati[8], A.Marini[1], E.Mauceli[1],
S.Merkowitz[1], Y.Minenkov[3], I.Modena[5,3], A.Moleti[5,3],
G.Modestino[1], G.V.Pallottino[4,2], M.Papa[1], G.Pizzella[5,1], G.Raffone[1],
F.Ronga[1], M.Schipilliti[5], R.Terenzi[6,3], M.Visco[6,3], L.Votano[1]

[1] *INFN - Laboratori Nazionali di Frascati, Frascati, Italy*
[2] *INFN - Sezione di Roma 1, Roma, Italy*
[3] *INFN - Sezione di Roma 2, Roma, Italy*
[4] *Università di Roma "La Sapienza", Roma, Italy*
[5] *Università di Roma "Tor Vergata", Roma, Italy*
[6] *CNR- Istituto di Fisica dello Spazio Interplanetario, Roma, Italy*
[7] *Università di L'Aquila, L'Aquila, Italy*
[8] *Kamerlingh Onnes Laboratory, Leiden University, Leiden, The Netherlands*

Abstract. We report on our progresses in the design and construction of the suspension system for a 3 kHz spherical detector. The plan for our SFERA project is to build a small cryogenic spherical detector by the year 2001.

The next generation of resonant-mass gravitational detectors will likely be of spherical shape. A single sphere is advantageous for many reasons. It is capable to detect waves coming from all directions and with any polarizations, and it can determine the direction and the tensorial character of the incident wave. Another great advantage is that a sphere has a larger mass and consequently a larger cross section with respect to a cylindrical bar working at the same frequency. Studies devoted to define a project of a large spherical detector of 40 to 100 tons in mass, cooled at temperature below 100 mK, are under way in Brazil, Italy, The Netherlands and USA [1].

We are designing a prototype spherical cryogenic detector made of copper-aluminum alloy having mass of about 1 ton and resonant frequency around 3 kHz.

An efficient mechanical isolation is one of the main characteristics required for a gravitational wave antenna. The vibration isolation is accomplished in two stages: a room temperature isolation stack, made of multiple alternating layers of metal and rubber and a low temperature stage made of multi-mode metallic isolators. The overall attenuation required at the resonance of our detector is around 350 dB. Most of this attenuation, 200 dB, should be produced by the low temperature isolators.

In our prototype we decided to decouple the suspension system from the cryogenics in order to avoid the noise due to the cooling apparatus (boiling of liquid helium etc.) and to introduce a thermal link mechanically soft, but with high thermal conductivity, between the cryogenics and the last stages of the attenuator.

Some preliminary tests were made on the suspension apparatus at room temperature using a TIGA-shaped 800 kg Al 5056 antenna. The last stage of attenuation is a circular rod supporting the antenna from its surface [2]. The other low temperature stages are made of steel cylinders (120 kg in mass, 304 mm in diameter and 200 mm thick) supported by steel cables (diameter 13.4 mm and length 80 mm). Tests were performed on a two stages system obtaining a transfer function close to that obtained by finite element simulation and an attenuation of 35 dB for each stage at the resonant frequency of the TIGA (Figure 1). These results make possible the design of a cryogenic suspension system for the 3 kHz Cu-Al spherical detector. Our plane is to built this detector by the year 2001.

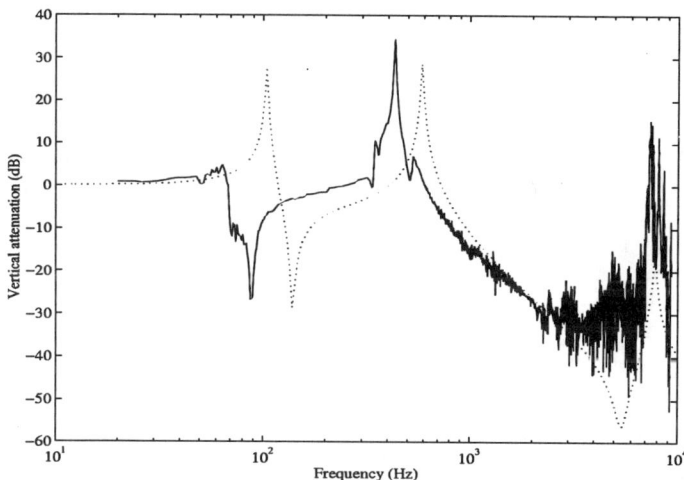

FIGURE 1. Attenuation of one stage, measurements(solid line) and model (dashed line)

REFERENCES

1. Editors Velloso W.F., Aguiar O.D. and Magalhães N.,*Proceedings of the first international workshop on Omnidirectional gravitational radiation observatory*, Singapore, World Scientific, 1997.
2. Merkowitz S.M., Coccia E., Fafone V., Raffone G., Schipilliti M. and Visco M., *Rev. Sci. Instr.* **70**, 1553-1560 1999.

SUSPENSIONS AND THERMAL NOISE

High-sensitivity measurement and control of thermal noise in a cavity

A. Heidmann, P.F. Cohadon, Y. Hadjar and M. Pinard

Laboratoire Kastler Brossel[1], 4 place Jussieu, 75252 Paris Cedex 05, France

Abstract. We performed an experiment with a movable mirror in a high-finesse cavity. We observed the Brownian motion of the mirror at room temperature with a very high sensitivity. We also used an auxiliary beam to cool the mirror by radiation pressure.

When a movable mirror is exposed to a laser beam, its position is coupled to the laser intensity via radiation pressure [1]. This optomechanical coupling plays an important role in quantum limits of high-precision optical measurements. This coupling can be enhanced by using a high-finesse optical cavity which is very sensitive to small mirror displacements. This device allows to study quantum effects of radiation pressure (quantum limits, quantum nondemolition measurement,...) or to observe the Brownian motion of the mirrors.

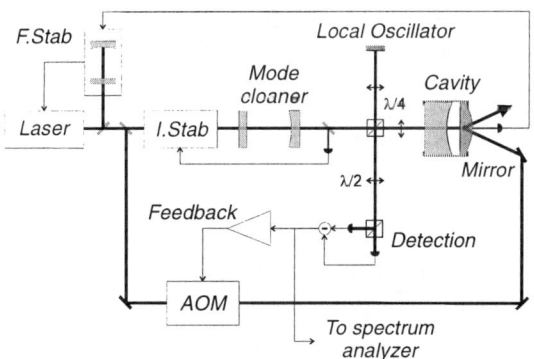

FIGURE 1. Experimental setup. The Brownian motion of the mirror is measured with a high-finesse cavity. A frequency (F. stab.) and intensity (I. stab.) stabilized laser beam is sent into the cavity and the phase of the reflected field is measured by homodyne detection. This signal is fed back to the mirror via the radiation pressure of a modulated auxiliary beam (AOM).

---

[1] Laboratoire de l'Université P. et M. Curie et de l'Ecole Normale Supérieure associé au C.N.R.S.

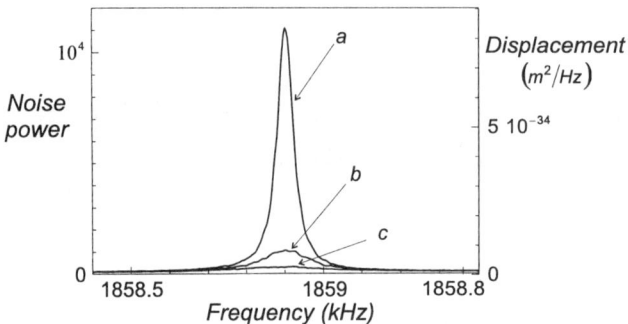

FIGURE 2. Phase noise spectrum of the reflected field normalized to the shot-noise level (scale on the left) and equivalent displacement (scale on the right) for a frequency span of 1 kHz around the fundamental resonance frequency of the mirror. The Brownian motion of the mirror without feedback (a) is reduced by the feedback for increasing gains (b and c).

In our experiment, a Titane-Sapphire laser is locked on a resonance of a single-ended high-finesse optical cavity (Figure 1). The totally reflecting back mirror is coated on a small plano-convex mechanical resonator. The phase of the field reflected by the cavity is measured by homodyne detection and reflects the mirror motion.

We observed the thermal motion of the moving mirror both at the mechanical resonance frequencies and far on the tails of the resonances. The sensitivity is limited by the quantum noise at a level of $2 \times 10^{-19} \text{m}/\sqrt{\text{Hz}}$ [2]. Optical excitation by an auxiliary laser beam reflected from the back on the moving mirror was used to study the mechanical response of the mirror to an applied force.

We have also cooled the mirror by the radiation pressure of light (Figure 2) [3]. The intensity of the auxiliary beam is modulated by an acousto-optic modulator driven by the signal of the homodyne detection. This beam exerts a radiation pressure whose phase can be adjusted in order to increase the damping of the resonator without adding any thermal fluctuations. Cooling up to a factor of 20 was obtained.

Such a cooling mechanism may be applicable to gravitational-wave interferometers to reduce the internal thermal noise of the mirrors [3]. This would increase the sensitivity of the interferometer in the intermediate frequency domain.

REFERENCES

1. Pinard M., Hadjar Y. and Heidmann A., *Eur. Phys. J. D* **7**, 107 (1999).
2. Hadjar Y., Cohadon P.F., Aminoff C.G., Pinard M. and Heidmann A., *Europhys. Lett.* **47**, 545 (1999).
3. Cohadon P.F., Heidmann A. and Pinard M., quant-ph/9903094, to be published in *Phys. Rev. Lett.* (1999).

The Maraging Steel Blades of the Virgo Super Attenuator

S. Braccini,[3] C. Casciano,[3] F. Cordero,[1] F. Corvace,[1] M. De Sanctis,[4] R. Franco,[1] F. Frasconi,[3] E. Majorana,[2] R. Passaquieti,[3] G. Paparo,[1] P. Rapagnani,[2] F. Ricci,[2] A. Solina,[4] and R. Valentini[4]

[1] Ist. di Acustica "O.M. Corbino", CNR Area di Ricerca Roma-Tor Vergata, Roma, Italy
[2] Dip. di Fisica, Univ. di Roma "La Sapienza" and INFN Sezione di Roma 1, Roma, Italy
[3] INFN, Sezione di Pisa, San Piero a Grado (PI), Italy
[4] Dip. Ingegneria Chimica e Scienza dei Materiali, Univ. di Pisa, Pisa, Italy

The Maraging is a steel with particularly high hardness and low creep under stress. Such mechanical properties are obtained through thermal treatments producing finely dispersed precipitate particles, which block the movement of the dislocations. This type of steel is presently used for constructing the blades of the Virgo Super Attenuator[1]. These are the elements that contribute to define the frequency cut off of the seismic noise. The main requirements for the blades are creep under stress and elastic energy loss coefficient as low as possible, in order to improve the stability and effectiveness of the mechanical filters to which the parts of the interferometer are suspended.

We have measured the Young's modulus, shear modulus, Poisson coefficient and the corresponding elastic energy loss coefficients in samples subjected to the same treatments as the parts of the interferometer Virgo[2]. It is found that, in the absence of plastic deformation, the elastic energy loss coefficient under flexural vibrations around 1 kHz can vary by more than one order of magnitude depending on the thermal treatments, and is dominated by the thermoelastic effect. The main reason for such strong variations is supposed to be the dependence of the thermal conductivity on the average sizes and distances between the precipitate particles[2].

We have performed also acoustic emission measurements on the blades of the suspension system of the Virgo. The term "Acoustic Emission" (AE) refers to the release of elastic energy as high frequency wave pulses on behalf of a solid, which undergoes mechanical stress. The detection of AE gives the possibility to study the phenomenon during its time evolution on structures in working conditions. The ultrasonic pulses were recorded while the sample was loaded up to 50 kg. Our measuring apparatus consists of a piezoelectric sensors coupled to the surface of the blade blocked between two wooden supports, on a table with two c-clamp. The signals are amplified (gain factor 40 dB and dc output) with stable maximum sensitivity peaked at 25 kHz and 150 kHz (depending on the frequency response of the piezo-electric sensor). Each single acoustic burst detected by the sensors is thus integrated into a pulse-like voltage signal with a rise time much shorter than 100 ms and then recorded as a time function.

In the figures we report AE data of a Maraging blade taken during two different runs started just after the blade loading. The data of the first plot were obtained with the blade stressed cyclically just few times after the thermal treatments. In the second plot the data were collected few months later after with the same blade that, in the mean time, experienced several stress cycles. By comparing the two plots, it appears a strong reduction of the amplitude and the rate of the AE events. The reason is that the influence of dislocations is considerably reduced during the second run of measurements, proving that the dislocations are blocked after the several cycles of applied stress. In conclusion the deep study of the mechanic and thermal properties of the Maraging blades of the Virgo Super Attenuator results in a definition of an optimised thermal treatment for this material that enhances its hardness and tensil strength and limit the dislocation motion

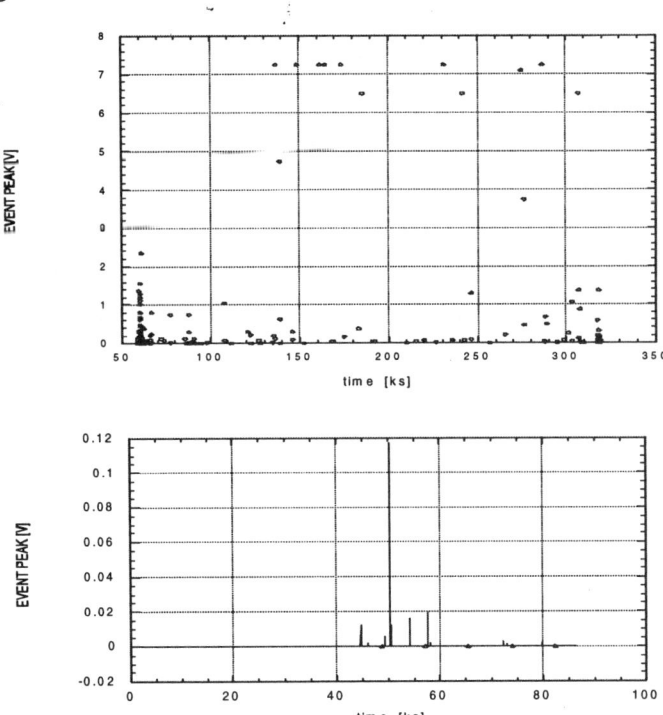

REFERENCES

1 M. Beccaria et al., Classical and Quantum Gravity, 15, (1998), 3339
2 S. Braccini et al., paper submitted to Meas. Scie. Tech.

The ANU Thermal Noise Experiment

Malcolm B. Gray, Bram J. J. Slagmolen,
Karl G. Baigent and David E. McClelland

*Department of Physics, Faculty of Science,
The Australian National University, A.C.T. 0200, Australia.*

The thermal noise experiment at the ANU is aimed at directly measuring the residual displacement noise of mirror suspension systems in order to both verify and optimise designs prior to installation in large scale Gravitational Wave Detectors, and to measure off-resonance thermal noise at frequencies from 10Hz to 10kHz. In addition we plan to investigate the use of squeezed and squashed light states in order to reduce optical readout noise in Fabry-Perot interferometers.

Here we consider four noise sources that will compromise the measurement of thermal noise: seismic noise, radiation pressure noise due to the laser intensity noise fluctuations, laser frequency noise, and shot noise. The seismic noise is isolated by use of a common suspension designed at the University of Western Australia. The radiation pressure noise in the test cavity is suppressed by an in-loop and out-of-loop intensity servo system. To over come the laser frequency noise we use a relatively long, high finesse mode cleaner cavity. Finally, shot noise is overcome by selecting sufficient test cavity power and finesse such that the resulting displacement noise is well below the desired thermal noise measurement level.

The isolation stack for this experiment is in the final development stage at the University of Western Australia. It uses both a horizontal and vertical pre-isolator followed by a 3 stage suspension/spring system. The whole isolation stack will be fitted into a vacuum tank at the ANU to isolate from both acoustic and seismic noise.

The laser intensity noise is suppressed using a two loop feedback system, measuring both out-of-loop and an in-loop laser intensity. The out-of-loop stabilisation measurement is made directly at the laser out and fed back to the laser diode pump current using a high gain servo (greater than 50dB at low frequencies in order to eliminate laser technical noise features). However as this measurement is performed on 20% of the laser output the shot noise of the measurement limits the 80% laser output to greater than 7dB above shot noise. By using a low gain in-loop servo as well, we place the final test cavity within this intensity servo. We can therefore benefit from sub-shot noise performance of this in-loop "squashed light" [1]. We estimate, based on the optical losses within the test cavity that the intra-cavity intensity noise can be suppressed to 2dB below shot noise. In order to achieve this we need only a gain of 7 - (-2) = 9dB on this in-loop servo.

Frequency stabilisation is performed by locking the laser to a reference/mode cleaner cavity. In our case a low thermal expansion high finesse ring cavity. Locking is achieved by using Tilt-Locking [2]; a simple, modulation free technique. In previous experiments [3] the frequency noise has been suppressed adequately across the signal bandwidth of DC to 100Hz. Above 100Hz the servo gain was insufficient to suppress frequency noise to the shot noise limit. In this experiment we will add a broadband phase modulator to increase high frequency gain and achieve shot noise limited performance up to 10kHz.

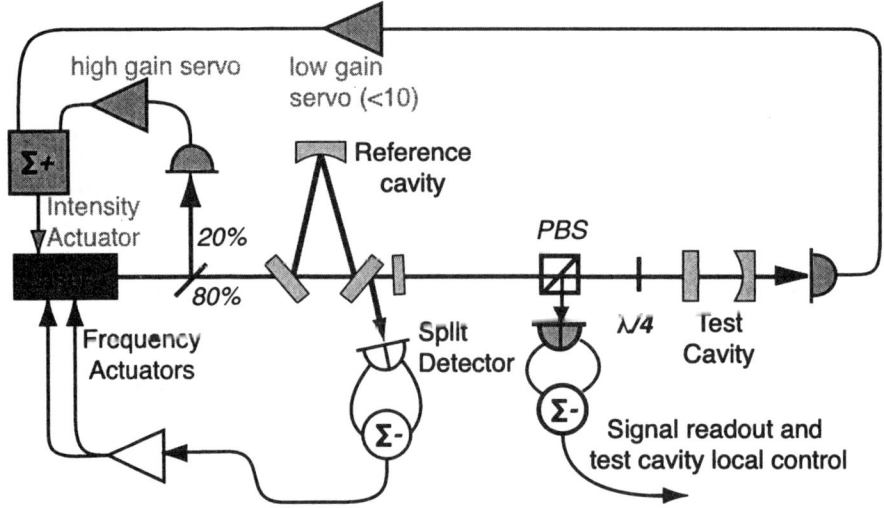

FIGURE 1. A schematic of the optical and servo system designed for our experiment.

In figure 1 we demonstrate our chosen configuration for both the optical system and the servo system. Note that it is necessary to use an in-loop/out-of-loop intensity control system in order to maintain high gain and stability while preserving the sub-shot noise benefit of the in-loop system alone.

REFERENCES

1. Buchler, B., Gray, M., Shaddock, D., Ralph, T., and McClelland, D., *Opt. Lett.* **24**, 259, 1998.
2. Shaddock, D., Gray, M., and McClelland, D., *Opt. Lett.* **24**, 1499 (1999).
3. Baigent, K. G., Shaddock, D. A., Gray, M. B., and McClelland, D. E., in press, General Relativity and Gravitation, 1999.

Design of an isolation system to a medium size spherical resonant gravitational wave detector

W.F. Velloso Jr.[2], J.L. Melo[1], O. D. Aguiar[1]

1. Instituto Nacional de Pesquisas Espaciais. Av. dos Astronautas 1758, 12227-010 - São José dos Campos, SP, Brazil
2. Instituto Astronômico e Geofísico - USP. Av. Miguel Stefano 4200, 04301- Sao Paulo, SP, Brazil[*]

Abstract: We describe the method we have proposed to design a mechanical isolation system to a medium size resonant gravitational wave detector we plan to construct in Brazil. The proposed isolation structure was numerically described and analyzed using a network of solid finite elements. Our results show that the mechanism could allow a damping factor of about 250 dB at 2100 Hz, which is adequate to obtain the sensitivity level we want.

THE ISOLATION SYSTEM DESIGNED FOR *NEWTON*.

The dynamical behavior of the system we designed was numerically simulated using the Finite Element Method (FEM) and the NASTRAN software. The FEM, particularly in the development of gravitational antennas, has been used[1,2] to model the modules used in the mechanical isolation systems. The basic idea in our design was to construct a multiple stage pendulum, capable of attenuating the mechanical vibrations in the spectral region of detection (around 2100 Hz for the NEWTON detector). Four cylinders joined together by hollow rods compose our system. The rods connect the upper face of a cylinder to the bottom face of the immediately below one. So the hollow rods can be long enough to have a relatively small elastic constant, resulting in low resonance frequencies. The basic constraints involved in the design were: 1) to use cylinders large enough to have the first internal resonance at a frequency larger than and far from the detection frequency; and 2) to choose the geometrical parameters of the hollow rod (internal and external radius and height), adjusting the elastic characteristics in a way to obtain a kind of spectral window, free of resonances, around the detection frequency.

THE FINITE ELEMENTS ANALYSIS

Using the above considerations we defined a first geometric set-up for the vibrational isolation system and constructed a FEM model. This model was analyzed in an iterative way, changing the mechanical parameters (external and internal radius of the hollow

---

1. [*] This work has been supported by the Brazilian agency FAPESP (proc. n. 96/2238-7, 97/14437 and 99/03050-0)

rods, etc.) in order to move the resonances out of the spectral region of interest for detection. After four design iterations we achieve a good set up, resulting in a 559 Hz large spectral window free of resonances (from 1861 Hz to 2420 Hz). Then we used the FEM again to calculate all the normal modes, from 0Hz to 3000Hz, in order to determine the dynamical behavior of the structure and to simulate its frequency response, using the modal approach.

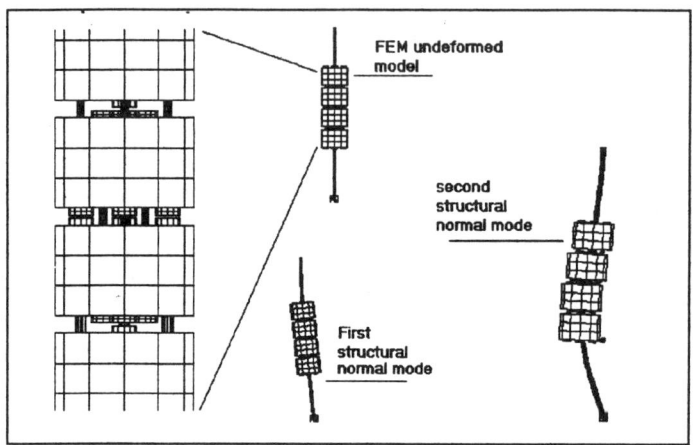

FIGURE 1. The Hooke system FEM undeformed model and the two lowest normal mode shapes.

CONCLUSION. THE HOOKE-4 FREQUENCY RESPONSE

Based on all the normal modes from 0 to 2500 Hz, the calculations showed that the attenuation at 2100Hz is about 250dB, in both X and Y directions, which is large enough to our proposals. However, in the Z direction the attenuation level is only about 100 dB, which is reasonable if we consider that the structure is much more rigid to movements in this direction than in the other ones. We will have to design the room temperature stages to complement the isolation needed in the vertical direction.

REFERENCES

[1] Merkowitz S. M., Coccia E., Fafone V., Raffone G., Schipilliti M., and Visco M., Rev. Sci. Instrum. **70**, 2 (1999).

[2] Aldcroft T. L., Michelson P. F., Taber R. C., and McLoughlin F. A., Rev. Sci. Instrum. **63**, 8 (1992).

Development of a Double Pendulum for Gravitational Wave Detectors

Mark A. Beilby, Gabriela Gonzalez, Michelle Duffy, Amber Stuver, Jennifer Poker

Department of Physics, Pennsylvania State University, University Park, PA 16802

Abstract. Seismic noise will be the dominant source of noise at low frequencies for ground based gravitational wave detectors, such as LIGO now under construction. Future interferometers installed at LIGO plan to use at least a double pendulum suspension for the test masses to help filter the seismic noise. We are constructing an apparatus to use as a test bed for double pendulum design. Some of the tests we plan to conduct include: dynamic ranges of actuators, and how to split control between the intermediate mass and lower test mass; measurements of seismic transfer functions; measurements of actuator and mechanical cross couplings; and measurements of the noise from sensors and actuators. All these properties will be studied as a function of mechanical design of the double pendulum.

INTRODUCTION

The next upgrade installed at LIGO (1) plans to use at least a double pendulum suspension for the test masses to help filter seismic noise. We are constructing a facility to be used as a test-bed for testing the mechanical and local control design of test mass suspension systems for interferometer gravitational wave detectors. The basic design that we are following is based on the GEO gravitational wave detector suspension design, which uses multiple pendulums (2). Our prototype is shown schematically in Figure 1. In addition to the double pendulum, which includes the test mass, a second identical double pendulum is just behind the test mass pendulum. This second pendulum acts as a reaction pendulum, which is used to hold the sensors and actuators used to control the test position of the test mass. Putting the sensors and actuators on an identical second pendulums allows a smaller relative motion between the actuators and the test mass. In addition to gaining extra seismic filtering of a double pendulum, the double pendulum also allows the possibility of controlling the test mass at the intermediate mass rather than from the test mass itself. Therefore, the magnets used for position control, which unfortunately ruin the high mechanical Q of the test mass, can be moved to the intermediate mass. The disadvantage of a double pendulum is that it is more complex to model and hence control, so that testing of actual configurations is a critical activity. The double pendulum is hung from cantilever spring, so that there is also vertical seismic isolation.

One feature of our test bed facility is that the design is made flexible so that a variety of test mass configurations can be tested, such as changing the positions of the sensors and actuators. Another feature in our test bed facility is the use of a high precision three-axis vibration shaker used for diagnostics.

The construction and assembly of the mechanical pieces of the initial single intermediate pendulum supported by cantilever springs is near completion. The first goal is to control this single pendulum in all six degrees of freedoms. Testing and calibration of position controller: sensors (LED shadow detectors) and actuators (magnets and coils) are now in progress. Testing and characterizing the three-axis vibration shaker has been in progress. The next phase will be to build a double pendulum with a dummy test mass of the size of the test masses currently being installed in LIGO. Finally, a double pendulum, with its corresponding double reaction mass pendulum will be built and tested.

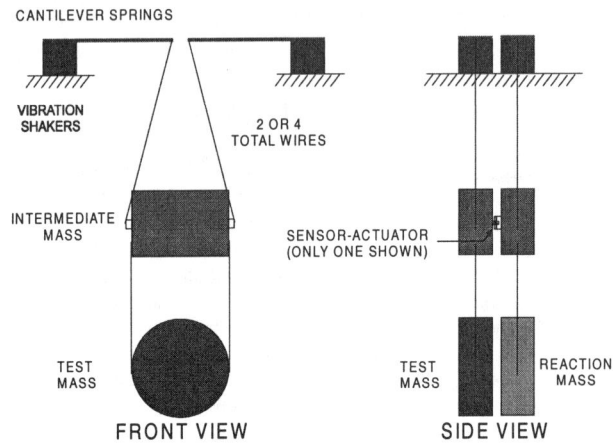

FIGURE 1. Prototype Double-Double Pendulum.

Some of the tests we plan to conduct include: dynamic ranges of actuators, used to control the position of the double pendulum masses and how to split control between the intermediate mass and lower test mass; measurements of seismic transfer functions of the double pendulum; measurements of actuator and mechanical cross couplings; and measurements of the noise from sensors and actuators. All these properties will be studied as a function of mechanical design of the double pendulum, such as two versus four wire suspension, wire attachment points (which determine the resonant frequencies of the pendulum), actuator and sensor placement, intermediate mass shape and size, cantilever spring design and number, and modal damping versus point to point damping.

ACKNOWLEDGMENTS

We would like to thank Dr. Mike Plissi of the GEO Project for his interest in this work. This research is supported by The Pennsylvania State University and NSF Grant No. PHY-9870032.

REFERENCES

1. A. Abromovici *et al.*, Science **256**, 325(1992).
2. M.V. Plissi *et al.*, Rev. Sci. Instrum. **80**, 235(1998).

Bi-filar pendulum mode Q factor for silicate bonded pendulum

K. Tokmakov,* V. Mitrofanov,* V. Braginsky*
S. Rowan,[†] J. Hough[‡]

*Dept. of Physics, Moscow State University, Moscow 119899, Russia
[†] Ginzton Laboratory, Stanford University, Stanford, CA 94305-4085, USA
[‡] Dept. of Physics and Astronomy, University of Glasgow, Glasgow, G12 8QQ, UK

Abstract. An all fused silica pendulum with record Q factor fabricated using the technique of hydroxide-catalysis bonding is described. Possible sources of loss which limited the measured Q are discussed.

INTRODUCTION

Thermal fluctuations of the mirror surface position in an interferometric gravitational wave detector are one of the principle limits to interferometer sensitivity. In order to reduce these fluctuations it is necessary to reduce the dissipation in the relevant modes of the mirror's suspension. Fused silica is a good material for the test masses (mirrors) and their suspension fibers because of the low mechanical losses of this material. The highest quality factors of pendulum and violin modes of suspensions have been obtained for monolithic fused silica pendulums with suspension fibers attached to the test masses and support structure by welding. Welding excludes the additional losses often associated with clamps. Experiments in Moscow (1) and at Glasgow (2) have demonstrated loss factors of $(1\pm0.3)\times10^{-8}$ for pendulums suspended by fused silica fibers with welded attachments. Improvements of the technique for fabricating monolithic fused silica pendulums and more perfect experimental set-up allowed the Moscow group to achieve the quality factor $Q \approx 1.7\times10^8$ for the torsional mode of a 2 kg bifilar pendulum, whose loss properties are similar to those of the simple pendulum mode.

An alternative method of attaching the fused silica suspension fibers to the test masses - hydroxide-catalysis bonding - was investigated at Glasgow (3). This method is more convenient for jointing suspension fibers to large test masses. Measurements give an estimate of the excess loss of $(3\pm1)\times10^{-8}$ associated with a hydroxide-catalysis bonded area of ≈ 0.8 cm^2 for the fundamental modes of fused quartz test masses (3). This paper describes a pendulum with record Q factor fabricated using hydroxide-catalysis bonding technique.

METHOD AND RESULTS

A 0.5 kg fused quartz cylinder was suspended by two fused silica fibers of length

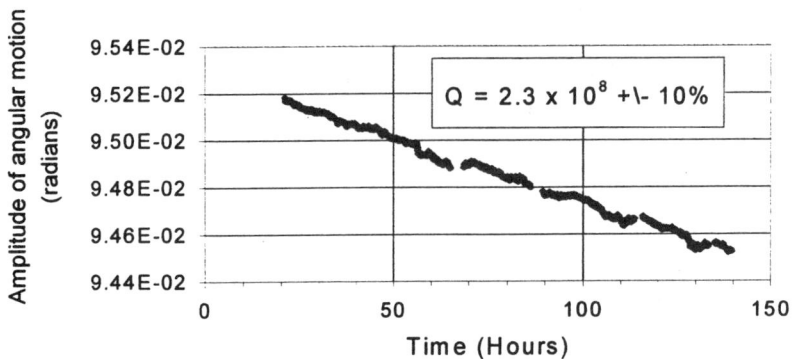

FIGURE 1. Decay in amplitude of angular motion of bi-filar fused silica pendulum

24 cm and diameter ~ 0.2 mm. In previous all-welded pendulums the fibers were welded to small bumps carved in a cylinder. In the pendulum described here the fibers were welded to the fused quartz cones which were attached to the cylinder using hydroxide catalysis bonding. One can find a detailed description of this technique in [3]. The top ends of the suspension fibers were welded to a fused silica disk which was attached via an indium gasket to the cover of a vacuum chamber fastened rigidly to a concrete wall. The chamber was then evacuated to a pressure of ~ 1×10^{-7} Torr. The construction was baked at 100-150 deg C for about 6 hours. The torsional mode of the bi-filar pendulum, with a resonant frequency of 1.17 Hz, was excited using a mechanical pusher. An optical sensor was used to monitor the amplitude of angular motion. Free decay of the amplitude is shown in Fig.1. The Q factor was found to be 2.3×10^8 +/- 10%.

This measured Q can be compared with the expected $Q \approx 1 \times 10^9$, obtained from calculations based on the pendulum dilution factor and the measured loss angle of $\phi \approx 1.4 \times 10^{-7}$ for the fused silica fibers. The discrepancy may be explained by incomplete elimination of possible sources of excess loss such as recoil losses, contact losses in indium gasket at mounting point, fiber surface contamination and others. But it is important that at achieved level of Q we did not see losses associated with hydroxide-catalysis bonding.

REFERENCES

1. Braginsky V.B., Mitrofanov V.P., Tokmakov K.V., *Phys.Lett. A* **218**, 164-166 (1996).
2. Rowan S., Twyford S.M., Hutchins R., Kovalik J., Logan J.E., McLaren A.C., Robertson N.A., Hough J., *Phys.Lett. A* **233**, 303-308 (1997).
3. Rowan S., Twyford S.M., Hough J., Gwo D.-H., Route R., *Phys.Lett. A* **246**, 471-478 (1998).

Prototype of the Suspension Last Stages for the Mirrors of the Virgo Interferometric Gravitational Wave Antenna

A. Bernardini[a], L. Brocco[a], E. Majorana[b], P. Puppo[b], P. Rapagnani[a,b], F. Ricci[a,b], G. Testi[a]

[a]*Dipartimento di Fisica, Università "La Sapienza", Roma, Italy*

[b]*I.N.F.N. Sez. di Roma, P. le A. Moro 5 - 00185 Roma, Italy*

The Virgo detector (1) differentiates itself from other long base-line interferometers, such as LIGO (2,3), by the use of a seismic noise Super-Attenuator (SA) system (4) that will allow to expand the detection bandwidth down to 4 Hz.

The Virgo Super Attenuator (SA) Last Stage is a branched system formed by three pendula. On the first pendulum mass, the "marionette", other two masses are hanged: the mirror ad its reaction mass. The use of a reaction mass to control a suspended mirror was an idea originally pursued in great detail by the GEO group (5).

The marionette can be controlled, by means of electromagnetic actuators, along four degrees of freedom, y, z, θ_x and θ_y, being z the optical axis and y the vertical one. Other electromagnetic actuators, set between the mirror and the reaction mass, allow direct control of the mirror along z, θ_x and θ_y for automatic alignment and locking purposes.

The choice of the final configuration is based on results of the system modeling and studies of the static and dynamic behaviour performed on a complete prototype assembled in our laboratory. In particular the structural study of the marionette and reaction-mass has allowed us to understand the thermal fluctuations influence of these elements on the mirror motion. Indeed, the seismic noise is filtered by the entire attenuation chain, but the Nyquist force acting on the last stage causes a mechanical noise at the output. Such a noise has a spectral density with peaks at the frequencies of the internal modes with amplitudes depending on each mode shape. Hence, in order to evaluate the contribution of all these mechanical noise sources to the Virgo sensitivity curve, a careful study of the internal modes of all the elements of the last stage suspension has been performed.

Using a detailed finite element analysis of marionette and reaction mass, we have determined frequency and shape of each mode, in the Virgo frequency range. We have checked the predictions of these models by measuring the frequency response of each element to impulsive calibrated excitations. Then, we have used these data to verify that the contribution of marionette and reaction mass thermal motion to the output signal is negligible in the Virgo bandwidth (6).

Using our prototype, we have also measured the electromechanical transfer function of the system. We have sent a white noise to each pair of marionette actuators and recorded the mirror position by means of a standard Virgo displacement monitor (7).

Such a monitor was modified to implement an automatic calibration giving the mirror coordinates in the laboratory reference frame.

As an example, in Figure 1 the transfer function measurement along the z axis is shown. The identification of the low frequency normal modes has been done using our "branched system" model, described above.

The results of the tests performed on the protype confirm that our system meets Virgo requirements (8). Further study is needed to characterize the system once it will be coupled to the Virgo Super Attenuator.

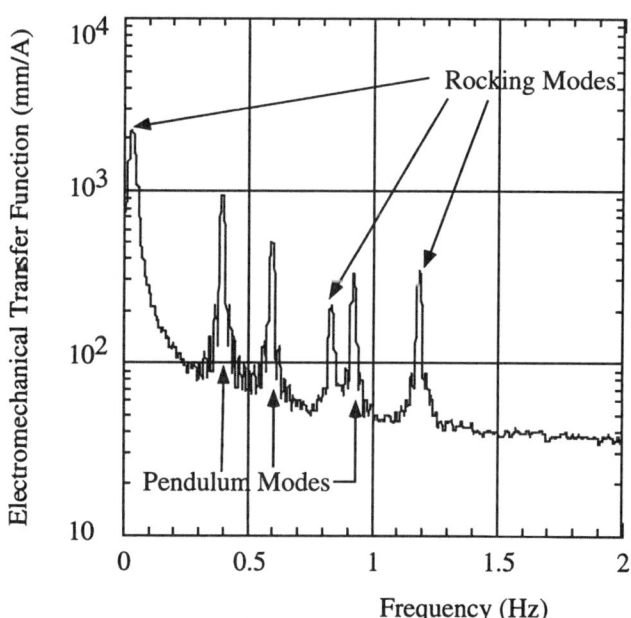

FIGURE 1. Transfer function measurement of the mirror motion along the z-axis, induced by a white noise current input into the coils.

References

1. Caron, B. et al., *Class. Quantum Grav.* **14**, 1461-1469 (1997).
2. Abramovici, A. et al., *Science* **256**, 325-333 (1992).
3. Brillet, A., Ricci, F., *Ann. Rev. Part. Sci.* **47**, 111-156 (1997).
4. Braccini, S. et al., *Rev. Sci. Inst.* **67**, 8, 2899 (1996).
5. Plissi, M.V. et al., *Rev. Sci. Inst.* **69**, 8, 3055 (1998).
6. Bernardini, A. et al., *Rev. Sci. Inst.*, **70**, 8, 3463 (1998).
7. Drezen, C., Heitmann H., *Rev. Sci. Inst.* **68**, 8 (1997).
8. Virgo Collaboration, *Virgo Final Design*, Pisa: E.T.S., 1997

SIGNAL PROCESSING
AND DATA ANALYSIS

Toward gravitational wave detection

L.S. Finn[1,2,3], G. Gonzalez[1,2], J. Hough[4], M.F. Huq[3),(1],
S. Mohanty[1,2], J. Romano[5], S. Rowan[6], P.R. Saulson[7] and
K.A. Strain[4]

[1] *Center for Gravitational Physics & Geometry, Penn State University;* [2] *Department of Physics, Penn State University;* [3] *Department of Astronomy and Astrophysics, Penn State University;* [4] *Department of Physics and Astronomy, The University of Glasgow;* [5] *Department of Physical Sciences, The University of Texas, Brownsvilllle;* [6] *Department of Applied Physics, Stanford University;* [7] *Department of Physics and Astronomy, Syracuse University*

Abstract. An overview of some tools and techniques being developed for data conditioning (regression of instrumental and environmental artifacts from the data channel), detector design evaluation (modeling the science "reach" of alternative detector designs and configurations), noise simulations for mock data challenges and analysis system validation, and analyses for the detection of gravitational radiation from gamma-ray burst sources.

I DETECTOR CHARACTERIZATION

Quality data analysis requires quality data. Part of the process of producing quality data is identifying and, as far as possible, removing instrumental and environmental artifacts. Here we illustrate, using data taken during November 1994 at the LIGO 40 M prototype, the identification and removal, through linear regression, of artifacts due to harmonics of the 60Hz power mains.

A power spectrum (psd) of the LIGO 40 M IFO_DMRO (interferometer differential-mode read-out; hereafter, "gravity-wave") channel shows a series of narrow spectral features at 60 Hz and its harmonics. Similar narrow spectral features are evident in the magnetometer channel, IFO_MAGX, which was recorded simultaneously with the gravity wave channel.

Focus attention, in both the magnetometer and gravity-wave channel, on a narrow band about one of the harmonics. We suppose that, in this narrow band, the gravity wave channel h is related to the magnetometer channel M through an expression of the form

$$h[k] = \frac{B(q)}{A(q)} M[k-n] + \frac{C(q)}{D(q)} e[k], \qquad (1)$$

where the index k indicates the sample number, the residual $e[k]$ is white and $A(q), B(q), C(q)$ and $D(q)$ are polynomials in the lag operator q^{-1},

$$q^{-1}M[k] := M[k-1]. \tag{2}$$

The ratio $B(q)/A(q)$ is a linear filter that can be thought of as the transfer function between the magnetometer and the gravity wave channel; similarly, the ratio $D(q)/C(q)$ can be thought as a filter that whitens that part of the gravity wave channel not explained by the magnetometer channel.

Using a small sample of data we find the "best" filters $B(q)/A(q)$ and $C(q)/D(q)$, where better choices yield smaller residuals and have fewer poles and zeros. (Fewer poles and zeros are desired because we don't want to over fit the data; smaller residuals are desired because we want to identify everything in h that can be explained by M.)

To illustrate, we focus on the 540 Hz harmonic in an approximately 2666 second continuous stretch of LIGO 40 M data taken on 19 November 1994. We mix this harmonic down to zero frequency and down-sample the data to a 4 Hz bandwidth. Using a 100 second segment of data from both the magnetometer and gravity wave channels, we find the filters $B(q)/A(q)$ and $C(q)/D(q)$ and the lag n. In this case the best filters have six zeros and one pole each. The quantity

$$h[k] - \frac{B(q)}{A(q)} M[k-n] \tag{3}$$

is then as free from the effects of the 540 Hz harmonic as we can make it, under the hypotheses of this model.

Figure 1 shows the effectiveness of this analysis. The top two panels show the gravity wave and magnetometer channel psd, with the 60 Hz harmonic features marked with asterisks. In the left bottom panel one curve shows one quadrature of the mixed and decimated gravity wave channel, a second shows the prediction that comes from applying the filter $B(q)/A(q)$ to the lagged magnetometer channel, and the third is their difference (cf. eq. 3). The final panel shows the psd of this difference, superposed with the psd of the original $h[k]$ and the magnetometer prediction. The magnetometer channel explains 40 dB of the contamination of the gravity wave channel by the 540 Hz harmonic.

The techniques described here operate in the time-domain and focus on the narrow-band of the artifact, minimizing possible contamination of the data stream by transients in the environmental channels. For a wide-band, frequency-domain approach to the problem of cross-talk see [1].

II BENCHMARKS FOR DETECTOR DESIGN

Gravitational wave detectors are built to detect gravitational waves. Better detectors do a better job of detecting gravitational waves. But, what are the relative

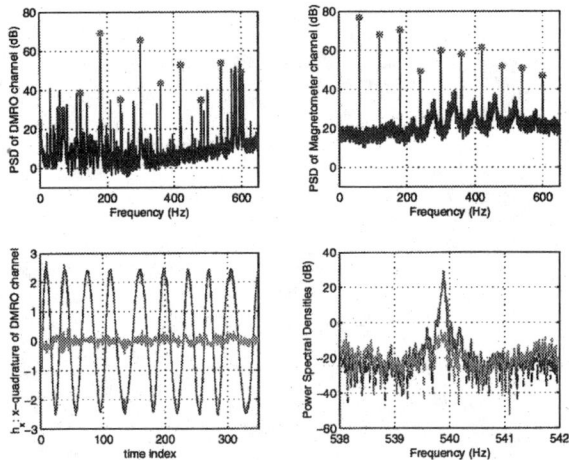

FIGURE 1. 60 Hz harmonics and their regression from the interferometer data stream. See section I for details.

advantages of, *e.g.*, better sensitivity in a narrow band as opposed to somewhat worse sensitivity, but over a broader band? How do we quantify better? To aid in answering this question we have developed a Matlab model, **bench**, that calculates different figures of merit, based on source science, for use in detector design and configuration trade studies.

An interferometer is described to **bench** in terms its laser, optical surface, suspension and substrate properties, since it is these that determine the dominant contributions to the detectors noise performance.[1] From this characterization **bench** determines the detectors expected thermal noise in the mirror suspensions and substrates, radiation pressure and laser shot noise. Using this idealized noise model **bench** calculates two different figures of merit: the first, an effective distance to which inspiraling binary neutron stars can be observed above a fixed threshold signal-to-noise, and the second, a measure of the upper-bound that can be placed on the intensity of a stochastic gravitational-wave background in the LIGO detector system, assuming identical interferometers installed at each observatory.

The **bench** model for the principal interferometer noise sources has the following features:

- Radiation pressure and laser shot noise expressions support interferometer configurations including power recycling and resonant sideband extraction through the specification of three mirror transmittances and associated losses in the optical system. Thermal lensing effects are estimated and a warning issued if the laser power on the beam splitter exceeds the bounds permitted

---

[1] Gross parameters, such as arm length, may also be varied.

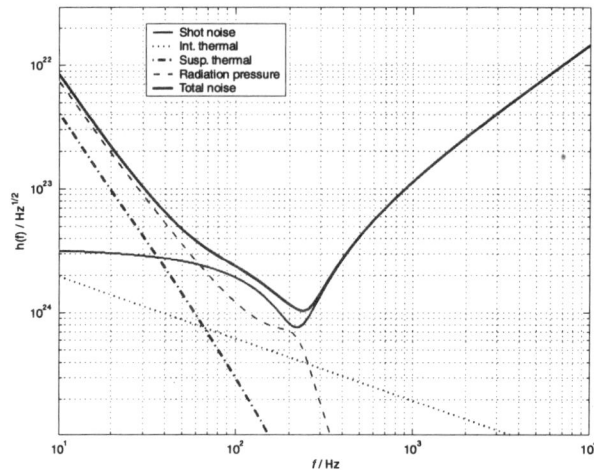

FIGURE 2. A noise model produced by bench, a tool for use in evaluating the science reach of different interferometric detector configurations or designs. An interferometer with this configuration can observe binary inspiral in an effective volume of radius 288 Mpc. See section II for details.

by the losses in the optical system.

- The suspension thermal noise model includes thermoelastic and structural damping for ribbon and cylindrical suspensions composed of different materials (and, for ribbon suspensions, different aspect ratios) [2,3];

- Thermal noise in the (cylindrical) mirror substrates depends on substrate dimensions, material properties (Young modulus, Poisson ratio, loss angle), and the incident (laser) beam radius [4].

The binary inspiral "effective distance" figure of merit is a distance r_0 such that the observed rate of inspiraling binary neutron star systems with S/N greater than 8 is equal to $4\pi r_0^3 \dot{n}/3$, where $\dot{n}$ is the rate per unit volume of inspiraling binary systems [5]. The stochastic signal sensitivity benchmark determines a threshold on the cross-correlation between two identical detectors (located and oriented like the LIGO detectors) such that, in the absence of a stochastic signal (or any other cross-correlated noise), the cross-correlation estimated using 1/3 yr data would exceed this threshold in only one of every one hundred trials. Other benchmarks are planned.

Figure 2 shows a sample noise model produced by bench for an interferometer whose parameters are those described in [6] as a possible LIGO II goal, but whose mirror reflectivities are optimized to maximize the distance to which neutron star binary inspirals could be observed. In this configuration, a single interferometer could survey an effective volume of radius 290 Mpc for neutron star binary inspirals: a volume large enough to expect an event rate of just less than one per month. The two LIGO interferometers operating in this configuration would observe an effective volume $2^{1/2}$ times large, with an expected event rate of one just less than once every two weeks.

III MOCK DATA FOR MOCK DATA CHALLENGES

Reliable analysis software is a prerequisite for reliable data analysis. Validating the performance of analysis system software will involve "Mock-Data Challenges" (MDCs). In a MDC, "mock data" — artificially generated time-series whose statistical character and signal content is known exactly — is passed through an analysis pipeline.

MDCs take two forms. In the first, idealizations of the detector noise, for which the pipeline response can be anticipated, are constructed and passed through the analysis pipeline. Agreement between the anticipated and actual system response validates the analysis system implementation. In the second form, more faithful simulations of detector noise are used to *calibrate* the analysis system: *i.e.*, determine, in a realistic but controlled environment, the detection efficiency and false alarm frequency as a function of the pipeline thresholds associated with the selections and data cuts.

In either form, mock data always includes the fundamental noise sources that contribute the greatest part of the detector noise power. In existing and planned interferometric detectors these fundamental contributions arise from radiation pressure noise, laser shot noise, suspension and substrate thermal noise. The thermal noise contributions have the character of structurally damped harmonic oscillators with small loss angles. The significant contribution from the substrate thermal noise arises from the low-frequency tail of the noise distribution, whose power spectral density (psd) is proportion to f^{-1}. The significant contribution from the suspension thermal noise arises from the high-frequency tail of the suspension pendulum mode, where the psd is proportional to f^{-5}. Additionally, the resonant peaks associated with the weakly damped suspension violin modes contribute important instrumental artifacts that must be part of a realistic noise simulation.

The general plan of our noise simulator is to find a combination of linear filters, acting in parallel on independent white noise sequences, whose sum gives rise to a sequence whose power spectral density (psd) has the desired form. The design of short, effective linear filters that capture either the odd-power dependence on f characteristic of the thermal noise tail of structurally damped oscillators, or the strong resonant peaks of the weakly damped systems, has been a stumbling block in this program. We have overcome those difficulties by developing a physical model of a structurally damped system whose noise psd has the desired in-band character. Arising from a physical model, the psd can be factored into a real, linear, zero-pole-gain filter that is stable and invertible (*i.e.*, has all of its poles in the left half-plane

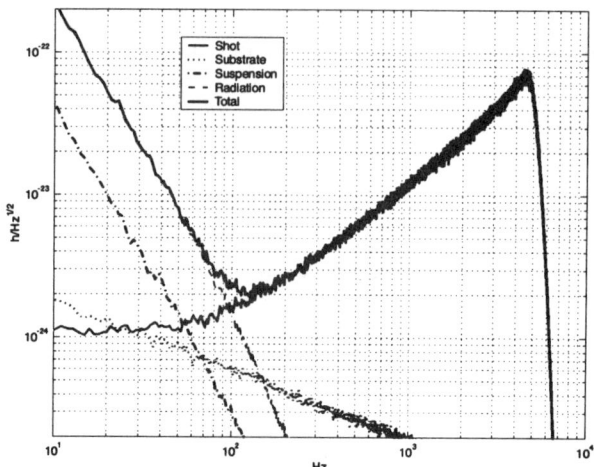

FIGURE 3. Simulations of the principle contributions to the overall noise of an interferometric gravitational wave detector. For more details see section III.

and zeros in the right half-plane), and with the required magnitude response. The filter's zeros, poles and gain are determined uniquely and directly by location and quality factor of the resonance, and the desired simulation bandwidth.

The first panel of figure 3 shows the psd of noise simulated to have the character of a structurally damped harmonic oscillator over three decades in frequency above and below the resonance. The simulation psd is overlaid with the spectrum of an idealized structurally damped harmonic oscillator with the same loss-angle and resonance frequency. This model involved thirteen poles and an equal number of zeros. They agree to better than 1% over the detector bandwidth. The second panel shows the noise psd of the other components of the simulation (radiation pressure, shot, internal thermal and pendulum mode suspension thermal noise) for a LIGO II like interferometer without resonant sideband extraction.[2]

IV GRAVITATIONAL WAVES AND γ-RAY BURSTS

Gamma-ray bursts (GRBs) are likely triggered by the violent formation of a solar mass black hole, surrounded by a debris torus, at cosmological distances. Given the distance, the violence of the formation event, and the range of possible progenitors,

---

[2] When simulating noise at the LIGO I sample rate of 16.384 KHz the simulator, currently implemented as an interpreted Matlab program, produces mock data at the rate of 81,920 samples per second, or 5× the real-time detectors sample rate, on a Sun Ultra-30 workstation. The inverse is the true, not amortized, cost per simulated sample and holds for any number of samples.

waveforms from events like these cannot be predicted *a priori,* nor the gravitational radiation associated with an individual burst detected directly.

Nevertheless, if GRBs are accompanied by gravitational wave bursts (GWBs) the correlated output of two gravitational wave detectors evaluated in the moments just prior to a GRB will differ from that evaluated at other times. This *difference* can be detected, with increasing sensitivity as the number of detector observations coincident with GRBs increases. Observations at the two LIGO observatories, operating at the anticipated LIGO I sensitivity and coincident with 1000 GRBs, can be used to set a 95% confidence upper limit of $h_{\text{RMS}} \sim 1.7 \times 10^{-22}$ on the gravitational waves associated with GRBs. (See [7] for more details.)

Consider the correlation X between the output h_1 and h_2 of two LIGO gravitational wave detectors:

$$X := \langle x_1, x_2 \rangle = \int \int_0^T dt\, dt'\, x_1(t) Q(|t-t'|) x_2(t'), \qquad (4)$$

where we have adjusted the origin of time in each detector so that plane gravitational waves from a direction $\vec{n}$ arrive "simultaneously" in the two detectors. Assuming that GWB signals from GRBs are broadband bursts, take the Fourier transform of Q to be $\tilde{Q}(f) = (S_1(|f|)S_2(|f|))^{-1}$, where $S_i(f)$ is the power spectral density (psd) of detector i, for f in the detector band, and 0 otherwise.

Every time a GRB occurs (say, at time t_0) adjust the origin of time so that $\vec{n}$ points towards the GRB and form X with the interval $(t_0 - T, t_0)$ of data from the two detectors. The duration T of this interval we choose large enough so that we are likely to have included in the interval any associated GWB. For current models of gamma-ray bursts this is no longer than several hundred seconds (where we have accounted for the cosmological redshift of these distant sources).

For each observed GRB we thus have an X. Collect these X into the *on-source* observation set $\mathcal{X}_{\text{on}}$. Similarly, we build an *off-source* observation set $\mathcal{X}_{\text{off}}$ following the same procedure but choosing random times t_0, not associated with any GRBs, and random directions $\vec{n}$ in the sky.

Assuming that GRB signals are weak compared to the detector noise, the sample sets $\mathcal{X}_{\text{off}}$ and $\mathcal{X}_{\text{on}}$ differ only in their means. This difference, $\bar{s}$ is just the average over the source population of $\langle h_1, h_2 \rangle$, where h_k is the GWB signal in detector k. For the two LIGO detectors h_1 and h_2 are, to a good approximation, identical and $\bar{s}$ is proportional to the mean-square amplitude of the wave of h over the source population.

We can test for the difference in the means of the two distributions $\mathcal{X}_{\text{on}}$ and $\mathcal{X}_{\text{off}}$ using Student's t-test, a standard test for determining if two distributions have different means. This provides a simple, yes/no answer to the question of whether GWBs are associated with GRBs.

Alternatively, we can use the value of the t-statistic to set an upper bound on $\bar{s}$. To assess the strength of the upper bound, assume that there is no gravitational radiation associated with GRBs. In this case the ensemble mean, median and mode of the t statistic is zero. Assuming that we actually observed t equal to zero we would obtain the 95% upper bound

$$h_{\text{RMS},95\%}^2 \leq \left[9.4 \times 10^{-22}\right]^2 \left(\frac{T}{500\,\text{s}} \frac{1000}{N_{\text{on}}}\right)^{1/2} \frac{S_0}{(3 \times 10^{-23}\,\text{Hz}^{-1/2})^2} \left(\frac{\Delta f}{100\,\text{Hz}}\right)^{3/2}. \quad (5)$$

where, for convenience, we have modeled the LIGO I detector noise as approximately constant with power spectral density S_0 over the bandwidth Δf, and much higher elsewhere. The value of T adopted here is consistent with external shock models of GRBs; if, on the other hand, it becomes clear that internal shock models are more appropriate (as is becoming more likely), then T will be reduced by a factor of 1000 and the limit will improve by a factor of nearly six.

This upper limit is remarkably strong, especially because it arises without assuming any model for the GWB source or waveform, or the detector noise.[3] Focusing on the difference in the population means has the important consequence that noise correlated between the detectors, but not associated with gravitational waves from GRBs, does not affect the difference in the means. Correspondingly, statistical tests built around the difference in the means are insensitive to noise correlated between the two gravitational wave detectors. Observations with this sensitivity will have important astrophysical consequences, either confirming or constraining the black hole model for GRBs, neither or which can be done with strictly electromagnetic observations.

We are grateful to the LIGO Laboratory for permitting the use of LIGO 40M prototype data in this work. This work was funded by the United States National Science Foundation awards PHY 98-00111, 99-6213, 98-00970, 96-02157, 96-30172; the University of Glasgow and the United Kingdom funding agency PPARC.

REFERENCES

1. B. Allen, W. Hua, and A. Ottewill, Automatic cross-talk removal from multi-channel data, gr-qc/9909083.
2. S. Rowan et al., Phys. Lett. A **227**, 152 (1997).
3. S. Rowan et al., Phys. Lett. A **233**, 303 (1997).
4. F. Bondu, P. Hello, and J.-Y. Vinet, Phys. Lett. A **246**, 227 (1998).
5. L. S. Finn, Phys. Rev. D **53**, 2878 (1996).
6. E. Gustafson, D. Shoemaker, K. Strain, and R. Weiss, Technical Report No. LIGO-T990080, LIGO, LIGO Laboratory, California Institute of Technology (unpublished).
7. L. S. Finn, S. D. Mohanty, and J. D. Romano, Detecting an association between Gamma Ray and Gravitational Wave Bursts, 1999, in press, Physical Review D.

---

[3] The t statistic is a robust one: *i.e.*, it is insensitive to the actual distribution of the X in $\mathcal{X}_{\text{on}}$ and $\mathcal{X}_{\text{off}}$. Additionally, since each X is a sum over many statistically independent random variables the distribution of the X is also, by The Central Limits Theorem, normal.

Separation of the gravitational-wave signals and the solar oscillation signals

Wei-Tou Ni and Xiaohui Xu

Center for Gravitation and Cosmology, Department of Physics, National Tsing Hua University, Hsinchu, Taiwan 30055, ROC.

Abstract. Astrodynamical missions and dedicated gravitational-wave missions using optical devices to map relativistic gravity of the solar system and to monitor gravitational waves have the capability to detect the solar p-mode and g-mode oscillations by observing their associated changing gravity field. We show that the low-l solar oscillation signals and the gravitational-wave background signals can be separated.

Efforts to see inside the Sun have stimulated important works in solar neutrino experiments and helioseismological observations. Recently, Schutz[1], Gough[2], and Cutler and Lindblom[3] suggested the possibility that solar oscillations might be observable by measuring their associated gravitational perturbations. With a detailed analysis, Cutler and Lindblom concluded that LISA may be confusion limited at the relevant frequencies due to the galactic background from short-period white dwarf binaries and present estimates of the number of these binaries would require the solar modes to have energies above about 10^{33} ergs to be observable by LISA.

ASTROD has better sensitivity in the frequency-band considered. In the TAMA Workshop on Gravitational Wave Detection, we reported that the confusion limit might be lifted for the gravitational detection of the solar oscillations for ASTROD due to higher strain sensitivity and its orbit configuration[4].

The change of external gravitational potential, $\delta U^{(s)}_{n\ell m}$, due to the $\binom{s}{n\ell m}$ mode oscillation of the sun can be written in the following simple form:

$$\delta U^{(s)}_{n\ell m} = -\frac{4\pi}{2\ell+1} \frac{\bar{\xi}^{(s)}_{n\ell m}}{R_\odot} \frac{GM_\odot}{R_\odot} \left(\frac{R_\odot}{r}\right)^{\ell+1} Y_{\ell m} e^{i\omega t}. \quad (1)$$

The surface helioseismology favors p-mode detection while gravitational helioseismology favors g-mode detection. Table 1 compiles some important parameters for selected solar oscillations.

TABLE 1. Frequencies, energies and surface velocity amplitudes of various $l=2$ modes of solar oscillations with $\bar{\xi}=1$mm [5]

Mode	Frequency (μHz)	Energy (ergs)	Surface radial velocity (mm/s)
g_3	220.4	2.82×10^{27}	0.183
g_2	254.0	3.93×10^{27}	0.308
g_4	192.2	4.08×10^{27}	0.138
p_1	381.6	5.13×10^{27}	1.15
p_2	514.4	3.56×10^{28}	8.20
p_3	663.6	1.96×10^{29}	35.5

For the ASTROD, the strain sensitivity for one year intergation for Signal-to-Noise-Ratio S/N = 5 threshold is 10^{-23} in the frequency range 100 μHz - 1 mHz. With this strain sensitivity, the $\ell = 2$ mode detection threshold is about 1 mm for ξ. This is about 2 orders of magnitude more sensitive than LISA. The time constants for solar oscillations are long — over 10^6 yr for low l g-mode oscillations[6] and over 2-3 months for low l p-mode oscillations[7]. The distance to the Sun of the inner ASTROD spacecraft varies from 0.77 AU to 1 AU and that of the outer ASTROD spacecraft varies from 1 AU to 1.32 AU. When the spacecraft move in solar orbits, the amplitude and direction of the solar oscillation signals receive deep modulations in addition to the modulations due to spacecraft motion and orientation. The time constant for the gravitational radiation (or orbit evolution) of the close white dwarf binaries (CWDB) is more than 10^6 yr, and hence the CWDB confusion background is steady in the inertial space. This background is modulated only by the orientations and motions of spacecraft, not by the distances and orientations of the spacecraft relative to the Sun. With this extra modulation — deep in magnitude and direction, the detectability of the solar oscillation signals reaches at least 5 orders lower than the confusion limit, i.e., to the instrumental noise floor.

REFERENCES

1. Schutz, B. F., in *Processdings of the First Amaldi Conference on Gravitational Wave Experiments*, edited by G. Pizzerlla and E. Coccia, (World Scientific, Singapore, 1995).
2. Gough, D., *Nature* **376**, 120 (1996).
3. Cutler, C., and L. Lindblom, *Phys. Rev. D* **54**, 1287 (1996).
4. Ni, W.-T., "ASTROD and Gravitational Waves", pp.117-129 in *Gravitational Wave Detection*, edited by K. Tsubono, M. -K. Fujimoto and K. Kuroda, 117 (Universal Academy Press, Tokyo, Japan 1997).
5. Ni, W.-T., A.-M. Wu and X. Xu, New Eye to See Inside the Sun: Gravitation Probe, to be submitted for publication.
6. Kumar, P., and E. J. Quataert, *The Astrophys. J.* **458**, L83 (1996).
7. Harvey, J. W., et al., *Science* **272**, 1284 (1996).

Data Archiving and Distribution of the Virgo Antenna for Gravitational Wave Detection

F. Barone[1,2], A. Eleuteri[2], F. Garufi[2], L. Milano[1,2]

[1]*Dipartimento di Scienze Fisiche, Università di Napoli "Federico II"*
Complesso Universitario di Monte S.Angelo - Edificio G - Via Cintia, I-80126 Napoli, Italia
[2]*Istituto Nazionale di Fisica Nucleare - Sez.Napoli,*
Complesso Universitario di Monte S.Angelo - Edificio G - Via Cintia, I-80126 Napoli, Italia

Abstract. The system consists of two sections: an acquisition and storage section (LynxOS based system with the disks directly connected to CPU slave boards) and a data management section (DEC-Unix Alpha Server - Data Server). The performances of these systems can be summarized as: 1- maximum raw data archiving sustained data flow of 10 *Mbyte/s* on DLT tapes (35/70 *Gbyte*); 2- up to 1.2 *Tbyte* disk capacity with a maximum sustained data acquisition flow of 25 *Mbyte/s*; 3- up to 10 *Mbyte/s* retrieval data flow for the on-line data distribution.

INTRODUCTION

The global architecture of VIRGO[1] data acquisition and storage is described in Fig.1. The raw data formatted in frames by the Frame Builder are sent to the Raw Data Archiving (RDA) System, that stores them on DLT tapes (35/70 *Gbyte*). The same frames are processed by the On-line Processing system that provides each frame with the reconstructed [*t, h*] pairs and other auxiliary information, selects the frames likely to contain a gravitational wave event and sends them to the Data Distribution System that archives them both on disks and on DLT tapes (Data Summary Tapes).

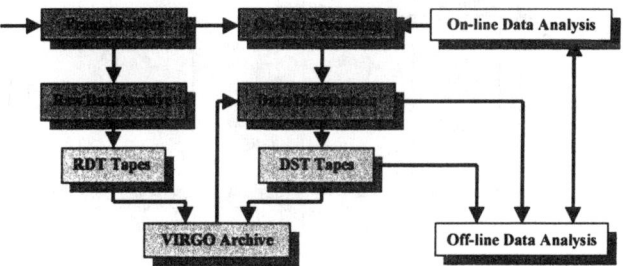

FIGURE 1. Virgo Data Flow.

RAW DATA ARCHIVE

The RDA[2] is organized according to a client/server architecture. In order to match the data flow to the DLT writing speed, we implemented a two stage modular storage procedure consisting in the parallel staging of the data on fast disks (writing speed >10 *Mbyte/s - 18 Gbyte* capacity), and the following copy of the data on DLT. The VIRGO RDA system that implements the above outlined structure consists of a 21 slot ELMA VME crate with a master CPU (VMPC4a PowerPC604e from Cetia) running LynxOS v3.0 operating system and linking the system to the Fast Ethernet network, two slave CPUs (VMPC4A) each provided with two 18 *Gbyte* SCSI disks (Cheetah from Seagate) and two 14 cartridge DLT autoloaders (DLTstor7114 from Quantum).

DATA DISTRIBUTION

The Data Distribution System[2] is made of two sections: acquisition and distribution (see Fig.2). This acquisition section is a VME based system made of a master CPU (VMPC4a from Cetia) running LynxOS and slave CPUs (same model) handling each 5 disks (18 *Gbyte* Chetaah from Seagate) connected to fast & wide SCSI interfaces. The master CPU acquires the frames via a Fast Ethernet network and sequentially distributes them to the slave CPUs (via VME bus) for disk storage. The distribution section manages the users requests following a standard networking procedure. The data server is a DECAlpha server 4100, running DEC/Unix and supporting several SCSI buses to connect disks. The link with the VME CPUs is ensured by a Fast Ethernet interface, but both the server and the CPUs may also support FDDI. The data retrieval is done via NFS mounting the disks on the Alpha server.

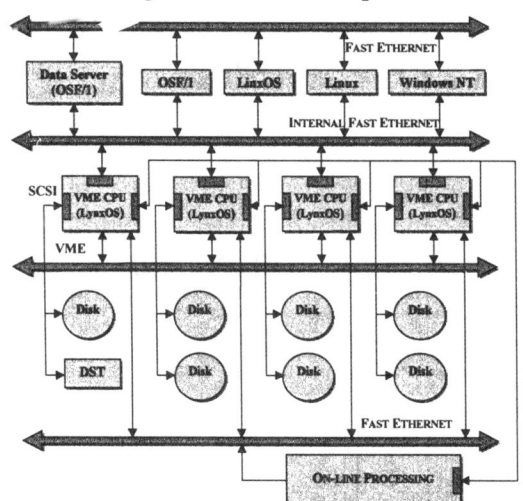

FIGURE 2. Data Distribution Architecture.

REFERENCES

1. The Virgo Project, Final Design of the Italian-French large base interferometric antenna VIRGO for gravitational wave detection (INFN, Italy, and CNRS, France, 1989, 1992, 1995).
2. Barone, F., Garufi, F., Milano, L., and Mours, B., Rev. Sci. Instrum. 68, 3907-3913 (1996).

The Environment Monitoring of the VIRGO antenna for gravitational wave detection

A.Anastasio[1], F.Barone[1,2], A.Eleuteri[1], F.Garufi[1], L.Milano[1,2]

[1]*Istituto Nazionale di Fisica Nucleare - Sez.Napoli,*
Complesso Universitario di Monte S.Angelo - Edificio G - Via Cintia, I-80126 Napoli, Italia
[2]*Dipartimento di Scienze Fisiche, Università di Napoli "Federico II"*
Complesso Universitario di Monte S.Angelo - Edificio G - Via Cintia, I-80126 Napoli, Italia

Abstract. The Environment Monitoring System of the VIRGO antenna is a full modular system that can be easily adapted and extended to fulfil the present and future Virgo needs. This system already started data taking of temperature, pressure and acoustic noise for their identification.

INTRODUCTION

The Environment Monitoring is the VIRGO system that monitors all the environmental quantities that may have effect on the interferometer output[1]. The environmental data flow together with the logical links among the Virgo systems is shown in Fig.1. In this figure it is possible to see that all the environmental data (including the quantities related to the buildings) are sent to the Frame Builder and the to the Data Distribution which provides also the historical monitoring.

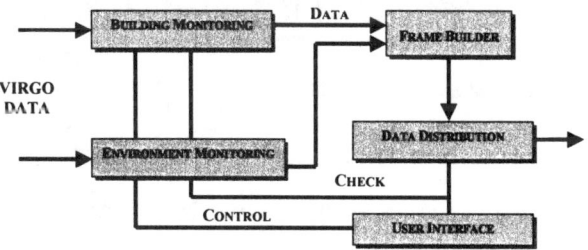

FIGURE 1. Environment Monitoring Data Flow.

ENVIRONMENT MONITORING ARCHITECTURE

A client/server architecture has been chosen for this system, so that an User Interface is used for the check and the control of the apparatus[2]. In Fig.2 the global architecture of the Environment Monitoring System is shown, including the Buildings Monitoring section, while in Fig.3 the server architecture is shown.

CP523, *Gravitational Waves: Third Edoardo Amaldi Conference,* edited by S. Meshkov
© 2000 American Institute of Physics 1-56396-944-0/00/$17.00

The system is able to acquire different environmental noise sources at different sampling rates (i.e. electromagnetic noise [20kHz], acoustic noise [20kHz], seismic noise [1 kHz], temperature [0.1 Hz], pressure [0.1 Hz], humidity [0.1 Hz], etc.). Concentrated and/or distributed acquisition is possible according to the VIRGO needs.

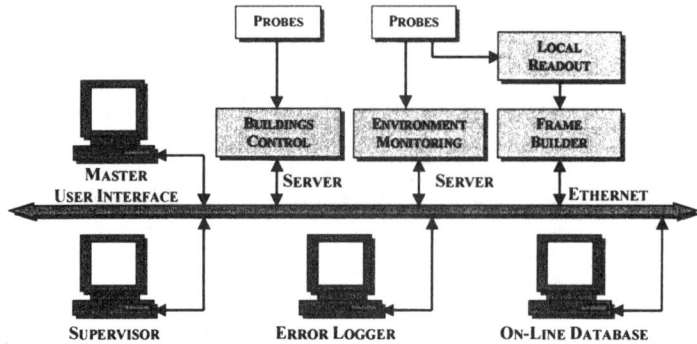

FIGURE 2. Environment Monitoring Architecture.

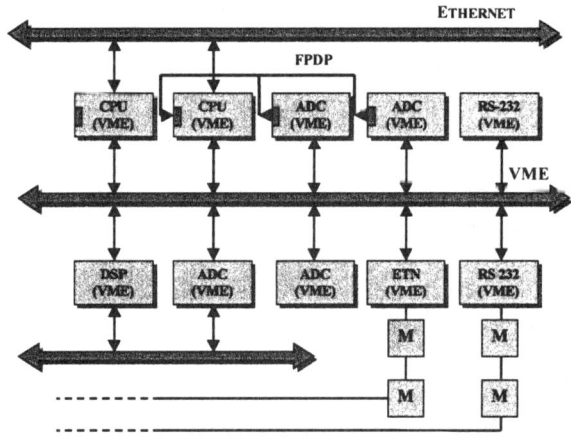

FIGURE 3. Environment Monitoring Server Architecture.

The system is now being intensively used for monitoring the temperatures of the Central Building, Mode Cleaner Building and of the mode cleaner suspension with a temperature accuracy already better than ± 0.2° C. Also an acoustic noise monitoring in the Central Building is started together with tests for seismic noise acquisition.

REFERENCES

1. The Virgo Project, Final Design of the Italian-French large base interferometric antenna VIRGO for gravitational wave detection (INFN, Italy, and CNRS, France, 1989, 1992, 1995).
2. Barone, F., Calloni, E., Di Fiore, L., Grado, A., Milano, L., and Russo, G., Rev. Sci. Instrum. 67, 4353-4359 (1996).

A Neural Network Approach for the Noise Identification and Data Quality of the VIRGO Antenna

F.Barone[1,2], A.Ciaramella[2,3], A.Eleuteri[2], F.Garufi[2], L.Milano[1,2], R.Tagliaferri[3]

[1]*Dipartimento di Scienze Fisiche, Università di Napoli "Federico II",*
Complesso Universitario di Monte S.Angelo - Edificio G - Via Cintia, I-80126 Napoli, Italia
[2]*Istituto Nazionale di Fisica Nucleare - Sez.Napoli,*
Complesso Universitario di Monte S.Angelo - Edificio G - Via Cintia, I-80126 Napoli, Italia
[3]*Dipartimento di Matematica ed Informatica, Università di Salerno*
via S.Allende, 84081 Baronissi (SA), Italia

Abstract. We are exploring the possibility of using neural networks for noise identification and extraction in connection with the environment monitoring and within the global architecture of Data Quality. We report here the very promising results of a test of real-time acoustic noise identification and extraction for a bench test Michelson interferometer.

INTRODUCTION

The identification of the environmental noise effects on the VIRGO[1] output signal is essential both for data analysis and system check. This task requires the development, implementation and test of suitable on-line and off-line noise identification techniques, fully adaptable to the interferometer and to the changes of the environmental conditions. Moreover, these techniques must be easily integrated within the Virgo Data Quality scheme (see Fig.1), that is the general scheme of the Virgo antenna in which data quality algorithms are running at the level of the macro-systems in which Virgo has been subdivided.

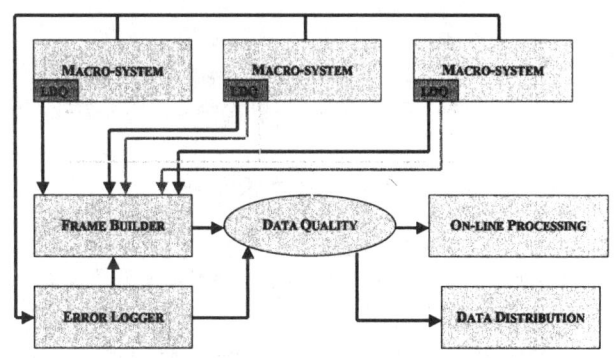

FIGURE 1. Virgo Data Quality Architecture.

CP523, *Gravitational Waves: Third Edoardo Amaldi Conference*, edited by S. Meshkov
© 2000 American Institute of Physics 1-56396-944-0/00/$17.00

In particular, neural network techniques can be easily matched to such dynamic architecture and are very powerful. We are exploring now if they can be really helpful for VIRGO and can be integrated with the classic ones[2,3]. Here, we report an example of a real-time acoustic noise identification with a neural network obtained combining a tapped delay line with an ADALINE applied to a bench test uncontrolled Michelson interferometer in air with a microphone along one of the two arms. The neural network was trained to identify the acoustic noise, so that when a perturbation was applied (in this case with a diapason) it was able to extract this noise directly from the Michelson photodiode output in real-time. We are now extending and improving these techniques also for the identification and study of other environmental noises.

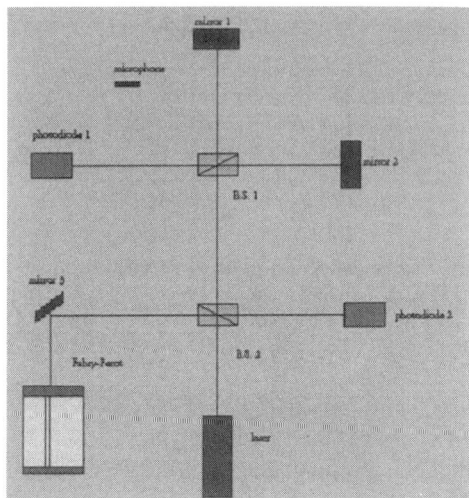

FIGURE 2. Michelson bench test interferometer of Napoli VIRGO lab.

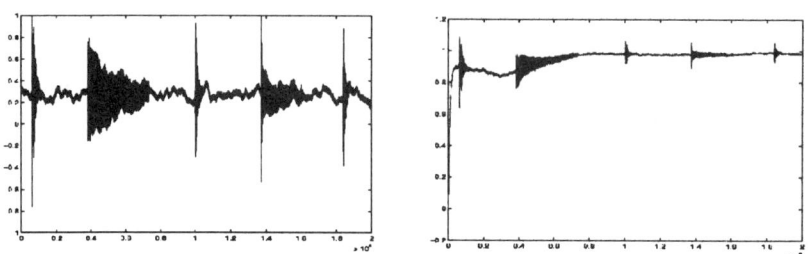

FIGURE 3. a) acoustic noise perturbation (diapason); b) acoustic noise extracted by the neural network.

REFERENCES

1. The Virgo Project, Final Design of the Italian-French large base interferometric antenna VIRGO for gravitational wave detection (INFN, Italy, and CNRS, France, 1989, 1992, 1995).
2. Barone, F., De Rosa, R., Eleuteri, A., Garufi, F., Milano, L., and Tagliaferri, R., "A neural network-based ARX Model of Virgo Noise", WIRN 99 Conference Proceedings, M.Marinaro and R.Tagliaferri eds., Vietri sul Mare, Italy, 20-22 May, 1999 (in press).
3. Barone, F., Ciaramella, A., Eleuteri, A., Garufi, F., Milano, L., and Tagliaferri, R., IEE Trans. Computer and Digital Techniques (1999) (submitted).

Gravitational Wave Signal Detection with Neural Networks for the VIRGO Antenna

F. Barone [1,2], A. Eleuteri [2], F. Garufi [2], L. Milano [1,2]
R. Tagliaferri [3,4,5]

[1] Universita` di Napoli "Federico II", Dipartimento di Scienze Fisiche, Edificio G-Complesso Universitario di Monte S.Angelo, Via cintia, I-80126 Napoli, Italia
[2] Istituto Nazionale di Fisica Nucleare, sez. Napoli, Edificio G-Complesso Universitario di Monte S.Angelo, Via Cintia, I-80126 Napoli, Italia
[3] Dipartimento di Matematica ed Informatica, Universita` di Salerno, via S.Allende, 84081 Baronissi (SA) Italia
[4] INFM unita` di Salerno, via S.Allende, 84081 Baronissi (SA) Italia
[5] IIASS E.R. Caianiello, Vietri s/m (SA), Italia

Abstract. In this paper we describe a neural network-based acoustic noise identification procedure. In particular, we have performed some experimental tests on a classic Michelson interferometer used as a microphone, that although different from the VIRGO[1] antenna provides us with global information on the performance of neural networks. The preliminary results appear to be very promising for the analysis of real VIRGO outputs.

INTRODUCTION

Gravitational Wave (GW) detection is a very complex problem [2]. Due to the interferometer sensitivity to environmental and internally generated noises, the not well known shapes of GW signals and their intrinsic weakness, the resulting s/n ratio is very poor, so it is important to develop robust models for the detection of GW signals in high noise environments. The GW signal can actually be considered as an anomaly in the signal output of the interferometer.

ANOMALY DETECTION BY NEURAL NETWORKS

In the performed experiments we used a multi-layer perceptron (MLP) neural network (NN) [3] to build a model of the dynamics of the acoustic noise process. The experiment we performed had the following purposes:

1. Show that it is possible to identify the acoustic noise contribution to the output of the interferometer using a NN trained on the acoustic noise.
2. The presence of an anomaly in the noise sequence can be detected analysing the prediction residuals, which show how the model fails in reproducing the dynamics of the modelled system (if we are confident that the model is capable of reproducing the system dynamics).

The approach we followed to build the model is based on sound theoretical foundations. First, we reconstructed the state space vector x(t) from the observed data s(t) via a *phase space reconstruction*[4] method. The vectors so obtained were then used to create a model by following a probabilistic Bayesian approach[3,4]; this allowed the exploration of a model space, rather than a single "optimal" model as usually happens in practice. By using Bayes' theorem and marginalizing the posterior distribution of model parameters to integrate out the hyperparameters[1] α and β (which are used only to determine the form of the distributions) we have:

$$p(\mathbf{w} \mid D) = \frac{1}{p(D)} \iint p(\mathbf{w} \mid \alpha) p(D \mid \mathbf{w}, \beta) p(\alpha) p(\beta) d\alpha d\beta. \tag{1}$$

The Bayesian learning framework has several advantages over traditional ones: *the model cannot overfit the data* also if it is very complex, and we can obtain *error bars* to assess the uncertainty in the predictions of the model.

RESULTS

The trained model had 23 hidden units with tanh activation function, 11 input units (corresponding to the embedding dimension of the system) and 1 output unit with linear activation function. Fourteen regularization parameters have been used: one for each group of connections from each input units, one for each layer's bias connections and one for the group of connections to the output unit. The model was trained for 200 epochs. The residuals show that the model can achieve good predictive performances, also if it not very complex. To test the detection capability of the model a synthetic anomaly has been added to the interferometer signal. This anomaly is a segment of a chaotic time series, namely the solution of the Mackey-Glass time-delay differential equation. The signal has been chosen because it mixes well with the interferometer output. At a glance, it is not possible to see the anomaly. However, since its amplitude is greater than the residuals, we expect the network to detect it, which indeed it does. The Bayesian error bars are useful to see how well our model can detect anomalies; if we do not trust the model enough, we could simply detect as anomalies those spikes which are greater than, say, 2 or 3 times the error bars. Note that the model could be made more reliable by using a greater training set, using more units and training for more epochs. Further experiments need to be done with more complex models and signals to verify the applicability of the procedure to VIRGO data analysis.

REFERENCES

1. The VIRGO Project, Final Design (INFN, Italy, and CNRS, France, 1989,1982,1995).
2. Blair, D. G., The Detection of Gravitational Waves, Cambridge:Cambridge University Press, 1991
3. Bishop, C. M., Neural Networks for Pattern Recognition, Oxford:Clarendon Press, 1995
4. Abarbanel, H. D. I., Analysis of Observed Chaotic Data, New York:Springer, 1995
5. Neal, R. M., Bayesian Learning for Neural Networks, Berlin:Springer, 1996

---

[1] The *alphas* are also termed *regularization coefficients*, since their use favours smooth mappings to be generated. We allow different *alphas* for different groups of parameters to give the model wider flexibility. The *beta* is the *precision* with which we realize the mapping.

LIGO End-to-End simulation Program

B. Bhawal*, G. Cella**, M. Evans*, S. Klimenko[†], E. Maros*, S.D. Mohanty[††], M. Rakhmanov*, R.L. Savage*, jr., H. Yamamoto*

*LIGO Laboratory, California Institute of Technology, Pasadena, CA 91125, USA
**Dipartmento di Fisica, Universita' di Pisa, piazza Torricelli 2, 56100 – Pisa, Italy
[†] Department of Physics, University of Florida, Gainsville, FL 32611, USA
[††] Department of Physics, Pennsylvania State University, PA 16802, USA.

Abstract. A time-domain simulation program has been developed to provide an accurate description of interferometric gravitational wave detectors. This is being utilized to build a model of LIGO with the aim of aiding in the shakedown and integration of the interferometer subsystems, and ultimately the optimization of detector sensitivity.

Computer simulation is expected to play a crucial role in understanding and improving the performance of the LIGO interferometers. With that aim a simulation package with various modelling tools has been developed at Caltech. This is called LIGO End-to-End (E2E) simulation program [1].

The E2E package simulates the time-evolution of fields, optics, mechanical structures and electronic and control systems. It is written in C++ and its modular design makes it possible to simulate wide variety of experimental configurations and processes using the same software without modifying the program. This package can be viewed as a software toolbox, like Matlab, and complex systems can be simulated by combining building blocks. The flexibility of the simulation environment makes it easy to add new physics or functionalities.

A graphical user interface, named Alfi, allows users to edit the description files for the simulation engine even for complex configurations in an easy way. Figure 1 shows the Alfi window for an example configuration of a Fabry-Perot (FP) cavity. For each module in the window (e.g., mirror, propagator, laser etc.), parameters can be set using the corresponding dialog box.

The evolution of optical fields and their interaction with optical media are calculated using the modal model [2]. The field is expanded using the Hermite-Gaussian eigenstate solutions of the paraxial wave equation. The E2E software can be used to simulate any planar optics configuration by combining mirrors and propagators without modifying the source code. This method is accurate and flexible but it suffers from speed penalty when the frequency of interest is much smaller than the inverse of the characteristic time-scale of the system. The simulation speed of

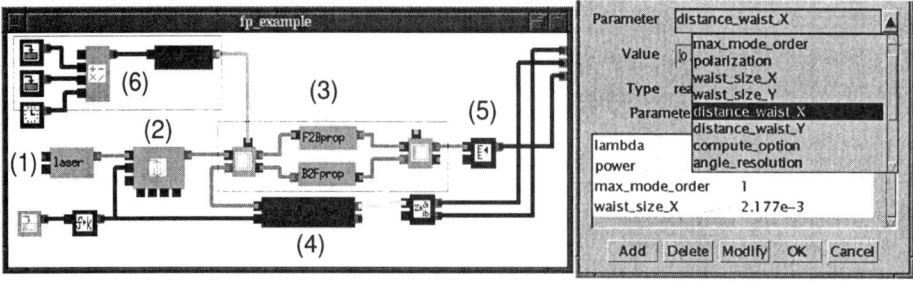

(a) GUI front-end of e2e   (b) parameter setting windows

FIGURE 1. Setup for Fabry-Perot cavity in E2E.

this kind of system can be improved by using approximate equations [3] of field-evolution for writing a module corresponding to a composite system like a cavity. At present three kinds of cavities have been implemented using this technique: FP cavity, triangular cavity, recycled Michelson cavity.

The simulation of mechanical systems involves: seismic ground motion, seismic isolation and the suspension systems. The power spectra of the measured ground motion and the transfer functions of the isolation system can be parametrized and simulated using digital filters [4]. LIGO-I suspension system uses a single pendulum; A few models [5,6] have been implemented and compared to validate each other. A new fully 3-D simulation code [6] called Mechanical Simulation Engine (MSE) is being implemented. This uses a basic set of fundamental objects (masses, beams etc.) which are combined to represent complex mechanical systems.

Modelling of the full 2-Km LIGO and its subsystems like Pre-Stabilized Laser and Input Optics and their validation are going on. Some open and closed loop transfer functions generated using this model are compared with measurements performed on the installed hardware and reasonable agreement is achieved.

Experiments on 2 Km FP cavity at the LIGO Hanford Observatory began in November, 1999. The E2E software was used to study length control and lock acquisition. Also it was used to understand various higher order harmonic signals caused by the non-linear response of the optics systems in the in-lock state.

REFERENCES

1. Bhawal B., Evans M., Maros E., Rakhmanov M., Yamamoto, H., LIGO documents T970193, T970194.
2. Hefetz Y., Mavalvala N., Sigg D., JOSA, B14, 1597 (1997).
3. Redding D., LIGO document T960171; Bhawal B., JOSA, A15, 120 (1998).
4. Daw E., LIGO document T990112.
5. Mohanty S., LIGO document T990014; Rakhmanov M., Ph.D. theses (Caltech, 2000).
6. Cella G., LIGO documents T990106, T990107, T990108.

The Logging and Data Retrieve System for the GW Detector AURIGA

A. Ortolan[#], L. Baggio[*], M. Cerdonio[*], V. Martinucci[§], G.A. Prodi[§], L. Taffarello[*], S.Vitale[§], G.Vedovato[#], J.P. Zendri[*]

[#]INFN Nat. Laboratory of Legnaro, Via Romea 4, Legnaro (PD) Italy.
[*]Department of Physics and INFN sez. of Padova, via Marzolo 8, Padova, Italy.
[§]Department of Physics and INFN gr. coll. Trento, Povo (TN) Italy.

Abstract. We have created a logging and data retrieve system for the AURIGA g.w. detector by means of the commercial software ORACLE RDBMS (Relational DataBase Management System). The system is able to manage and correlate efficiently many years of analyzed data relative to g.w, candidate events and to the detector noise estimates and calibrations.

AURIGA is a gravitational wave detector[1] located at INFN Nat. Lab. of Legnaro designed to search for millisecond bursts from the Local Group of Galaxies: to this aim a coordinate coincidence program with similar detectors (NAUTILUS, EXPLORER, ALLEGRO, NIOBE), to form the first gravitational wave observatory, has been started too (see the IGEC agreement at http://igec.lnl.infn.it/igec). The performances and capabilities of AURIGA in searching g.w. events depend on many experimental factors, including environmental conditions, hardware setting, working temperature of the bar, etc. and these parameters often interact in unexpected ways. In addition, AURIGA data acquisition system produces ~ 3 Gbytes per day of raw data (including the auxiliary data which ensure that the detector is working properly) which are first reduced by the on-line analysis to 20 Mbytes per day of flat files[2]. These data, together with the detector calibrations and data diagnostics, have to be efficiently stored for years in order to recover any information necessary to the search of g.w. signals. To manage rapidly and correlate efficiently such a large amount of data we have created a logging and data retrieve system by mean of the commercial software ORACLE RDBMS. A dedicated UNIX workstation runs the Oracle Server 8; suitable tools (ProC, ODBC) allow to insert data into the database and to fetch them for application programs (Paw++, Excel, etc.) for their graphical representation. For the database design an Entity/Relationship (E-R) diagram was used. In this diagram are represented the types of data produced by the on-line analysis (entities), the relations among them and the relevant information they can provide. Each type of data is characterized by the set of attributes used. In fig. 1 we report a simplified version of the complete AURIGA E-R diagram. Entities (rectangles) represent a collection of homogeneous data with the same attributes. Relationships (rhombs) point out the relation among entities. We put in parentheses the relation cardinality, i.e. the maximum and minimum number of objects belonging to the entities in relation. To give an example, an event can be in relation with almost one calibration but a

CP523, *Gravitational Waves: Third Edoardo Amaldi Conference,* edited by S. Meshkov
© 2000 American Institute of Physics 1-56396-944-0/00/$17.00

calibration is in relation with many events. Data and Setup are generalized entities, i.e. they consist of entities that can be aggregated. The Events entity is the main entity as it collects the gravitational wave candidate events, which are described by amplitude, arrival time, signal-to-noise ratio, χ^2 and other attributes.

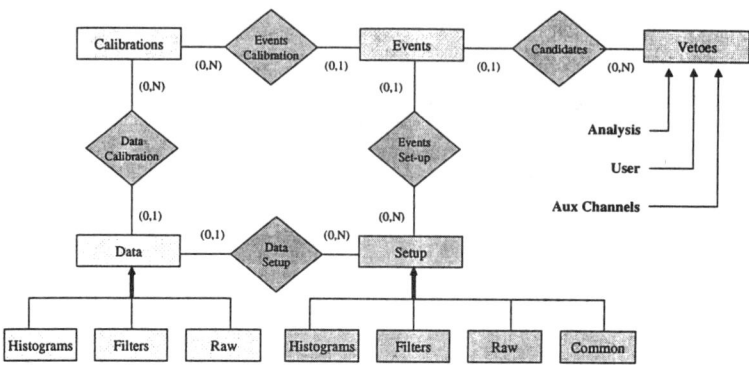

FIGURE 1: Entity/Relationship diagram for the AURIGA database.

When we retrieve a gravitational wave candidate event, we have to recover the experimental setup and the detector calibration as well to check if the detector was working properly in that period of time (non-vetoed event). The data produced by the on-line analysis are also related to the entities Calibration and Setup; they have been grouped according to their characteristic as Histograms data, Filtered data and Raw data. Each data type is produce by the on-line analysis by means of the corresponding set-up and by the Global setup parameters. The Vetoes entity stores the time information about the detector faulty operation. The vetoed time intervals can be due to detector cryogenic maintenance or to the failure of self-consistency tests of the on-line data analysis (see L. Baggio *et al.* in these proceedings). Other vetoed time intervals could be determined from the auxiliary channels, which monitor the environmental activity. As a consequence of the long term operation of the AURIGA detector we decided to organize the database in space units (each one containing one year of data) that can be taken off-line and recovered independently. The annual tablespaces have the same logical structures and are composed of a set of tables containing the relevant information about the detector output, the hardware and software set up, and the auxiliary instrumentation channels. This data organization occupies about 80 Mbytes per year for storage and management. It is worth to notice that the database keeps track also of the location of the flat files produced by the on-line analysis allowing the recover of the original information stored in these files.

REFERENCES

1. G. A. Prodi *et al*, in Proc. of the 2nd E. Amaldi Conference on Gravitional Waves, E. Coccia G. Veneziano an G. Pizzella Eds. (World Scientific 1998 Singapore) p. 148.
2. A. Ortolan *et al*, in Proc. of the 2nd E. Amaldi Conference on Gravitional Waves, E. Coccia G. Veneziano an G. Pizzella Eds. (World Scientific 1998 Singapore) p. 204.

An Efficient Matched Filtering Algorithm for the Detection of Continuous Gravitational Wave Signals

Peter R. Williams and Bernard F. Schutz

MPI for Gravitational Physics, Albert Einstein Institute, Am Mühlenberg 1, D-14476 Golm, Germany

INTRODUCTION

Neutron stars are perhaps the most promising class of gravitational wave (GW) sources, and searches for such GW signals is particularly suited to the characteristics of the GEO600 detector (see Schutz "Getting Ready for GEO600 Data" gr–qc9910033). However, the instantaneous GW frequency of such a source will evolve due to both intrinsic spindown effects and Doppler modulations induced by the motion of the Earth. Thus because of the large parameter space of likely signals, directly implemented optimal matched filtering is not computationally feasible.

In response to this problem, Schutz and Papa have developed an alternative strategy: the Hough–Hierarchical search algorithm (see Schutz and Papa "End-to-End Algorithm for Hierarchical Area Searches for Long–Duration GW Sources for GEO600" gr–qc9905018). In order to carry out a blind search over a range of intrinsic GW frequencies, the following three stages must be calculated for each point in the parameter space of sky positions and intrinsic spindown parameters:

Stage I: Calculate demodulated Fourier transforms (DeFTs) on an intermediate time baseline (of order 1 day) by combining FFTs of short durations (approximately 30 minutes) of the time series data. In this context demodulated means that if there is a source at the sky position in question, and with the intrinsic spindown parameters in question, then all spindown and modulatory effects will have been correctly removed from the DeFTs: all signal power will be confined to one and the same frequency bin in each DeFT. This frequency is the intrinsic frequency of the source measured at the start of the observing time. It is expected that the total observing time will be of order 4 months, and thus roughly 120 of these DeFTs will be calculated for each point in parameter space.

Stage II: In general source parameters will not coincide exactly with those searched for, and residual frequency evolution and modulation will remain in the DeFTs. Thus, the peak in power associated with a given source may change frequency bins from DeFT to DeFT. Because of the relatively small time baseline of these DeFTs and the resultant poor signal–to–noise of any expected continuous GW signal, this evolution will be not directly apparent in the DeFTs, but can be recovered statistically using the Hough Transform algorithm.

Stage III: Calculate DeFTs for candidate sources with the full frequency resolution of the total observation time, by combining the intermediate baseline DeFTs produced in stage I.

Thus, during stage II, regions of the parameter space in which it is statistically unlikely that there are GW sources are eliminated from the search. Thereby, in stage III, the most computationally expensive part of the algorithm, the long time baseline DeFTs are calculated over only a very small fraction of parameter space and over a very small range of frequencies.

In this paper we outline the methods used in the first and third stages of this algorithm in constructing a longer time baseline DeFT from a number of shorter time baseline FFTs or DeFTs.

THE METHOD

Consider a time series x_a of total duration T, which has been divided into M short time series, each having N data points. Then the DeFT for a signal with a time independent amplitude and phase $2\pi\Phi_{ab}(\vec{\lambda})$ is

$$\hat{x}_b(\vec{\lambda}) = \sum_{a=0}^{NM-1} x_a e^{-2\pi i \Phi_{ab}(\vec{\lambda})} = \sum_{\alpha=0}^{M-1}\sum_{j=0}^{N-1} x_{\alpha j} e^{-2\pi i \Phi_{\alpha j b}(\vec{\lambda})}, \quad (1)$$

where the time indices are related by $N\alpha + j = a$, and b is a long time baseline frequency index. In the following discussion Latin indices j,k,l always sum over N, while Greek indices sum over M. Note that $\Phi_{ab}(\vec{\lambda})$ is dependent on a vector $\vec{\lambda}$ of parameters which characterize the signal one is searching for. In searching for GW signals from neutron stars these will include intrinsic spindown parameters, and the position of the source in the sky. If $\tilde{x}_{\alpha k}$ is the matrix formed by carrying out Fourier transforms along the short time index j in $x_{\alpha j}$, then equation 1 can be written as

$$\hat{x}_b(\vec{\lambda}) = \sum_{\alpha=0}^{M-1}\sum_{k=0}^{N-1} \tilde{x}_{\alpha k}\left[\frac{1}{N}\sum_{j=0}^{N-1} e^{-2\pi i\left(\Phi_{\alpha j b}(\vec{\lambda}) - \frac{jk}{N}\right)}\right] = \sum_{\alpha=0}^{M-1} Q_\alpha(b,\vec{\lambda})\sum_{k=0}^{N-1}\tilde{x}_{\alpha k} P_{\alpha k}(b,\vec{\lambda}), \quad (2)$$

where the product $Q_\alpha(b,\vec{\lambda})P_{\alpha k}(b,\vec{\lambda})$ is defined by the terms in square brackets, and $Q_\alpha(b,\vec{\lambda})$ contains all parts of the square brackets independent of the short time index j and short frequency index k.

In equation 2 we have effectively re-written equation 1, a long time baseline DeFT in the time domain, as a sum (α index) of short time baseline DeFTs in the frequency domain (k index), where $Q_\alpha(b,\vec{\lambda})P_{\alpha k}(b,\vec{\lambda})$ are these frequency domain filters. In the presence of stationary noise with a flat spectrum, equation 2 is the optimal detector. However, through applying various approximations, the detector can be made "acceptably sub-optimal", in the sense that only a small fraction of power from a signal is lost in comparison to the optimal case, while achieving vast savings in computational cost.

To illustrate these mathematical approximations it is instructive to discuss equation 2 for a specific case of $\Phi_{\alpha jb}(\vec{\lambda})$: a linearly varying frequency model, i.e. in the continuum limit $\Phi(t, f_0, \dot{f}_0) = f_0 t + \dot{f}_0 t^2$, where f_0 and $\dot{f}_0$ are the intrinsic frequency and spindown of the source respectively, and t is time. In the case of an actual search for GW signals from pulsars, $\Phi_{\alpha jb}(\vec{\lambda})$ will not be so simple. However, this model is sufficiently complex to effectively demonstrate all of the approximations to be discussed here.

In discrete form, $\Phi(t, f_0, \dot{f}_0)$ can be written as $\Phi_{\alpha j\beta l}(\gamma) = (\beta+Ml)(N\alpha+j)/NM + \gamma(N\alpha+j)^2/N^2M^2$, where the long time baseline frequency index $b = \beta + Ml$. The chosen discretization of the spindown parameter $\dot{f}_0 \equiv \gamma/T^2$ is not practically appropriate. However, in an actual search, a grid of points in spindown parameter space will be chosen to ensure an acceptable loss of power from unresolved signals. Thus, in the following discussion, only searches for resolved $\dot{f}_0$ parameters will be considered.

Approximation 1: By Taylor expanding the model phase function $\Phi(t)$ about the middle of each short duration time series (i.e. about $j = N/2$) and discarding terms of order $(j/N)^2 \equiv t^2$ and higher, in the limit $N \to \infty$ the function $P_{\alpha k}(\beta, l, \gamma)$ is Re $P_{\alpha k}(\beta, l, \gamma) = \text{sinc } x$ and Im $P_{\alpha k}(\beta, l, \gamma) = (1 - \cos x)/x$. In the phase model considered here $x = -2\pi(\beta/M + l + (2\alpha + 1)\gamma/M^2 - k)$ and $Q_\alpha(\beta, l, \gamma) = \exp\{-2\pi i(\alpha\beta/M + \alpha\gamma^2/M^2)\}$.

Approximation 2: Consider the case where the short time baseline is chosen such that the instantaneous model frequency $f(t) = \dot{\Phi}(t, f_0, \dot{f}_0, \ddot{f}_0, \ldots)$ does not move by more than one short time baseline frequency bin over the duration of a short time baseline data set, i.e. in the model discussed here $|\dot{f}_0|T/M < M/T$. Then for a given α, the function $P_{\alpha k}(b, \vec{\lambda})$ will be peaked in power about the model frequency averaged over the duration of time associated with the αth short data set, i.e. about $x = 0$ (the first three terms in the above definition of x are the index of this average model frequency). Thus only a few terms around this model frequency will contribute significantly to the summation over k in equation 2.

Approximation 3: The semi-periodic nature of $P_{\alpha k}(b, \vec{\lambda})$ means that this function can be efficiently evaluated from a look-up table of values containing the periodic parts, and three further operations: to calculate one instance of $P_{\alpha k}(b, \vec{\lambda})$ will require only 8 floating point operations.

Approximation 4: If one approximates the model frequency parameter β in the calculation of $P_{\alpha k}(\beta, l, \gamma)$ as a fixed value, for example with $\beta = \beta_0$, equation 2 can be calculated as an FFT, i.e.

$$\hat{x}_{\beta l}(\gamma) = \sum_{\alpha=0}^{M-1} \left[Q'_\alpha(\gamma) \sum_{k}^{n_{term}} \tilde{x}_{\alpha k} P_{\alpha k}(\beta_0, l, \gamma) \right] e^{-2\pi i \frac{\alpha \beta}{M}}, \qquad (3)$$

where n_{term} relates to approximation 2, $P_{\alpha k}(\beta, l, \gamma)$ is defined above, and for the phase model discussed here $Q'_\alpha(\gamma) = \exp\left\{-2\pi i \left(\alpha^2 \gamma/M^2 - \gamma/4M^2\right)\right\}$. Thus for values of β sufficiently near to β_0, the loss in power due to this approximation will be small. To obtain $\hat{x}_{\beta l}(\gamma)$ for other values of β, the calculation must be repeated using another β_0.

RESULTS AND DISCUSSION

Numerical tests have shown that if one chooses 10% as an acceptable loss in power in comparison to the optimal case, then $N_{FFT} = 8$ and $n_{term} = 16$ are the preferred parameter combination, if the short time baseline T/M is chosen such that in the phase model discussed here $|\dot{f}_0|T/M < M/T$. If one decides that only a 5% loss in optimal power is acceptable, then this can be achieved with the same parameters, but choosing T/M such that $|\dot{f}_0|T/M < M/2T$.

The computational cost of calculating one DeFT in stage I in floating point operations is

$$C_{DeFT} \simeq 5.3 \times 10^{10} \left(\frac{B}{300 \text{ Hz}}\right) \left(\frac{T}{1 \text{ day}}\right) \left(\frac{N_{FFT}}{8}\right) \left(\frac{n_{term.}}{16}\right), \qquad (4)$$

where B is the bandwidth of the search. This is comparable to the computational cost of the corresponding steps in the Hierarchical Stack / Slide algorithm of Brady and Creighton ("Searching for Periodic Sources with LIGO: Hierarchical Searches" gr-qc9812014). The Hough–Hierarchical search algorithm also has a number of computational advantages. To calculate a given bandwidth of a DeFT requires only the FFT data from this bandwidth and an additional small overlap. Thus the algorithm can be easily parallelized by distributing data and work by bandwidth; and no communication between processors is required. Also, the complete three stage algorithm can be arranged in such a way that once a bandwidth of FFT data is read from disk by a processor, all computation required on this data can be carried out while this data is held in memory, thus time spent reading data from disk is a negligible fraction of the total computational time: each processor will need to read roughly 40 Mb from disk once every two weeks. Furthermore, little additional memory is required as workspace for stages I and III: less than 100 kb.

The GEO600 data analysis team are currently working on coding this algorithm in a computationally optimal manner, as well as integrating this with the Hough Transform part of the procedure.

A Likelihood Based Scheme for Coincidence Analysis

Soma Mukherjee, Soumya D. Mohanty

Center for Gravitational Physics and Geometry, the Pennsylvania State University, University Park, PA 16802, U.S.A.

Abstract. A fresh approach to coincidence analysis is initiated. The coincidence window size and the rates of source and background that best explain a given list of event times from two detectors is found using Maximum Likelihood Estimation. This approach is transparent and provides a unified scheme for a wide variety of situations.

I MOTIVATION

A statistically significant excess in the number of coincidences in arrival times of events in two detectors indicates a common source. Very often, however, either the rate of background events or the correct coincidence window size or both are not known *a priori*. For instance, what coincidence window size to use for the gravitational wave strain channel and a magnetometer? How do gaps in the data and environmental variability affect the significance of the results? Usually, such issues are addressed in a very data specific manner with the data itself influencing some of the assumptions made in the analysis. This makes it very difficult to understand the result and its statistical significance. These problems will become more serious for the large datasets that will be produced by the upcoming interferometers.

Maximum Likelihood Estimation (MLE) [1] appears to be a promising avenue for addressing the concerns raised above. As a demonstration, we consider here the simplest problem : two detectors producing data that consists of only event arrival times. Both background and source events are assumed to occur as Poisson processes (constant rates and no gaps in data collection). A procedure that can handle both gaps and variable rates is under development [2].

II AN OUTLINE OF THE METHOD

The data consists of two lists of event times, $L_1 = \{t_1, \ldots, t_{n_1}\}$ and $L_2 = \{\tau_1, \ldots, \tau_{n_2}\}$, from detectors 1 and 2 respectively. The likelihood, $P(L_1, L_2 | r_s, r_1, r_2, \Delta)$, of the data is its probability *given* source rate r_s, background rates r_1 and r_2 in detector 1 and 2 and the coincidence window size Δ. The values of r_s, r_1, r_2 and Δ at which $P(L_1, L_2|r_s, r_1, r_2, \Delta)$ is maximized is the MLE estimate for the best fit values of these parameters.

Note that no prior estimation of source or background rates is obtained from the data. By allowing Δ to be a free parameter we have added an extra flexibility which will be useful when comparing, for instance, two environmental monitors.

Monte Carlo simulation: Several realizations of L_1 and L_2 were generated with $r_1 = 20$ events/day, $r_2 = 30$ events/day, $r_s = 50$ events/**year** and $\Delta = 1$ min and the MLE values of r_s, r_1 and r_2 were obtained for each. Fig. 1 shows the histograms for r_s, r_1 and r_2. We did not search over Δ here because computational requirements multiply with every additional parameter in the Monte Carlo simulation.

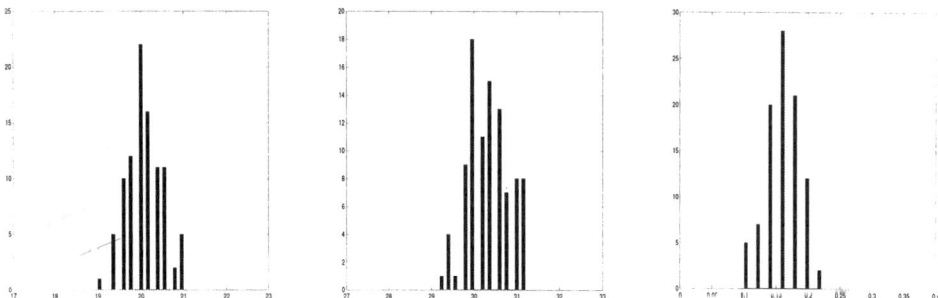

FIGURE 1. Histograms of best fit values. Left panel : r_1. Middle panel : r_2. Right panel : r_s (events/day).

Future work — Our final aim is to construct a code that implements MLE taking into account variable event rates, gaps and also extends coincidence to quantities such as event amplitude. Methods for handling gaps (such as Bayesian Blocks [3]) and variable rates are under exhaustive study. The detection problem, as opposed to estimation, is also being addressed.

ACKNOWLEDGEMENT

This work is supported by NSF awards PHY 98-00111 and PHY 99-96213 to the Pennsylvania State University.

REFERENCES

1. Stuart, A. and Ord, J. K., *Kendall's Advanced Theory of Statistics*, Edward Arnold, 1991.
2. Collaborative project between IGEC and Soma Mukherjee, Soumya D. Mohanty.
3. Jeffrey D. Scargle, Astrophys. J., **504**, 405 (1998).

A robust test for detecting non-stationarity

Soumya D. Mohanty

*Center for Gravitational Physics and Geometry,
Pennsylvania State University, University Park, PA 16802, U.S.A.*

Abstract. Tests of stationarity that require a statistical model for ambient noise may not be appropriate for huge data sets since, in most cases, the models would have to be estimated from the data itself. An alternative is presented in the form of a *robust* test : its false alarm rate is almost independent of the statistical nature of the ambient noise. Apart from robustness this test is also suitable as an online monitor of stationarity.

I THE PROBLEM

Tests for detecting transient non-stationarity [1,2], or *bursts*, that require a statistical model for the ambient stationary noise present a paradox : In practice, the noise model must be obtained from the data itself. But in order to use a portion of data for this purpose, it is first necessary to certify it as stationary ! Though an iterative procedure may get around this problem, automating it would prove to be difficult. Hence, it may not work for the humungous amount of data that interferometric detectors will produce and especially for the auxillary channels since ambient terrestrial noise is not well understood statistically.

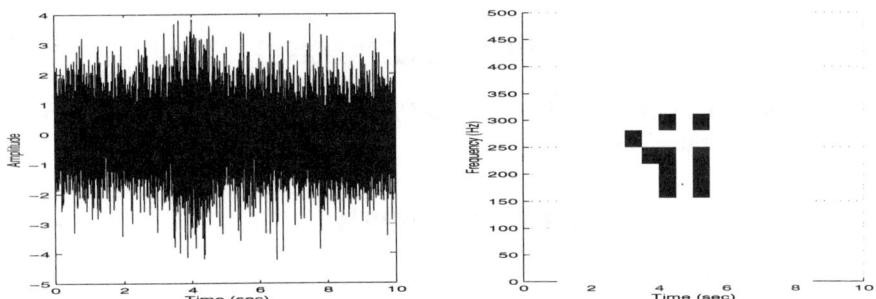

FIGURE 1. Left panel : White Gaussian noise (variance $\sigma^2 = 1$) plus broadband burst (peak amplitude 3σ). Right panel : Output of the test. The black cluster corresponds to the burst. The white areas represent stationarity.

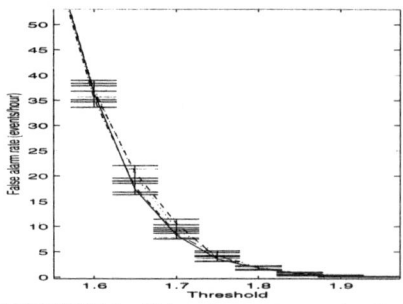
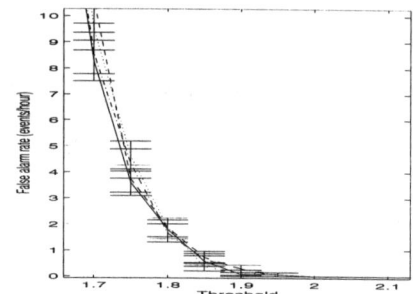

FIGURE 2. False alarm versus threshold curve (right panel: detail from left). Solid : GWN with variance $\sigma^2 = 1$. (G: Gaussian, W: white N: Noise.) Dashed : GWN ($\sigma = 10$), Dotted : Exponential WN ($\sigma = 1$) and Dashed-Dotted : Coloured GN with LIGO-I PSD ($\sigma = 1$).

II A SOLUTION

Use a *robust* test for which the *false alarm rate* at a given threshold depends weakly on the nature of ambient noise. We present such a test which is also computationally trivial and unambiguously provides the time of arrival and bandwidth of a burst. The basic idea behind this test is the detection of a temporary change in the Power Spectral Density (PSD) of the input data that should occur, in general, during a burst. Sample PSDs are computed for two disjoint segments of the data. Student's t-test [3] is used to test for a statistically significant difference in the PSDs at each frequency. Fig. 1 shows the output, a time-frequency plot.

The false alarm rate versus threshold curve for this test is estimated for several different types of noise. An example, shown in Fig. 2, demonstrates the robustness of this test (the error bars are at 1σ). The test is surprisingly powerful at detecting a wide variety of bursts. For instance, at a false alarm rate of 1 /hour, bursts centered at 200 Hz with a bandwidth of 20 Hz and embedded in noise with a LIGO-I PSD can be detected with probability 0.8 when their peak amplitude is only 1.6× the r.m.s. of the background ambient noise.

ACKNOWLEDGEMENT

This work was supported by NSF awards PHY 98-00111 and PHY 99-96213 to the Pennsylvania State University.

REFERENCES

1. Arnaud, N., *et al*, Phys. Rev. D **59**, 082002 (1999).
2. Flanagan, E. E. and Hughes, S. A., Phys. Rev. D **57**, 4535 (1998).
3. Snedecor, G. W. and Cochran, W. G., *Statistical methods*, Iowa State Univ. Press, 1980, ch. 6.

PROGRAM

Monday, July 12: AM
Introductions
Chair: A. Giazotto

9:00	B. Barish
	Welcome
	D. Baltimore
	Remarks
9:15	U. Amaldi
	A Retrospective on Edoardo Amaldi
9:45	S. Phinney
	Astrophysical GW Sources
10:20	Discussion
10:30	Coffee break
11:00	A. Rüdiger
	Interferometry Today
11:35	Discussion
11:45	E. Coccia
	Bars In Action
12:20	Discussion
12:30	Session adjourns

Monday PM
Sources
Convenor: K. Thorne

2:00	V. Kalogera
	Compact Binary Mergers and Accretion-Induced Collapse: Event Rates and Mass and Spin Distributions
2:30	J. Houser
	Commentary: Simulation of Accretion-Induced Collapse
2:35	Discussion
2:45	R. Price
	Gravitational Waves from Black-Hole and Neutron-Star Mergers: Status Report

3:05	Discussion
3:15	B. Owen
	Gravitational Waves from the r- Modes of Rapidly Rotating Neutron Stars: Status Report
3:35	Discussion
3:45	Coffee break
4:00	G. Ushomirsky
	Gravitational Waves from Accreting Neutron Stars in Low-Mass X-ray Binaries: Status Report
4:25	Discussion
4:35	S. Hughes
	Gravitational Waves from Inspiral into Supermassive Black Holes: Status Report
4:55	Y. Mino
	Commentary: Radiation Reaction Force on a Compact Body Spiraling into a Supermassive Black Hole - I
5:00	L. Burko
	Commentary: Radiation Reaction Force - II
5:05	Discussion
5:15	B. Allen
	Gravitational Waves from the Early Universe: Status Report
5:30	C. Ungarelli
	Commentary: Are Pre-Big Bang Models Falsifiable by Gravitational Wave Experiments?
5:35	O. Aguiar
	Commentary: Is There a Signature in Gravitational Waves from Structure Formation of the Universe?
5:40	Discussion

Monday Evening

8:00	LIGO Research Community General Meeting
	Chair: R. Stebbins

Tuesday, July 13: AM
Status of Interferometers
Chair: R. Weiss

8:30	M. Coles
	LIGO
9:05	Discussion
9:15	F. Marion
	VIRGO
9:50	Discussion
10:00	Coffee Break
10:15	H. Lück
	GEO 600
10:50	Discussion
11:00	K. Tsubono, M. Ando
	TAMA
11:35	Discussion
11:45	D. McClelland
	ACIGA
12:20	Discussion
12:30	Session adjourns

Tuesday PM
Overviews
Convenor: A. Brillet
Chair: V. Braginsky

2:00	S. Braccini
	Suspensions
2:20	Discussion
2:30	L. Gammaitoni
	Thermal and Mechanical Noise
2:50	Discussion
3:00	Coffee Break
3:15	K. Strain
	Interferometer Configurations
3:35	Discussion
3:45	F. Khalili
	Energetic Quantum Limitations in the Antennae on Free Masses
4:05	Discussion

Tuesday PM
Workshop - Advanced Configurations
Convenors: P. Fritschel, D. McClelland

4:30	G. Heinzel
	Dual Recycling at the Garching 30 m Interferometer
4:55	Discussion
5:00	M. Gray
	A Power Recycled, Fabry-Perot Arm Cavity Michelson Interferometer with Resonant Sideband Extraction
5:25	Discussion
5:30	J. Mason
	Length Sensing and Control of an RSE Interferometer for LIGO II
5:45	Discussion
5:47	J. Camp
	Arm Cavity Resonant Sideband Control for Gravitational Wave Detection
6:02	Discussion
6:04	P. Beyersdorf
	The Polarization Sagnac Interferometer as a Candidate Configuration for an Advanced Detector
6:19	Discussion
6:21	M. Rakhmanov
	Dynamics of Fields in 3-Mirror Coupled Cavity
6:36	Discussion
6:38	Session Adjourns

Tuesday Evening

8:00	Poster Session

Wednesday, July 14: AM
Workshop - Lasers and Optics
Convenors: W. Winkler, E. Gustafson

8:30	R. Byer
	Diode Pumped Solid State Lasers
9:00	Discussion
9:10	P. Veitch
	ACIGA High Power Nd:YAG Laser Development
9:30	Discussion
9:40	B. Willke
	GEO Stabilized Laser
10:00	Discussion
10:10	Coffee Break
10:30	B. Feigelson
	History of Materials' Development Program: Lithium Niobate
10:55	Discussion
11:05	R. Route
	Crystal Growth: Prospects for Future Core Optics
11:30	Discussion
11:40	L. Ju
	Optical Properties of Sapphire
12:00	Discussion
12:10	Lunch

Wednesday PM
LISA Mission Definition Team Meeting

1:00-5:00	NASA Project Status, ESA Project Status, Technology Status, Time-Domain Signal Processing, Gravitational-Wave Backgrounds

Wednesday Evening
Public Lecture
Chair: B. Barish

8:00	K. Thorne
	Probing Black Holes and the Dark Side of the Universe with Gravitational Waves

Thursday, July 15: AM
LISA
Convenors: K. Danzmann, R. Stebbins

8:30	K. Danzmann
	Status of LISA
9:00	Discussion
9:10	S. Vitale, M. Peterseim
	Advances in LISA Technology
9:40	Discussion
9:50	Coffee Break
10:10	A. Vecchio
	Deep Surveys of Massive Black Holes with Space-Borne Interferometers
10:40	Discussion
10:50	O. Blaes
	Scienice of Massive Black Holes
11:20	Discussion
11:30	W. Folkner
	Flight Tests
12:00	Discussion
12:10	Session adjourns

Thursday PM
Workshop - Bar Antennae
Convenors: M. Cerdonio, W. Hamilton

2:00	M. Cerdonio, W. Hamilton
	IGEC Status: Poster Highlights
2:20	Discussion
2:30	W. Johnson
	Bar Technology for Interferometric Detectors
2:55	Discussion
3:05	Coffee Break
3:20	L. Conti
	An Optical Transduction Chain for the AURIGA
3:40	Discussion
3:50	A. de Waard
	MiniGRAIL: A Small (60cm) Spherical Antenna

4:10	Discussion
4:20	G. Modestino
	Detection of Cosmic Rays by NAUTILIS
4:40	Discussion
4:50	D. Blair
	NIOBE: Improved Noise Temperature Performance and Background Noise Suppression
5:10	Discussion
5:20	Session Adjourns

Friday, July 16: AM
Workshop - Suspensions and Thermal Noise
Convenors: G. Gonzalez, J. Hough, K. Kuroda

8:30	V. Braginsky
	Thermal and Nonthermal Noise in the Suspension (Recent Results of MSU Group)
8:50	Discussion
8:55	S. Rowan
	Mechanical Loss Factors of Material and Suspension Systems for Advanced Gravitational Wave Detectors
9:15	Discussion
9:20	J. Giaime, B. Lantz
	Active Isolation Systems
9:40	Discussion
9:45	G. Harry
	Role of Surfaces in Thermal Noise Budgets
10:05	Discussion
10:10	K. Kuroda
	Reduction of Thermal Noise by Cooling Mirror Itself
10:30	Discussion
10:35	Coffee break
10:50	N. Robertson
	Suspension Design for GEO 600
11:10	Discussion
11:15	R. DeSalvo, G. Losurdo
	Aspects of Low-Frequency Suspensions

11:45	Discussion
11:50	D. Blair
	Ultralow Residual Motion Vibration Isolation
12:10	Discussion
12:15	J. Kovalik, M. Punturo
	Research in Suspension Design for VIRGO
12:45	Discussion
12:50	S. Augst, R. Drever
	Experiments on Levitated Paramagnetic Crystals
1:00	Session adjourns

Friday PM
Signal Processing and Data Analysis
Convenors: B. Allen, S. Vitale

2:00	G. Prodi
	Validation of Data in Operating Resonant Detectors
2:30	Discussion
2:40	J. Creighton
	Observational Limit on Gravitational Waves From Binary Neutron Stars in the Galaxy
3:10	Discussion
3:20	J. Armstrong
	The Cassini Gravity Wave Experiment
3:40	Discussion
3:50	Coffee Break
4:05	K. Scholberg
	The SN Neutrino Network
4:25	Discussion
4:35	S. Mukherjee
	Simultaneous Dynamical Tracking and Removal of Multiple Violin Modes
4:55	Discussion
5:05	G. Pizzella
	Search for Gravitational Radiation with the Allegro and Explorer Detectors
5:35	J. Hough
	Summary
6:05	Conference Adjourns

AUTHOR INDEX

A

Aguiar, O. D., 94, 413, 417, 441
Alexandrovski, A. L., 293, 395
Amaldi, U., 5
Anastasio, A., 463
Ando, M., 128
Andrade, L. A., 417
Astone, P., 275, 369, 425, 430
Aufmuth, P., 119
Augst, S. J., 338

B

Baggio, L., 345, 421, 423, 471
Baigent, K. G., 439
Baker, M., 140
Barish, B. C., 3
Barone, F., 461, 463, 465, 467
Bassan, M., 275, 369, 425, 427, 430
Beausoleil, R., 385
Beilby, M., 443
Benabid, F., 140, 222
Benacquista, M., 405
Bernardini, A., 447
Bertolini, A., 320, 409
Beyersdorf, P. T., 200, 385
Bhawal, B., 469
Bildsten, L., 65
Blaes, O., 248
Blair, D., 283
Blair, D. G., 140, 222, 325
Bonaldi, M., 419, 421
Bonifazi, P., 275, 369, 425, 430
Braccini, S., 153, 437
Braginsky, V. B., 180, 445
Brocco, L., 447
Brozek, O. S., 119, 215, 389, 393
Burko, L. M., 86
Byer, R. L., 200, 385

C

Cagnoli, G., 119, 162, 293, 313
Carelli, P., 275, 369, 425, 430
Casciano, C., 437
Casey, M., 119
Cella, G., 320, 469
Cerdonio, M., 261, 345, 419, 421, 423, 471
Charlton, P., 140
Ciaramella, A., 465
Clubley, D. A., 391
Coccia, E., 32, 275, 369, 425, 430
Cohadon, P. F., 435
Coles, M. W., 101
Conti, L., 261, 421
Cordero, F., 437
Corvace, F., 437
Cosmelli, C., 369
Crivelli Visconti, V., 345, 419, 421
Cutler, C., 65

D

D'Ambrosio, E., 320
D'Antonio, S., 275, 425, 430
Danzmann, K., 119, 215, 389, 393
de Andrade, L. A., 413
de Araújo, J. C. N., 94, 413
DeBra, D., 300
De Rosa, M., 261
DeSalvo, R., 320, 409
De Sanctis, M., 437
de Waard, A., 268
Dolesi, R., 231
Drever, R. W. P., 338
Duffy, M., 443

E

Eleuteri, A., 461, 463, 465, 467
Evans, M., 469

F

Fafone, V., 275, 369, 425, 430
Falferi, P., 419, 421
Fallnich, C., 215, 389, 393

Fejer, M. M., 200, 293, 385, 395
Fidecaro, F., 409
Finn, L. S., 362, 451
Fortini, P. L., 421
Frajuca, C., 413, 417
Francesconi, M., 409
Franco, R., 437
Frasca, S., 275, 369
Frasconi, F., 437
Freise, A., 119
Freitag, I., 393
Frossati, G., 268, 413, 430
Furtado, S. R., 413

G

Gammaitoni, L., 162
Garufi, F., 461, 463, 465, 467
Geng, K., 369
Giaime, J., 300
Gonzalez, G., 443, 451
Gorodetsky, M. L., 180
Goßler, S., 119, 215
Grado, A., 119
Gray, M. B., 140, 193, 439
Greenwood, D., 140
Gretarsson, A. M., 306
Grote, H., 119
Gustafson, E. K., 293, 385

H

Hadjar, Y., 435
Hamilton, M. W., 140
Hamilton, W. O., 369
Hardham, C., 300
Harry, G. M., 306, 415
Heidmann, A., 435
Hellings, R. W., 255
Heng, I. S., 283
Hollitt, C., 140
Hough, J., 119, 293, 313, 376, 445, 451
Houser, J. L., 51
How, J., 300
Hughes, S. A., 76
Huq, M. F., 451
Husman, M. E., 119, 313

I

Ivanov, E. N., 283

J

Jin, I., 415
Johnson, W. W., 369
Ju, L., 140, 222

K

Kalogera, V., 41
Kawabe, K., 119
Khalili, F. Y., 180
Klimenko, S., 469
Kovalik, J., 162

L

Lantz, B., 300
Lobo, J. A., 411
Locke, C. R., 283
Losurdo, G., 332
Lück, H., 119, 215

M

Magalhães, N. S., 413, 417
Majorana, E., 437, 447
Marchesoni, F., 162
Marin, F., 261
Marinho Jr., R. M., 413
Marini, A., 275, 425, 430
Marion, F., 110
Maros, E., 469
Martinucci, V., 345, 421, 423, 471
Mason, J., 208
Mattioli, A., 419
Mauceli, E., 275, 369, 425, 430
Mazzitelli, G., 275
McClelland, D. E., 140, 193, 439
McHugh, M. P., 369
McIntosh, S., 293, 313
McNamara, P., 119
Melo, J. L., 413, 441

Merkowitz, S. M., 369, 411, 430
Mezzena, R., 419, 421
Milano, L., 461, 463, 465, 467
Minenkov, Y. F., 275, 369, 425, 427, 430
Mino, Y., 82
Miranda, O. D., 94, 413
Mitrofanov, V., 445
Modena, I., 275, 369, 425, 430
Modestino, G., 275, 369, 425, 430
Mohanty, S. D., 451, 469, 477, 479
Moleti, A., 275, 369, 425, 430
Moore, T. A., 255
Morse, A., 369
Mossavi, K., 119, 215
Mudge, D., 140
Mukherjee, S., 362, 477
Munch, J., 140

N

Newton, G. P., 119, 391
Ni, X.-T., 459
Notcutt, M., 222

O

Oliveira Jr., N. F., 413
Ortolan, A., 345, 421, 423, 471
Ostermeyer, M., 140
Ottaway, D., 140
Owen, B. J., 55

P

Paik, H. J., 415
Pallottino, G. V., 275, 369, 425, 430
Palmer, D., 313
Pampaloni, V., 275
Papa, M. A., 275, 369, 425, 430
Paparo, G., 437
Passaquieti, R., 437
Penn, S. D, 306
Peterseim, M., 393
Pinard, M., 435
Pizzella, G., 275, 369, 425, 430
Plissi, M. V., 119, 313

Poker, J., 443
Prodi, G. A., 261, 345, 419, 421, 423, 471
Punturo, M., 162
Puppo, P., 447

Q

Quetschke, V., 119, 215, 389

R

Raffone, G., 430
Rakhmanov, M., 204, 469
Rapagnani, P., 437, 447
Ribeiro, K. L., 413, 417
Ricci, F., 437, 447
Richman, S., 300
Roberts, P. C. E., 399
Robertson, D. I., 119, 313
Robertson, N. A., 119, 313
Romano, J., 451
Ronga, F., 275, 425, 430
Route, R. K., 293, 385, 395
Rowan, S., 119, 293, 313, 445, 451
Rüdiger, A., 119

S

Salles, K. B. M., 413
Sandeman, R. J., 140
Sannibale, V., 320, 409
Sathyaprakash, B. S., 119
Saulson, P. R., 306, 451
Savage Jr., R. L., 469
Schiller, J. J., 306
Schilling, R., 119
Schipilliti, M., 430
Scholberg, K., 355
Schutz, B. F., 119, 473
Scott, S. M., 140
Shaddock, D. A., 140, 193
Sintes, A. M., 403
Skeldon, K. D., 119, 391
Slagmolen, B. J. J., 140, 439
Sneddon, P., 293, 313
Solina, A., 437

Solomonson, N., 369
Soranzo, G., 421
Startin, W. J., 306
Stebbins, R., 300
Stevenson, T. R., 415
Strain, K. A., 119, 173, 313, 451
Stuver, A., 443

T

Taffarello, L., 261, 345, 419, 421, 423, 471
Tagliaferri, R., 465, 467
Takamori, A., 320, 409
Taylor, I., 119
Terenzi, R., 275, 369, 430
Testi, G., 447
Thorne, K. S., 180
Tobar, M. E., 283
Tokmakov, K., 445
Torrie, C. I., 119, 313
Traeger, S., 385
Tsuboni, K., 128

U

Ungarelli, C., 90
Ushomirsky, G., 65

V

Valentini, R., 437
Vecchio, A., 90, 238, 403
Vedovato, G., 345, 42i, 423, 471
Veitch, P. J., 140
Velloso Jr., W. F., 413, 441

Vinante, A., 421
Visco, M., 275, 369, 425, 430
Vitale, S., 231, 261, 345, 419, 421, 423, 471
Votano, L., 275, 425, 430

W

Waluschka, E., 401
Ward, H., 119, 313
Welling, H., 215, 389, 393
Wellstood, F. C., 415
Whiting, B. J., 140
Willems, P., 208
Williams, P. R., 473
Willke, B., 119, 215, 389
Winkler, W., 119
Winterflood, J., 140, 325

X

Xu, X., 459

Y

Yamamoto, H., 320, 469

Z

Zawischa, I., 215, 389
Zendri, J. P., 261, 345, 419, 421, 423, 471
Zhao, C., 119
Zhou, Z. B., 140, 325
Zhu, N., 369